Remote Sensing Handbook for Tropical Coastal Management

Edmund P. Green, Peter J. Mumby
Alasdair J. Edwards and Christopher D. Clark

Edited by Alasdair J. Edwards

UNESCO PUBLISHING

Department for International Development

DFID is the British Government department responsible for promoting development and the reduction of poverty. The policy of the Government was set out in a White Paper on International Development published in November 1997. The central focus of the policy is a commitment to the internationally agreed targets to halve the proportion of people living in extreme poverty by 2015 together with the associated targets including basic health care provision and universal access to primary education by the same date. DFID seeks to work in partnership with governments which are committed to the international targets and also seeks to work with business, civil society and the research community to encourage progress which will help reduce poverty. DFID also works with multilateral institutions including the World Bank, United Nations agencies and the European Commission.

United Nations Educational, Scientific and Cultural Organization

UNESCO promotes international cooperation in natural and social sciences, culture, communication and education in an effort to 'build the defenses of peace in the minds of humanity'. Such co-operation benefits from the Organization's capacity to integrate action from these different domains, for example, utilizing its Coastal Regions and Small Islands (CSI) initiative, launched in 1996. On the CSI platform, over 20 intersectoral pilot projects have been initiated involving some 60 countries, uniting decision-makers, local communities, cultural heritage experts, scientists and others. Interlinked UNESCO Chairs foster interdisciplinary training and capacity building for environmentally sustainable, socially equitable and culturally appropriate coastal development. On the basis of lessons learnt so far, a preliminary set of 'wise practices' are being elaborated and widely disseminated. One such practice is to provide easy access to useful resource material; this publication is a prime example.

Published in 2000 by the United Nations Educational, Scientific and Cultural Organization
7, place de Fontenoy, 75352 Paris 07 SP

Cover design: Jean-Francis Chériez
Printed by Corlet, Imprimeur, S.A., 14110 Condé-sur-Noireau
N° d'Imprimeur : 45801 - Dépôt légal : mai 2000

ISBN 92-3-103736-6

Foreword

Interactions between the coastal environment and the people who use its resources are still poorly understood. In the developing countries many of these people are poor.

Demographic changes and economic development have increased demands on coastal resources. More people than ever are generating at least part of a livelihood from activities which directly affect, or are affected by, changes in the coastal environment.

As these pressures grow, resource-use conflicts in coastal regions could result in increased environmental degradation and social inequality. In such conditions the livelihoods of poor people will be vulnerable. Extractive uses for food, income generation, medicines and building materials and non-extractive uses for tourism continue to degrade coastal ecosystems. The widespread death of corals following global bleaching events in 1998 have further aggravated these pressures.

Although the present decade has witnessed substantial investment in environmental institutions, coastal management planning has remained largely empirical with little detailed analysis of environmental components and their reaction to natural variation or resource use. The absence of effective monitoring systems to generate baseline data against which the impacts of natural and anthropogenic changes can be measured has been a significant constraint to effective planning.

The challenges presented by these factors are huge and they will only be tackled through working in effective partnerships at all levels, but particularly between the development agencies.

This *Handbook* is one of the products of a successful partnership between UNESCO and the United Kingdom Department for International Development (DFID). It attempts to match user need to the level of technology required for planning and monitoring. The information presented seeks to close the gap between managers objectives/expectations and what can be realistically achieved in an operational context. It provides clear guidance on the reliability, accuracy and cost of applications. Effective recommendations are provided to inform coastal management initiatives and projects supporting sustainable livelihoods in coastal communities.

We hope this *Handbook* will be a valuable reference and provide practical guidance for all who work towards the goal of the sustainable and wise use of resources in coastal regions.

ANDREW J. BENNETT
Chief Natural Resources Adviser
Department for International Development
London

DIRK G. TROOST
Chief, Environment and development
in coastal regions and in small islands
UNESCO
Paris

Editor's note

The field of remote sensing is fast evolving, with new satellites being launched and new sensors developed every year. Inevitably any book on the subject is slightly out-of-date as soon as it is written. The *Handbook* summarises the state-of-play in terms of operational use of remote sensing in coastal management applications in 1998/1999. Since the bulk of it was written, several important satellites, which will influence the application of remote sensing to coastal management problems, have been launched. These include Landsat 7 carrying the Enhanced Thematic Mapper (ETM+), the Sea-viewing Wide Field-of-view Sensor (SeaWiFS) for ocean colour mapping, SPOT 4 with the new High Resolution Visible and Infra-Red (HRVIR) sensor, IKONOS 2 with its 1 m panchromatic and 4 m multispectral image data (www.ersi.bc.ca/ikonos.html), and the Indian Remote Sensing satellite IRS-1D with improved panchromatic spatial resolution of around 5 m. Apart from IKONOS 2 which was launched in late September 1999, these sensors are all discussed in the text but the implications of recent sensor developments and changes in pricing have yet to be evaluated fully for coastal management applications. However, the protocols used to evaluate the range of sensors discussed in the *Handbook* provide guidelines as to how the usefulness and cost-effectiveness of the new sensors may be assessed.

Although specific costs (e.g. those stated in Chapter 5 for various remote sensing products) may change, the type of information that needs to be gathered and compared in order to decide which sensor may be most appropriate for user objectives, remains the same. A key aim of the *Handbook* is to allow users to take the methodologies outlined here and apply these in their own countries for their own objectives and thus be less reliant on outside inputs. With the rapid development of PC computing power and decrease in the cost of appropriate hardware and software, this is becoming a more achievable goal every year. To allow users to stay abreast of new developments in the field we have included many URLs to key web sites both in the text and in Appendix 1.

The authors hope that the *Handbook* proves a valuable sourcebook to all involved in managing coastal resources or planning remote sensing campaigns to assist coastal management initiatives. Whilst every attempt has been made to ensure the accuracy of the information presented, in such a wide ranging book some errors or omissions are bound to have crept in, and I apologise in advance if such are found.

ALASDAIR EDWARDS

Acknowledgements

This *Handbook* would not have been possible without the generous help of the following people and organisations and we are extremely grateful to them for their time, advice, help or useful discussions. We thank: the Ministry of Natural Resources of the Turks and Caicos Islands for logistical assistance during our fieldwork and in particular Christie Hall, Chris Ninnes, Paul Medley, Perry Seymore and John Ewing; the Environment Research Programme of the UK Department for International Development for funding the research; Hugo Matus, Janet Gibson and David Vousden, UNDP/GEF Coastal Zone Management Project, Belize; Herb Ripley and Ariel Geomatics Incorporated; Angie Ellis; School for Field Studies, South Caicos; Alastair Harborne, Peter Raines, Jon Ridley and Coral Cay Conservation Ltd; all respondents to questionnaire on use of remote sensing for tropical coastal zone management; the Belize Fisheries Department; John Turner, University of North Wales; Charles Sheppard, University of Warwick; John McManus, International Center for Living Aquatic Resources Management, Manila; Bob Clarke, Plymouth Marine Laboratory; The Groundbusters Corporation; Jean-Pierre Angers, Centre University Saint-Louis-Maillet; Clive Anderson and Coomaren Vencatasawmy, University of Sheffield (for statistical advice); Ian Sotheran and Rob Walton, SeaMap, University of Newcastle; Agneta Nilsson and Lieven Bydekerke UNEP EAF/14 Programme and other attendees of the UNEP-IOC Seagrass Assessment Workshop in Zanzibar, 1997; Mick Sharp for his patient reformatting of the book from Word into QuarkXPress; Judy Preece for preparing some thirty figures.

The book was pageset by Mick Sharp in QuarkXPress and Figures 1.1, 3.2-3.5, 5.7-5.8, 5.10, 5.15-5.17, 6.3-6.7, 7.1, 7.3, 7.5, 8.1, 9.1-9.4, 10.3, 10.6-10.7, 11.2, 11.9-11.10, 15.1, 16.1 and 19.1 were drawn by Judy Preece based on the sources acknowledged.

Contributors

Production of the *Handbook* was a team effort by the four authors. Alasdair Edwards had overall editorial control and oversaw production of the figures and plates. The lead authors on each section are listed in the Table of Contents along with the other members of the project team who had a significant input to those sections. Any queries on individual sections should be directed to the lead author of the section, at the addresses listed below.

The book should be cited as:

Green, E. P., Mumby, P.J., Edwards, A. J., Clark, C. D., (Ed. A. J. Edwards), 2000. Remote Sensing Handbook for Tropical Coastal Management. *Coastal Management Sourcebooks 3*, UNESCO, Paris. x + 316 pp.

Addresses

EDMUND P. GREEN
Centre for Tropical Coastal Management Studies,
Department of Marine Sciences and Coastal Management,
University of Newcastle, Newcastle upon Tyne NE1 7RU, UK.
Present address: World Conservation Monitoring Centre,
219 Huntingdon Road, Cambridge CB3 0DL, UK.

PETER J. MUMBY
Sheffield Centre for Earth Observation Science,
Department of Geography, University of Sheffield,
Sheffield S10 2TN, UK.
Present address: Department of Marine Sciences and Coastal Management, University of Newcastle,
Newcastle upon Tyne NE1 7RU, UK.

ALASDAIR J. EDWARDS
Centre for Tropical Coastal Management Studies,
Department of Marine Sciences and Coastal Management,
University of Newcastle, Newcastle upon Tyne NE1 7RU, UK.

CHRISTOPHER D. CLARK
Sheffield Centre for Earth Observation Science,
Department of Geography, University of Sheffield,
Sheffield S10 2TN, UK.

Table of Contents

How to Use this *Handbook*

The *Remote Sensing Handbook for Tropical Coastal Management* is aimed primarily at practitioners, that is those in government, non-governmental organisations (NGOs), universities, research institutes and consulting who are responsible for or directly involved in managing the coastal resources of tropical nations to ensure sustainable and wise use. It is also a unique compendium of information from diverse sources which will be invaluable to undergraduate and graduate students and research workers studying remote sensing in general and its application to coastal management or marine resource conservation in particular. Remote sensing is a rapidly developing subject and, although every effort has been made to ensure that information is accurate at the time of going to press, readers may wish to check out recent technological advances.

Linked to the *Handbook* is a computer-based remote sensing distance-learning module on *Applications of Satellite and Airborne Image Data to Coastal Management* produced for UNESCO which is available on CD-ROM or over the World Wide Web. This consists of image processing software (*Bilko for Windows* v. 2.0), digital images and training materials which teach practitioners how to carry out a range of key coastal management applications.

The *Handbook* is not intended to be read from cover to cover but to be dipped into and treated as a reference work. It outlines the ways in which remote sensing instruments carried by satellites and aircraft can practically assist the achievement of coastal management objectives with particular attention to the costs involved and the accuracy of outputs such as habitat maps. Cost and accuracy are two factors crucial to practitioners but factors which are all too often neglected by enthusiastic researchers intent on exploring the theoretical potential (i.e. what could be done given levels of resourcing, technical support and time seldom available to practitioners) of the new and exciting remote sensing technologies continually being developed.

With regard to accuracy, the research worker may be primarily interested in establishing a relationship between remotely sensed image data and, for example, a habitat type or fishery resource. However, such a worker will be less concerned as to whether the strength of the relationship and consequent accuracy of habitat/resource maps are sufficient for decision making, or with the likely economic consequences of decisions on resource management or habitat conservation. With regard to costs, a knowledge of the likely costs of inputs (images, processing, field survey) and the accuracies of outputs achievable for given levels of inputs is essential to practitioners in deciding whether a remote sensing approach is appropriate (i.e. will it accomplish objectives within the available budget), and, if it is, which technologies are most cost-effective. Given the importance of these factors to practitioners, they are emphasised throughout the *Handbook*.

Technical chapters present information at three levels to cater for readers with different levels of technical expertise and differing needs. At the start of each chapter is a layman's introduction to the techniques or processes being covered. This provides a concise background to the topic then gives an intuitive discussion of the concepts, objectives and practical importance of the techniques in non-technical language. These chapter introductions are likely to be all that a decision maker, project manager or funding agency personnel might need to gain a grasp of the key issues. For those with greater technical involvement, they provide a clear overview of the topic and indications of further sections it may be advantageous or necessary to study.

At the next level, the technical reader is guided in a practical step-by-step fashion through the techniques or processes under discussion. Flow diagrams are used extensively to allow the practitioner a clear view of the goals and the often complex route to achieving them. Also practical hints based on our experience are scattered through the text (flagged with the symbol ☞). So as not to interfere with the process of leading the practitioner through the techniques or processes, the mathematical and theoretical background is kept to a bare minimum in the main text. However, this background information may be included in boxes so those readers wishing to explore the essential mathematical foundation and theory behind techniques can do so. For those wishing to delve even deeper into techniques, each chapter has its own bibliography. Where appropriate, chapters also examine the sensitivity of outputs to increased sophistication of image processing and the costs and effort required, so that practitioners can assess the importance of techniques in the context of their project objectives.

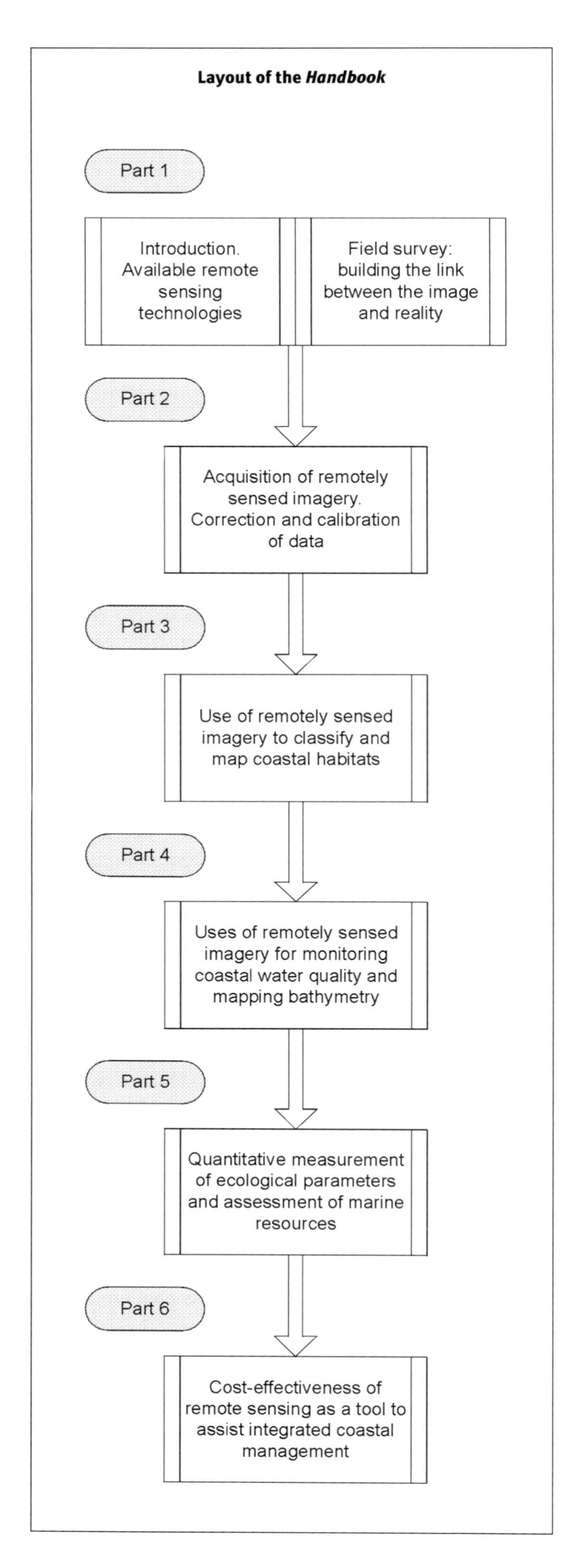

Layout of the *Handbook*

A synopsis of the *Handbook* is given in an initial section entitled *Guidelines for Busy Decision Makers* which provides readers with a one page *Executive Summary* and a nine page overview of key findings.

The main *Handbook* is divided into six parts, each comprising a number of chapters. **Part 1** forms an introduction to remote sensing for coastal managers. Following a general introduction to the remote sensing of coastal environments in Chapter 1, the precise objectives of coastal managers in using remote sensing are explored in Chapter 2. Objectives of remote sensing campaigns are quite often poorly defined (with a 'we used the technology because it was there' type of approach) and we suspect that many past remote sensing studies of tropical coasts would either not have been undertaken or would have been radically redesigned if objectives (and the inputs required to meet them) had been clearly defined. A major aim of the *Handbook* is that Chapter 2 and later chapters assist users in clear and achievable objective setting to avoid this problem. In Chapter 3 the plethora of sensors on satellites and aircraft which are available to assist in meeting management related objectives are reviewed and the key specifications of the sensors and the platforms carrying them are discussed. Constraints on sensor operation such as cloud cover and suspended sediments in coastal waters, and the effects of these constraints in practice, are outlined.

Chapter 4 deals with the crucial process of building the links between the remotely sensed image and the objects of interest on the Earth's surface with a practical guide to the planning of field surveys on tropical coasts and the assessment of the accuracy of outputs such as habitat maps.

In summary, Part 1 introduces the reader to the coastal management objectives achievable using remote sensing, the remote sensing technologies available to meet these objectives, and the need for concomitant field survey.

Part 2 of the *Handbook* covers the acquisition, correction and calibration of remotely sensed data. Firstly, the processes involved in acquiring satellite imagery, digital airborne imagery and aerial photographs of the type routinely used for coastal management applications are outlined in Chapter 5. Digital imagery is seldom directly useable for coastal management applications in the form in which it is supplied; the succeeding three chapters describe how to make it useable. The techniques for correcting the supplied image data are described in Chapters 6–8. These techniques seek to correct: i) spatial distortions of the image (Chapter 6), ii) scattering of radiation and other effects of the atmosphere lying between the sensor and the Earth's surface (Chapter 7), and for underwater features, iii) the confounding effects of the depth of seawater on the signal recorded from the seabed (Chapter 8). Without these corrections, outputs are not likely to contribute effectively to management objectives. By giving clear guidance as to how to carry out atmospheric and water column correction we hope that these will be more routinely carried out by practitioners and the quality of management information thereby improved.

In summary, Part 2 takes the reader through how to obtain remotely sensed image data then prepare those data so that they are optimised for achieving coastal management objectives.

Part 3 is perhaps the core of the *Handbook* and describes how the corrected image data can be used for the classification and mapping of coastal habitats. Chapter 9 looks at four approaches to defining and classifying habitats on the ground and their advantages and disadvantages in terms of different project objectives. Chapter 10 then describes how to classify images and match image classes with field survey data to produce habitat maps of known accuracy. The three succeeding chapters focus on three major types of coastal habitats – coral reefs and macroalgal assemblages, seagrass beds, and mangroves – and explore the particular problems and practical solutions of mapping these. In each chapter, the critical question of the accuracy of outputs achievable and the type of field survey required is explored in some detail.

In summary, Part 3 describes how remote sensing is used routinely to support coastal management initiatives and what can be readily accomplished in the way of mapping coastal habitats.

Part 4 describes how image data can be used for the mapping of water quality and bathymetry. Chapter 14 reviews the ways in which remote sensing can be used to monitor coastal water quality, particularly poor water quality resulting from pollution. Uses and limitations of remote sensing for monitoring sediment loadings, oil pollution, sewage discharges, toxic algal blooms, eutrophication, industrial wastes and thermal discharges are reviewed. Chapter 15 describes how the bathymetry of shallow water areas (<25 m deep) can be crudely mapped using passive optical imagery in areas where the water is relatively clear. Alternative approaches to mapping bathymetry using LIDAR and acoustic technologies are outlined briefly.

In summary, Part 4 reviews how remote sensing is used for coastal management applications other than habitat mapping.

Part 5 focuses on the next step in data analysis, which is to move beyond the mapping and classification of habitats to some sort of quantitative measurement of ecological parameters of the ecosystems supporting coastal production (Chapters 16 and 17) and assessment of key marine resources (e.g. fish, seaweeds, conch and *Trochus*) associated with particular habitats or oceanographic features (Chapter 18). Such studies involve the use of detailed field survey data to construct empirical models which establish relationships between image attributes and ecological parameters such as seagrass standing crop, mangrove canopy structure (e.g. leaf area index or percentage canopy closure), or the abundance of a resource. This type of assessment is relatively novel and requires the deployment of considerable technical and human resources in field survey. It does, however, provide outputs directly relevant to coastal management and is an area where remote sensing appears to offer great potential to resource managers. Whether this potential is likely to be achievable is discussed in some detail.

In summary, Part 5 takes the practitioner beyond the routine uses to applications where the greatest potential may exist to support management.

Part 6 consists of a single chapter which reviews the crucial question of the cost-effectiveness of remote sensing technologies in helping coastal managers meet resource management objectives. This is divided into two sections. The first of these briefly explores the objectives and yardsticks which can be used to assess 'cost-effectiveness' and how in the long term one might expect remote sensing campaign outputs to contribute to sustainable economic development of coastal areas (i.e. what precisely are the benefits). The second section is much more pragmatic and summarises the relative cost-effectiveness of a range of airborne and satellite sensors used to map habitats and assess resources in the Turks and Caicos Islands at meeting each of a number of objectives.

The *Handbook* ends with a brief look at the future prospects for remote sensing in the context of coastal management. Considerable useful reference information is given in a series of appendices.

Remote sensing technologies are evolving rapidly but we hope that this handbook provides a concise, informative and easily used companion for those involved in coastal management so that they can both use the existing technologies to full advantage and evaluate new ones as they become available.

Associated training module

A computer-based remote sensing distance-learning module on *Applications of Satellite and Airborne Image Data to Coastal Management* has been written for the UNESCO Coastal Regions and Small Islands (CSI) unit to link with the *Handbook*. It consists of eight lessons which relate to topics covered in *Handbook* chapters as follows:

1. Visual interpretation of images with the help of colour composites: getting to know the study area (relates to Chapters 4 and 10).

2. The importance of acquiring images of the appropriate scale and spatial resolution for your objectives (relates to Chapters 11–13).
3. Radiometric correction of satellite images: when and why radiometric correction is necessary (relates to Chapter 7).
4. Crude bathymetric mapping using Landsat TM satellite imagery (relates to Chapter 15).
5. Compensating for variable water depth to improve mapping of underwater habitats: why it is necessary (relates to Chapter 8).
6. Mapping the major inshore marine habitats of the Caicos Bank by multispectral classification using Landsat TM (relates to Chapters 9–11).
7. Predicting seagrass standing crop from SPOT XS satellite imagery (relates to Chapters 12 and 16).
8. Assessing mangrove leaf-area index (LAI) using CASI airborne imagery (relates to Chapters 13 and 17).

The image processing software (*Bilko for Windows* v. 2.0), digital images and training materials are available on CD-ROM or via the World Wide Web (http://www.unesco.bilko.org/). The training module allows practitioners to develop their image processing and interpretation skills on their desktop PC, guides them through fairly complex procedures such as radiometric and water column correction, and introduces them to exciting new resource assessment techniques such as mapping seagrass standing crop and mangrove leaf area index from satellites and digital airborne imagery. Once the techniques have been mastered they can be applied to local image data.

The module has been field tested at a workshop in East Africa under the auspices of the Intergovernmental Oceanographic Commission (IOC) and United Nations Environment Programme (UNEP) and has benefited from feedback from the regional scientists who participated.

Companion reading

We have found two books to be particularly useful and recommend them to practitioners as companion volumes to the *Handbook*. The first is *Computer Processing of Remotely-Sensed Images: An Introduction,* by Paul Mather, which provides an excellent and very readable introduction to the principles of remote sensing and techniques of image analysis, enhancement and display. A very useful feature of this book is the provision of a CD-ROM containing software and image data sets. The second is the *ERDAS Field Guide* produced by ERDAS Inc.†, the company which produces the popular *ERDAS Imagine* image analysis and GIS software. This guide naturally focuses on how to do image processing tasks using *ERDAS Imagine* but the wealth of detailed, very clearly explained hands-on advice and information is extremely useful to practitioners whatever software they are using.

References

Mather, P.J., 1999, *Computer Processing of Remotely-Sensed Images: An Introduction*. 2nd Edition. (Chichester: John Wiley & Sons). ISBN 0-471-98550-3 (paperback).

ERDAS Inc., 1994, *ERDAS Field Guide*. Third Edition. (Atlanta: ERDAS Inc.).

† Mention of a company name does not constitute endorsement of a particular product by the authors.

Guidelines for Busy Decision Makers

Executive summary

1. The purpose of this *Handbook* is to enhance the effectiveness of remote sensing as a tool in coastal resources assessment and management in tropical countries by promoting more informed, appropriate and cost-effective use.
2. A primary objective of the *Handbook* is to evaluate the cost-effectiveness (in terms of accuracy of resource estimates or habitat maps) of a range of commonly used remote sensing technologies at achieving objectives identified by the user community. The research activities reported were focused on the Turks and Caicos Islands with comparative work in Belize but outputs are seen as applicable in all countries where water clarity permits optical remote sensing.
3. Sensor technologies evaluated included: Landsat Multispectral Scanner (MSS), Landsat Thematic Mapper (TM), both multispectral (XS) and panchromatic (Pan) SPOT High Resolution Visible, Compact Airborne Spectrographic Imager (CASI) and aerial photographs. Spatial resolution of the digital imagery ranged from 1–80 m, and spectral resolution from 1–16 wavebands in the visible and infra-red.
4. For habitat mapping, which is the primary objective of users, the achievable accuracy of outputs depends on: i) the level of habitat discrimination required, ii) the type of sensor used, iii) the amount of ground-survey carried out, and iv) the image processing techniques used. Accuracies of outputs are evaluated for different levels of marine or mangrove habitat discrimination for each sensor with varying ground-survey and image processing inputs (costs).
5. Satellite-mounted sensors are only able to provide information on reef geomorphology and broadscale ecological information such as the location of coral, sand, algal and seagrass habitats with an accuracy ranging from 55–70%. For this type of information, the most cost-effective satellite sensors for habitat mapping are Landsat TM (for areas greater than one 60 x 60 km SPOT scene) and SPOT XS for areas within a single SPOT scene. Both these sensors can deliver overall accuracies of about 70%. The remaining satellite sensors tested (SPOT Pan and Landsat MSS) could not achieve 60% overall accuracy even for coarse-level (only 4 marine habitat classes being distinguished) habitat discrimination .
6. For fine-scale habitat mapping (9+ habitat classes) only colour aerial photography and airborne multispectral digital sensors offer adequate accuracy (around 60% or better overall accuracy). The most accurate means of making detailed reef or mangrove habitat maps involves use of digital airborne multispectral instruments such as CASI (Compact Airborne Spectrographic Imager). Using CASI we were able to map sublittoral marine habitats to a fine level of discrimination with an accuracy of over 80% (compared to less than 37% for satellite sensors and about 57% for colour aerial photography). Similar accuracy was achieved for mangrove habitats using CASI. Comparison of the costs of using CASI and colour aerial photography indicates that CASI is cheaper as well as producing more accurate results.
7. Additional field studies were carried out of the capabilities of remote sensing to map bathymetry and assess mangrove resources (Leaf Area Index (LAI) and percentage canopy closure) and seagrass standing crop ($g.m^{-2}$). Remote sensing offers very cost-effective rapid assessment of the status of the plant resources but passive optical remote sensing of bathymetry is considered too inaccurate for most practical purposes. The practicalities of using remote sensing technologies for i) coastal resources (fisheries, conch (*Strombus gigas*), *Trochus*, and seaweed resources) assessment, and ii) monitoring of coastal water quality (including sediment loadings, oil, thermal discharges, toxic algal blooms, eutrophication and other pollution) are reviewed.

Background

Habitat maps derived using remote sensing technologies are widely and increasingly being used to assess the status of coastal natural resources and as a basis for coastal planning and for the conservation, management, monitoring and valuation of these resources.

Despite the fact that Earth resources data from some satellites has been routinely available for over 25 years, there has until now been almost no rigorous assessment of the capacity of the range of operational remote sensing technologies available to achieve coastal management-related objectives.

Digital sensors commonly used for coastal management applications have spatial resolutions ranging from about 1 m to 80 m on the ground and spectral resolutions ranging from a single panchromatic band (producing an image comparable to a black and white photograph) to around 16 precisely defined wavebands which can be programmed for specific applications. Costs of imagery range from about £250 for a low-resolution (80 m pixel size) satellite image covering 35,000 km^2 to around £80,000 for a high-resolution (3 m pixel size) airborne multispectral image covering less than half this area. In addition, high-resolution analogue technologies such as colour aerial photography are still in routine use.

Coastal managers and other end-users charged with coastal planning and management and the conservation and monitoring of coastal resources require guidance as to which among this plethora of remote sensing technologies are appropriate for achieving particular objectives. A study of reports of the use of remote sensing in developing countries and research of other available literature indicated that there is wide use of the various technologies but no clear idea of the extent to which different technologies can achieve objectives. In addition, objectives are often poorly defined.

Data which would allow clear guidance to be given have not been available, with most remote sensing research being devoted to potential applications of an often experimental nature and little attention being paid to the operational realities or the costs of applications. This *Handbook*, based on three years research funded by the UK Department for International Development (DFID), seeks to fill the gap and provide practical guidance to end-users, particularly those in developing countries.

The Caicos Bank was chosen as the site for the research to discover what coastal management objectives are realistically achievable using remote sensing technologies. It offered almost ideal clear-water conditions and a broad mix of coastal habitat types, a very large area (>10,000 km^2) of shallow (<20 m deep) coastal waters which would be amenable to the technologies. Results from the site can thus be considered as best-case scenarios. To test the generic applicability of the outputs, additional work was carried out in Belize with the UNDP/GEF Belize Coastal Zone Management Project and Coral Cay Conservation. Of particular concern was the quantification of the accuracies achievable for all outputs, since habitat or resource maps of poor or unknown accuracy are of little use as a basis for planning and management. The key research findings are listed hereafter.

Key findings

Uses of remote sensing in the coastal zone

Chapter 2 of the *Handbook* examines the objectives of coastal managers in using remote sensing. Coastal managers and scientists around the world were asked a) to identify what they saw as the primary applications of remote sensing and b) to prioritise the usefulness to them of various levels of information on coastal systems.

The most in-demand applications of remotely sensed data were to provide background information for management planning and to detect coastal habitat change over time (Figure 1). The term 'background information' reflects the vagueness with which habitat maps are often commissioned and indicates a need for objectives of remote sensing surveys to be defined more rigorously. Some 70% of respondents who were using remote sensing for change detection were concerned with mangrove assessment and/or shrimp farming. The primary uses made of the habitat/resource maps are shown in Figure 1.

Matching available technologies with managers' objectives

Analysis of the questionnaire responses revealed that managers sometimes had unrealistic expectations of what remote sensing could achieve, notably in the management of seagrass and coral reef habitats. However, for mangroves the outputs which managers considered most useful (mapping boundaries, clearance and density) are readily achievable using current technologies. The *Handbook* seeks to close the gap between managers' expectations and what can be realistically achieved with remote sensing in an operational as opposed to research context.

The need to integrate field survey and remote sensing

Remote sensing is often (erroneously) considered as an alternative to field survey. It should be seen as a complementary technology which makes field survey more cost-effective. A solely field-survey based approach to habitat mapping is shown to be extremely cost-inefficient. For example, it would cost about £0.5 million excluding staff salaries to produce a coarse-level habitat map of the *ca* 15,000 km^2 Caicos Bank to reasonable accuracy without remote sensing inputs and take a team of three more than eight years! On the other hand, a remote sensing approach without extensive field survey is shown to be too inaccurate. For example, <50% overall accuracy is achievable for unsupervised classification (i.e. one without field survey inputs) of a four habitat class (reef, macroalgae, sand, seagrass) Landsat TM image subjected to the highest level of processing. The integrated approach

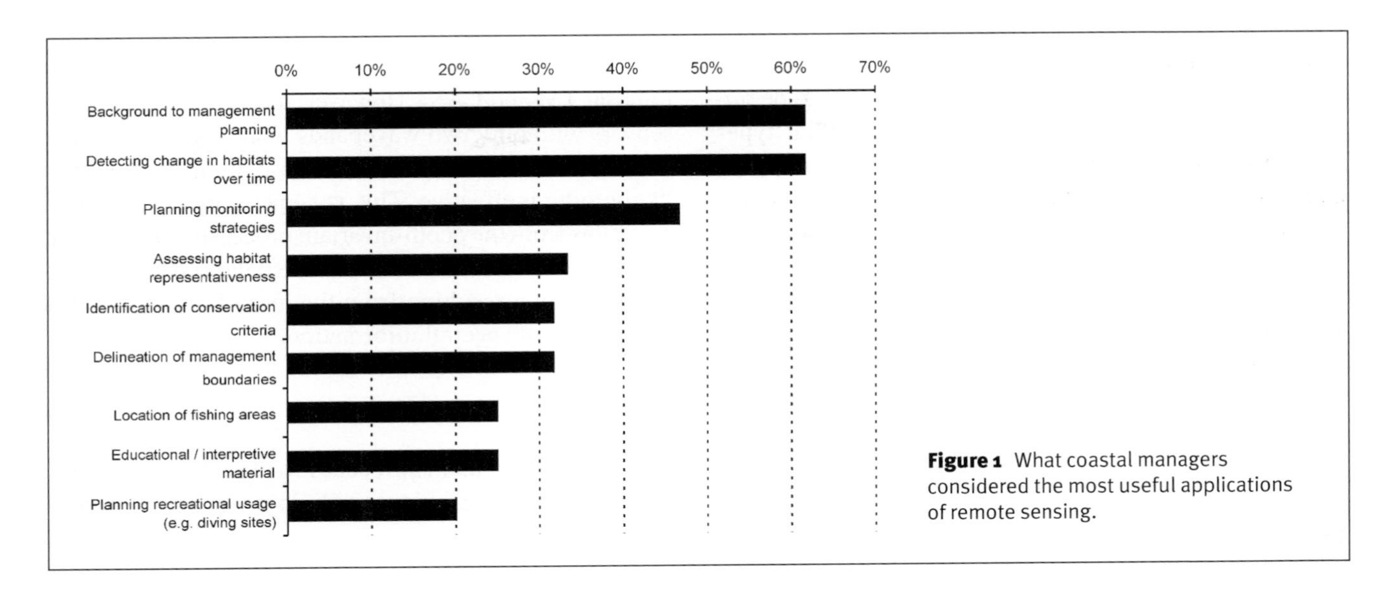

Figure 1 What coastal managers considered the most useful applications of remote sensing.

is summarised in the following box:

- Remote sensing is very good at indicating the extent of habitats and location of boundaries between habitats;
- Field survey identifies what the habitats are;
- Digital image processing then extends the field survey coverage to the whole area of management interest.

The end-user community has too often failed to appreciate this; the *Handbook* (particularly Chapters 9 and 10) seeks to remedy this misapprehension, emphasising the need for an integrated field and image-based approach.

Cloud cover is a key constraint

Most projects requiring remotely sensed imagery have to be completed within a timeframe of one to three years. The question of whether appropriate imagery is likely to be obtainable in such a timeframe is sometimes overlooked at the planning stage. A primary constraint which project planners need to consider when using optical imagery is cloud cover. This constraint does not apply to radar imagery such as that from synthetic aperture radar (SAR) sensors, which can operate through cloud. However, radar does not penetrate water and thus is of limited use for coastal management.

The extent to which cloud cover is a constraint to image acquisition was analysed for a range of developing nations from those with arid desert coastlines to those comprised of high islands in the humid tropics (Chapter 5 of the *Handbook*). Data sources were both archived imagery and local weather stations. In the humid tropics there was a 4–10% chance (depending on location) of obtaining a satisfactory image (defined as one with ≤12.5% cloud cover) on a given satellite overpass, equivalent to a 41–77% chance of at least one good image per year for SPOT sensors (if not specially programmed) or a 58–91% chance per year for Landsat sensors, assuming these are recording at the time of every overpass. However, sensors are often inactive over tropical areas (particularly small-island states) because of low commercial demand and this is reflected in the relatively low numbers of archived images available for such areas. This means that there is often a less than even chance of obtaining a reasonable image in a year. Planners and managers need to be aware that this can present a significant constraint to applying satellite remote sensing technologies in developing countries, particularly where projects have short timeframes or require up-to-date imagery.

Maps of unknown accuracy are of little value to managers

The manager or other end-user needs to know the accuracy of the habitat maps or other remote sensing products which are to be used in planning and management (Plate 1). We found that only a quarter of 86 tropical studies reviewed included any assessment of accuracy. Although additional field survey costs will be incurred to enable an accuracy assessment of a habitat map to be carried out, one has to ask what value a map of poor or unknown accuracy has as a basis for management and planning? The *Handbook* emphasises throughout the need for accuracy assessment and describes the statistical and field survey methodologies required (Chapter 4). Such assessment must be an integral part of any remote sensing survey if the results are intended for more than mere wall decoration.

The importance of rigorous methodologies for defining habitats

In order for remote sensing studies to realise their full usefulness, particularly for the key objective of change detection identified by end-users (Figure 1), the field survey methodologies used to define habitats need to be rigorous and preferably quantitative so that habitat types defined in one study can be clearly identified in subsequent ones. *Ad hoc* methods, though quick and cheap, produce outputs of very limited use and are a false economy. Chapter 9 reviews which methodologies are suitable for different objectives and advocates a hierarchical approach to allow a) descriptive resolution (i.e. the level of habitat discrimination) to be tailored to management objectives and sensor capabilities, and b) habitat categories to reflect natural groupings appropriate to the area under study.

The need for atmospheric and water column correction

Most remotely sensed images will be supplied already geometrically corrected, so that they can be used like a map (Chapters 5 and 6). For most coastal applications the images will also require radiometric and atmospheric correction. These corrections take into account sensor peculiarities and the effects of atmospheric haze, the angle of the sun at the time the image was acquired, etc. which would otherwise make it impossible to compare one image with another in any meaningful way. For underwater habitat mapping, water column correction is also recommended to compensate for the effect of water depth (due to attenuation of light in the water column) on the signal received by the sensor. If this is not done, then it is difficult to distinguish marine habitats reliably because depth effects dominate the image.

Remarkably, water column correction was carried out in only 9% of 45 underwater habitat mapping studies reviewed, often leading to unnecessarily low accuracies in habitat classification and thus inefficient use of resources. The main reason for the poor uptake of the technique appears to be that users have found the rather mathematical research papers describing the techniques daunting and difficult to implement. The techniques are thus underutilised despite being essential for cost-effective achievement of many common management objectives. Chapters 7 and 8 seek to present these image processing techniques in a more accessible and intuitive way so that they will be more routinely adopted.

For change detection (one of the two principal objectives of remote sensing identified by the user-community) radiometric and atmospheric correction are essential if information on habitat spectra generated by the supervised classification of one image are to be made use of in classifying subsequent or neighbouring images. Not using such information is very wasteful of resources.

Water column correction significantly improves the accuracy of habitat maps for multispectral sensors with more than two wavebands that penetrate water, such as Landsat TM and CASI. However, for those multispectral sensors with only two wavebands which penetrate water (SPOT XS and Landsat MSS) water column correction is likely to be ineffective. This is because in these sensors any gains from the depth-invariant processing to carry out water column correction seem to be balanced by reduced discrimination resulting from the loss of one of the two dimensions of spectral information.

Image classification

There are three main approaches to classifying remotely sensed images (assigning pixels in an image to habitat types). These are:

- photo-interpretation/visual interpretation,
- unsupervised classification of multispectral imagery, and
- supervised (using field survey information) multispectral image classification.

These approaches are described and evaluated in the *Handbook* for a range of applications (Chapters 10–13). All have their uses although supervised multispectral classification based on extensive field survey was found to be most effective and generally produced the most accurate habitat maps. Visual interpretation of remotely sensed imagery (identification by eye of habitat types based on their colour, tone, texture and context within the imagery) can often reveal considerable detail on the nature and distribution of habitats. However, the subjectivity and operator dependence of visual interpretation is a major drawback, particularly if comparisons over time are envisaged.

Unsupervised classification (i.e. a computer-based classification or one based upon a good local knowledge of the habitats, local maps and field experience but without use of field survey data from known positions) produced maps with unacceptably low overall accuracies which for Landsat TM were less than 50% at coarse habitat resolution (compared to over 70% using supervised classification).

Habitat maps produced using supervised multispectral image classification can be significantly improved by additional image processing such as water column correction (see above) and contextual editing (the addition of context-dependent decision rules into the image classification process – done automatically by the brain in visual interpretation). Gains in accuracy ranged from 6–17% per day's processing effort. Use of contextual editing and water column correction together significantly improved the accuracy of habitat maps derived from all the multispectral sensors tested.

Operational versus theoretical spatial resolution of aerial photography

Colour aerial photographs at a photographic scale of 1:10,000 to 1:25,000 can offer potential ground resolution of the order of 0.5–1 m (the minimum detection unit). However, conventional mapping techniques do not make use of this resolution because to define polygons at such fine spatial scales would be prohibitively time-consuming and practically difficult. Thus minimum mapping units at the photographic scales above are in the order of 10–20 m in diameter, equivalent to the spatial resolution of SPOT satellite sensors. At least one order of magnitude of spatial resolution is therefore lost during preparation of habitat maps using standard aerial photographic interpretation (API) techniques. However, the high resolution does provide extra detail on the texture of habitats, which facilitates the visual interpretation process and thus allows more accurate assignment of habitats than would be true if the minimum detection unit were the same as the minimum mapping unit.

Mapping coral reefs and macroalgae

Satellite-mounted sensors are able to provide useful information on reef geomorphology and broadscale ecological information such as the location of coral, sand, algal and seagrass habitats with accuracies of around 70% for the best satellite sensors (Landsat TM and SPOT XS). However, even the best sensors could not provide medium to finescale ecological classifications of habitats at better than 50% to 35% accuracy respectively and, unless such low accuracies are acceptable, we do not recommend using satellite imagery for such purposes.

The most cost-effective satellite sensors for broadscale habitat mapping (e.g. four submerged habitat classes) are Landsat TM (for areas greater than one SPOT scene) and SPOT XS for areas within a single SPOT scene (i.e. <60 km in any direction).

Colour aerial photography can resolve more detailed ecological information on reef habitats but for general purpose mapping (4–6 reef habitat classes) satellite imagery is more effective (i.e. slightly greater accuracy and much faster to use). However, at finer levels of habitat discrimination (9–13 reef habitats) aerial photography performs significantly better than the best satellite sensors.

The most accurate means of making detailed reef habitat maps involve the use of airborne multispectral instruments such as CASI (Compact Airborne Spectrographic Imager). In the Caribbean, CASI can map coral reef habitats at a fine level of discrimination with an accuracy of over 80% (compared to less than 37% for satellite sensors and about 57% for colour aerial photography).

Mapping seagrass beds and assessing seagrass standing crop

The location and extent of seagrass beds can be mapped reasonably accurately with satellite imagery (about 60% user accuracy). Landsat TM appears to be the most appropriate satellite sensor although some mis-classification may occur between seagrass habitats and habitats with similar spectral characteristics (e.g. coral reefs and algal dominated areas). Contextual editing significantly improves the accuracy of seagrass habitat maps. Airborne multispectral digital imagery (e.g. CASI) allows seagrass beds to be mapped with high user accuracies (80–90%) and permits changes of a few metres in seagrass bed boundaries to be monitored. Similarly, the interpretation of colour aerial photography (API) can generate maps of reasonably high accuracies (around 60–70%) but map-making may be time-consuming where seagrass patches are small. Although API is excellent for mapping high-density seagrass (standing crop = 80–280 $g.m^{-2}$), it is poor at mapping low to medium density seagrass (standing crop = 5–80 $g.m^{-2}$). A non-destructive visual assessment technique for standing crop was found to be very cost-effective and is recommended for ground-truthing (Chapter 12).

Remote sensing is well suited to mapping the standing crop of seagrass of medium to high density and SPOT XS, Landsat TM and CASI performed similarly overall. To map seagrass standing crop, 'depth-invariant bottom-index' images are created (using the water column correction techniques discussed above) and those which show the closest correlation between bottom-index and seagrass standing crop for ground-truthed pixels are used to predict standing crop over the whole area of interest.

Accurate predictions of low standing crop are difficult due to confusion with the substrate and increased patchiness but, at higher biomasses, measurement of standing crop by remote sensing compares favourably with direct quadrat harvest and, unlike the latter, has no adverse impact on the environment (see Figure 16.7). Chapter 16 describes how to map seagrass standing crop.

Detecting seagrass degradation is a key application. Allowing for known errors in image rectification and classification we estimate that satellite sensors can detect changes in seagrass bed boundaries in the order of 20–90 m (2–3 pixel widths) with confidence. These are huge changes! Airborne digital sensors are capable of more accurately identifying seagrass habitats and providing spatially-detailed maps of biomass with spatial errors <10 m. This allows monitoring of change in seagrass standing crop in two dimensions at a more useful scale and would, for example, allow the impacts of a point source of pollution (e.g. a thermal discharge) on a seagrass bed to be monitored.

Mapping mangroves and assessing the status of mangrove resources

Mangrove areas are often difficult to reach and equally difficult to penetrate. Thus, field survey of mangroves is logistically demanding particularly where the areas are large. Remote sensing offers a very cost-effective method of extending limited field survey to map large areas of mangroves. Mangrove habitat maps are primarily used for three management applications: resource inventory, change detection and the selection and inventory of aquaculture sites. Chapters 13 and 17 analyse a range of remote sensing approaches to mangrove mapping and the quantitative assessment of mangrove resources using satellite and airborne imagery, and make recommendations as to the most effective options.

Image processing techniques appropriate for mangrove mapping can be categorised into five main types: 1) visual interpretation, 2) vegetation indices, 3) unsupervised classification, 4) supervised classification, and 5) principal components analysis of band ratios. The image processing method (5) based on taking ratios of different red and infrared bands and using these as inputs to principal components analysis (PCA) generated the most consistently accurate maps of mangroves.

In the Turks and Caicos Islands mangrove vegetation could be distinguished from terrestrial vegetation (thorn scrub) with high accuracies (~90%) using Landsat TM or CASI data. However, it is not always readily distinguishable from adjacent habitats. SPOT XS imagery could not separate mangrove from thorn scrub in the Turks and Caicos Islands, although it has been used to map mangrove successfully elsewhere. Similarly, problems of distinguishing mangroves from rainforest have been reported for Landsat MSS and TM sensors in Australia and Indonesia. At coarse levels of habitat resolution (only two types of mangrove habitat being distinguished), Landsat TM data were found to be more cost-effective than CASI. However, CASI data were greatly superior when it came to mapping mangroves at fine levels of habitat discrimination (nine habitat classes) where Landsat TM had an overall accuracy of only 30%. Using CASI, finescale mangrove habitats of the Turks and Caicos Islands could be mapped to an overall accuracy of 83% (Figure 10.9, Plate 11).

Thus, in the context of qualitative mapping, satellite imagery may achieve little more than mapping mangrove occurrence and distinguishing mangrove from non-mangrove vegetation in areas where mangroves are poorly developed. Where mangroves are well-developed and extensive (e.g. Sunderbans, Niger delta, etc.) satellite imagery may be more effective and allow mapping of major zonation. By contrast, digital airborne imagery can provide detailed and accurate maps of the zonation of mangrove ecosystems even where poorly developed. However, in a quantitative context remote sensing has more to offer.

A number of vegetation indices which can be derived from satellite and airborne digital imagery of mangrove areas show a close correlation with the leaf area index (LAI) and percentage canopy closure of mangroves. LAI can be used to predict growth and yield and to monitor changes in canopy structure due to pollution and climate change. It is thus a good measure of the status and productivity of mangrove ecosystems. Similarly, canopy closure can been used as a measure of tree density. LAI and percentage canopy closure can be predicted with reasonable confidence (Table 1) from Landsat TM, SPOT XS (using the TM image to mask out non-mangrove vegetation) and CASI airborne images using the correlation between Normalised Difference Vegetation Index (NDVI) images and these parameters. Other image-derived indices including the Global Environment Monitoring Index (GEMI) and Angular Vegetation Index (AVI) gave equally good correlations.

Table 1 The accuracy of LAI and percentage canopy closure prediction using three image types.

	Landsat TM	SPOT XS	CASI
Leaf area index	71%	88%	94%
Canopy closure	65%	76%	80%

Higher spatial resolution SPOT XS imagery (20 m pixel size) was more successful than Landsat TM (30 m pixel size), with the 1 m resolution CASI being best of all. Thus, both satellite and airborne digital imagery can provide useful forestry management information on mangroves which would be very difficult to obtain by field survey.

Monitoring coastal water quality

Chapter 14 of the *Handbook* reviews how remote sensing may aid coastal managers in assessing and monitoring coastal water quality. Pollution of coastal waters (leading to poor water quality) by, for example:

- sediment run-off as a result of deforestation or unsustainable agricultural development,
- fertilisers and pesticides from agriculture,
- oil spills from land-based or marine installations,
- thermal discharges from power-stations or other industry requiring cooling water,
- discharges of insufficiently treated sewage,
- toxic and other industrial wastes, and
- nutrient-rich effluents from coastal aquaculture ponds,

are just some of the issues which coastal managers may be called upon to assess or monitor. Such pollutants (loosely defined here as substances in the water column which are likely to have detrimental effects on the environment at the levels observed – see below for a more formal definition) may have adverse impacts on coastal ecosystems and the economically important resources that these support. Thus, monitoring of 'water quality' is in practice about detecting, measuring and monitoring of pollutants.

Pollution means the introduction by man, directly or indirectly, of substances or energy into the marine environment (including estuaries) resulting in such deleterious effects as harm to living resources, hindrance to marine activities including fishing, impairment of quality for use of seawater and reduction of amenities (GESAMP, 1991).

Water quality is far more ephemeral and dynamic than habitat status, often changing dramatically over a tidal cycle, or after a monsoon downpour, or following the flushing of an aquaculture pond or the discharge of a tanker's ballast water. Thus, the usefulness of water quality measurements at a single point in time (as available from a single satellite image) is likely to be far less than, say, knowledge of the habitat type at a particular location.

Since most incidents of pollution are likely to be short-lived (hours to days) and relatively small in size (tens to hundreds of metres across) one needs high temporal resolution and moderately high spatial resolution in order to monitor them. Further, since the pollutants can only be detected and measured if they can be differentiated from other substances in the seawater, one usually also needs a sensor of high spectral resolution.

Such a combination of characteristics is at present only available from airborne remote sensors. Satellite sensors are thus only likely to be useful in coastal water quality assessment where mapping large scale sediment plumes from estuaries or huge oil spills such as that released during the Gulf War of 1990/1991. This means that application of effective remote sensing techniques in pollution monitoring is likely to be expensive (costing one thousand to several thousand dollars a day) because of the costs of hiring sensors and aircraft to fly them and the probable need for repeat surveys to establish the dynamics of the pollution in space and time. These high costs, the relatively small areas which need to be investigated (particularly where point-sources are involved), and the need to collect field survey data for calibration (establishing a quantitative relationship between the amount of pollutant and some characteristic measured by the remote sensor) and legislative purposes anyway, may make the cost-effectiveness of using remote sensing to aid pollution monitoring questionable.

Remote sensing can, at present, only directly detect some of the pollutants listed above, for example, oil, sediment and thermal discharges. The others can in some cases be detected by proxy but only via ephemeral site-specific correlations between remotely sensed features and the pollutants in question. Thus, fertilisers, pesticides, nutrients and colourless toxic and industrial wastes (containing, for example, heavy metals, aromatic hydrocarbons, acid, etc.) cannot be detected but in some instances their concentrations (measured by field survey at the time of image acquisition) may be correlated with levels of substances which can be detected. Such experimentally demonstrated ephemeral relationships are still far from allowing operational detection and monitoring of these pollutants in a cost-effective way.

Whilst remembering these shortcomings, the primary benefits from adding a remote sensing perspective are:

- rapid survey for pollutants over large or remote areas which could not feasibly be surveyed from the surface,
- a synoptic overview of the pollution and clear demarcation of its boundaries, and,
- once reliable calibration has been achieved, the potential for measurement of the pollutant without the need for more field survey.

The *Handbook* critically reviews the operational use of remote sensing for the study of sediment loadings in coastal waters, oil pollution, sewage discharges, toxic algal blooms and eutrophication, and thermal discharges.

Mapping bathymetry

Optical remote sensing has been suggested to offer an alternative to traditional hydrographic surveys for measuring water depth, with the advantage that data are collected synoptically over large areas. However, bathymetry can only be derived from remote sensing to a maximum depth of about 25 m in the clearest water, and considerably less in turbid water. Bathymetric mapping is also often confounded by variation in seabed albedo (e.g. change from white sand to dark seagrass).

Chapter 15 of the *Handbook* describes and investigates three commonly used methods for mapping bathymetry from satellite imagery, with worked examples from the Turks and Caicos Islands. The accuracy of depth prediction of each method was tested against field data. Predicted depth correlated very poorly with actual depth for all but one method. Even this method does not produce bathymetric maps suitable for navigation since the average difference between depths predicted from imagery and ground-truthed depths ranged from about 0.8 m in shallow water (<2.5 m deep) to about 2.8 m in deeper water (2.5–20 m deep).

By contrast, airborne LIDAR can produce very accurate (±30 cm) and detailed bathymetric charts to 50–60 m in clear waters and about 2.5 times the depth of penetration of passive optical remote sensing technologies. Boat-based acoustic surveys using single or multibeam depth sounders can produce bathymetric maps of similar accuracies and can operate to depths in excess of 500 m. The drawback of such boat-based techniques is that it may not be feasible to survey shallow water less than 2–3 m deep because of sounder saturation and/or the draught of the survey boat. If bathymetric charts are an objective then airborne LIDAR or boat-based acoustic techniques (depending on depth range required) appear to be the remote sensing technologies to use. However, there may be instances where crude bathymetric maps which indicate major depth contours are useful (see Figure 15.10, Plate 17).

Assessment of marine resources

With rare exceptions (such as some near-surface schooling fish and marine mammals, which can be remotely sensed from aircraft, and seaweeds, which can be mapped from both satellites and aircraft) most marine resources cannot be assessed directly using remote sensing. However, there is potential for assessing some stocks indirectly using a combination of field survey and remote sensing technologies. Chapter 18 of the *Handbook* briefly considers how remotely sensed characteristics of coastal waters, such as chlorophyll concentration and sea surface temperature, and the ability to discriminate habitats, may help in assessing potential levels of economic resources such as finfish and shellfish.

The progression from mapping of habitats or ocean colour to assessment of living aquatic resources requires the establishment of a significant quantitative relationship between some feature which the remote sensor can reliably detect and levels of the living resource being studied. Establishing such a relationship requires a large amount of field survey data and thus any attempts to assess resources using remote sensing are likely to be costly.

In terms of cost-effectiveness, the question which needs to be asked is whether the costs of the addition of a remote sensing component to the stock assessment study are outweighed by the benefits. Remote sensing does not replace traditional methods of stock assessment, it only augments existing methods and may allow a) a more accurate determination of stocks, and b) some reduction in the need for field survey (once a clear relationship between a feature which can be distinguished on the imagery and levels of the stock being assessed has been established).

The *Handbook* examines these issues in relation to the assessment of:

- phytoplankton and primary production,
- coastal fisheries resources (using both direct and indirect remote sensing approaches),
- Queen Conch (*Strombus gigas*) resources,
- *Trochus* resources, and
- seaweed resources.

Cost-effectiveness of remote sensing

The cost-effectiveness of a remote sensing survey is perhaps best assessed in relation to alternative means of achieving the same management objectives. The three primary objectives of remote sensing surveys identified by end-users were: providing a background to management planning, detecting change in habitats over time, and planning monitoring strategies (Figure 1). In all cases the expected outputs are habitat/resource maps of varying detail. Once the political decision has been taken that coastal resource management needs to be strengthened and that as part of this process an inventory of coastal ecosystems is required, the question becomes: 'Does a remote sensing approach offer the most cost-effective option to achieve this objective?' Chapter 19 of the *Handbook* examines both this question and that of which remote sensing technologies are most cost-effective for which objectives.

The only real alternative to using remote sensing to map the marine and shoreline habitats is to carry out the mapping using boat-based and land-based survey alone. Such surveys would involve inordinate amounts staff time as well as boat and vehicle expenses. We calculate that, even with staff salaries excluded, mapping a large area (e.g. the Caicos Bank) using an integrated remote sensing/field survey approach using Landsat TM for coarse-level habitat discrimination would cost less than 3% of the costs of using field survey alone. Even for a small area of 16 km^2 (the median size of marine protected areas), a detailed survey of marine habitats using boat-based surveys would cost nearly double that of hiring CASI in hire charges, not to mention involving an extra 16 person-months of staff time just to collect the data!

The savings achieved from an integrated remote sensing/field survey approach rely on reducing the amount of costly field survey needed and using the imagery to extrapolate reliably to unsurveyed areas. For example, the project collected field survey data from 1100 km^2 of the Caicos Bank, about 7% of the area mapped, at a boat-survey cost (excluding staff time) of approximately £2700 over about 6 weeks.

So, although almost 70% of questionnaire respondents considered the cost of remote sensing as a main hin-

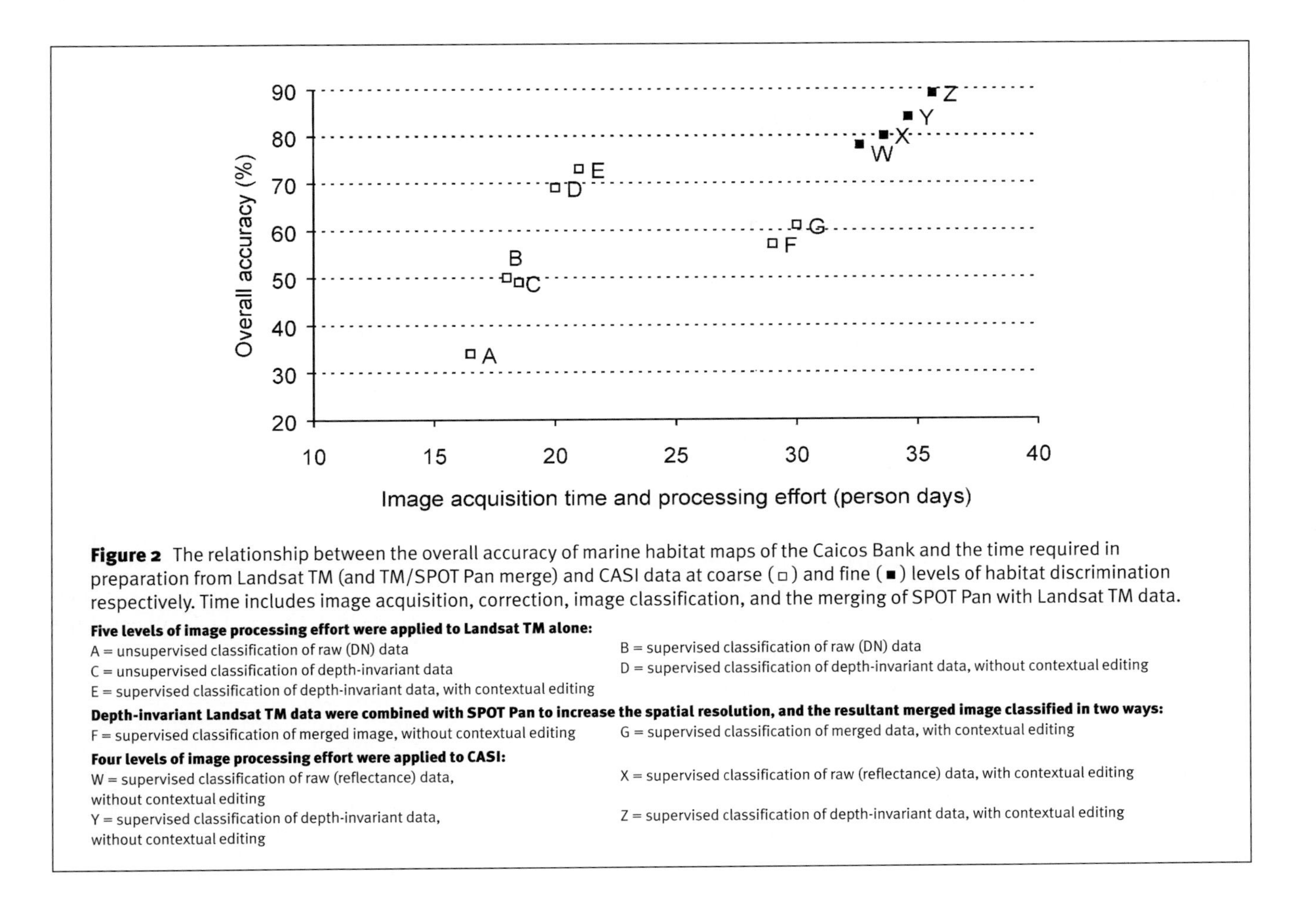

Figure 2 The relationship between the overall accuracy of marine habitat maps of the Caicos Bank and the time required in preparation from Landsat TM (and TM/SPOT Pan merge) and CASI data at coarse (□) and fine (■) levels of habitat discrimination respectively. Time includes image acquisition, correction, image classification, and the merging of SPOT Pan with Landsat TM data.

Five levels of image processing effort were applied to Landsat TM alone:
A = unsupervised classification of raw (DN) data
B = supervised classification of raw (DN) data
C = unsupervised classification of depth-invariant data
D = supervised classification of depth-invariant data, without contextual editing
E = supervised classification of depth-invariant data, with contextual editing

Depth-invariant Landsat TM data were combined with SPOT Pan to increase the spatial resolution, and the resultant merged image classified in two ways:
F = supervised classification of merged image, without contextual editing
G = supervised classification of merged data, with contextual editing

Four levels of image processing effort were applied to CASI:
W = supervised classification of raw (reflectance) data, without contextual editing
X = supervised classification of raw (reflectance) data, with contextual editing
Y = supervised classification of depth-invariant data, without contextual editing
Z = supervised classification of depth-invariant data, with contextual editing

drance to using it for coastal habitat mapping, the issue is not that remote sensing is expensive but that habitat mapping is expensive. Remote sensing is just a tool which allows habitat mapping to be carried out at reasonable cost.

Four types of cost are encountered when undertaking remote sensing: (1) set-up costs, (2) field survey costs, (3) image acquisition costs, and (4) the time spent on analysis of field data and processing imagery. The largest of these are set-up costs, such as the acquisition of hardware and software, which may comprise 40–72% of the total cost of the project depending on specific objectives. However, it should be borne in mind that most of the institutional set-up expenditure (computer hardware and software, maps/charts, almanacs, remote sensing textbooks and training courses, etc.) needed to establish a remote sensing capability is likely to be necessary anyway if an institution is to make effective use of the output habitat maps.

For coarse-level habitat mapping with satellite imagery, the second-most important cost is field survey, which accounts for about 25% of total costs and over 80% of costs if a remote sensing facility already exists (i.e. in the absence of set-up costs). Field survey is a vital component of any habitat mapping programme and may constitute approximately 70% of the time spent on a project (e.g. 70 days for mapping more than 150 km^2 of reef with satellite imagery).

The next two sections show how time invested in image processing and in field survey can dramatically improve the accuracy of outputs.

Investment in image processing is time well spent

The relationship between accuracy of outputs and amount of image processing input is examined in Figure 2.

Two key points emerge from Figure 2. Firstly, to achieve acceptable accuracy a considerable investment in staff time is required to carry out image acquisition, correction and subsequent processing; approximately one month for Landsat TM (and other satellite imagery such as SPOT XS), and about 1.5 months for CASI or similar airborne imagery. Secondly, increased image processing effort in general leads to increasing accuracy of outputs. This said, merging Landsat TM data with SPOT Pan to increase spatial resolution, despite producing visually pleasing outputs (see Figure 19.5, Plate 24), produced no benefits in terms of improved accuracy (in fact reducing it). Given the costs of inputs such as field survey data and set-up costs, extra effort devoted to image processing is clearly very worthwhile.

Accurate habitat mapping is dependent on adequate field survey

Field survey represents a major component of total costs. To assess how increasing field survey affects the accuracy of outputs, 25%, 50%, 75% and 100% of field survey data (from 157 sites) were used as inputs to direct supervised classification of a Landsat TM image. The accuracy of the resulting habitat maps was then assessed using independent field data. Figure 3 shows that the amount of field survey used to direct supervised classification profoundly influenced the accuracy of outputs although the increase in accuracy between 75% and 100% of field survey inputs was not significant. As a goal when extrapolating to large areas (several hundred km^2) we recommend that around 30 sites per class be surveyed in detail for use in directing supervised classification of multispectral imagery. To assess the accuracy of outputs we recommend that around 50 independent accuracy-assessment sites be visited per class. Using 75% of the signature file in Figure 3 would give about 30 sites per class at coarse-level (four habitat classes). For smaller study areas fewer survey and accuracy-assessment sites will be needed.

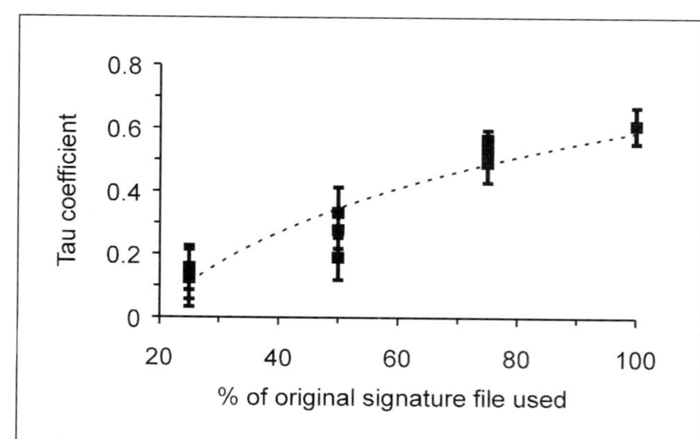

Figure 3 The effect of increasing amounts of fieldwork on classification accuracy (expressed as the tau coefficient) for Landsat TM supervised classification (coarse-level habitat discrimination). The vertical bars are 95% confidence limits. Three simulations were carried out at each level of partial field survey input (25%, 50% and 75% of signature file). A tau value of 0.6 equates to about 70% overall accuracy.

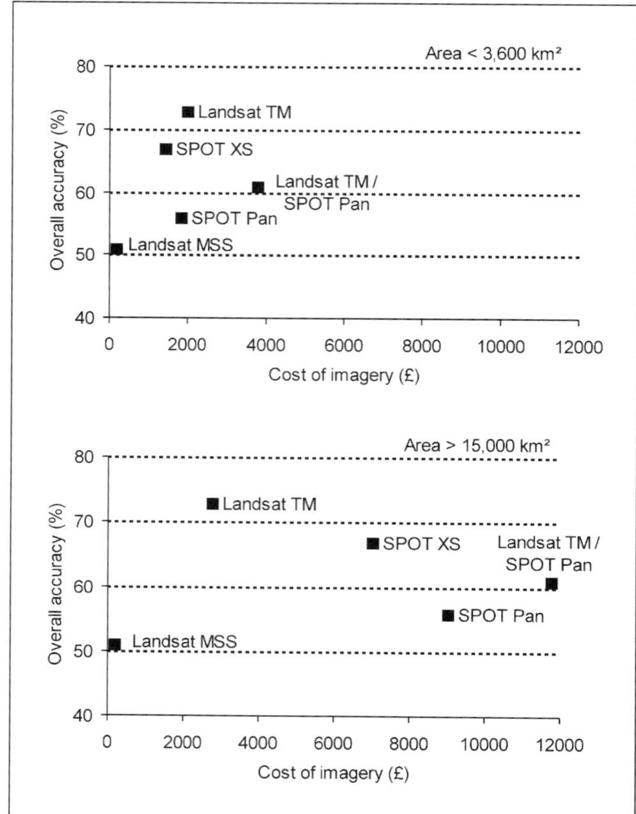

Figure 4 Cost-effectiveness of various satellite sensors for mapping coastal habitats of the Caicos Bank with coarse detail. Upper figure shows imagery costs for a relatively small area, such as would be contained within a single SPOT scene (3,600 km^2). Lower figure shows costs for a larger area (about the size of the Caicos Bank), which would fall within a single Landsat scene but would require five SPOT scenes.

Conclusions

Since the set-up costs, field survey costs and analyst time do not vary appreciably between satellite sensors, the selection of cost-effective methods boils down to map accuracy and the cost of imagery, the latter depending on the size of the study area and choice of sensor. For coarse-level habitat discrimination, Landsat TM is generally the most cost-effective option, allowing habitat maps of over 70% user accuracy to be created over large areas at relatively modest cost. However, for areas contained within one 60 x 60 km SPOT scene, SPOT XS is likely to be more cost-effective, the small decrease in accuracy being offset by the lower cost of a single SPOT image (see Figure 4). Airborne remote sensing will allow maps of greater accuracy to be produced but it would be a waste of the imagery's capabilities to use it for coarse habitat mapping alone. SPOT Panchromatic images may be useful for preparing base maps of areas because of its relatively high spatial resolution (10 m) but is not a cost-effective option for habitat mapping. Landsat MSS is very cheap but to achieve an accuracy in excess of 50%, the cost of ground-truthing and processing are likely to be 10–20 times the cost of the imagery, thus buying it may be a false economy (Figure 4).

At fine descriptive resolution, only digital airborne scanners (e.g. CASI) and interpretation of colour aerial photography can deliver overall accuracies in excess of 50% (the former providing accuracies of around 80% and the latter of 57%). Thus, detailed habitat mapping should only be undertaken using these techniques. The cost of commissioning the acquisition of such imagery can be high (£15,000–26,000 even for small areas of 150 km^2) and may constitute 33–45% of total costs (64–75% if set-up costs are excluded). For a moderate-sized study larger than *ca* 150 km^2, about a month will be required to derive habitat classes from field data and process digital image data – irrespective of the digital data used (though, if staff need to be trained, this will increase time requirements considerably).

The relative cost-effectiveness of digital airborne scanners and aerial photography is more difficult to ascertain because they depend on the specific study. Where possible, it is recommended that professional survey companies be approached for quotes. In our experience, however, the acquisition of digital airborne imagery such as CASI is more expensive than the acquisition of colour aerial photography. However, this must be offset against the huge investment in time required to create maps from aerial photograph interpretation (API). If habitat maps are needed urgently, say in response to a specific impact on coastal resources, API might take too long and therefore be inappropriate. For small areas of say 150 km^2, a map could be created within 120 days using CASI but might take almost twice this time to create using API. We estimate that API is only cheaper if the staff costs for API are less than £75 per day. If consultants are used, this is unlikely to be the case. Furthermore, as the scope of the survey increases in size, the cost of API is likely to rise much faster than that arising from digital airborne scanners, making API less cost-effective as area increases. Where the costs of API and digital airborne scanners are similar, the latter instruments (digital scanners) should be favoured because they are likely to yield more accurate results than API. Possible drawbacks are the large data volumes which must be handled and the relative novelty of the technology, which means that institutions are less likely to be equipped to handle it and transfer of the technology may be less easy.

In summary, the costs of airborne imagery are only justified if finescale habitat maps are required and, if they are required, only airborne imagery (CASI or API) can deliver outputs of adequate accuracy. If broadscale habitat maps are required then multispectral satellite imagery (either Landsat TM or SPOT XS depending on the area to be covered) provides the most cost-effective option.

Background to the Research behind the *Handbook*

Why the research was needed

Habitat maps derived using remote sensing technologies are widely and increasingly being used to assess the status of coastal natural resources, as a basis for coastal planning, and for the conservation, management, monitoring and valuation of these resources. Despite the fact that Earth resources data from some satellites have been routinely available for 25 years, there had been almost no rigorous assessment of the capacity of the range of operational remote sensing technologies available to achieve coastal management-related objectives.

Digital sensors commonly used for coastal management applications have spatial resolutions ranging from about 1–80 m and spectral resolutions ranging from a single panchromatic band to around 16 precisely defined wavebands which can be programmed for specific applications. Costs of imagery range from about £250 for a low-resolution (80 m pixel) satellite image covering 35,000 km^2 to around £80,000 for a high-resolution (3 m pixel) airborne multispectral image covering less than half this area. In addition, high-resolution analogue technologies such as colour aerial photography are still in routine use. Coastal managers and other end-users charged with coastal planning and management and the conservation and monitoring of coastal resources require guidance on which of this plethora of remote sensing technologies are appropriate for achieving particular objectives. A study of reports of the use of remote sensing in tropical countries and other available literature indicated that there was wide use of the various technologies but no clear idea of the extent to which different technologies could achieve objectives.

In 1994, data that would allow clear guidance to be given to managers wishing to use remote sensing were not available. Most remote sensing research was devoted to potential applications of an often experimental nature and little attention was being paid to the operational realities of applications. This lack of practical information on which remote sensing technologies were appropriate for different coastal habitat mapping, management and planning objectives in tropical countries and at what cost, was identified as a key constraint to coastal management. The project therefore set out, firstly, to carry out applied research to rectify this and, secondly, to distil the outputs into a *Handbook* to disseminate the guidance and recommendations resulting to target end-users.

The Caicos Bank in the Turks and Caicos Islands (Figure 5) was chosen as a test site to discover what coastal management objectives were realistically achievable using different remote sensing technologies. This site offered clear water conditions, a broad mix of coastal habitat types, and a very large area (> 10,000 km^2) of shallow (< 20 m deep) coastal waters which would be amenable to the technologies. To test the generic applicability of the outputs, additional work was carried out in Belize with the UNDP/GEF Belize Coastal Zone Management Project and Coral Cay Conservation. Of particular concern was the quantification of the accuracies achievable for all outputs, since habitat or resource maps of poor or unknown accuracy are of little use as a basis for planning and management.

The UK Department for International Development supported the necessary research over three years via their Environment Research Programme. The ultimate purpose of the research was to enhance the effectiveness of remote sensing as a tool in coastal resources assessment and management in tropical countries by promoting informed, appropriate and cost-effective use. We hope this *Handbook* and the associated computer-based learning module on *Applications of Satellite and Airborne Image Data to Coastal Management* produced for UNESCO go some way towards achieving that purpose.

Research activities

The research was carried out from September 1994 to August 1997 and was designed to allow:

- better information to be provided to both practitioners and decision makers on what can be achieved with remote sensing technologies in the area of coastal management,
- advice to be given on best-practice for the most common applications, and

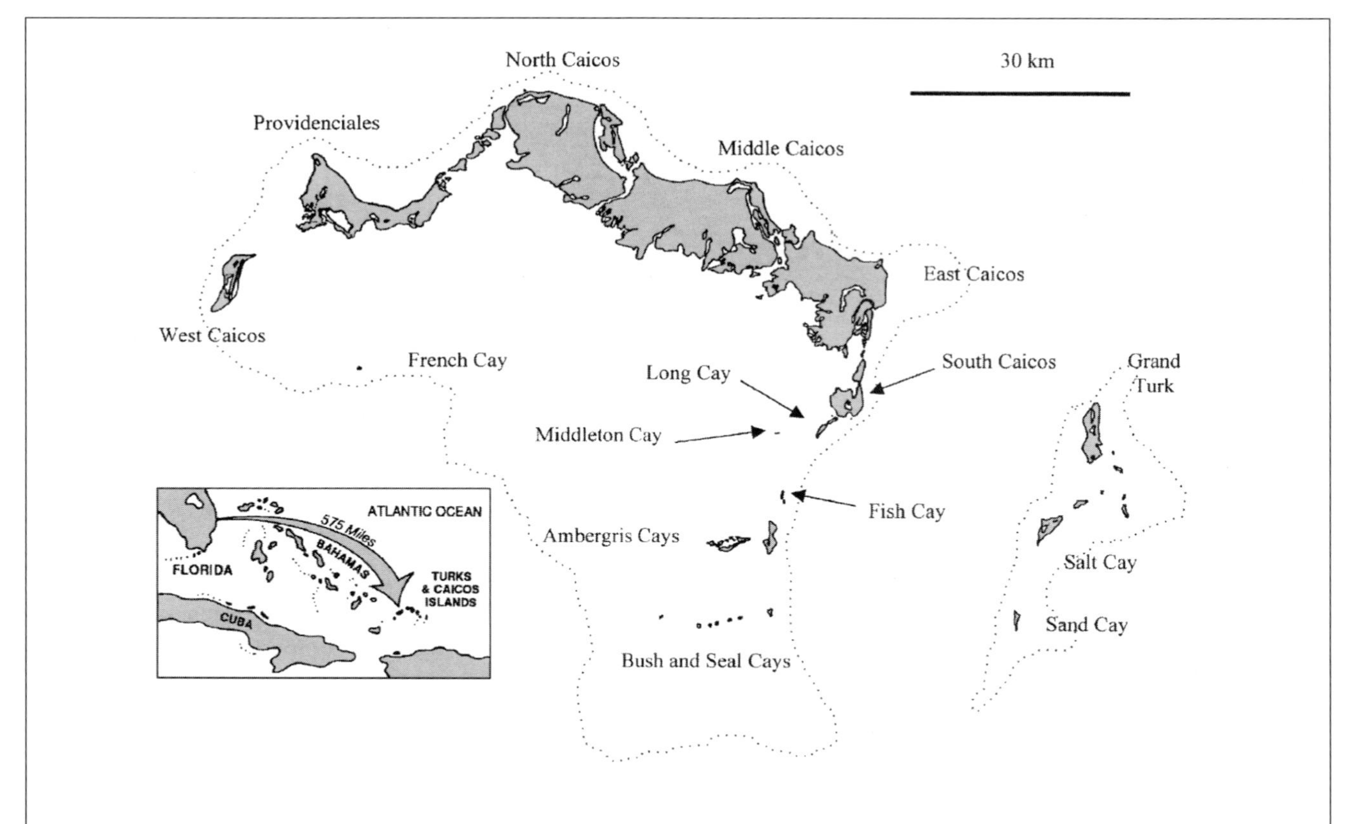

Figure 5 The Turks and Caicos Islands are situated some 575 miles from Miami, at the end of the Bahamas chain. The main islands, certain places mentioned in the text and the approximate position of the edge of the drop-off into deep-water (dotted lines) are shown.

- evaluation of the cost-effectiveness (gains in accuracy for increasing inputs/costs) of different methodological approaches to these applications.

For practitioners who may be interested, the project's research activities are outlined below:

1. Acquisition of imagery (see Chapters 3 and 5 for full details). Landsat Multispectral Scanner (MSS: 80 m spatial resolution, 28/06/92) and Thematic Mapper (TM: 30 m resolution, 11/11/90), and SPOT (Satellite Pour l'Observation de la Terre) multispectral (XS: 20 m resolution, 27/03/95) and panchromatic (Pan: 10 m resolution, 27/03/95) imagery of the Caicos Bank were acquired. Satisfactory Landsat images were available from archive but no SPOT imagery with < 30% cloud cover was available and so these images had to be specially requested from SPOT Image in Toulouse.

   Compact Airborne Spectrographic Imager (CASI) digital airborne multispectral imagery was flown successfully at 1 m and 3 m spatial resolution with 8 and 16 wavebands respectively using a local Cessna 172N with a specially modified door in July 1995. Ariel Geomatics Incorporated provided the sensor and operator and pre-processed the imagery, carrying out roll and radiometric correction.

   This imagery represented a cross-section of the commercially available satellite imagery that has been commonly used in coastal management applications. We would have liked also to evaluate Indian Remote Sensing satellite (IRS) imagery, which offers spatial and spectral resolution well suited to coastal management applications and is widely used on the Indian subcontinent, but were unfortunately unable to obtain images at the start of the project.

   A variety of airborne and satellite scenes and subscenes of the Caicos Bank are used as examples to illustrate specific remote sensing applications or processes in the *Handbook*. To orientate the reader, the locations of many of these are shown in Figures 6 and 7. All imagery was geometrically, radiometrically and water column corrected as appropriate (see Chapters 6, 7 and 8 for details).
2. A comprehensive and critical review of remote sensing studies of coastal resources was undertaken and published (Green *et al.* 1996).
3. Requirements and objectives of remote sensing studies in tropical coastal areas were analysed and prioritised by reference to the literature and by means of a ques-

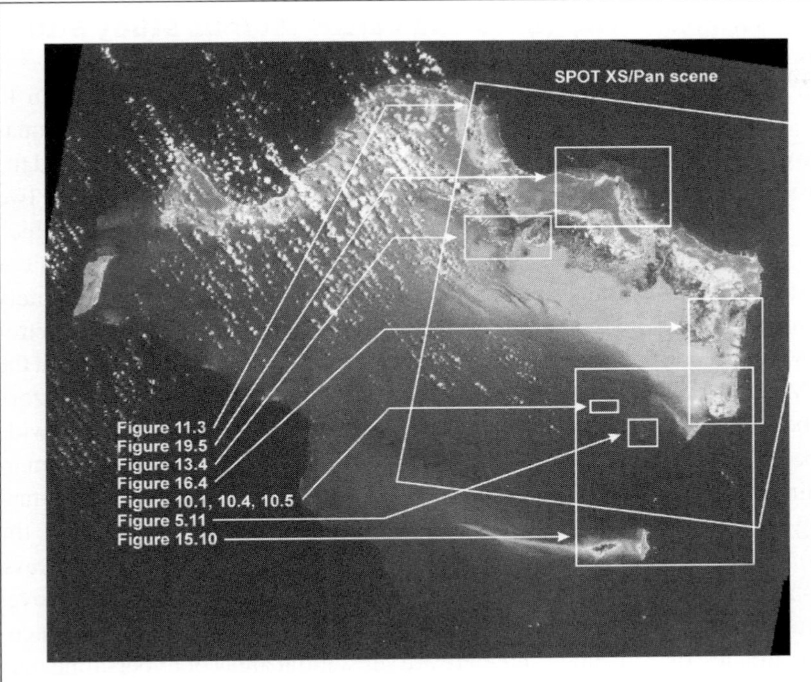

Figure 6 To orientate the reader, the approximate locations on the Caicos Bank of the imagery used in figures in the *Handbook* are superimposed on a greyscale representation of a colour composite Landsat MSS image of the Bank. The location of airborne digital imagery from small areas around South Caicos is illustrated in Figure 7. Narrow bands of cloud obscure much of the north-west quadrant of the MSS image. A single scene of Landsat Thematic Mapper (TM) or Multispectral Scanner (MSS) data covers an area of roughly 185 x 185 km. The Caicos Bank falls within a single Landsat scene, from which a subset was extracted to cover the area of interest. A single scene of SPOT multispectral or panchromatic data covers only 60 x 60 km, and so data were acquired for an area including Middle, East and South Caicos islands and Fish Cay (Figure 11.3).

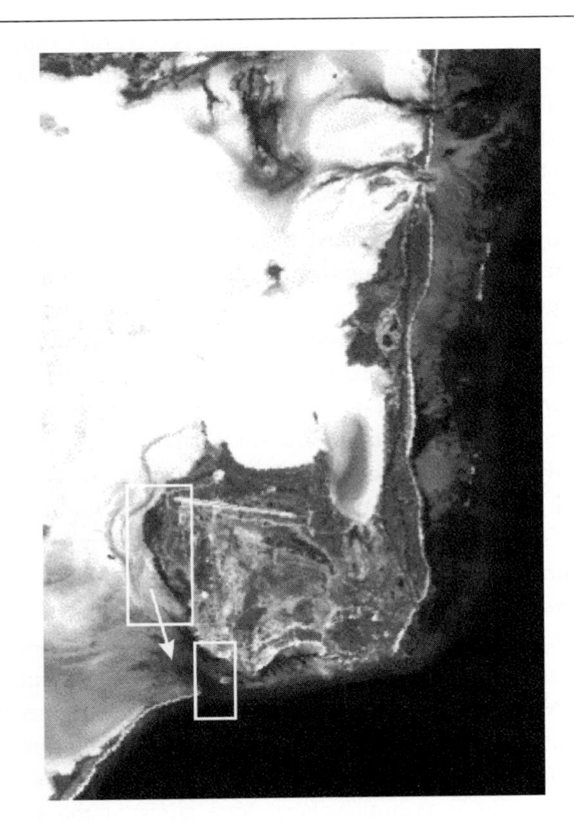

Figure 7 Landsat Thematic Mapper imagery of South Caicos, from an area which approximates to Figures 15.3, 16.4 and 19.3. The oblique aerial photograph in Figure 5.9 was taken looking in the direction of the white arrow and the Compact Airborne Spectrographic Imager (CASI) imagery in Figures 10.8 and 10.9 was taken over an area represented by the upper box. The CASI imagery shown in Figures 8.4 and 8.6 was from an area approximately defined by the lower box.

tionnaire distributed worldwide to some 140 end-users. 60 responses were obtained (see Chapter 2 for full details).

4. Ground-truthing surveys of the Caicos Bank were designed based a) on maps, charts and other existing data sources ('habitat-directed'), and b) on preliminary processing (unsupervised classification) of the satellite imagery ('image-directed'). These served to guide the initial field-survey programme in 1995. The effect of increasing amounts of field-survey input on the accuracy of outputs was later examined for four levels of inputs. General aspects of planning a fieldwork campaign are outlined in Chapter 4, methods for defining natural habitats in Chapter 9 and details of fieldwork specific to different applications of remote sensing in Chapters 11–13.
5. Field surveys were carried out over the Caicos Bank from June–August 1995 with additional accuracy assessment surveys in March 1996. A total of 780 marine field sites were surveyed with the assistance of the Ministry of Natural Resources of the Turks and Caicos Islands. Of these, 180 sites were surveyed in detail to allow i) an objective hierarchical habitat classification scheme to be developed (see Chapter 9), ii) mapping of these habitat classes at three levels of detail (see Chapter 11), iii) direction of supervised multispectral image classification (see Chapter 10), and iv) assessment of seagrass standing crop from imagery (see Chapter 16). At the remaining 600 marine sites the habitat type was recorded; data from these sites were used for assessing the accuracy of images.

In addition, a total of 202 shore sites were surveyed to allow i) an objective hierarchical habitat classification scheme to be developed for mangroves (see Chapter 13), ii) mapping of mangrove types (see Chapter 13), and iii) the assessment of mangrove leaf area index and percentage canopy closure from imagery (see Chapter 17).

6. A range of image classification techniques were tested with varying levels of ground-truthing input to evaluate which techniques produced the most accurate outputs and at what cost (Chapters 11 and 19). Multi-sensor data fusion was carried out with Landsat TM and SPOT Pan imagery but unfortunately could not be attempted between optical and ERS synthetic aperture radar imagery as the latter could not be obtained for the Caicos Bank. Three bathymetric mapping algorithms were tested and evaluated (see Chapter 15).
7. The accuracy of the outputs of the studies above were tested using the second phase of ground-truthing accuracy data (see point 5 above) and user accuracies of habitat maps (see Chapters 11–13), seagrass standing crop maps (see Chapter 16), and mangrove leaf area index and percentage canopy closure maps (see Chapter 17) were assessed.
8. The methodologies developed were also applied in Belize, as part of the UNDP/GEF Belize Coastal Zone Management Project, by Mr Hugo Matus. A detailed assessment of the accuracy of the National Marine Habitat Map of the Belize Barrier Reef and associated environments showed concordance with the results obtained for the Turks and Caicos (see Table 2 below), indicating the general applicability of the Turks and Caicos Islands study results, at least on a regional basis.
9. Using the results from the project and costs in staff time, training, capital equipment and consumables of achieving key objectives identified by end-users, the cost-effectiveness of different approaches and techniques was compared for a range of objectives (summarised in Chapter 19).

Table 2 Comparison of thematic map accuracy derived from Landsat Thematic Mapper imagery for the Turks & Caicos Islands (TCI) and Belize (Southwater Cays).

Thematic map accuracy	TCI study	Belize study
Detailed mapping of reef habitats	37%	42%
Broadscale mapping of reef habitats	73%	70%

Papers arising from the research project are listed at the end of this chapter in order of publication.

The Turks and Caicos Islands study site

The Turks and Caicos Islands are a group of more than 40 islands lying at the southeastern extremity of the Bahamas distributed over two archipelagos, the Turks Bank and the Caicos Bank (Figures 5 and 6). The margins of these two limestone platforms are defined by a 'drop-off' which plunges steeply into water of abyssal depth. The Caicos Bank is the larger platform and covers approximately 15,000 km^2. Water depth ranges from a few centimetres along the inland coasts of the Caicos Islands to 30 m at the top of the drop-off and clarity is typically good (horizontal Secchi distance of 30–50 m at 0.5 m depth). A wide range of habitats occur across the Caicos Bank. The margins are dominated by coral, algal and gorgonian (sea-fan) communities growing on hard substrate whereas the middle of the bank is typically covered by sparse seagrass, calcareous green algae and bare oolitic sand. Mangroves grow in fringes along the inland margins of the Caicos Islands and there are substantial areas of hypersaline mud dominated by salt-marsh plants.

Local impacts on reefs are at present limited because the population is low (~14,000) and only six islands are inhabited. Furthermore, the climate is too arid to support agriculture, there is negligible terrestrial run off, there is no industry and large construction projects have been mostly restricted to the island of Providenciales (the location of the international airport). Consequently the reefs in the Turks and Caicos Islands show few signs of being adversely affected by human activity. They are similar to the Bahamas with a deep fore reef dominated by gorgonians and the coral *Montastraea annularis* with live hard coral cover rarely exceeding 25%. Seaweeds are abundant on the fore reef, especially *Laurencia* spp., *Microdictyon* and *Lobophora*. The main impact on the fore reef comes from intense scuba diving activity, especially on the north coast of Providenciales, West Caicos and the western drop-off on Grand Turk. Large patch reefs 50–250 m in diameter occur in the shallow sheltered waters of the Caicos Bank, with the corals *Montastraea annularis*, *Porites porites* and *Acropora cervicornis* being the dominant species. These patch reefs are the main habitat for the spiny lobster *Panulirus argus* and are therefore the site of intense fishing during the lobster season. Adverse effects of this fishery are relatively limited, although the use of bleach to startle the lobsters from their holes is commonplace. Vast areas of the Caicos Bank are covered by bare sand, fleshy and calcareous algae and seagrass. These habitats are crucially important as nursery grounds for conch (*Strombus gigas*) and lobster but because of the size of the areas in question (1000s of km^2), their remoteness from population centres and fisheries management measures, they are under little threat at present.

References

Green, E.P., Mumby, P.J., Edwards, A.J., and Clark, C.D., 1996, A review of remote sensing for the assessment and management of tropical coastal resources. *Coastal Management*, **24**, 1–40.

Clark, C.D., Ripley, H.T., Green, E.P., Edwards, A.J., and Mumby, P.J., 1997, Mapping and measurement of tropical coastal environments with hyperspectral and high spatial resolution data. *International Journal of Remote Sensing*, **18** (2), 237–242.

Green, E.P., Mumby, P.J., Clark, C.D., Edwards, A.J., and Ellis, A.C., 1997, A comparison between satellite and airborne multispectral data for the assessment of mangrove areas in the eastern Caribbean. *Proceedings of the Fourth International Conference on Remote Sensing for Marine and Coastal Environments*, Florida, 17–19th March 1997, **1**, 168–176.

Green, E.P., Mumby, P.J., Edwards, A.J., Clark, C.D., and Ellis, A.C., 1997, Estimating leaf area index of mangroves from satellite data. *Aquatic Botany*, **58**, 11–19.

Green, E.P., Mumby, P.J., Edwards, A.J., and Clark, C.D., 1997, Mapping reef habitats using remotely sensed data: exploring the relationship between cost and accuracy. *Proceedings of the 8th International Coral Reef Symposium*, Panama, June 24–29, 1996, **2,** 1503–1506.

Mumby, P.J., Green, E.P., Clark, C.D., and Edwards, A.J., 1997, Reef habitat assessment using (CASI) airborne remote sensing. *Proceedings of the 8th International Coral Reef Symposium*, Panama, June 24–29, 1996, **2**, 1499–1502.

Mumby, P.J., Green, E.P., Edwards, A.J., and Clark, C.D., 1997, Measurement of seagrass standing crop using satellite and digital airborne remote sensing. *Marine Ecology Progress Series*, **159**, 51–60.

Mumby, P.J., Edwards, A.J., Green, E.P., Anderson, C.W., Ellis, A.C., and Clark, C.D., 1997, A visual assessment technique for estimating seagrass standing crop. *Aquatic Conservation: Marine and Freshwater Ecosystems*, **7**, 239–251.

Mumby, P.J., Green, E.P., Edwards, A.J., and Clark, C.D., 1997, Coral reef habitat mapping: how much detail can remote sensing provide? *Marine Biology*, **130**, 193–202.

Mumby, P.J., Clark, C.D., Green, E.P., and Edwards, A.J., 1998, Benefits of water column correction and contextual editing for mapping coral reefs. *International Journal of Remote Sensing*, **19** (1), 203–210.

Green, E.P., Mumby, P.J., Edwards, A.J., Clark, C.D., and Ellis, A.C., 1998, The assessment of mangrove areas using high resolution multispectral airborne imagery. *Journal of Coastal Research*, **14** (2), 433–443.

Mumby, P.J., Green, E.P., Clark, C.D., and Edwards, A.J., 1998, Digital analysis of multispectral airborne imagery of coral reefs. *Coral Reefs*, **17** (1), 59–69.

Green, E.P., Mumby, P.J., Clark, C.D., Edwards, A.J., and Ellis, A.C., 1998, Remote sensing techniques for mangrove mapping. *International Journal of Remote Sensing*, **19** (5), 935–956.

Mumby, P.J., Edwards, A.J., Clark, C.D., and Green, E.P., 1998, Managing tropical coastal habitats. Should I use satellite or airborne sensors? *Backscatter*, May 1998, 22–26.

Green, E.P., Mumby, P.J., Edwards, A.J., and Clark, C.D., 1998, Cost-effective mapping of coastal habitats. *GIS Asia Pacific*, August/September 1998, 40–44.

Edwards, A.J., (Ed.) 1999, Applications of airborne and satellite image data to coastal management. *Coastal Region and Small Island Papers,* No. 4. UNESCO, Paris, vi + 185 pp.

Mumby, P.J., Green, E.P., Edwards, A.J., and Clark, C.D., 1999, The cost-effectiveness of remote sensing for tropical coastal resources assessment and management. *Journal of Environmental Management*, **55**, 157–166.

Part 1

Remote Sensing for Coastal Managers

An Introduction

1

Introduction to Remote Sensing of Coastal Environments

Summary *This introduction is aimed primarily at decision makers and managers who are unfamiliar with remote sensing. It seeks to demonstrate to non-specialists the essential differences between digital images, such as those generated by most satellite sensors, and photographic pictures. Firstly, what is meant by remote sensing of tropical coastal environments is briefly defined. Key reference books which give a grounding in the fundamental theoretical concepts of remote sensing are then listed. Finally, electromagnetic radiation (EMR), the interactions between it and objects of coastal management interest, and multispectral digital imagery are introduced using a hypothetical image of a small atoll as an example.*

Introduction

Remote sensing of tropical coastal environments involves the measurement of electromagnetic radiation reflected from or emitted by the Earth's surface and the relating of these measurements to the types of habitat or the water quality in the coastal area being observed by the sensor. Some unconventional definitions of remote sensing culled from a recent conference are given below. Although tongue-in-cheek, there are perhaps grains of truth in many of them.

Remote sensing is . . .

- advanced colouring-in.
- the art of dividing up the world into little multi-coloured squares and then playing computer games with them to release unbelievable potential that's always just out of reach.
- the most expensive way to make a picture.
- seeing what can't be seen, then convincing someone that you're right.
- teledeception.
- being as far away from your object of study as possible and getting the computer to handle the numbers.
- astronomy in the wrong direction.

One of the aims of this *Handbook* is to make sure that the more derogatory definitions cannot be applied in earnest!

The theoretical fundamentals of remote sensing are very well-covered in several textbooks and in this introductory chapter we will not attempt to duplicate the many excellent expositions that already exist on this topic. Some key reference books, each of which provides a good grounding in the theoretical concepts underlying remote sensing, are listed in Table 1.1 (overleaf).

In the section on *How to Use this Handbook* we singled out Mather (1999) and the *ERDAS Field Guide* as companion volumes for use with the *Handbook* but any one of the other texts listed in Table 1.1 would provide an adequate replacement for Mather (1999).

The aims of this introduction are to outline in simple terms to coastal managers or decision makers, unfamiliar with remote sensing, such fundamental concepts as what we mean by electromagnetic radiation (EMR), the interaction of EMR and objects of coastal management interest on the Earth's surface (e.g. mangrove forests, houses, shrimp ponds, roads, coral reefs, beaches), and how digital images are generated from these interactions by sensors on aircraft or satellites. Analogue images such as colour aerial photographs are not discussed here because most people are already sufficiently familiar with cameras to understand these techniques intuitively.

Table 1.1 Key reference books on remote sensing listed in reverse order of date of publication. Costs and ISBN numbers are for paperback editions where these exist and for cloth editions otherwise. Prices are for mid-1998 in pounds sterling or US dollars (except for Mather).

Reference	ISBN	Cost
Mather, P.M. (1999). *Computer Processing of Remotely Sensed Images: An Introduction.* Second Edition. New York: Wiley (Includes CD-ROM)	0-471-98550-3	£29.99
Sabins, F.F. (1996). *Remote Sensing. Principles and Interpretation.* (Third Edition). San Francisco: Freeman	0-7167-2442-1	£32.95
Wilkie, D.S. and Finn, J.T. (1996). *Remote Sensing Imagery for Natural Resources Monitoring: A Guide for First-time Users.* New York: Columbia University Press	0-231-07929-X	£22.00
Jensen, J.R. (1995). *Introductory Digital Image Processing. A Remote Sensing Perspective.* (Second Edition). Englewood Cliffs: Prentice-Hall	0-13-489733-1	$66.67
Richards, J.A. (1995). *Remote Sensing Digital Image Analysis: An Introduction.* (Second Edition). New York: Springer-Verlag	0-387-58219-3	$59.95
Lillesand, T.M. and Keifer, R.W. (1994). *Remote Sensing and Image Interpretation.* (Third Edition). New York: Wiley	0-471-57783-9	£24.95
Barrett, E.C. and Curtis, L.F. (1992). *Introduction to Environmental Remote Sensing.* (Third Edition). London: Chapman and Hall	0-412-37170-7	£35.00
Cracknell, A.P. and Hayes, L.W. (1990). *Introduction to Remote Sensing.* London: Taylor and Francis	0-85066-335-0	£18.95
Harrison, B.A. and Judd, D.L. (1989). *Introduction to Remotely Sensed Data.* Canberra: Commonwealth Scientific and Industrial Research Organisation	0-643-04991-6	£43.95
Open Universiteit (1989). *Remote Sensing. Course Book and Colour Images Book.* Heerlen: Open Universiteit. (Available from The Open University, Milton Keynes, UK as PS670 study pack)	90-358-0654-9 90-358-0655-7	£55.00
Curran, P.J. (1986). *Principles of Remote Sensing.* London: Longman	0-582-30097-5	£20.99

Electromagnetic radiation (EMR)

EMR ranges from very high energy (and potentially dangerous) radiation such as gamma rays and X-rays, through ultra-violet light which is powerful enough to give us sunburn, visible light which we use to see, infra-red radiation part of which we can feel as heat, and microwaves which we employ in radar systems and microwave cookers, to radio waves which we use for communication (Figure 1.1). All types of EMR are waveforms travelling at the speed of light and can be defined in terms of either their wavelength (distance between successive wave peaks) or their frequency (measured in Hertz (abbreviated Hz), equivalent to wavelengths travelled per second). Shorter wavelength radiation (infra-red or shorter) tends to be described in terms of its wavelength; thus visible light sensors will have their wavebands listed in nanometres or micrometres (see Table 1.2). For example, blue light has wavelengths of 455–492 nm, red has wavelengths of 622–780 nm, and thermal infra-red has wavelengths of 3–15 μm (Figure 1.1). Longer wavelength radiation (e.g. microwave radiation such as radar, and radio frequencies) is often described in terms of its frequency, for example, the synthetic aperture radar on the European Remote-Sensing Satellite operates at 5.3 GHz).

Various types of sensor can measure incoming EMR in different parts of the electromagnetic spectrum but for most coastal management applications we are primarily interested in sensors measuring visible or near infra-red light reflected from the Earth's surface or emitted thermal infra-red radiation, although radar wavelengths are providing increasingly useful data.

Table 1.2 Some common prefixes for standard units of measurement.

Prefix	Symbol	Multiplier	Common name
giga	G	10^9	billion (US)
mega	M	10^6	million
kilo	k	10^3	thousand
centi	c	10^{-2}	hundredth
milli	m	10^{-3}	thousandth
micro	μ	10^{-6}	millionth
nano	n	10^{-9}	thousand millionth

Visible and near infra-red sensors measure the amount of sunlight reflected by objects on the Earth's surface, whereas thermal infra-red sensors detect heat emitted from the Earth's surface (including surface waters). These sensors are known as passive sensors, as the radiation sources are the Sun and Earth respectively

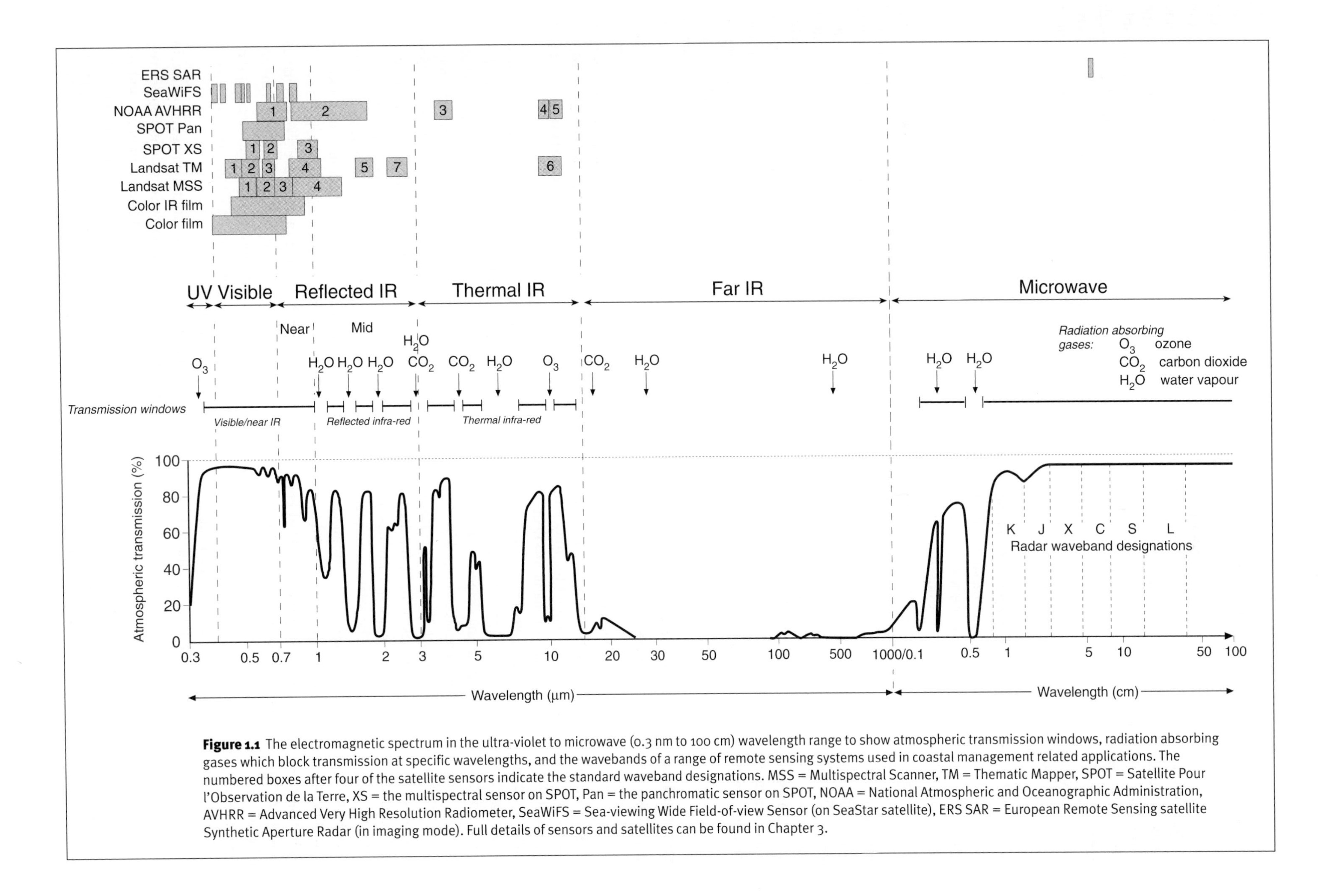

Figure 1.1 The electromagnetic spectrum in the ultra-violet to microwave (0.3 nm to 100 cm) wavelength range to show atmospheric transmission windows, radiation absorbing gases which block transmission at specific wavelengths, and the wavebands of a range of remote sensing systems used in coastal management related applications. The numbered boxes after four of the satellite sensors indicate the standard waveband designations. MSS = Multispectral Scanner, TM = Thematic Mapper, SPOT = Satellite Pour l'Observation de la Terre, XS = the multispectral sensor on SPOT, Pan = the panchromatic sensor on SPOT, NOAA = National Atmospheric and Oceanographic Administration, AVHRR = Advanced Very High Resolution Radiometer, SeaWiFS = Sea-viewing Wide Field-of-view Sensor (on SeaStar satellite), ERS SAR = European Remote Sensing satellite Synthetic Aperture Radar (in imaging mode). Full details of sensors and satellites can be found in Chapter 3.

and the sensor does no more than passively measure the reflected or emitted radiation. Microwave (radar) sensors, by contrast, send out radar pulses (with wavelengths usually in the centimetre to tens of centimetre range, equivalent to frequencies in the GHz to tens of GHz range) and measure the echoes reflected back from the Earth's surface and objects on it. Such sensors are termed active sensors as they also provide the source of the radiation being measured.

The atmosphere lets visible radiation through, except when cloudy, but water vapour, carbon dioxide and ozone molecules absorb EMR, particularly in the infra-red part of the spectrum (hence these molecules are 'greenhouse' gases), making the atmosphere opaque to certain wavelengths (Figure 1.1). Consequently satellite sensors are designed to sense in so-called transmission 'windows' where the atmosphere allows radiation through. Note in Figure 1.1 how the Landsat satellite's Thematic Mapper sensors for bands 5, 6 and 7 make use of three such windows in the infra-red part of the spectrum. Microwave radiation has the advantage over visible light that it can be sensed through cloud but the disadvantage for coastal applications that it does not penetrate water.

Interactions of electromagnetic radiation and the Earth's surface

To illustrate how a digital image is created we will consider a hypothetical sensor observing a small imaginary atoll in the Indian Ocean and consider the interactions of visible light with five types of habitat/object on the Earth's surface (Figure 1.2, Plate 1). The image covers an area of 800 x 800 m. The hypothetical sensor is similar to our own eyes in that it has detectors which sense light in three different wavebands: blue, 450–480 nm; green, 530–560 nm; and red, 625–675 nm. For colour vision our eyes have three types of light-sensitive nerve cells called cones which have peak sensitivities at about 450 nm, 550 nm and 610 nm. The relative stimulation of each type of cone by reflected or emitted light allows us to perceive colour images of the world around us.

On our imaginary atoll each different type of habitat/object reflects or absorbs incident sunlight in a different way and will appear different to the blue, green and red detectors (Figure 1.2). For example, the deeper water appears relatively bright in the blue waveband, much darker in the green and very dark in the red. This is because blue light penetrates water best and is scattered back to the sensor whereas red light is rapidly absorbed in the top few metres. By contrast, the reddish roofed buildings appear bright in the red and very dark in both the green and blue wavebands. The red paint basically absorbs green and blue light but is a powerful reflector of red light. The whitish coral sand beach appears bright in all three wavebands. The green leaves of the trees in the forested area reflect a lot of green radiation and some in the blue but almost no red wavelengths. The shallow sandy lagoon area reflects considerable amounts of blue and green light but is still deep enough that most red light is absorbed in the water column, giving it a characteristic turquoise colour in the colour composite image.

Digital images

The hypothetical sensor is a multispectral scanner which has a field-of-view such that at any moment in time it is viewing a 2 x 2 m square on the Earth's surface and recording how much blue, green and red light is being reflected from that square. Each square is known as a picture element or pixel. The majority of current multispectral sensors have for each waveband a line of detectors, each viewing one pixel (in this case one 2 x 2 m square). Our hypothetical sensor views a 800 m (400 pixel) wide swath of the Earth's surface in each of three wavebands and would thus need three rows of 400 detectors, one row measuring reflected blue light, one for green and one for red. As the platform carrying the sensor (which may be an aircraft or satellite) moves over the atoll it records the amount of light being reflected by line after line of 2 x 2 m squares which together make up the image (or picture). Each of the images in Figure 1.2 (Plate 1) is thus made up of 400 x 400 pixels, or 160,000 picture elements.

For each pixel the amount of light reflected in the blue, green and red wavebands is recorded by detectors as a digital number (DN) which, for this sensor, is a number between 0 and 255, with zero (displayed as black) indicating no detectable reflectance and 255 (displayed as white) indicating the maximum level of reflectance recordable by the detector. Many satellite sensors use the same range of brightness levels (radiometric resolution) although others may use less (e.g. 64 brightness levels) or more (e.g. 1024 brightness levels). In Figure 1.3 (Plate 2), groups of 5 x 5 pixels have been merged together to allow the values of pixels in the inset area to be displayed more easily. The merged pixels are thus 10 x 10 m in size and the inset area covers a 130 x 130 m block.

The advantages of the digital image (i.e. an image made up of numbers) is that we can define objectively the spectral characteristics of different habitats and process the images using computers. For our hypothetical image the unique spectra which distinguish the five types of surface can be obtained from Figure 1.3 (Plate 2) and are given in the associated table. By empirically relating such spectra to known types of habitat at known positions during field survey, a computer can be used to classify all pixels in an image, creating a habitat map. In real life, an entire image may cover tens or hundreds of kilometres, which means that remote sensing effectively extends the results of limited field survey to much larger areas. In reality the spectra will be much less clear-cut than those shown in Figure 1.3 and there will be considerable variation in pixel DN values within individual habitat types; however, the essential concept holds true and forms the basis of image classification (Chapter 10). You will notice that four of our merged pixels in Figure 1.3 are composed of mixtures of two different habitats and do not fall into the five main classes. Such pixels are known as 'mixels' and are a particular problem with low spatial resolution digital images where individual pixels may cover two or more habitat types.

Conclusion

This brief chapter has sought to introduce you to the basic concepts underlying remote sensing and digital images. These concepts, where needed for practical remote sensing of tropical coastal environments, will be expanded on in subsequent chapters but if you would like to delve deeper into the theoretical background you are referred to the texts listed in Table 1.1. This *Handbook* will concentrate on the practicalities of remote sensing of tropical coastal areas, as these have been relatively neglected; the theory is well-served elsewhere. The take-home message is that digital images are made up of numbers and to unlock their potential one needs to process the raw data on computers. Purchasing photographic prints derived from these digital images negates 90% of the advantages gained from being able to utilise such imagery.

2

Remote Sensing Objectives of Coastal Managers

Summary *Remote sensing has been applied to many management and research issues. The most common objectives include the production of coastal habitat maps and the identification of change in habitat cover, usually as a result of development. Monitoring of mangrove loss and aquaculture development is a particular concern.*

Sixty coastal managers and scientists evaluated the usefulness of various types of information on coastal ecosystems. The overall response for coral and seagrass ecosystems was that all types of information are more or less equally useful and that specific information requirements are dependent on study or management objectives. However, for mangrove ecosystems, users considered that the most important information was (i) the location of boundaries, (ii) areas of clearance and (iii) mangrove density. Although potentially very useful features of mangroves (such as percentage canopy closure and leaf area index) can be fairly readily mapped using remotely sensed imagery, these were considered of lower priority.

The need for managers to consider carefully the precise objectives of using remote sensing and to explore fully alternatives in terms of costs and effectiveness is stressed.

Introduction

Remote sensing is used to address a wide variety of management and scientific issues in the coastal zone. This chapter aims to summarise the applications of remote sensing in a tropical coastal management context and, in so doing, provide practitioners with a practical perspective on which management objectives might feasibly be addressed using remote sensing.

Two approaches are taken to summarise the applications of remote sensing. First, a summary of coastal management applications is given based on a review of published literature (Green *et al.* 1996). However, since a great deal of remote sensing work is not published, one hundred and fifty coastal managers and scientists were asked to complete a questionnaire. The questionnaire asked for respondents' opinions on four key issues:

1. Which applications have remote sensing actually been used for (i.e. what are the 'real' as opposed to 'potential' applications)?
2. What is the relative importance of different types of information on coastal habitats?
3. To what degree is remote sensing considered a cost-effective tool?
4. What are the main hindrances to remote sensing?

The results of the questionnaire provide a second perspective on the current utilisation of remote sensing for tropical coastal management. We would like to express our thanks to the 60 individuals who took the time to return our questionnaire.

Remote sensing objectives – perspectives from the published literature

The objectives of using remote sensing could be grouped into 14 primary categories (Table 2.1). Each of these is discussed briefly overleaf with examples although many

Table 2.1 Summary of remote sensing applications for tropical coastal zone management listing the number of papers reviewed in each category. API = Aerial Photographic Interpretation. For actual citations see Green *et al.* (1996).

Management Use	Landsat MSS	Landsat TM	SPOT XS	SPOT XP	Airborne MSS	API
1. CARTOGRAPHIC BASE MAPPING	2	1	2	1		2
2. RESOURCE INVENTORY AND MAPPING						
Coral reef	11	7	7	1		9
Seagrass	3	13	7		2	2
Mangrove	6	6	6	1		
Algae		4	2			
3. CHANGE DETECTION						
Tidal flats/salt affected land	1	1				
Extent of cyclone damage (reefs)		3				
Change in reef cover	1	2				1
Change in seagrass cover			1			1
Deforestation of mangrove	5	3	3			
4. BASELINE ENVIRONMENTAL MONITORING						
Coastal fisheries habitat		2	4			
Seasonal variation in reefal cover						3
Sea-level change						1
5. ENVIRONMENTAL SENSITIVITY MAPPING						
Exposure (storm vulnerability)	1					
Site assessment for marine parks	1					
Mangrove conservation areas	2	1				
Planning snorkel activity						1
Oil spill – sensitive mangrove areas			2			
6. MAPPING BOUNDARIES OF MANAGEMENT ZONES	2	2		1		1
7. BATHYMETRIC CHARTING						
Mapping shipping hazards	2	2	1		1	
Mapping transport corridors	2					
Update/augment existing charts	12	3	5			
Planning hydrographic surveys	3		2			
Mapping sediment accumulation/loss					2	
Help interpretation of reef features	2	1				
8. PLANNING FIELD SURVEYS						
Coral reef	2	1	1			
Seagrass		1	1			
Mangrove	1					
Algae		1	1			
9. PRODUCTIVITY MEASUREMENT						
Phytoplankton chlorophyll-*a*		1			1	
Biomass of seagrass ($g.m^{-2}$)		2				
10. STOCK ASSESSMENT						
Biomass of *Trochus niloticus*			2			
11. AQUACULTURE MANAGEMENT						
Inventory of aquaculture sites			5			
Shrimp aquaculture site selection		2	3			
Aid to the location of fish cages		1				
Monitoring pearl culture activity			3			
Site assessment for mariculture	2	1	1			
12. EROSION AND SHORELINE DRIFT	1			2	4	
13. IDENTIFICATION OF WATER BODIES / CIRCULATION	2	1	1			
14. LOCATION OF SUSPENDED SEDIMENT PLUMES	5	2	2		1	

of the applications are discussed in more detail in succeeding chapters.

1. Cartographic base mapping

Cartographic mapping is primarily concerned with the accurate delineation of land masses and water ways and is usually based on stereoscopic pairs of aerial photographs, each of which have sufficient ground control to produce a detailed and accurate map of the area of interest. Unfortunately, these techniques may be unsuitable for mapping offshore islands because obtaining adequate ground control is likely to be problematical (Friel and Haddad 1992, Mumby *et al.* 1995), and if the area to be mapped is large, it may be too expensive to use aerial photography (Jupp *et al.* 1981). Satellite images cover much larger areas than aerial photographs and require much less ground control per unit area and so can be employed as an alternative to conventional cartographic techniques. In Australia, 24 Landsat MSS scenes were used to map the entire Great Barrier Reef Marine Park with a root mean square accuracy of 64 m. This level of accuracy is considered to be satisfactory for producing maps at a scale of 1:250,000 (Jupp *et al.* 1985, Kenchington and Claasen 1988).

2. Resource inventory and mapping

Of the papers reviewed, 74% of applications were associated with assessing coastal resources and most of these (37%) concern their general inventory and mapping. In terms of coastal management, the fate of a resource inventory is not always clear. The resource map may be used to identify representative areas of distinct habitats which may be selected for protection (e.g. McNeill 1994) or used to demarcate and map the boundaries to management zones (Kenchington and Claasen 1988, Danaher and Smith 1988). Perhaps more commonly, the resource map may simply serve as background materials for a management plan. Provided the spatial and classification accuracy of the map are known, it can serve as a valuable baseline for future comparisons. For further details, refer to chapters on specific ecosystems (Part 3). Landsat TM appeared the most commonly used sensor in this category (34%), closely followed by SPOT XS and Landsat MSS (25% and 23% respectively).

3. Change detection

Intuitively, remote sensing appears ideally suited to the task of assessing changes in the coastal environment. For example, Landsat TM imagery is theoretically available for all tropical areas at time intervals of 16 days (see Chapter 3). In theory then, future changes in marine resources can be monitored almost twice-monthly or historic changes assessed by reference to archived imagery. In practice, however, change detection is complicated by several issues which are discussed throughout this book. The main limitations are the cost of obtaining multiple sets of imagery (Part 6), the need to standardise images for different atmospheric conditions (Chapter 7), cloud cover, (Chapters 3 and 5), the sensors' ability to detect change reliably, which is partly a function of pixel size (Chapter 3), and the fact that most satellite data are not available in real time, which can create a delay between (say) a major impact and acquisition of new imagery (see Chapter 5).

There have been some successful change detection studies, such as the impact of tropical cyclones on reefs of the Aitutaki atoll, Cook Islands (Loubersac *et al.* 1989), the decline of dense coral reefs over seven years in Bahrain (Zainal *et al.* 1993) and the extent of mangrove deforestation in the Pulang Redang Marine Park, Malaysia (Ibrahim and Yosuh 1992). Hopley and Catt (1989) have demonstrated the potential application of high resolution digital aerial photography for assessing the recovery of shallow reef flats from predation by the crown of thorns starfish (*Acanthaster planci*). Change detection analysis has become increasingly viable with the advent of satellite-borne synthetic aperture radar (SAR), which can acquire data on the Earth's surface despite the presence of clouds. However, SAR cannot penetrate water so its use is largely restricted to change on land (e.g. mangrove loss).

4. Baseline environmental monitoring

The success of remote sensing in providing baseline environmental monitoring seems to be restricted to large scale studies which generally have not been performed on a routine basis. For example, several papers clearly outline the potential of using low-altitude colour and infra-red aerial photography for mapping coral and algal cover (Catt and Hopley 1988, Hopley and Catt 1989, Thamrongnawasawat and Catt 1994). At this stage, however, it seems that the use of these techniques for routine management is confined to the Surin Marine National Park in Thailand (see Thamrongnawasawat and Hopley 1995).

5. Environmental sensitivity mapping

Several of the remote sensing studies reviewed in Table 2.1 have extremely clear objectives which stem from a particular management requirement. Biña *et al.* (1980) set out the selection criteria for mangrove conservation areas in the Philippines. Criteria which were detectable using Landsat MSS included locations (i) near a river mouth, (ii) near urban centres, (iii) offering a natural protection from hazards, (iv) comprising primary dense young forest of greater than 10% small islands. Similarly, Biña (1982) described how Landsat MSS products were used to highlight potential sites for marine reserve status. The selection criteria included: (i) clusters of reefs that occurred within 300 km^2, (ii) atolls and fringing reefs greater than 10 km in length, (iii) a diverse range of habitats both within and adjacent to reefs, and (iv) areas not situated near potential threats.

In the Surin Marine National Park (Thailand) efforts are currently under way to manage snorkelling activity using large scale digital aerial photography (Thamrongnawasawat and Hopley 1995). Analysis of these photographs reveals the presence of living corals and areas of dead coral and sand. The authors suggest that such data will allow planners to lay interpretative snorkelling routes for tourists and estimate the reefs' carrying capacity for snorkelling activity based on the ratio of massive to branching coral (assuming that massive corals are less vulnerable to snorkelling-induced damage).

Environmental sensitivity mapping has also been carried out for mangrove ecosystems. By using mangrove canopy cover as a surrogate measure of tree density, SPOT data were used to map mangrove density in Florida (Jensen *et al.* 1991). The authors suggest that such information will be important in identifying which mangrove areas are most susceptible to oil spill (assuming that oil dispersal is inversely proportional to tree density).

6. Mapping boundaries of management zones

Maps of coastal habitats have been used to demarcate the boundaries of management zones (Kenchington and Claasen 1988, Danaher and Smith 1988). This may include the delineation of fisheries-exclusion zones, marine park areas and so on.

7. Bathymetric charting

A great deal has been published on the application of remote sensing techniques for mapping bathymetry. Chapter 15 discusses the theory and methods behind bathymetric mapping but the fundamental principle is described here. Different wavelengths of light penetrate water to varying degrees: red light attenuates rapidly in water and does not penetrate deeper than 5 m or so, whereas blue light penetrates much further and, in clear water, the seabed will reflect enough light to be detected by a satellite sensor even when the depth of water approaches 30 m. The depth of penetration depends on the wavelength of light and the water turbidity. Suspended sediment particles, phytoplankton and dissolved organic compounds will all affect the depth of penetration because they scatter and absorb light. Taking a Landsat MSS example from the Great Barrier Reef, it is known that green light (band 1, 0.50–0.60 μm) will penetrate to a maximum depth of about 15 m, red light (band 2, 0.60–0.70 μm) to 5 m, near infra-red (band 3, 0.70–0.80 μm) to 0.5 m and infra-red (band 4, 0.80–1.1 μm) is fully absorbed (Jupp 1988). Appropriate image processing enables the analyst to derive a 'depth of penetration' image which is segmented into depth zones. Each zone represents a region in which light is reflected in one band but not the next. For example, if light is detected in band 2 but not in band 1, the depth should lie between 5 m and 15 m. Using this approach, three zones can be derived from a Landsat MSS image: 5–15 m, 0.5–5 m and 0–0.5 m (zero being the surface). Taken a stage further, each depth zone can be subdivided into discrete depth contours. This is achieved using a small amount of field work and assumptions about the homogeneity of substrate, water quality and atmospheric conditions (see Jupp 1988). Sensors with a greater number of spectral bands will clearly allow the analyst to generate a greater number of depth zones (e.g. Airborne MSS, see Danaher and Cottrell 1987). Chapter 15 explains several of these methods in more detail.

Depth of penetration type images have been used to augment existing charts (Bullard 1983, Pirazolli 1985), assist in interpreting reef features (Jupp *et al.* 1985) and map ship transportation corridors (Kuchler *et al.* 1986). However, they have not been used as a primary source of bathymetric data for navigational purposes (e.g. mapping shipping hazards). The major limitation is inadequate spatial resolution. For example, an emergent coral outcrop that is much smaller than the sensor pixel will remain undetected and therefore the image is not detailed enough for safe navigation. In addition, where the substratum is variable there may be too much uncertainty attached to specific depth measurements for their safe use in navigation.

Bathymetric measurements are feasible in areas of clear water less than 20 m deep with a uniform substratum. If these conditions are met then bathymetry can be measured to an accuracy of about ± 10% (Warne 1978, Danaher and Smith 1987, Lantieri 1988). If there is reason to believe that there are ephemeral patches of turbid water, images should be taken at different times of year, with little cloud, calm seas and a high sun elevation (UNESCO 1986).

Radar images may also be used to map depth changes. Water is an excellent reflector of microwaves and so the characteristics of a radar image are controlled by the nature of the sea surface. Variations in centimetre-scale roughness are strongly correlated with changes in water depth of up to approximately 100 m but only if a significant current is flowing across a fairly smooth seabed. SAR imagery (Chapter 3) has provided information on bathymetry, although the highly specific conditions of sea state and currents (Lodge 1983, Hesselmans *et al.* 1994) necessary for such analysis have prevented general application.

Surprisingly, Landsat MSS was the most popular sensor for this type of study with 55% of bathymetric studies using it, despite its relatively poor spatial resolution and limited water penetration.

8. Planning field surveys

Remote sensing products can play an important role in planning detailed field studies, particularly where charts or maps are poor. Biña *et al.* (1980) used Landsat MSS imagery of the Philippines to plan detailed aerial surveys of intact mangrove areas. The MSS data revealed which areas of mangrove had already been denuded and allowed the survey planners to target the surveys with an estimated 50% saving in fuel and air time. Benny and Dawson (1983) suggested that Landsat products would be a cost-effective way to plan hydrographic surveys in shallow water (< 30 m depth).

9. Productivity measurement

Armstrong (1993) used Landsat TM to map seagrass biomass in the Bahamas and Mumby *et al.* (1997) conducted similar tests for a variety of sensors in the Turks and Caicos Islands (Chapter 16). Greenway and Fry (1988) made anecdotal remarks relating the colour of aerial photographs and airborne MSS data to seagrass biomass but, unfortunately, the relationship was not assessed quantitatively.

10. Stock assessment

Several authors have attempted to conduct stock assessment of finfish and shellfish, a subject that is reviewed in detail in Chapter 18. For example, Bour (1989) estimated the biomass of the commercial mollusc *Trochus niloticus* in New Caledonia. This was essentially a two-stage process. Firstly, field data were used to identify the density of *Trochus* in various marine habitats. Secondly, since these habitats could be mapped using satellite imagery (in this case, SPOT XS), an estimate of total biomass was made by multiplying *Trochus* density (and thus biomass) by the total area of the habitat.

11. Aquaculture management

Over the past decade, tropical shrimp production has doubled every two to three years, fuelling a progressive search for new aquaculture sites (Meaden and Kapetsky 1991). SPOT imagery has been used to satisfy the following site selection criteria for shrimp ponds (Loubersac and Populus 1986): location, bathymetry, possible pond shape and pattern, site area, potential for expansion, access routes, practicalities of pumping seawater and the identification of possible sources of pollution. This type of information has been central to the development of shrimp farming in New Caledonia (IFREMER 1987). SPOT data have also been used successfully to monitor on-going aquaculture activity. For example, active (water-filled) ponds can be easily distinguished from redundant (empty) ponds (IFREMER 1987, Populus and Lantieri 1991). Pearl oyster (*Pinctada margaratifera*) beds in French Polynesia can be distinguished from undisturbed sea bed (Hauti 1990, Chenon *et al.* 1990). SPOT XS was the most widely used sensor (67%) for aquaculture applications.

12. Erosion and longshore drift

Several authors have attempted to use satellite imagery to examine coastal erosion (Welby 1978) or longshore drift (Kunte and Wagle 1993). Since both applications involve detecting change, these studies are subjected to the limitations outlined earlier such as pixel size and temporal resolution. For example, the minimum shift in shoreline that Landsat TM could detect would be in the order of 30 m (1 pixel width) and when geometric correction errors are considered (Chapter 6), the practical detectable change might be as large as several pixel widths (about 100 m). However, Danaher and Smith (1988) overcame some of these problems by using a high resolution airborne scanner (pixel size 2.5 m) to measure bathymetry of the Gold Coast, Australia. The results were used to understand the dynamics of sand accumulation/loss although the authors concede that the use of an airborne scanner was expensive.

13. Identification of water bodies / water circulation

Most of the published oceanographic applications of remote sensing (e.g. measurement of sea surface temperature and chlorophyll concentration) are confined to oceanic systems and temperate coastal areas. The status and relevance of these studies to coastal management are reviewed in Chapter 14.

14. Location of suspended sediment plumes

Chapter 14 reviews the use of remote sensing for estimating suspended sediment concentration but we present a single case study here for illustrative purposes.

Enhanced discharge of suspended sediment plumes due to excessive land clearance is one of the major threats to coral reefs world-wide (Wilkinson 1992). In management terms, it is useful to identify the pathways of coastal suspended sediments and land based run-off, since this will highlight regions in which particulate matter or solutes may directly influence the community. This was carried out in a pro-active sense using Landsat MSS in the Philippines where mining activities were terminated because mining-induced sediments were found to be reaching coral reefs (Biña, 1982).

Remote sensing objectives – a survey of coastal managers and scientists

Aims of the questionnaire

The questionnaire was designed to assess the current utilisation of remote sensing for tropical coastal resources assessment and management. The specific objectives of the questionnaire were to:

1. identify the actual (as opposed to potential) uses of remote sensing,
2. assess the relative importance of different types of information on coastal habitats,
3. ascertain the degree to which remote sensing was considered cost-effective, and
4. identify the main hindrances to remote sensing.

The results of the first two objectives are presented in this chapter. Cost-effectiveness (objective 3) is discussed in Part 6 and hindrances to remote sensing (objective 4) are outlined in Chapter 3.

Respondents to the questionnaire

A questionnaire was completed by 60 managers and scientists throughout the world including the Caribbean, Central America, South America, the USA, the Mediterranean, the Indian Ocean, South East Asia, Micronesia, Australia, West Africa and East Africa. Of 60 respondents, 28 worked for government departments, 23 for educational/research institutes, 5 for non-governmental organisations or private companies and 4 for the United Nations (UNEP, UNDP, UNESCO).

A broad spectrum of management and research mandates were represented (Table 2.2) although the responsibilities of many individuals fell under several categories, usually owing to an integrated approach to coastal zone management and research. Approximately half (29) of respondents considered themselves to be primarily research-orientated (scientists) and the other half (31) described themselves as coastal managers.

Table 2.2 Distribution of management and research mandates represented by respondents to the questionnaire

Management or Research Mandate	Number
Water quality	8
Land use planning	11
Environmental impact studies	4
Coral reef management	20
Mangrove/wetland management	36
Seagrass management	8
Marine parks	8
Management of coastal fisheries	16
Beach erosion and suspended sediments	15
Aquaculture and mariculture	5
Physical oceanography	8
Cartography	6
Bathymetry	3
Extractive activities	3

Actual uses of remote sensing by coastal managers and scientists

The most common applications of remote sensing were to provide background information for management planning and the detection of change in coastal resources (Figure 2.1). The term 'background information' reflects the vagueness with which habitat maps are often used. In many cases, the habitat map simply provides the location

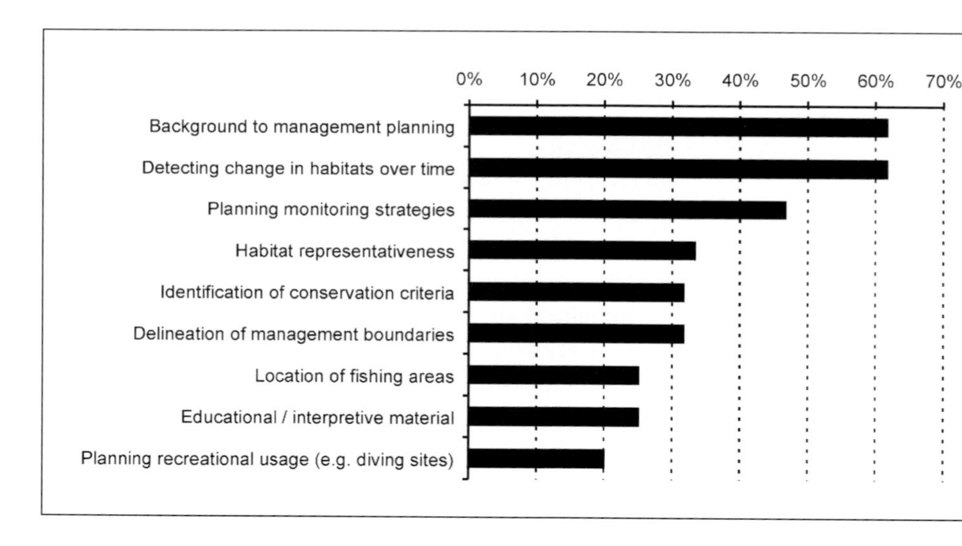

Figure 2.1 Responses of 60 coastal managers and scientists to a questionnaire asking them to identify what they considered the primary applications of remote sensing for tropical coastal management.

of coastal habitats but such information might not be used explicitly (e.g. to identify representative habitats for inclusion in marine parks). This result mirrors that of the literature review above.

Approximately 70% of the respondents who used remote sensing to detect changes in coastal areas were concerned with mangrove assessment and/or shrimp farming (Plate 3). Maps of mangrove are used to identify deforested areas such as shrimp farms and urban expansions. This does not necessarily suggest that a time sequence of images has been used to identify changes in coastal resources because deforested areas can usually be identified visually in a single image due to the contrast against forested areas and the presence of linear boundaries (e.g. near roads and settlements).

Almost half of the respondents used habitat maps to plan monitoring strategies (e.g. locating seagrass beds for a seagrass monitoring programme). Habitat maps were also used to provide information on conservation criteria, such as habitat representativeness, rarity and diversity.

Respondents to the questionnaire highlighted the use of habitat maps for educational and interpretative material. This application did not arise in the literature review and marks the importance of habitat maps to a wide range of users including owners of small boats, recreational fishermen and tourists.

Informational requirements of coastal managers and scientists

Respondents were asked to prioritise the usefulness of various types of information on coastal resources for coral reef, seagrass and mangrove ecosystems. The interval scale for 'usefulness' was: U (unknown), 1 (poor), 2 (fairly useful), 3 (moderately useful) and 4 (extremely useful). The results are summarised hereafter for each type of ecosystem. In each case, Kruskal–Wallis tests (Zar 1996) were used to identify significant differences in the usefulness of various types of information (e.g. whether live coral cover was considered to be significantly more useful than coral reef boundaries). The analyses were not conducted separately for coastal managers and scientists because the distinction is somewhat arbitrary and we wished to represent the views of a broad range of remote sensing users.

Coral reefs

Five types of information on coral reefs were listed in the questionnaire (Table 2.3). In increasing order of detail (and difficulty to obtain using remote sensing) these were: coral reef boundaries, geomorphological zones of coral reefs, coral density, individual coral colonies and live coral cover.

Table 2.3 Usefulness of various types of coral reef information to respondents of questionnaire.

Objectives	Mean usefulness	
	Numeric	Interpreted
Coral reef boundaries	2.7	fair–moderate
Geomorphological zones	2.5	fair–moderate
Coral density	2.9	moderate
Coral colonies	2.7	fair–moderate
Live coral cover	3.1	moderate

The mean usefulness of these data ranged from 2.5 (geomorphology) to 3.1 (live coral cover) – i.e. fairly to moderately useful. Usefulness was not found to differ significantly between different types of information ($P = 0.56$, $n = 31$, $df = 4$). As is shown in Chapter 11, coral reef boundaries and geomorphological zones can be mapped readily with both satellites and airborne instruments but coral density, coral colonies and live coral cover can only be evaluated in shallow clear water with airborne sensors.

Seagrass

Usefulness was assessed for seagrass boundaries, exposed versus submerged seagrass, seagrass density and seagrass biomass (Table 2.4). The averaged responses suggested that data were of fair to moderate use but that, as for coral reefs, the usefulness of various types of information did not vary significantly ($P = 0.49$, $n = 32$, $df = 3$). All these objectives can be achieved using both satellite and airborne sensors (Chapter 12).

Table 2.4 Usefulness of various types of seagrass information to respondents of questionnaire.

Objectives	Mean usefulness	
	Numeric	Interpreted
Seagrass boundaries	2.5	fair–moderate
Exposed/submerged seagrass	2.6	fair–moderate
Seagrass density	2.8	moderate
Seagrass biomass	2.8	moderate

Mangrove

In terms of usefulness (Table 2.5), mangrove parameters fell into two groups which differed significantly ($P < 0.001$, $n = 40$, $df = 6$). The most useful information was considered to be mangrove boundaries, cleared mangrove and mangrove density. The remaining parameters were considered to be fairly useful.

As discussed in Chapter 13, the first five objectives can be achieved using both satellite and airborne imagery, whereas mangrove species composition normally requires airborne instruments. However, mangrove biomass cannot realistically be mapped at present.

Table 2.5. Usefulness of various types of mangrove information to respondents of questionnaire.

Objectives	Mean usefulness	
	Numeric	Interpreted
Mangrove boundaries	3.0	moderate
Cleared mangrove	2.9	moderate
Mangrove density	3.0	moderate
Mangrove height	2.0	fair
Canopy cover of mangrove	1.8	poor–fair
Mangrove species composition	2.0	fair
Mangrove biomass	2.1	fair

The difference in perceived usefulness of the different types of mangrove information was surprising and may reflect the current availability of information on mangrove parameters. For example, canopy closure has only recently been measured using remote sensing (Jensen *et al.* 1991, Green *et al.* 1997). The application of synoptic measures of canopy closure (and leaf area index) are only now becoming clear. Leaf area index can be related to the net photosynthetic primary production of mangrove (English *et al.* 1994) and it is now possible to map this parameter throughout an entire mangal (Green *et al.* 1997). This may have implications for forestry management and help prioritise the selection of sites for protection (e.g. to protect the most productive mangrove in a particular ecosystem). Once the full implications of such information are realised, managers and scientists might assign greater importance to these data in future.

Conclusions

Most types of information on coastal resources are useful for one application or another. This chapter provides a general overview and it must be borne in mind that the usefulness of any particular type of information will depend on the specific objectives of research or management. For example, if we need to assess a reef's response to sewage run-off, live coral cover is likely to be a more informative measure than (say) a map of reef geomorphology.

Above all, managers need to think clearly about the precise objectives they wish to achieve using remote sensing and fully explore alternatives in terms of costs and effectiveness. They also need to be aware that some management objectives will not be achievable using remote sensing. This volume seeks to allow managers to make more informed decisions when matching remote sensing methods to survey objectives.

References

Armstrong, R.A., 1993, Remote sensing of submerged vegetation canopies for biomass estimation. *International Journal of Remote Sensing*, **14**, 10–16.

Benny, A.H., and Dawson, G.J., 1983, Satellite imagery as an aid to bathymetric charting in the Red Sea. *Cartography Journal*, **20**, 5–16.

Biña, R.T., 1982, Application of Landsat data to coral reef management in the Philippines. *Proceedings of the Great Barrier Reef Remote Sensing Workshop*, Townsville, May 5th 1982, (Townsville: James Cook University), pp. 1–39.

Biña, R.T., Jara, R.B., and Roque, C.R., 1980, Application of multi-level remote sensing survey to mangrove forest resource management in the Philippines. *Natural Resources Management Center Research Monograph No. 2* (Manila: NRMC), pp. 1–5.

Bour, W., 1989, SPOT images for coral reef mapping in New Caledonia. A fruitful approach for classic and new topics. In *Proceedings of the 6th International Coral Reef Symposium*, edited by J.H. Choat *et al.* (Townsville: Sixth International Coral Reef Symposium Committee, Townsville) **2**, 445–448.

Bullard, R.K., 1983, Detection of marine contours from Landsat film and tape. In *Remote Sensing Applications in Marine Science and Technology*, edited by A.P. Cracknell, (Dordrecht: D. Reidel), pp. 373–381.

Catt, P., and Hopley, D., 1988, Assessment of large scale photographic imagery for management and monitoring of the Great Barrier Reef. *Proceedings of the Symposium on Remote Sensing of the Coastal Zone, Gold Coast, Queensland.* (Brisbane: Department of Geographic Information). pp. III.1.1–1.14.

Chenon, F., Varet, L., Loubersac, L., Grand, S., and Hauti, A., 1990, SIGMA POE RAVA a GIS of the Fisheries and Aquaculture Department. A tool for a better monitoring of public ownership and pearl oyster culture. *Proceedings of the International Workshop on Remote Sensing and Insular Environments in the Pacific: Integrated Approaches*, (Noumea: ORSTOM/IFREMER), pp. 561–572.

Danaher, T.J., and Smith, P., 1988, Applications of shallow water mapping using passive remote sensing. *Proceedings of the Symposium on Remote Sensing of the Coastal Zone, Gold Coast, Queensland*, September 1988 (Brisbane: Department of Geographic Information), pp. VII.3.2–8

Danaher, T.J., and Cottrell, E.D., 1987, Remote sensing: a cost–effective tool for environmental management and monitoring. *Pacific Rim Congress*, Gold Coast Australia, 1987 pp. 557–560.

English, S., Wilkinson, C., andBaker, V., 1994, Survey manual for tropical marine resources. ASEAN–Australia Marine Science Project: Living Coastal Resources, (Townsville: AIMS).

Friel, C., and Haddad, K., 1992, GIS manages marine resources. *GIS World*, **5**, 33–36.

Green, E.P., Mumby, P.J., Edwards, A.J., and Clark, C.D., 1996, A review of remote sensing for the assessment and management of tropical coastal resources. *Coastal Management*, **24**, 1–40.

Green, E.P., Mumby, P.J., Ellis, A.C., Edwards, A.J., and Clark, C.D., 1997, Estimating leaf area index of mangroves from satellite data. *Aquatic Botany*, **58**, 11–19.

Greenway, M., and Fry, W., 1988, Remote sensing techniques for seagrass mapping. *Proceedings of the Symposium Remote Sensing Coastal Zone, Gold Coast, Queensland*, (Brisbane: Department of Geographic Information), pp. VA.1.2–1.11.

Hauti, A., 1990, Monitoring marine concessions for pearl aquaculture. Problematics and first solutions. *Proceedings of the International Workshop on Remote Sensing and Insular Environments in the Pacific: Integrated Approaches*, (Noumea: ORSTOM/IFREMER), pp. 547–560.

Hesselmans, G.H.F.M., Wensink, G.J., and Calkoen, C.J., 1994, The use of optical and SAR observations to assess bathymetric information in coastal areas. *Proceedings of the 2nd Thematic Conference Remote Sensing for Marine Coastal Environments, New Orleans*, (Michigan: ERIM), **1**, 215–224.

Hopley, D., and Catt, P.C., 1989, Use of near infra-red aerial photography for monitoring ecological changes to coral reef flats on the Great Barrier Reef. *Proceedings of the 6th International Coral Reef Symposium*, Townsville, **3**, 503–508.

Ibrahim, M., and Yosuh, M., 1992, Monitoring the development impacts on the coastal resources of Pulau Redang marine park by remote sensing. In *Third ASEAN Science & Technology Week Conference Proceedings*, edited by L.M. Chou and C.R. Wilkinson (Singapore: University of Singapore), Vol. 6 Marine Science Living Coastal Resources. pp. 407–413.

IFREMER (Institut français de recherche pour l'exploitation de la mer), 1987, The preselection of sites favourable for tropical shrimp farming. *France Aquaculture*.

Jensen, J.R., Ramsey, E., Davis, B.A., and Thoemke, C.W., 1991, The measurement of mangrove characteristics in south-west Florida using SPOT multispectral data. *Geocarto International*, **2**, 13–21.

Jupp, D.L.B., 1988, Background and extensions to depth of penetration (DOP) mapping in shallow coastal waters. *Proceedings of the Symposium on Remote Sensing of the Coastal Zone, Gold Coast, Queensland*, September 1988, (Brisbane: Department of Geographic Information), pp. IV 2.1–19.

Jupp D.L.B., Mayo, K.K., Kuchler, D.A., Claasen, D.R., Kenchington, R.A., and Guerin, P.R., 1985, Remote sensing for planning and managing the Great Barrier Reef Australia. *Photogrammetria*, **40**, 21–42.

Jupp, D.L.B., Mayo, K.K., Kuchler, D.A., Heggen, S.J., and Kendall, S.W., 1981. Remote sensing by Landsat as support for management of the Great Barrier Reef. *2nd Australasian Remote Sensing Conference: Papers & Programme* pp. 9.5.1–9.5.6.

Kenchington, R.A., and Claasen, D.R., 1988, Australia's Great Barrier Reef – management technology. *Proceedings of the Symposium on Remote Sensing of the Coastal Zone, Gold Coast, Queensland*, (Brisbane: Department of Geographic Information) pp. KA.2.2–2.13.

Kuchler, D.A., Jupp, D.L.B., Claasen, D.R., and Bour, W., 1986, Coral reef remote sensing applications. *Geocarto International*, **4**, 3–15.

Kunte, P.D., and Wagle, B.G., 1993, Determination of net shore drift direction of central west coast of India using remotely sensed data. *Journal of Coastal Research*, **9**, 811–822.

Lantieri, D., 1988, Use of high resolution satellite data for agricultural and marine applications in the Maldives. Pilot study on Laamu Atoll. *FAO Technical Report TCP/MDV/4905. No. 45.*

Lodge, D.W.S., 1983, Surface expressions of bathymetry on SEASAT synthetic aperture radar images. *International Journal of Remote Sensing*, **4**, 639–653.

Loubersac, L., Dahl, A.L., Collotte, P., LeMaire, O., D'Ozouville, L., and Grotte, A., 1989, Impact assessment of Cyclone Sally on the almost atoll of Aitutaki (Cook Islands) by remote sensing. *Proceedings of the 6th International Coral Reef Symposium*, Townsville, **2**, 455–462.

Loubersac, L., and Populus, J., 1986, The applications of high resolution satellite data for coastal management and planning in a Pacific coral island. *Geocarto International*, **2**, 17–31.

McNeill, S.E., 1994, The selection and design of marine protected areas: Australia as a case study. *Biodiversity and Conservation*, **3**, 586–605.

Meaden, G.J., and Kapetsky, J.M., 1991, Geographical information systems and remote sensing in inland fisheries and aquaculture. *FAO Fisheries Technical Paper No. 318.*

Mumby, P.J., Gray, D.A., Gibson, J.P., and Raines, P.S., 1995, Geographic Information Systems: a tool for integrated coastal zone management in Belize. *Coastal Management*, **23**, 111–121.

Mumby, P.J., Green, E.P., Edwards, A.J., and Clark, C.D., 1997, Measurement of seagrass standing crop using satellite and digital airborne remote sensing. *Marine Ecology Progress Series*, **159**, 51–60.

Pirazzoli, P.A., 1985, Bathymetry mapping of coral reefs and atolls from satellite. *Proceedings of the 5th International Coral Reef Symposium* Tahiti, **6**, 539–545.

Populus, J., and Lantieri, D., 1991, Use of high resolution satellite data for coastal fisheries. *FAO RSC Series No. 58*, Rome.

Thamrongnawasawat, T., and Catt, P., 1994, High resolution remote sensing of reef biology: the application of digitised air photography to coral mapping. *Proceedings of the 7th Australasian Remote Sensing Conference*, Melbourne pp. 690–697.

Thamrongnawasawat, T., and Hopley, D., 1995, Digitised aerial photography applied to small area reef management in Thailand. In *Recent Advances in Marine Science and Technology 94*, edited by O. Bellwood, J.H. Choat, N. Saxena, (Townsville: James Cook University of N. Queensland), pp. 385–394.

UNESCO, 1986, The application of digital remote sensing techniques in coral reef, oceanographic and estuarine studies. Report on a regional Unesco/COMAR/GBRMPA Workshop. *Unesco Reports in Marine Science* 42 (Paris: UNESCO).

Warne, D.K., 1978, Landsat as an aid in the preparation of hydrographic charts. *Photogrammetric Engineering and Remote Sensing*, **44**, 1011–1016.

Welby, C.W., 1978, Application of Landsat imagery to shoreline erosion. *Photogrammetric Engineering and Remote Sensing*, **44**, 1173–1177.

Wilkinson, C.R., 1992, Coral reefs of the world are facing widespread devastation: can we prevent this through sustainable management practices? *Proceedings of the 7th International Coral Reef Symposium*, (Guam: University of Guam), **1**, 11–21.

Zainal, A.J.M., Dalby, D.H., and Robinson, I.S., 1993, Monitoring marine ecological changes on the east coast of Bahrain with Landsat TM. *Photogrammetric Engineering and Remote Sensing*, **59**, 415–421.

Zar, J.H., 1996, *Biostatistical Analysis*. (New Jersey: Prentice-Hall).

3

Satellite and Airborne Sensors Useful in Coastal Applications

Summary *Choice of image type is of critical importance, as users must be certain that images are suitable for a particular application before large amounts of money and time are invested in purchasing and processing the data. To assist in matching types of imagery to the particular application, this chapter provides details on the characteristics of the main forms of remotely sensed data that have been used routinely in tropical coastal management or have clear potential.*

Fundamental characteristics, such as spatial, spectral, radiometric and temporal resolutions, are outlined and the main limitations are discussed. Sensors useful for remote sensing of tropical coasts are considered in three groups according to their spatial resolutions (>100 m, 10–100 m, and <10 m) and the technical specifications of each sensor outlined. This chapter, in combination with information on the costs of different data (Chapter 5), the coastal habitat mapping capabilities of each image type (Chapters 11–13) and their cost-effectiveness (Chapter 19), provides readers with a foundation on which to base the choice of image types appropriate for specific applications.

Certain types of imagery with a spatial resolution > 100 m are useful for measuring sea surface temperature and phytoplankton biomass, information which is useful for fisheries management (Chapter 18), mapping toxic algal blooms (Chapter 14) and mapping the progress of warm water anomalies associated with El Niño such as those that caused coral bleaching around the tropics in 1998. Imagery with a spatial resolution of 10–100 m has been most commonly used in coastal zone management and is the type most widely discussed in later chapters. Imagery with a spatial resolution < 10 m has been used less frequently, partly because the development of digital airborne sensors is comparatively recent. However, these high resolution data appear to be the most useful to managers and are being used increasingly for coastal applications.

Introduction

A remote sensing system is composed of sensors carried on a platform. A platform may be either a satellite or some type of aircraft (most frequently a light aeroplane), for example the Thematic Mapper sensor is carried on the Landsat satellite. The nature of the sensor-platform system determines the characteristics of the image data.

The sensor (more than one can be carried on a single platform) is a device which gathers energy and converts it to a signal. Sensors carried on satellites and aircraft gather information about interactions of electromagnetic radiation with the Earth's surface. When electromagnetic radiation falls upon a surface, some of the energy is absorbed, some is transmitted through the surface and some is reflected. Surfaces also naturally emit radiation which at the temperature of the Earth is mostly in the form of thermal infra-red radiation (heat). The upwelling reflected and emitted radiation is recorded on either a photographic film or digitally by the sensor. Since the intensity and wavelengths of this radiation are a function of the surface in question, each surface possesses a characteristic spectrum or spectral signature. If different spectra can be distinguished, it is possible to map the extent of such surfaces. Along tropical coastlines specific surfaces such as mangrove canopy, seagrass beds, sand, fringing reef, etc. usually have distinctive enough spectra to enable them to be distinguished and therefore mapped (Figure 3.1).

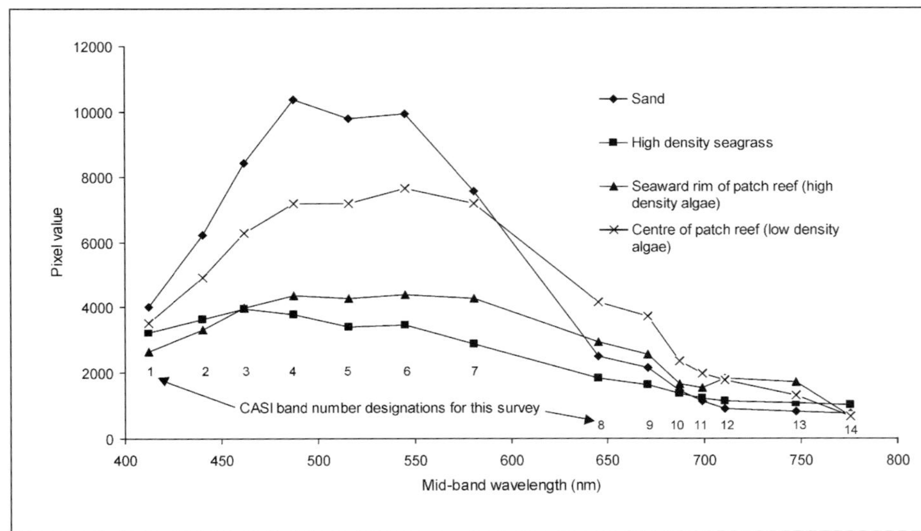

Figure 3.1 Spectra of four different bottom types taken from a 14 band CASI image (see Table 5.9 for precise settings used). The spectra of low density algae, high density algae, sand and high density seagrass differ most in bands 1–7 (but note that high density seagrass and algae are very similar in bands 2 and 3) but are all fairly similar in bands 10–14 where the dominant factor affecting reflectance is absorption of far red and near infra-red light in the water column. The habitats should be separable by careful selection of a combination of bands 1–9 to provide distinct spectral envelopes.

Types of sensor

Sensors can be categorised as passive or active. Passive sensors receive natural radiation, such as reflected sunlight: active sensors receive radiation specifically generated by the sensor, such as radar.

Passive sensors

Passive sensors collect electromagnetic radiation (EMR), usually in the visible (0.4–0.7 µm) and infra-red (0.7–14 µm) part of the electromagnetic spectrum, which is reflected or emitted from the Earth. Therefore passive sensors collect colour, infrared and temperature information. They use the Sun or Earth as a source of radiation. This has two implications. Firstly, and of great practical importance, the collection of light in the visible and reflected infra-red part of the electromagnetic spectrum is restricted to daylight hours with clear skies (see Chapter 5 for the practical implications of cloud cover). Secondly, as solar radiation passes through the atmosphere it is absorbed and scattered so that the signals received at the sensor are a distorted version of the true signal (the upwelling radiance just above the surface in question). Absorption in the atmosphere is wavelength specific and increases with the concentration of certain atmospheric gases, especially water vapour, carbon dioxide and ozone. These can be highly variable across geographical regions or around urban areas. Scattering also varies with the concentration of atmospheric aerosols (e.g. those released by volcanic eruptions). The combined effect is to introduce uncertainty into the measurements. Under certain circumstances this can be ignored but careful correction using information about the composition of the atmosphere is often required (see Chapter 7).

Active Sensors

Active sensors generate their own radiation (e.g. radar, laser) and measure that which is reflected or scattered back to the sensor. Radar radiation (microwave wavelengths 0.1–10 cm) will penetrate cloud and, as it is gen-

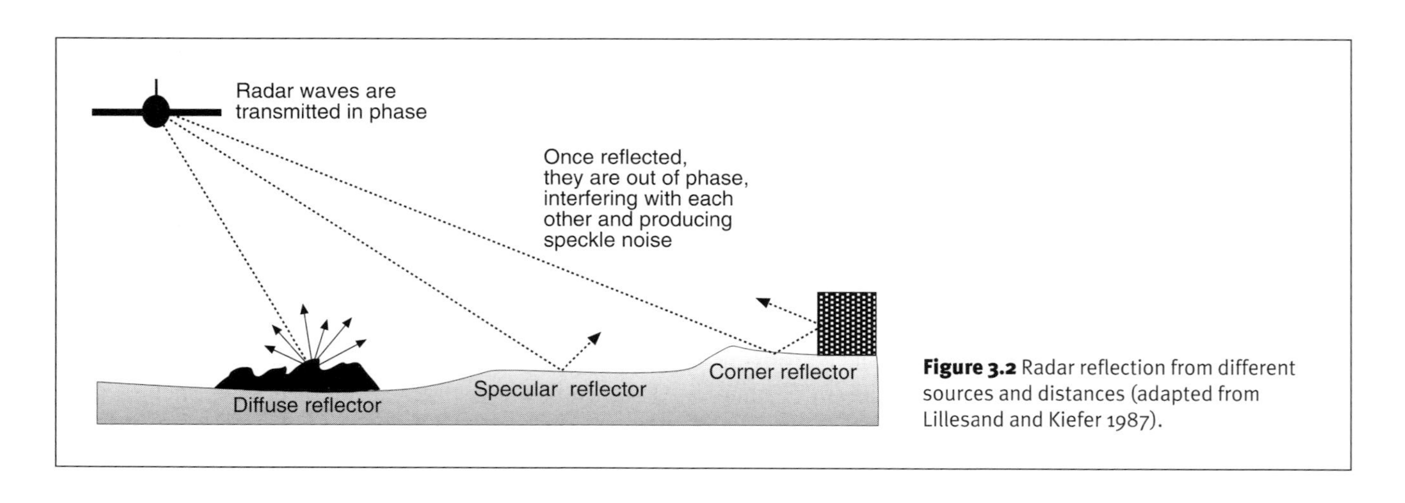

Figure 3.2 Radar reflection from different sources and distances (adapted from Lillesand and Kiefer 1987).

erated by the sensor, it is independent of solar energy and hence can be used at night. Energy at these wavelengths is not attenuated by the atmosphere, except in the case of heavy rain clouds, so there is no need for the atmospheric corrections that are usually required for visible and near infra-red wavelengths.

Owing to the much longer wavelengths used in radar remote sensing, the information content of the signal is fundamentally different to that from visible and infra-red wavelengths. For coastal applications, the most important difference is that the energy does not penetrate the sea surface and so very little is revealed about underwater habitats (although some information on seafloor topography can be deduced from patterns of surface roughness). For this reason, radar images are generally only used for the terrestrial environments of mangrove, sand dunes, wetlands and beach. The amount of energy returned (backscattered) to the sensor is dependent upon the slope, roughness and dielectric properties (principally controlled by moisture) of the surface in question and therein lies the information content of the signal.

In comparison with visible and infrared remote sensing, radar has yet to be fully explored in terms of mapping and measurement of coastal environments. This is in part due to our poorer understanding of radar interactions with surfaces, the more complex nature of this interaction and the fact that reliable radar data from satellites has only been available since 1991. There has been some success using radar data to map mangrove extent and in the future it may prove possible to obtain quantitative information on biophysical properties of mangrove forests (see Aschbacher *et al.* 1995, Pons and Le Toan 1994). At present, the greatest potential for radar lies in the realm of change detection. As radar is unhindered by cloud cover and nightfall, imagery can be collected with high frequency when necessary. Furthermore, because the atmosphere does not distort the measured signal, we get reliable measurements of backscatter from the surfaces in question and can therefore detect any changes through time. Such changes may arise from differences in moisture or canopy geometry but, more usefully in a management context, they may arise from major changes such as destruction of an area of mangrove. In this way it is possible to use radar images to map areas of change that require further investigation. Radar imagery, for example, is being used with great success to map deforestation (Rignot *et al.* 1996, Conway *et al.* 1996).

Another difference between visible/IR and radar remote sensing is that the latter type of image contains 'speckle'. Speckle can be considered to be noise that reduces the visibility of features in an image. It is similar in appearance to a television picture with poor reception, and is infuriating in the sense that the more an image is enlarged, the harder it becomes to distinguish features clearly. Speckle appears in Synthetic Aperture Radar (SAR) images because of the coherent processing of the returned signal. The pulse of radiation produced by an active radar sensor is transmitted and travels in phase; in other words waves interact minimally on their way to the target. Reflection from different types of surface disrupt this and cause the waves to become out of phase (Figure 3.2) and so interfere destructively or constructively. Destructive interference produces dark areas on the image, constructive interference bright. These combine to create speckle. For many applications it is necessary to reduce the effect of speckle, which can be achieved through use of a number of special filtering techniques.

Further reading

More information on active and passive sensors can be found in any remote sensing text book but Kramer (1994) is especially recommended.

Terms used to describe image data

Images obtained from digital sensors are an ordered set of numbers, each value being related to the amount of radiance coming from a ground area represented by a single cell or pixel (pixel is an amalgamation of the words 'picture element', the blocks which make up an image). All image types can be characterised by the following terms:

- *Spatial resolution* is a measure of the area on the ground covered by each sampling unit of the sensor and is dependent on altitude and sensor design. Many complex definitions of spatial resolution exist but for practical purposes it can be thought of as the size of each pixel (Plate 4) and so is commonly measured in units of distance (m or km). A more or less equivalent term is instantaneous field of view (IFOV) which is the angle θ in Figure 3.3, and measured in degrees or radians. A whole image may comprise of many millions of pixels, which partly accounts for the large data volumes often encountered in remote sensing. Altitude and spatial resolution remain more or less constant for satellites (at least theoretically) but the spatial resolution of airborne sensors depends upon the altitude of the aircraft e.g. the Compact Airborne Spectrographic Imager (see sensors with spatial resolution less than 10 m) flown at 840 m has a spatial resolution of 1 m; at an altitude of 3000 m it is about 3 m.
- *Swath width* is the width of the area on the Earth's surface covered by the image (the width swept by the sensor on one overpass). Spatial resolution and swath width determine the degree of detail that is revealed by the sensor and the area of coverage. Swath width can be thought of as constant for satellites (e.g. the swath

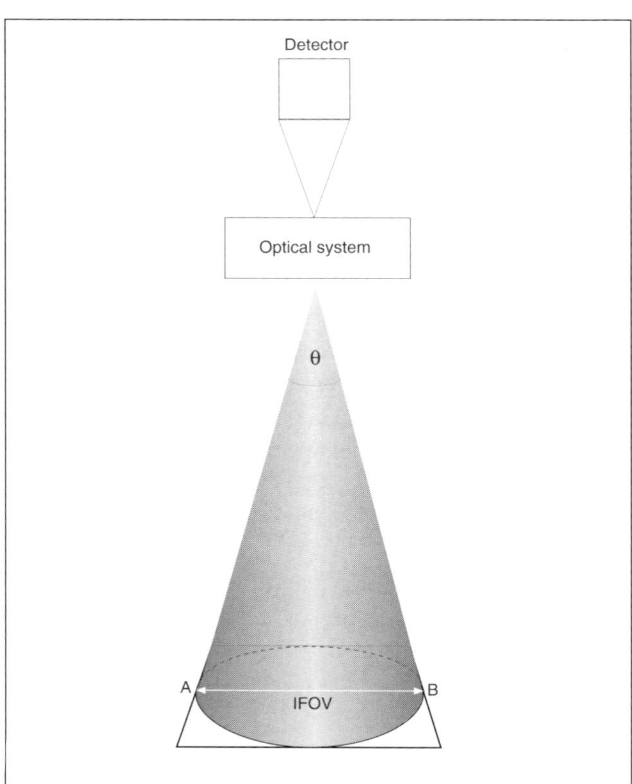

Figure 3.3 The instantaneous field of view (IFOV) is the area on the Earth's surface viewed by a sensor from a given altitude at a given moment in time. IFOV may be described as the angle viewed, θ, (expressed in milliradians or degrees) or the width (AB) on the ground of the area viewed. Note that AB is the diameter of a circle but the recorded radiance is displayed in an image as a square (pixel) of side AB.

width of SPOT XS is 60 km) but variable for airborne sensors, dependent on altitude of the aircraft. For example, CASI swath width is ~ 0.5 km at 840 m and ~1.5 km at 3000 m.

For airborne sensors the width of the ground swath observed varies with aircraft altitude. For these sensors swath width is seldom stated and, if used, will be stated for a given altitude. Instead, the field of view (FOV) of the sensor is stated (see Table 3.2). This is the angle viewed during a single scan. Using simple trigonometry, for a given altitude the swath width can be calculated from the FOV of the sensor.

- *Temporal resolution* or *repeat time*. Most satellites are sun-synchronous, which means that, for any given point on Earth, the satellite will pass overhead at the same time of day. For example, Landsat and SPOT satellites are sun-synchronous, crossing the equator at approximately 0945 h and 1030 h local sun time. This is subject to slight variation due to orbital perturbations. Both satellites pass overhead earlier in the day north of the equator and later to the south. Temporal resolution therefore is the time interval between consecutive overpasses by a satellite of a fixed point. Landsat TM images for example can be obtained every 16 days, weather permitting. Generally, it is the maximum frequency at which imagery of any area covered by the satellite's orbit can be obtained. The angle of view for sensors on the SPOT and RADARSAT satellites can be altered to allow the collection of off-nadir (oblique) images either side of the satellite's path. This can increase the temporal resolution (revisit time) of the satellites considerably.

BOX 3.1

AVIRIS (Airborne Visible/Infrared Imaging Spectrometer)

AVIRIS is an imaging spectrometer which simultaneously acquires 224 images at 20 m spatial resolution and 10 nm spectral resolution (Porter and Enmark 1987, Kruse *et al.* 1993). This allows mapping to be performed based on the spectroscopic properties of materials (Goetz *et al.* 1985, Boardman *et al.* 1995). AVIRIS has been used to study coastal geology, sediment plumes and algal blooms (Kruse *et al.* 1997, Richardson and Ambrosia 1997). The practicalities of using remote sensing for monitoring coastal water quality is covered in Chapter 14.

☞ **Temporal resolution has little meaning for airborne sensors because the frequency at which imagery can be obtained is highly variable and is largely determined by the weather.**

- *Spectral resolution*. Sensors are designed to record a specific portion of the electromagnetic spectrum. Sensors which record radiation over a wide part of the visible spectrum in a single waveband are called 'panchromatic'. However many sensors record radiation in several distinct bands (or channels) and are thus called multispectral. Spectral resolution refers to the number and width of bands. A sensor with high spectral resolution has numerous bands of narrow range. Hyperspectral sensors have many (tens of) narrow bands. The development of airborne hyperspectral sensors with more and narrower bands has been a recent trend in remote sensing, but as yet there are none in space. Box 3.1 provides some details on AVIRIS, an airborne hyperspectral sensor which has been used to study coastal phytoplankton dynamics.
- *Radiometric resolution* refers to the number of digital levels used to express the data collected by the sensor and is therefore a measure of its sensitivity. The radiometric resolution of Landsat TM is such that the level of light intensity recorded for each pixel – in each wave band – can have a value between 0 (no reflectance) and 255 (100% reflectance), each value being referred to as a digital number or DN. Landsat TM data is therefore 8-bit (2^8) data; 16-bit (2^{16}) data would express the reflectance in each band on a scale of 0-65535.

☞ **The size of an image file is determined by the spatial and spectral resolution. Imagine an area 1 km x 1 km. A sensor with a spatial resolution of 10 m would produce an image of 10,000 pixels; a sensor with a spatial resolution of 5 m, an image of 40,000 pixels. Now consider spectral resolution: if the first sensor was recording radiation in 7 bands then the computer would have to store 7 digital numbers for each pixel, a total of 70,000 DNs. If the 5 m sensor was panchromatic then only a single DN would be stored for each pixel; if it was a multispectral sensor with, say, eight bands the image would consist of 320,000 DNs. Higher spatial and spectral resolution mean more information, which requires larger files and more computing capacity to store and manipulate them.**

In terms of computer file sizes an 8-bit number forms the basic unit, called a byte, such that an image consisting of 320,000 DNs would have a file size of at least 320,000 bytes. If the data had been collected with greater sensitivity, for example, if it had been recorded as 16-bit, this would require 2 bytes to record the radiance of each pixel, doubling the file size to 640,000 bytes. A Landsat TM image has a high data volume (about 270 million bytes or 270 Mb) because of its relatively high spatial resolution, large area of coverage, and seven wavebands.

☞ **In general terms there is an inverse relationship between area covered by a sensor and spatial resolution. This is because there is a maximum amount of data that can be recorded at the satellite and a maximum rate at which those data can be transmitted and received at Earth receiving stations. Therefore, imagery tends to be either high resolution from relatively small areas, or relatively low resolution from larger areas.**

Limitations of remote sensing of tropical coasts

A number of authors agree that the capability of remote sensing technology may have been somewhat oversold in the past (Currey *et al.* 1987, Young and Green 1987, Meaden and Kapetsky 1991). Thus, by failing to live up to expectations, a degree of disillusionment may have arisen among potential users. It is vitally important that the uses of remote sensing are understood in relation to its inherent limitations. Fundamental limitations to remote sensing such as the limited penetration of light in water and areas of heterogeneous substrate within a single pixel ('mixels') remain facts of life. Some, such as the variable attenuation of light in the water column and atmosphere, can be compensated for (see Chapters 8 and 7 respectively).

BOX 3.2

Descriptive resolution of sensors

Descriptive resolution is a term referring to the level of detail to which a habitat can be described and mapped, and so is a function of spatial, spectral and radiometric resolution. For example, a sensor with coarse descriptive resolution would only be able to separate, say, coral from sand. In contrast, a sensor with fine descriptive resolution might be able to distinguish coral zonation, coral density and sand. Descriptive resolution is a concept which should help readers decide whether a particular type of imagery will enable them to fulfil a given management objective. The descriptive resolution of different sensors for the mapping of coral reef, seagrass and mangrove habitats is discussed in Chapters 11–13.

The limitations of remote sensing can be broadly divided into: (i) practical limitations, in the sense that they affect the application of remote sensing, and (ii) user limitations, in the sense that they complicate the incorporation of remote sensing into a coastal management plan. Practical limitations arise from sensor specifications and many can be expected to be progressively overcome as new sensors are developed. User limitations arise from the difficulties in assessing the descriptive resolution of sensors (Box 3.2), the true costs of deriving the necessary resource information and the difficulty of making rigorous accuracy assessments. User limitations are discussed in detail with reference to specific applications in Chapters 10–18.

Practical limitations of sensors

In an effort to present a balanced assessment of the utilisation of remote sensing, the major practical limitations cited in more than 150 scientific papers and articles (reviewed in Green *et al.* 1996) have been summarised in Table 3.1. The constraints listed are ranked in terms of frequency of citation; this does not necessarily indicate importance. The importance of each issue depends on the objective of the remote sensing study and differs from sensor to sensor. Indeed, it is fair to suggest that some of the constraints listed may not be regarded as undesirable under particular circumstances. For example, data from sun-synchronous orbital satellites (e.g. SPOT, Landsat) will be taken at the same time of day at any given site. If the objective of the study requires data from different times of day or similar tidal conditions, then this clearly presents a problem. However it may be desirable to record data at the same time of day and therefore with a similar sun angle.

Several of the general problems represented in the table are the result of specific physical limitations which

Table 3.1 Potential constraints to the practical use of remote sensing for applications in the tropical coastal zone. Constraints are listed in descending order of the frequency of citation and not necessarily by importance. Readers should note that these constraints are not discrete; the more general issues described here are the result of several technical constraints which are listed separately (see text for examples). n= number of citations.

Constraint	n
1 Cloud obscuring area of interest	24
2 Spatial resolution of the sensor too coarse	13
3 Turbidity and variations in water colour may confuse classification and bathymetry	12
4 Variations in depth and tide complicate image comparisons and classification	9
5 Spectral resolution too coarse	7
6 Substrate heterogeneity can result in 'mixels' and confuse bathymetric measurements	5
7 Overall depth of penetration inadequate/limited and varies tidally	5
8 Backscatter from atmosphere/sunglint from water confuses image interpretation	5
9 Temporal resolution too infrequent	4
10 Difficulty of obtaining ground control for image rectification	4
11 High cost of imagery	4
12 Inadequate user training	4
13 Satellites do not collect data in areas of interest	3
14 Remote sensing generally provides geomorphological rather than biological information	3
15 Disparity between dates of image and field work complicates image interpretation	3
16 Poor sea state may increase backscatter and reduce depth of light penetration	3
17 High cost of hardware and software	3
18 Error in locating pixels on the ground (for verification and ground-truthing)	2
19 Limited radiometric resolution	1
20 Banding (a sensor fault) requires pre-processing, reducing the integrity of image data	1
21 Periodicity of satellites means data is collected at the same time of day at every site	1
22 Most remotely sensed data are not available in real time	1
23 Overall dimensions of aerial imagery are small and thus expensive for large areas	1
24 Poor knowledge of remote sensing applications on the part of managers and politicians	1
25 Data related issues (e.g. large size of image files leading to storage problems)	1
26 Satellite sensors cannot be programmed to provide data on a routine basis	1
27 The need to make radiometric corrections when using multiple images – the magnitude of atmospheric parameters like aerosol density is often unknown	1

are also listed. For example, remote sensing tends to provide geomorphological rather than ecological information on reef structure (Table 3.1: 14). This is a result of the limited spectral and spatial resolution of the sensors (Table 3.1: 2, 5) and factors that confound image interpretation such as turbidity and variation in water depth (Table 3.1: 3, 4).

Cloud cover is probably the most serious problem in the humid tropics since it can significantly reduce the number of suitable images available and obscure the area of interest within an individual image. It is discussed in detail in Chapter 5.

Factors that contribute to satellite imagery not being available (Table 3.1: 1, 9, 13, 20) can have severe synergistic consequences. While this has not been rigorously quantified (to the best of our knowledge) we can illustrate the problem with a recent attempt to obtain Landsat TM data of the Turks and Caicos, a group of dry islands in the British West Indies (average annual rainfall of 720 mm). A search of archived imagery conducted in 1995 revealed that the most recent image with less than 20% cloud cover was dated in 1990. This has two implications. Firstly, no change detection is possible in the cloud-covered areas using Landsat TM over the past five years. Secondly, a gap of five years between the acquisition of imagery and field data may cause problems when trying to relate image and field data (Table 3.1: 15).

Financial costs of hardware, software and imagery do not rank highly as problems on a frequency of citation basis. This is largely due to the dominance of technical papers in the scientific literature. Naturally, authors of these papers generally discuss the problems of their research in terms of technical rather than economic constraints. The important issues of cost are discussed in Chapter 19.

Sensors useful for remote sensing of tropical coasts

Sensors with spatial resolution of more than 100 m

The benefit of this type of sensor lies in the large area of coverage and high temporal resolution of the data. Data are typically acquired every few days (or daily and twice daily in some cases) over thousands of kilometres. Spatial resolution however is coarse, generally about 1 km^2. Therefore, these sensors are most useful for monitoring environmental parameters such as sea-surface temperature, chlorophyll concentration and ocean colour over regional scales.

AVHRR sensor on NOAA 9–14 satellites

Primary coastal applications: Sea surface temperature, ocean colour.
Platforms: NOAA POES (Polar Orbiting Operational Environmental Satellites).
Sensor: AVHRR (Advanced Very High Resolution Radiometer).
Further information is available from:
http://edcwww.cr.usgs.gov/landdaac/1KM/avhrr.sensor.html or Kidwell (1995).
Operation: 1984 – present.
Order from remote sensing centres e.g. NRSC, CCRS (see Appendix 2).
Spatial resolution: 1.1 km.
Temporal resolution: 12 hours.
Radiometric resolution: 10 bit.
Wavebands: Band 1 = 580–680 nm, Band 2 = 725–1100 nm, Band 3 = 3550–3930 nm, Band 4 = 10.3–11.3 µm, (10.5–11.5 µm on NOAA-10), Band 5 = 11.4–12.4 µm (not on NOAA-10).
Image dimensions: 2399 km.
Products available: Photographic and digital images.
Cost per image: Free if receiving station is owned (it is possible to receive data directly from the satellites), otherwise media costs and postage only.

CZCS sensor on the NIMBUS-7 satellite

Primary coastal applications: measurement of temperature, chlorophyll concentration sediment distribution, and Gelbstoff concentrations as a salinity indicator.
Platform: NIMBUS 7 satellite.
Sensor: CZCS (Coastal Zone Color Scanner)
Operation: 1978–1986.
Further information is available at:
http://daac.gsfc.nasa.gov/CAMPAIGN_DOCS/OCDST/OB_Outline.html or a special edition of the *Journal of Geophysical Research* (Volume 99, April 1994) entitled *Ocean Color From Space: A Coastal Zone Color Scanner Retrospective*.
The entire CZCS archive may be browsed at:
http://daac.gsfc.nasa.gov/CAMPAIGN_DOCS/OCDST/CZCS/czcs_browser.html; Search criteria may be sent by electronic mail to ocdst_comments@daac.gsfc.nasa.gov;
Fax: +1 301 286 0268.
Spatial resolution: 825 m.
Temporal resolution: 6 days.
Radiometric resolution: 8 bit.
Swath width: 1566 km.
Wavebands (nm): Band 1 = 433–453, Band 2 = 510–530, Band 3 = 540–560, Band 4 = 660–680, Band 5 = 700–800, Band 6 = 10,500–12,500.

SeaWiFs sensor on the SeaStar satellite

Primary coastal applications: ocean colour and chlorophyll-*a* measurement.
Platforms: SeaStar satellite.
Sensor: SeaWiFS (Sea-viewing Wide Field-of-view Sensor)
Operation: Launched in July 1997.
Further information is available from:
http://seawifs.gsfc.nasa.gov/SEAWIFS.html or by writing to SeaWiFS Project, NASA, Goddard Space Flight Centre, USA.
Spatial resolution: 1.1 km.
Swath width: 2801 km.
Wavebands (nm): Band 1 = 402–422, Band 2 = 433–453, Band 3 = 480–500, Band 4 = 500–520, Band 5 = 545–565, Band 6 = 600–680, Band 7 = 745–785, Band 8 = 845–885.
Cost per image: One month-old data will be free. Other options available.

ATSR sensor on the ERS satellites

Primary coastal applications: sea state, sea surface winds, ocean circulation and coastal vegetation cover
Platforms: European Remote Sensing satellites (ERS-1 and ERS-2).
Sensor: ATSR-I (imaging Along Track Scanning Radiometer).
Operation: 1991 – present.
Further information is available from: http://www.esrin.esa.it:80/ (the European Space Agency).
Spatial resolution: 1 km.
Temporal resolution: 3 days.
Swath width: 785 km.
Wavebands (nm): Band 1 = 1580–1640, Band 2 = 3550–3930, Band 3 = 10,400-11,300, Band 4 = 11,500–12,500.
Cost: 80–150 ECU

OCTS sensor on the ADEOS satellite

Primary coastal applications: measurement of temperature, chlorophyll concentration sediment distribution, and Gelbstoff concentrations as a salinity indicator.
Platform: Advanced Earth Observing Satellite (ADEOS).
Sensors: OCTS (Ocean Color and Temperature Scanner).
Operation: 17 August 1996 – 30 June 1997. ADEOS II to be launched in 1999.
Further information is available from:
http://www.eorc.nasda.go.jp/ADEOS/UsersGuide/.
Some OCTS data are currently also held by NASA and can be browsed at:
http://seawifs.gsfc.nasa.gov/seawifs_scripts/octs_browse.pl

Spatial resolution: 700 m.
Temporal resolution: 3 days.
Swath width: 1400 km.
Radiometric resolution: 10 bit.
Wavebands (nm): Band 1 = 402–422, Band 2 = 433–453, Band 3 = 480–500, Band 4 = 510–530, Band 5 = 555–575, Band 6 = 655–675, Band 7 = 745–785, Band 8 = 845–885, Band 9 = 3550–3880, Band 10 = 8250–8800, Band 11 = 10,300–11,400, Band 12 = 11,400–12,500.
Cost: Currently, processed data are available via ftp, see: http://www.eorc.nasda.go.jp/ADEOS/UsersGuide/.

Sensors with spatial resolution of 10–100 m

Data from these sensors have the advantage that their spatial resolution is more appropriate to biological habitats. Therefore, they can be used to map different ground (and benthic) habitat types, as well as providing quantitative data on the spatial distribution of environmental information (e.g. seagrass biomass). However, the area of coverage is less as a result of the higher spatial resolution.

Multispectral Scanner on Landsat satellites

Primary coastal applications: resource inventory, cartographic mapping, baseline environmental monitoring, change detection, bathymetry. Note: MSS is no longer being acquired.
Platforms: Landsat 1–5.
Sensor: MSS (Multispectral Scanner).
Operation: 1972 – 1993.
Further information is available from: http://geo.arc.nasa.gov/sge/landsat/daccess.html.
Spatial resolution: approximately 80 m.
Temporal resolution: 16 days for Landsat 4-5.
Radiometric resolution: 6 bit, resampled and stored as 8 bit.
Swath width: 185 km.
Wavebands (nm): Band 1 = 500–600, Band 2 = 600–700, Band 3 = 700–800, Band 4 = 800–1100 (for MSS on Landsat 4 and 5).
Cost: See Chapter 5.

Thematic Mapper on Landsat satellites

Primary coastal applications: resource inventory, cartographic mapping, baseline environmental monitoring, change detection, bathymetry.
Platforms: Landsat 4 and 5.
Sensor: TM (Thematic Mapper).
Operation: 1982 – present.
Further information is available from: http://geo.arc.nasa.gov/sge/landsat/daccess.html.
Spatial resolution: 30 m.
Temporal resolution: 16 days.
Radiometric resolution: 8 bit.
Swath width: 185 km.
Wavebands (nm): Band 1= 450–520, Band 2 = 520–600, Band 3 = 600–690, Band 4 = 760–900, Band 5 = 1550–1750, Band 6 = 10,400–12,500, Band 7 = 2080–2350.
Cost: See Chapter 5.

Two other sensors have been carried on Landsat platforms but have been little used and are mentioned here only in passing. The RBV (Return Beam Vidicon camera) had a spatial resolution of 40 m, spectral range 505–750 nm and ground image size of 98 x 98 km.

The ETM (Enhanced Thematic Mapper) was carried on Landsat 6 but was lost with that satellite. Another is on board Landsat 7 (launched in April 1999). ETM+ has the seven TM bands plus an additional high resolution (15 m) panchromatic band (500–900 nm). Images will be made available in three modes, Mode 1: all seven spectral bands, Mode 2: panchromatic band with bands 4, 5, and 6, and Mode 3: panchromatic band with bands 4, 6, and 7. See http://landsat.gsfc.nasa.gov/.

HRV and HRVIR sensors on SPOT satellites

Primary coastal applications: resource inventory, cartographic mapping, baseline environmental monitoring, change detection, bathymetry. SPOT 1-3 had the same HRV sensor (two per satellite); SPOT 4 launched in March 1998 has the HRVIR sensor with an extra short-wave infra-red band (useful for coastal vegetation monitoring) and a blue rather than green band as band 1 (giving better water penetration).
Platforms: SPOT 1–4 (Satellite Pour l'Observation de la Terre). SPOT 2 and 4 are currently operational.
Sensor: HRV (Haute Résolution Visible, High Resolution Visible sensor, in English). HRVIR (High Resolution Visible and Infra-Red) on SPOT 4.
Operation: 1986 – present.
Contact: SPOT Image
http://www.spotimage.fr/home/contact/welcome.htm
Fax: +33 5 62 19 40 54. Tel: +33 5 62 19 40 89.
Further information is available from: http://www.spotimage.fr/
Spatial resolution: Multispectral Mode (XS) 20 m, Panchromatic Mode (XP) 10 m.
Temporal resolution: 26 days (but has off-nadir viewing).
Radiometric resolution: 8 bit.
Swath width: 60 km.
Wavebands for SPOT 1–3 (nm): XS mode Band 1 = 500–590, Band 2 = 610–680, Band 3 = 790–890. XP (panchromatic) mode = 510–730. Wavebands for SPOT 4 (nm): Blue band = 430–470, Red band = 610–680, Near-IR band = 780–890, Short-wave IR band = 1580–1750.
Cost: See Chapter 5.

LISS sensor on IRS Satellites

Primary coastal applications: resource inventory, cartographic mapping, baseline environmental monitoring, change detection. LISS data appears to have primarily been used by Indian scientists. Increased ease of availability should widen the application of these data.
Platforms: IRS 1A –1D (Indian Remote Sensing satellite series).
Sensors: LISS (Linear Imaging Self-Scanning Sensor). IRS 1A and 1B carried LISS-I and II sensors. IRS 1C and 1D carry LISS-III.
Operation: 1988 – present.
Contact: ISRO (Indian Space Research Organisation) NRSA (National Remote Sensing Agency) Data Centre, Hyderabad, India. Tel: +91 662-5790345 for image products. Space Imaging EOSAT, are the exclusive provider of IRS imagery outside of

India. Customer Service can be contacted by telephone on +1 301 552 0537 or by e-mail at info@spaceimage.com. See http://www.spaceimage.com/ for further details.

Spatial resolution: LISS-I 73 m, LISS-II 36.5 m, LISS-III 23.5 m

Temporal resolution: LISS-I and II 22 days, LISS-III 24 days.

Radiometric resolution: 7 bit.

Swath width: LISS-I: 148 km; LISS-II: 146 km; LISS-III: 142 km (band 1–3), 148 km (band 4).

Wavebands (nm): LISS-I and II: Band 1 = 450–520, Band 2 = 520–590, Band 3 = 620–680, Band 4 = 770–860; LISS-III: Band 1 = 520–590, Band 2 = 620–680, Band 3 = 770-860, Band 4 = 1550–1750.

Three sensors make up the payload of IRS-1C and IRS-1D: PAN (Panchromatic Camera), LISS-III (see above) and WiFS (Wide Field Sensor). PAN specifications are: one spectral band: 500–900 nm; spatial resolution: 5.8 m; swath width: 70.5 km; radiometric resolution 6 bit; temporal resolution 24 days (off-nadir revisit is possible, just like the HRV on SPOT, in 5 days: off-nadir viewing capability ± 26^0). WiFS specifications are: spectral bands (nm): Band 1 = 620–680, Band 2 = 770–860; spatial resolution: 188 km; swath width: 770 km; radiometric resolution 7 bit; temporal resolution 5 days.

Cost: See Chapter 5.

OPS sensor on the JERS-1 satellite

Primary coastal applications: resource inventory, cartographic mapping, baseline environmental monitoring, change detection.

Platform: JERS-1 (Japanese Earth Resources Satellite).

Sensor: OPS (Optical Sensor).

Operation: 1992 – present

Contact: National Space Development Agency of Japan at: http://www.nasda.go.jp/index_e.html
E-mail: www-admin@yyy.tksc.nasda.go.jp

Spatial resolution: 18.3 m.

Temporal resolution: 44 days.

Radiometric resolution: 8 bit.

Swath width: 75 km.

Wavebands (nm): Band 1 = 520–600, Band 2 = 630–690, Band 3 and 4 = 760–860 (for stereoscopic images), Band 4 = 760–860 (forward viewing), Band 5 = 1600–1710, Band 6 = 2010–2120, Band 7 = 2130–2250, Band 8 = 2270–2400.

Cost: See Chapter 5.

MESSR sensor on the MOS satellite

Primary coastal applications: resource inventory, cartographic mapping, baseline environmental monitoring, change detection.

Platform: MOS-1 (Marine Observation Satellite).

Sensor: MESSR (Multispectral Electronic Self-Scanning Radiometer).

Operation: 1987 – present

Contact: National Space Development Agency of Japan at e-mail: www-admin@yyy.tksc.nasda.go.jp, or via the WWW on: http://www.nasda.go.jp/index_e.html, or via the European Space Agency.

Spatial resolution: 50 m.

Temporal resolution: 17 days.

Radiometric resolution: 8 bit.

Swath width: 200 km.

Wavebands (nm): Band 1 = 510–590, Band 2 = 610–690, Band 3 = 730–800, Band 4 = 800–1100.

Cost: Only available on CCT. 275 ECU for one scene, or two scenes for 504 ECU.

SAR sensor on ERS satellites

Primary coastal applications: Terrestrial habitat mapping including mangrove extent, and biophysical parameters (LAI, cover). A particular advantage is that sensing can be made through cloud cover, fog and at night. Well-suited for change-detection studies. However, note that SAR is not capable of underwater sensing.

Platforms: ERS-1 and ERS-2 (European Remote Sensing satellites).

Sensor: SAR (Synthetic Aperture Radar).

Operation: 1991 – present.

Contact: ERS Help Desk, ESRIN, Via Galileo Galilei, 00044 Frascati, ITALY. Tel: +39 6 941 80 600.

Spatial resolution: 25 m.

Temporal resolution: 3 days.

Radiometric resolution: 16 bit.

Swath width: 100 km.

Wavebands: microwave, (C-band). Single band at 5.6cm wavelength (5.3 Ghz) with vertical transmit and receive polarisation (VV).

Products available: photographic and digital images. Range of processing levels from raw through to fully geo- and terrain corrected. Fast delivery product also available. Most investigators use the precision level of processing (PRI). Image data volume: 131Mb.

Cost: See Chapter 5.

RADARSAT

Primary coastal applications: Terrestrial habitat mapping including mangrove extent, and biophysical parameters (LAI, cover). A particular advantage is that sensing can be made through cloud cover, fog and at night. Well-suited for change-detection studies. Note that radar is not capable of underwater sensing. Advantages over ERS (outlined above) include a variety of incidence angles, spatial resolutions and image sizes. Because of this range of options it is usual to programme the satellite to collect data to your specifications rather than to take imagery from the archive.

Platform: RADARSAT.

Sensor: SAR (Synthetic Aperture Radar).

Operation: 1995 – present.

Further information is available from: http://www.ccrs.nrcan.gc.ca/ccrs/tekrd/radarsat/rsate.html.

Contact: Radarsat International (RSI) 3851 Shell Road, Suite 200, Richmond, British Columbia, V6X 2W2, Canada, Tel: 604 244 0400, Fax: 604 244 0404, E-mail: caspden@rsi.ca, http://www.rsi.ca.

Spatial resolution: Varies from 10 m up to 100 m, dependent on beam mode.

Temporal resolution: variable.

Radiometric resolution: 16 bit.

Swath width: 50 x 50 km to 500 x 500 km dependent on beam mode.

Wavebands (nm): microwave, (C-band). Single band at 5.6 cm wavelength (5.3 GHz) with horizontal transmit and receive polarisation (HH). Incidence angle can be varied between 20–60° dependent on beam mode.

Products available: photographic and digital images. Range of processing levels from raw through to fully geo- and terrain corrected. Most investigators use the Path Image level of processing (SGF). Image data volume: 131Mb (for SGF).

Cost: See Chapter 5.

Sensors with spatial resolution of less than 10 m

Imagery with a spatial resolution of less than 10 m is rare, for the good reason that coverage at this level of detail for large areas would quickly produce unmanageable volumes of data. There is however a growing archive of satellite data, primarily of high resolution Russian sensors down to a resolution of 5 m, although the availability of data for tropical coasts appears exceedingly limited.

Data from airborne multispectral imagers typically has very high spatial (they are operated more often nearer 1 m than 10 m resolution) and spectral resolution (tens or hundreds of potential bands). This enables fine detail to be revealed within images, detail which would be impossible from satellite imagery. A further advantage is that the aircraft may be flown and the data collected at times chosen by the operators (weather permitting). The users of airborne imagery are not restricted to data collection at fixed times such as satellite overpasses and can make opportunistic use of favourable weather windows. However, relative to satellites, airborne imagery covers tiny areas, the volume of data collected can be enormous and acquisition is very expensive (Chapter 5). Operating sensors from aircraft also has implications for geometric correction (see Chapter 6).

Such has been the explosion of interest over the last decade in airborne multispectral (and more recently, hyperspectral) imagery that many different sensors have been developed. Table 3.2 summarises briefly the main characteristics of nearly forty airborne sensors. Many of these sensors have more than one fixed configuration (or are fully programmable by the operator in the number and width of bands used). Furthermore the spatial resolution of the imagery can be varied by flying the sensor at different heights above sea level. Together these two factors provide the user of airborne imagery with an enormous variety of possible spatial/spectral resolutions, and great flexibility in tuning the sensor to a particular application. For practical purposes the selection of sensors is undoubtedly not as wide as Table 3.2 would suggest, since some of these sensors may not be available for deployment outside their country of origin. Having said that however, many will be commercially available and all have specifications which render them appropriate for the mapping of habitats, both submerged and terrestrial.

Space and time precludes a detailed description of a range of airborne multispectral sensors and so we describe one of the most popular and versatile systems, the CASI (Compact Airborne Spectrographic Imager). This is particularly appropriate for tropical coastal applications because of its programmable wavebands and its lightweight and compact configuration (Plate 3), which permits it to be transported easily to any field area and installed on non-survey aircraft if necessary (see Chapter 5). CASI is used as a detailed example and this section provides much of the technical detail necessary for a complete understanding of the application of CASI described in future chapters.

☞ **Daedalus ATM (Sensys Technologies Inc.) is another versatile airborne sensor, although it has to be mounted in a survey aeroplane. ATM specifications are given in Table 3.2: CASI and ATM costs are compared in Chapter 5.**

CASI (Compact Airborne Spectrographic Imager)

The CASI instrument consists of five modules: sensor head, instrument control unit, keyboard, power supply module, and monitor. Total instrument weight is 55 kg. Power requirements are 110 volts at 2.4 amps, and with a suitable inverter the CASI can be operated from the 28 volts DC power supply found on many aircraft. Designed to be compact enough to be flown on light aircraft the CASI has been flown from models such as the Piper Aztec and Cessna Citation (Plate 3).

CASI has a nominal spectral range of 391 nm to 904 nm with a spatial resolution of 512 pixels across the 35° field of view. Ground resolution depends on the aircraft altitude and ranges from one to ten metres. The spectral resolution is nominally 2.5 nm, with 288 spectral channels centred at 1.8 nm intervals. This bandwidth increases with wavelength. Data are recorded on a built-in digital tape recorder (EXABYTE) which uses 8 mm cassettes. This standardised data storage medium greatly facilitates post processing of the data. Each tape can store up to 1 GB of data or, depending on the frame rate, up to one hour of imagery. A representative value for the frame rate under typical conditions is 20 frames (lines) sec^{-1} for eight spectral channels in imaging mode. Due to the high data rate of the CASI sensor three operating modes have been developed (the user may select the most appropriate). Each mode maximises the information content while keeping the data rate at a manageable level. The three operating modes are:

- *Spatial mode*: full spatial resolution of 512 spatial pixels across the 35 degree swath. Channel wavelengths and bandwidths are user-specified (up to 21 bands). This mode is also called 'imager mode' (Figure 3.4).

Table 3.2 The characteristics of some airborne multispectral and hyperspectral imagers. The acronyms used for sensor names, agencies and companies in this table are defined in full in Table 3.3. The development of this type of sensor has been extremely rapid over the past six years: as a result this table is probably not complete. It should, however, serve as a basis for further investigation. FWHM = full width at half maximum, IFOV = instantaneous field of view (in milliradians), FOV = field of view in degrees (angle viewed in a single scan).
Note: Full width at half maximum (FHWM) is a commonly used measure of spectral resolution that can be applied to hyperspectral data. It is a better measure of spectral resolution than sampling interval. The response to a monochromatic target should be a peak whose (spectral) width measured at a level of half of the peak height is the FWHM.

Sensor	Agency/Company	Number of bands	Spectral coverage (nm)	FWHM (nm)	IFOV (mrad)	FOV(°)	Period of operation
AAHIS	SAIC	288	433–832	6.0	1.0 x 0.5	11.4	since 1994
AHS	Daedalus (Sensys)	48	440–12700	20–1500	2.5	86	since 1994
AISA	Karelsilva Oy	286	450–900	1.56–9.36	1.0	21.0	since 1993
AMS	Daedalus (Sensys)	10	320–12500	30–4000	1.25/2.5	86	since 1980
AMSS	GEOSCAN	32 8 6	490–1090 2020–2370 8500–12000	20.0–71.0 60.0 550–590	2.1 x 3.0	92.0	since 1985
ATM	Daedalus (Sensys)	10	420–2350	30–300	1.25/2.5	86	since 1989
ARES	Lockheed	75	2000–6300	25.0–70.0	1.17	3x3	since 1985
ASAS	NASA/GSFC	62	400–1060	11.5	0.8	25.0	since 1992
ASTER Simulator	GER	1	700–1000	300	1.0	28.8	since 1992
AVIRIS	NASA/JPL	224	400–2450	9.4–16.0	1.0	30.0	since 1987
CAMODIS	Chinese Academy of Sciences	64 24 1 2	400–1040 2000–2480 3530–3940 10500–12500	10 20 410 1000	1.2 x 3.6 1.2 x 1.8 1.2 x 1.8 1.2 x 1.2	80	since 1993
DAIS–7915	GER/DLR/JRC	32 8 32 1 6	498–1010 1000–1800 70–2450 3000–5000 8700–12300	16 100 15 2000 600	3.3 2.2 or 1.1	78	since 1994
DAIS–16115	GER	76 32 32 6 12 2	400–1000 1000–1800 2000–2500 3000–5000 8000–12000 400–1000	8.0 25 25 16 333 333	3	78	since 1994
DAIS–3715	GER	32 1 2 1 1	360–1000 1000–2000 2175–2350 3000–5000 8000–12000	20 1000 50 2000 4000	5.0	90	since 1994
FLI/PMI	MONITEQ	288	430–805	2.5	0.66/0.80	70	since 1990
GERIS	GER	24 7 32	400–1000 1000–2000 2000–2500	25.4 120 16.5	2.5, 3.3 or 4.5	90	since 1996
HIS	SAIC	128	400–900	4.3	0.14 x 1.0	8.0	until 1994
HYDICE	Naval Research Laboratory	206	400–2500	7.6–14.9	0.5	8.94	since 1995
ISM	DES/IAS/OPS	64	800–1700	12.5	3.3	40.0	since 1991

(cont'd)

Table 3.2 (cont'd)

Sensor	Agency/Company	Number of bands	Spectral coverage (nm)	FWHM (nm)	IFOV (mrad)	FOV(°)	Period of operation
MAS	Daedalus	9	529–969	31–55	2.5	85.92	since 1993
		16	1395–2405	47–57			
		16	2925–5325	142–151			
		9	8342–14521	352–517			
MAIS	Chinese Academy of Sciences	32	450–1100	20	3	90	since 1990
		32	1400–2500	30	4.5		
		7	8200–12200	400–800	3		
MEIS	Mc Donnell Douglas	>200	350–900	2.5	2.5	–	since1992
MISI	RIT	60	400–1000	10	1	±45	since 1996
		1	1700	50			
		1	2200	50			
		3	3000–5000	2000			
		4	8200–14000	2000			
MIVIS	Daedalus	20	433–833	20	2	70	since 1993
		8	1150–1550	50			
		64	2000–2500	8			
		10	8200–12700	400/500			
MUSIC	Lockheed	90	2500–7000	25–70	0.5	1.3	since 1989
		90	6000–14500	60–1400			
ROSIS	MBB GKSS DLR	84	430–850	4.0/12.0	0.56	16.0	since 1993
		30					
RTISR	Surface Optics Corp.	20 or 30	400–700	7.0–14.0	0.2–2.0	29.0 x 22.0	since 1994
SMIFTS	University of Hawaii	75	1000–5200	100 cm^{-1}	0.6	6.0	since 1993
		35	3200–5200	50 cm^{-1}			
TRWIS–A	TRW	128	430–850	3.3	1.0	13.8	since 1991
TRWIS–B	TRW	90	430–850	4.8	1.0	13.8	since 1991
TRWIS–II	TRW	99	1500–2500	11.7	0.5/1.0	6.9/13.8	since 1991
TRWIS–III	TRW	396	400–2500	5.0/6.25	0.9	13.2	since 1991
Hybrid VIFIS	University of Dundee	30	440–640	10–14	1.0	31.5	since 1994
		30	620–890	14–18	1.0	31.5	
WIS–FDU	Hughes SBRC	64	400–1030	10.3	1.36	10.0/15.0	since 1992
WIS–VNIR	Hughes SBRC	17	400–600	9.6–14.4	0.66	19.1	since 1995
		67	600–1000	5.4–8.6			

- *Spectral mode*: full spectral resolution of 288 channels for up to 39 look directions across the 35 degree swath. Look direction spacing and location are user-specified to sample the array. This sampling produces an image rake or comb. A single channel, full spatial scene recovery channel (FSS channel) can be selected. This mode is also called 'multispectrometer mode' (Figure 3.5).
- *Full-frame mode*: this mode, sometimes called 'calibration mode', outputs all the 288 spectral channels for all 512 spatial pixels (i.e. the whole array). This mode requires long data readout times, in the order of one second or more. In airborne operation the first two modes are typically used in successive flights over the same target area. The full-frame mode is used for calibration and ground measurements.

CASI data are calibrated to radiance after flight. Data processing can be done on a PC-type computer directly

Table 3.3 Definition of the acronyms used for sensors, agencies and companies in Table 3.2.

Acronym	Definition
AAHIS	Advanced Airborne Hyperspectral Imaging Spectrometer (Science Application International Corporation (SAIC) USA)
AHS	Airborne Hyperspectral Scanner (Daedalus, USA)
AIS	Airborne Imaging Spectrometer (National Aeronautics and Space Administration (NASA)/JPL, USA)
AISA	Airborne Imaging Spectrometer for Different Applications (Karelsilva Oy, Finland)
AMS	Airborne Multispectral Scanner (Sensys Technologies Inc. (formerly Daedalus Enterprises), USA)
AMSS	Airborne Multispectral Scanner (GEOSCAN Pty Ltd., Australia)
ARES	Airborne Remote Earth Sensing (Lockheed, Palo Alto Research Laboratory, USA)
ASAS	Advanced Solid–state Array Spectroradiometer (NASA/GSFC, USA)
ASTER	Advanced Spaceborne Thermal Emission Reflection Radiometer (built by Geophysical and Environmental Research Simulator (GER) Corporation)
ATM	enhanced Airborne Thematic Mapper (Sensys Technologies Inc. (formerly Daedalus Enterprises), USA)
AVIRIS	Airborne Visible/Infrared Imaging Spectrometer (NASA/JPL, USA)
CAMODIS	Chinese Airborne MODIS (Shanghai Institute of Technical Physics, Chinese Academy of Sciences, China)
CHRISS	Compact High Resolution Imaging Spectrograph Sensor (SAIC, USA)
DAIS	Digital Airborne Imaging Spectrometer (GER Corporation, USA)
FLI/PMI	Fluorescence Line Imager/Programmeable Multispectral Imager (Moniteq, Canada)
GERIS	Geophysical and Environmental Research Imaging Spectrometer (GER, USA)
HSI	Hyperspectral Imaging (Science Applications International Corporation (SAIC), USA)
HYDICE	Hyperspectral Digital Imagery Collection Experiment (built by Hughes Danbury Optical Systems Inc., USA for Naval Research Laboratory, USA)
ISM	Infrared Spectro–Imager (Department d'Etude Spatiale (DES)/Institute d'Astrophysique Spatiale (IAS)/Observatoire Paris–Mendon (OPS), France)
MAS	MODIS Airborne Simulator (built by Daedalus, USA for NASA/AMES, USA)
MAIS	Module Airborne Imaging Spectrometer (Shanghai Institute of Technical Physics, Chinese Academy of Sciences)
MEIS	Multispectral Environmental Imaging Sensor (MacDonnell Douglas Technologies Inc., USA)
MISI	Modular Imaging Spectrometer Instruments (Rochester Institute of Technology, USA)
MIVIS	Multispectral Infrared and Visible Imaging Spectrometer (Daedalus, USA; now Sensys Technologies Inc.)
MUSIC	Multi–Spectral Infrared Camera (Lockheed, Palo Alto Research Laboratory, USA)
ROSIS	Reflective Optics System Imaging Spectrometer (Messerschmidt–Bölkow–Blohm (MBB)/GKSS Research Centre/German Aerospace Establishment (DLR), Germany)
RTISR	Real Time Imaging Spectroradiometer (Surface Optics Corporation, USA)
SFSI	SWIR Full Spectral Imager (CCRS, Canada)
SMIFTS	Spatially Modulated Imaging Fourier Transform Spectrometer (University of Hawaii, USA)
TRWIS	TRW Imaging Spectrometer (TRW, USA)
VIFIS	Visible Interference Filter Imaging Spectrometer (University of Dundee, UK)
WIS–FDU	Wedge Imaging Spectrometer – Flight Demonstration Unit (Hughes Santa Barbara Research Centre (SBRC), USA)
WIS–VNIR	Wedge Imaging Spectrometer – Visible and Near–Infrared (Hughes SBRC, USA)
WIS–SWIR	Wedge Imaging Spectrometer – Short–Wave Infrared (Hughes SBRC, USA)

from the EXABYTE tapes. Calibration information and software have been supplied by the instrument manufacturer: calibration may be further supplemented by the operators. Processing of the data can be accomplished in as little as several hours after the return of the aircraft.

Geometric correction of data: CASI has a roll and pitch correction system installed. A vertical gyro provides real time aircraft attitude data to the CASI data stream written to tape. This information is used in the post processing of the data to produce geo-referenced images (see Chapter 6).

☞ **These are August 1997 specifications but CASI is continually being developed. Readers requiring up-to-date information are advised to contact the manufacturers, ITRES, directly at Suite 155, East Atrium, 2635–37 Avenue N.E., Calgary, Alberta, Canada, T1Y 5Z6, Tel: +1 403 250 9944, Fax: +1 403 250 9916 or to consult their web page at http://www.itres.com/casi/casi.html.**

SPIN-2

SPIN-2 imagery products are derived from the camera systems on Russian Cosmos spacecraft. The orbiting satellites carry two cameras, acquiring panchromatic imagery (510–760 nm) at 2 m and 10 m spatial resolution. Images are supplied geometrically corrected. The two metre imagery is produced by the KVR 1000 camera system and individual scenes cover a usable area of 40 km x 160 km. A spatial accuracy of 10 m (RMS error, see Chapter 6) is claimed and this presumably would be improved with the use of ground control points (see Chapter 6). This is the only 2 m imagery commercially available and is the

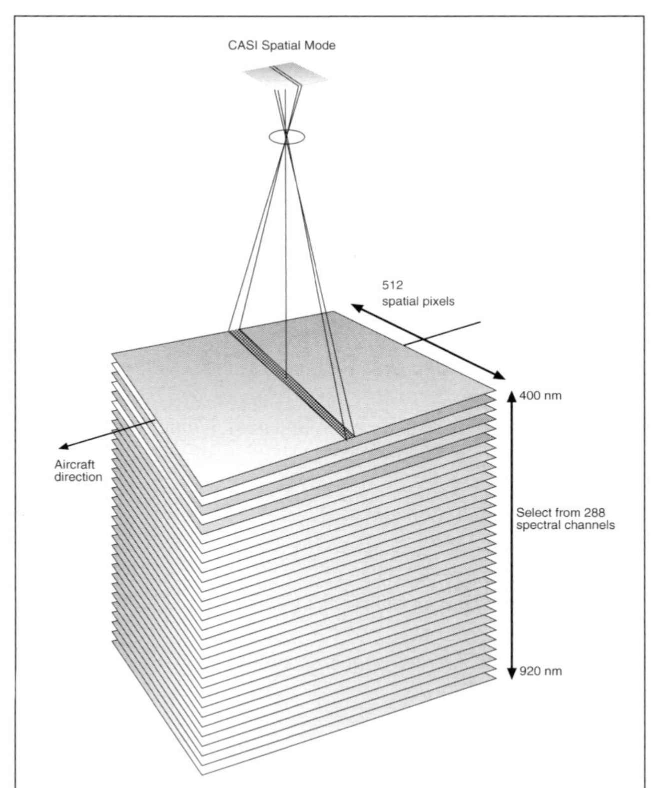

Figure 3.4 Compact Airborne Spectrographic Imager (CASI) spatial mode. The instrument scans lines 512 pixels wide with pixel size dependent on the altitude (1 m pixels at about 840 m altitude, or 3 m pixels at about 2500 m); 8 to 21 spectral channels are programmable between wavelengths of 400 nm and 920 nm, each 2.6 nm or greater in bandwidth. The minimum pixel size is determined by the scan rate and is currently about 1 m². CASI can be set up to simulate Landsat TM (bands 1–4) and SPOT XS.

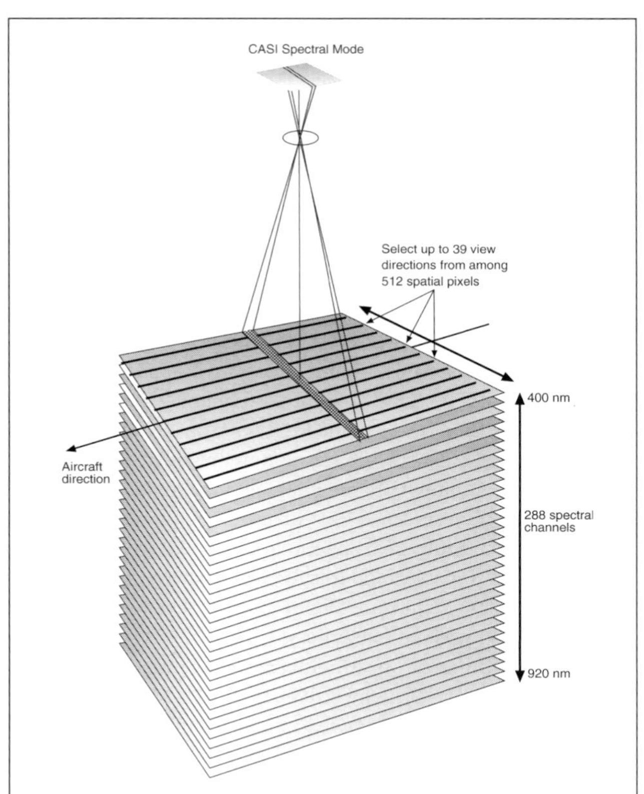

Figure 3.5 Compact Airborne Spectrographic Imager (CASI) spectral mode. In spectral mode, the full spectral resolution of 288 channels can be utilised for up to 39 view directions across the 35 degree swath width. Look direction spacing and location are user specified to sample the array. This sampling produces an image rake or comb. A single channel, full spatial scene recovery channel (FSS channel) can also be selected.

highest resolution satellite imagery on the market. Although untested for coastal management applications in theory 2 m SPIN-2 data should be comparable to aerial black and white photography with the considerable advantages of covering large areas without the need to mosaic separate photographs. Pre-1993 imagery is sold at US$30 km^{-2} and post-1993 imagery at US$40 km^{-2}, with a minimum order of 10 x 10 km (i.e. US$3000 for pre-1993 or US$4000 for post-1993 data).

The ten metre imagery is produced by the TK-350 camera system. A scene covers an area of 200 x 300 km. A spatial accuracy of 20–25 m RMS is claimed, but again ground control pints should improve this. Digital Elevation Models (DEMs) can be produced from the TK-350 because successive frames overlap by 80% producing stereo pairs. Ten metre imagery is sold at US$1 km^{-2} with a minimum order of 2500 km^2 or US$2500.

The SPIN-2 archive can be browsed at http://www.TerraServer.com and technical details reviewed at http://www.spin-2.com/. The main drawback to SPIN-2 data would appear to be that coverage seems to be mostly of the USA and Europe, with few tropical areas in the archive. However, interested individuals should contact the marketing company (SPIN-2 Marketing, 2121 K. Street NW, Suite 650, Washington, DC 20037, Fax: + 1 202 293 0560, E-mail: sales@spin-2.com) because special acquisitions can be arranged.

Conclusions

Certain types of imagery with a spatial resolution >100 m are useful for measuring sea-surface temperature and phytoplankton biomass, information which is useful for fisheries management (Chapter 18), mapping toxic algal blooms (Chapter 14) and mapping the progress of warm water anomalies associated with El Niño, such as those that caused coral bleaching around the tropics in 1998.

Imagery with a spatial resolution of 10–100 m has been most commonly used in coastal zone management. Data at this resolution are appropriate for mapping many biological habitats and man-made structures (Plate 4). For some sensors, such as Landsat MSS and TM, there is a large archive of past imagery, although cloud cover frequently limits the number of useful images available.

Imagery with a spatial resolution of less than 10 m has been used less frequently, partly because the development of airborne sensors is comparatively recent. However, these data are being used increasingly for coastal applications.

The process of selecting suitable data is critically important but not straightforward. This *Handbook* is structured to facilitate that decision as far as possible. Chapter 5 provides advice on acquiring imagery and on the cost of different data. Chapters 11–13 and 15–17 illustrate the use of data from five sensors (Landsat MSS, Landsat TM, SPOT XS and Panchromatic, CASI) for various coastal management applications. Chapter 19 discusses issues of cost-effectiveness which are essential when planning a remote sensing campaign.

References

Aschbacher, J., Ofren, R.S., Delsol, J.P., Suselo, T.B., Vibulsresth, S., and Charrupat, T., 1995, An integrated comparative approach to mangrove vegetation mapping using remote sensing and GIS technologies, preliminary results. *Hydrobiologia*, **295**, 285–294.

Boardman, J.W., Kruse, F.A., and Green, R.O., 1995, Mapping target signatures via partial unmixing of AVIRIS data. *Fifth Jet Propulsion Laboratory Airborne Earth Science Workshop*, Jet Propulsion Laboratory Publication, **95-1**, 1, 23–26.

Conway, J., Eva, H., and Dsouza, G., 1996, Comparison of the detection of deforested areas using the ERS-1 ATSR and the NOAA-11 AVHRR with reference to ERS-1 SAR data – a case study in the Brazilian Amazon. *International Journal of Remote Sensing*, **17**, 3419–3440.

Currey, B., Fraser, A.S., and K.L. Bardsley. 1987, How useful is Landsat monitoring? *Nature,* **328**, 587–589.

Goetz, A.F.H., Vane, G., Solomn, J.E., and Rock, B.N., 1985, Imaging spectrometry for earth remote sensing, *Science*, **228**, 1147–1153.

Green, E.P., Mumby, P.J., Edwards, A.J., and Clark, C.D., 1996, A review of remote sensing for the assessment and management of tropical coastal resources. *Coastal Management*, **24**, 1–40.

Kidwell, K.B., 1995, NOAA Polar Orbiter Data Users Guide (Washington: NOAA/NESDIS).

Kramer, H. J., 1994, *Observation of the Earth and its Environment: Survey of Missions and Sensors*. (Berlin: Springer-Verlag).

Kruse, F.A., Lefkoff, A.B., Boardman, J.B., Heidebrecht, K.B., Shapiro, A.T., Barloon, P.J., and Goetz, A.F.H., 1993, The spectral image processing system (SIPS) – interactive visualisation and analysis of imaging spectrometer data. *Remote Sensing of Environment*, **44**, 145–163.

Kruse, F.A., Richardson, L.L., and Ambrosia, V.G., 1997, Techniques developed for geologic analysis of hyperspectral data applied to near-shore hyperspectral ocean data. *Proceedings of the Fourth International Conference on Remote Sensing for Marine and Coastal Environments*, Orlando, Florida, 17–19 March 1997, **1**, 233–246.

Meaden, G.J. and Kapetsky, J.M., 1991, Geographical information systems and remote sensing in inland fisheries and aquaculture. *FAO Fisheries Technical Paper*, **318**, 1–262.

Pons, I., and Le Toan, T., 1994, Assessment of the potential of ERS-1 data for mapping and monitoring a mangrove ecosystem. *Proceeding of the First ERS-1 Pilot Project Workshop*, Toledo, Spain, June 1994 (European Space Agency SP-365), 273–282.

Porter, W.M., and Enmark, H.E., 1987, System overview of the Airborne Visible/Infrared Imaging Spectrometer (AVIRIS). *Proceedings of the Society of Photo-Optical Instrumentation Engineers (SPIE)*, **834**, 22-31.

Richardson, L.L., and Ambrose, V.G., 1997, Remote sensing of algal pigments to determine coastal phytoplankton dynamics. *Proceedings of the Fourth International Conference on Remote Sensing for Marine and Coastal Environments*, Orlando, Florida, 17–19 March 1997, **1**, 75–81.

Rignot, E., Salas, W.A., and Skole, D.L., 1997, Mapping deforestation and secondary growth in Rondonia, Brazil, using imaging radar and thematic mapper data. *Remote Sensing of Environment*, **61**, 179–180.

Young, J.A.P., and Green, D.R., 1987, Is there really a role for RS in GIS? In Advances in Digital Image Processing, *Proceedings of the Thirteenth Annual Conference of the Remote Sensing Society*, University of Nottingham, UK, 309–317.

CASI literature

1) Use of CASI for tropical coastal management:

Clark C.D., Ripley H.T., Green E.P., Edwards A.J., and P.J. Mumby, 1997, Mapping and measurement of tropical coastal environments with hyperspectral and high spatial resolution data. *International Journal of Remote Sensing*, **20**, 237–242.

Borstad, G., Brown, L., Cross, W., Nallee, M., and Wainwright, P., 1997, Towards a management plan for a tropical reef-lagoon system using airborne multispectral imaging and GIS. *Proceedings of the Fourth International Conference on Remote Sensing for Marine and Coastal Environments*, Orlando, Florida, 17–19 March 1997, **2**, 605–610.

Green, E.P., Mumby P.J., Edwards A.J., Clark, C.D., and Ellis, A.C., 1997, A comparison between satellite and airborne multispectral data for the assessment of mangrove areas in the eastern Caribbean. *Proceedings of the Fourth International Conference on Remote Sensing for Marine and Coastal Environments*, Orlando, Florida, 17-19 March 1997, **1,** 168–176.

Green, E.P., Mumby, P.J., Edwards, A.J., Clark, C.D., and Ellis, A.C., 1998, The assessment of mangrove areas using high resolution multispectral airborne imagery. *Journal of Coastal Research*, **14** (2), 433–443.

Green, E.P., Clark, C.D, Mumby, P.J., Edwards, A.J., and Ellis, A.C., 1998, Remote sensing techniques for mangrove mapping. *International Journal of Remote Sensing*, **19** (5), 935–956.

Mumby P.J., Green E.P., Clark C.D., and Edwards A.J., 1997, Reef habitat assessment using (CASI) airborne remote sensing. *Proceedings of the 8th International Coral Reef Symposium*, Panama City, Panama, June 1996, **2**, 1499–1502.

Mumby P.J., Green E.P., Clark C.D., and Edwards A.J., 1998, Digital analysis of multispectral airborne imagery of coral reefs. *Coral Reefs*, **17** (1), 59–69.

Mumby P.J., Green E.P., Edwards A.J., and Clark C.D., 1997, Coral reef habitat mapping: how much detail does remote sensing provide? *Marine Biology*, **130**, 193–202.

Mumby P.J., Green E.P., Edwards A.J., and Clark C.D., 1997, Measurement of seagrass standing crop using satellite and digital airborne remote sensing. *Marine Ecology Progress Series*, **159**, 51–60.

2) Aquatic applications of CASI:

Borstad, G.A., 1992, Ecosystem surveillance and monitoring with a portable airborne imaging spectrometer system. *Proceedings of the First Thematic Conference on Remote Sensing for Marine and Coastal Environments,* New Orleans, Louisanna, **1**, 883–892.

Borstad, G.A., Kerr, R.C., and Zacharias, M., 1994, Monitoring near shore water quality and mapping of coastal areas with a small airborne system and GIS. *Proceedings of the Second Thematic Conference on Remote Sensing for Marine and Coastal Environments*, New Orleans, Louisana, **2**, 51–56.

Borstad, G.A., and Hill, D.A., 1989, Using visible range imaging spectrometers to map ocean phenomena. *Proceedings of the SPIE Conference on Advanced Optical Instrumentation for Remote Sensing of the Earth's Surface from Space*, Paris, France, **1129**, 130–136.

Gower, J.F.R., and Borstad, G.A., 1993, Use of imaging spectroscopy to map solar-stimulated chlorophyll fluorescence, red tides and submerged vegetation. *Proceedings of the 16th Canadian Symposium on Remote Sensing, Sherbrooke*, Quebec, 95–98.

Jupp, D., Kirk, J.T.O., and Harris, G.P., 1994, Detection, identification and mapping cyanobacteria – using remote sensing to measure the optical quality of turbid inland waters. *Australian Journal Marine and Freshwater Research*, **45**, 801–828.

MacLeod, W., Stanton-Gray, R., Dyk, A., and Farrington, G., 1992, Discrimination of substrate type using airborne remotely sensed data, Bay of Quinte, Ontario, Canada. *Proceedings of the 15th Canadian Symposium on Remote Sensing*, Toronto, Ontario, 13–16.

Matthews, A.M., and Boxall, S.R., 1994, Novel algorithms for the determination of phytoplankton concentration and maturity. *Proceedings of the Second Thematic Conference on Remote Sensing for Marine and Coastal Environments*, New Orleans, Louisiana, **1**, 173–180.

Pettersson, L.H., Johannessen, O.M., and Frette, O., 1993, Norwegian remote sensing spectrometry for mapping and monitoring of algal blooms and pollution. *89' Proceedings of an European Space Agency Joint Research Council workshop ESA-SP*, 360.

Porter, W.M., and Enmark, H.E., 1987, System overview of the Airborne Visible/Infrared Imaging Spectrometer (AVIRIS). *Proceedings of Photo-Optical Instrumentation Engineers (SPIE)*, **834**, 22–31.

Saunders, J.F., Jupp, D.L.B., Harris, G.P., Hawkins, P.R., Byrne, G., and Hutton, P.G., 1994, Mapping optical water quality of the Hawkesbury River using casi airborne spectrometer data. *7th Australasian Remote Sensing Conference*, Melbourne, Australia. 57–69.

Stone, D.J., Freemantle, J.R., Shepherd, P.R., and Miller, J.R., 1993, Total water vapour amount determined from casi imagery. *Proceedings of the 16th Canadian Symposium on Remote Sensing*, Sherbrooke, Quebec, 805–809.

Zacharias, M., Niemann, O., and Borstad, G., 1992, An assessment and classification of a multispectral bandset for the remote sensing of intertidal seaweeds. *Proceedings of the Canadian Journal of Remote Sensing*, Toronto, Ontario, **18**, 4, 263–274.

4

Field Survey: Building the Link between Image and Reality

Summary *Field survey is essential to identify the habitats present in a study area, to record the locations of habitats for multispectral image classification (i.e. creation of habitat maps) and to obtain independent reference data to test the accuracy of resulting habitat maps. Field surveys should be planned to represent the range of physical environments within a study area and it is recommended that the locations of approximately 80 sites be recorded for each habitat type (30 for guiding multispectral classification and an additional 50 for accuracy assessment). The importance of accuracy assessment should not be under-estimated: a carefully produced habitat map does not guarantee high accuracy and inaccurate information can mislead decision makers.*

The costs of field survey may be divided into fixed and variable cost categories; an overview of these cost considerations is provided. Global Positioning Systems (GPS) are a vital component of field survey and the selection of an appropriate system will depend on the mapping objectives and type of remotely sensed imagery being used.

Several complementary methods of assessing the accuracy of habitat maps are available. These are described and their relative advantages listed.

Introduction

Almost every remote sensing exercise will require field surveys at some stage. For example, field surveys may be needed to define habitats, calibrate remotely sensed imagery (e.g. provide quantitative measurements of suspended sediments in surface waters), or for testing the accuracy of remote sensing outputs. This chapter aims to describe some of the key generic issues that must be borne in mind when planning a field survey. Specifically, the chapter sets out the general considerations involved in surveying coastal habitats, describes the importance of recording the positions of survey sites using Global Positioning Systems (GPS), and gives an introduction to the costs of field survey (costs are explored further in Chapter 19). The importance of assessing the accuracy of remote sensing outputs is stressed and guidance given on appropriate statistical methods for calculating the accuracy of habitat maps. Specific coral reef, seagrass and mangrove field survey methods (Plate 5) are too varied to include here and are discussed in Chapters 11, 12 and 13 respectively.

The need for field survey

Before the need for field survey is discussed, it is worth briefly reviewing the concept of remote sensing. Remote sensing provides a synoptic portrait of the Earth's surface by recording numerical information on the radiance measured in each pixel in each spectral band of the image being studied. To create a habitat map, the operator must instruct the computer to treat certain reference pixels as belonging to specific habitats. The computer then creates a 'spectral signature' for each habitat and proceeds to code every other pixel in the image accordingly, thus creating a thematic map.

Historically, some researchers have looked upon remote sensing as a means of mapping without the need to conduct field work. Whether this is an appropriate tenet depends on the objective of the study and familiarity of the operator with the study site. On a general basis, most people can view a satellite image or aerial photograph and easily distinguish different features according to their colour, contrast, pattern, texture and context. In some instances, this may be all that is required to make use of the imagery. For example, visual interpretation is usually sufficient to delineate the shape of coastlines. In the majority of studies, however, the objective is more sophisticated (e.g. mapping submerged habitats) and the thematician may not be able to draw on visual interpretation and background knowledge to identify each habitat type. In fact, the thematician is unlikely to be aware of the variety of habitat types in the image. Our own experience supports this view (see Chapter 9): even when moderately familiar with an area (the Caicos Bank), the overall accuracy of the final map was low if field surveys were not conducted (e.g. 15–30%).

The aims of field survey are three-fold. Firstly, to identify each feature of interest (e.g. each habitat type). Secondly, to locate representative areas of each feature in order to generate spectral signatures (spectra) from the imagery. Thirdly, to generate adequate additional data to test the quality or accuracy of the image classification (i.e. habitat map). This latter consideration is extremely important for any mapping exercise. In a coastal management context, imagine the legal problems in suggesting that a developer had cleared a particular mangrove area if the accuracy of mangrove maps were unknown. Taken a step further, where do decision makers stand legally if offenders are fined according to the extent of habitat that they have illegally destroyed? Legal problems may not be the only consequence. In biological terms, management initiatives based on a habitat map of unknown accuracy could lead to unnecessary or inappropriate action, although it is difficult to predict or generalise specific problems arising from such circumstances. Surprisingly though, accuracy assessments are fairly scarce in the context of mapping tropical coastal resources. Green *et al.* (1996) found that only a quarter of papers reviewed included an assessment of accuracy. The apparent scarcity of such assessments is understandable, although hardly acceptable. To test a classification rigorously, further field data are required which must be independent of the field data used to classify the imagery in the first place. It is often suggested that an adequate accuracy assessment is not possible on financial grounds. Such arguments may be countered by asking what the value is of habitat maps of unknown accuracy. The extra expenditure will clarify the degree to which the map can be 'trusted' for planning activities and should avert inappropriate management action on the basis of poor information. For example, the map might have to be disregarded for planning some areas whereas other sites might be well-represented. Methods of accuracy assessment will be discussed later in this chapter.

Planning field surveys

Field surveys must be planned carefully and due consideration must be given to the objectives of the study and the nature of habitats being surveyed. These issues will dictate most aspects of survey design, such as the sampling strategy, sampling technique, sampling unit, amount of replication, time to survey (i.e. weather conditions, date of image acquisition), ancillary data (e.g. depth, water turbidity) and the means of geographically referencing data. Specific considerations on methods, sampling units and ancillary data are described in the relevant chapters of this handbook (i.e. for mapping coral reefs, seagrass beds and mangroves) but more general comments are made here.

Most habitat mapping projects aim to represent the full range of relevant habitats. A helpful starting point is to conduct an unsupervised classification (Chapter 10) of imagery prior to conducting any field work. The unsupervised classification will provide a general guide to the location, size and concentration of habitat types (although it will not necessarily identify them correctly!). There are four main considerations when planning a field survey.

1. Represent all habitats

The physical environment (e.g. depth, wave exposure, aspect) will, to a large extent, control the distribution and nature of habitats. Relative depths can be inferred from most optical imagery (deeper areas are darker) and if the prevailing direction of winds and waves are known, the area can be stratified according to the major physical environments (e.g. Plate 12). If possible, each of these should be included in the survey.

2. Represent the range of environmental conditions for each habitat

Coastal areas will often possess gradients of water quality and suspended sediment concentration. Changes in these parameters across an image can lead to spectral confusion during image classification and misassignment of habitat categories. To mitigate this effect, surveys should represent a wide cross-section of each physical environment. This will provide further field data to train the image classification and provide data for accuracy assessment (to highlight the extent of the inaccuracies where they occur). As an example, an unsupervised classification of Landsat TM imagery of the Caicos Bank identified a specific habitat type on both sides of the Bank (some 40 km apart). Surveys near the field base identified this habitat as seagrass and it would have been easy to assume that all similar-looking habitats in the imagery fell into this class.

However, field surveys at the opposite side of the Bank identified a very different habitat type (organic deposits on sand), thus reinforcing the need for extensive field work.

3. Choose a sampling strategy

To ensure that all habitats are adequately represented by field data, a stratified random sampling strategy should be adopted (Congalton 1991). The unsupervised image classification and map of main physical environments can be used to stratify sampling effort. A similar number of sites should be obtained in each area. Truly random sampling within each stratum (area) is likely to be prohibitively time-consuming because the boat would have to visit randomly selected pairs of coordinates, thus incurring wasteful navigation time. In practice, driving the boat throughout each area with periodic stops for sampling is likely to be adequate. The main limitation to any field survey is cost/time. While every attempt is made to obtain the maximum amount of data, Congalton (1991) recommends that at least 50 sites of each habitat be surveyed for accuracy assessment purposes. We feel that an additional 30 sites should be visited for use in image classification (Chapter 19).

4. Estimate costs of field survey

Field surveys are expensive and not all of the costs incurred in gathering field data and relating it to remotely sensed data are immediately obvious. However, a full analysis of field costs is vital when designing a remote sensing campaign to ensure that realistic budgets and work schedules are planned. A generalised discussion of costs is presented here. Detailed advice on planning a remote sensing field campaign in terms of cost and the actual costs incurred in mapping the habitats of the Turks and Caicos Islands are given in Chapter 19.

Fixed costs

Fixed costs are defined as costs which are independent of the type of imagery used, the duration of fieldwork and the number and distribution of sites. As such they represent single payments, usually for equipment which can be used repeatedly for the field component of any remote sensing campaign. Of the costs being considered here, computing equipment constitutes the largest fixed cost. The following practical guidance points may be helpful:

- Several image processing packages are available as versions for high specification PCs but a UNIX workstation is still the most flexible, reliable and powerful platform on which to carry out image processing.
- A full Landsat TM scene is around 270 Mbyte in size. Image processing often generates a number of intermediate images which cannot be deleted before completion. It is also frequently useful to be able to display, compare and analyse more than one image simultaneously. Therefore, a large amount of free disk space (e.g. 2 Gbyte) is necessary to store original images and to allow the generation of intermediates. A large-capacity external hard drive is recommended.
- Remotely sensed images are delivered on EXABYTE tapes, Computer Compatible Tapes (CCTs) or Compact Disk (CD). A tape drive or CD-ROM drive is necessary to import these files. Data may also be written to tape or CD so a tape drive or CD writer will also allow work to be saved and retrieved (one 8 mm EXABYTE tape can store up to 14 Gbyte of data; one CD can store 650 Mbyte of data).
- Visual examination of raw and classified images is an essential part of image processing and therefore, it is advisable to purchase the best quality monitor that budgets allow. For example, we purchased a 20 inch (51 cm) colour monitor capable of displaying 24-bit graphics (approximately 16 million colours).
- A good-quality colour printer is necessary for hard copy output, preferably one capable of large format printing.
- Image processing software varies greatly in price from simple systems costing a few hundred pounds (e.g. IDRISI) to sophisticated packages costing several thousands of pounds (e.g. ERDAS Imagine, PCI). Universities and other educational institutions usually receive a discount on licences through software sharing schemes. Government departments normally pay a higher price but this is usually lower than the cost to commercial organisations. In addition to image processing software, a mapping package is useful for producing presentation-quality hard copy (e.g. MapInfo and Vertical Mapper), a spreadsheet package is useful for data management (e.g. Excel, Lotus) and a statistical package (e.g. Minitab, PRIMER) for analysis of field data.
- For field work, a Global Positioning System (GPS) is an essential piece of equipment to estimate the surveyor's location on the Earth's surface. There are two principal types of GPS, which differ in their cost and locational accuracy. The cheaper, less accurate unit is known as a 'stand-alone GPS' and the more expensive option, 'differential GPS' (DGPS), which provides accuracy to a higher order of magnitude (see section on GPS later in this chapter). The positional accuracy of a stand-alone GPS is perfectly adequate for use with imagery of coarse spatial resolution such as Landsat MSS. However, imagery with greater spatial resolution (e.g. SPOT XS and Pan) justifies the use of higher-specification DGPS.

Variable costs
Variable costs are defined as those which vary with the type of imagery used, the duration of fieldwork and the number and distribution of sites. Personnel salaries are the major cost in this category; they will be directly related to time costs and for this reason all time costs are expressed in person-days (see next section). Fuel and oil for boats is another cost which will depend on the amount of fieldwork undertaken and the distances covered. The last main variable cost is that of the imagery itself; this is discussed in detail in the following chapter.

The time required to undertake a field survey will depend on the following considerations:

- *The amount of fieldwork undertaken*: fieldwork time will be related to the number of survey and accuracy sites visited
- *The area over which survey sites are spread*: are survey and accuracy sites concentrated in a small area of an image or do they cover the entire scene?
- *The survey methods employed*: these determine the rate of site survey (number of sites per day). SCUBA surveys take much longer than snorkel surveys which, in turn, are considerably more time intensive than shipboard surveys using glass-bottom buckets. The time costs of SCUBA surveys are further increased for safety reasons (e.g. the need to have pairs of divers in the water, limited number of dives in a working day). Rapid visual assessment methods can convey considerable savings in time for larger surveys (see Chapter 11).
- *The data being collected*: species level data takes more time to collect than phylum level, for example. Survey data typically take longer to collect than accuracy data (when a site is usually being assigned to a particular category on an already well-defined habitat classification scheme).
- *Accessibility of survey areas*: (i) depth: deeper areas may require SCUBA surveys, very shallow areas can be inaccessible by motorised boats and alternative transport may be necessary, (ii) exposure: areas open to prevailing winds may be impossible to survey except on calm days, (iii) ease of access: areas with high concentrations of natural hazards, such as patch reefs or sand banks, may be difficult to navigate through. The interior of mangrove forests is difficult to penetrate.
- *The habitats themselves*: complex habitats like coral reefs take considerably longer to survey than simpler habitats like bare sand.

Global Positioning Systems (GPS)

In a remote sensing context, GPS has two major applications:

1. to measure the position of prominent features on an image *in situ* which can be used to provide ground control points for geometric correction (see Chapter 6),
2. to assign positions to field data. These field data can then be correlated with spectral information at the same point on a geometrically corrected image. Conversely, a group of image pixels of particular interest can be surveyed in the field by using a GPS to navigate to that location.

GPS was developed by the US Department of Defense as a global navigation system for military and civilian use. It is based on a constellation of 24 Navstar satellites which orbit the Earth at a height of 17,500 km and a period of 24 hours. These satellites act as fixed reference points whose distance from any point on the Earth's surface can be measured accurately. Each transmits a unique radio signal which serves to identify the satellite. The time between transmission and reception of the radio signal enables the distance between the satellite and any receiving point on the Earth's surface to be calculated. The signal is received by the GPS, which is typically either battery-powered and hand-held (or carried in a back-pack) or mounted on a boat and powered from the onboard supply. The receivers are highly mobile and positions can be taken at any location which allows radio reception (for example, there may be problems receiving satellite signals under dense forest canopies).

In order to compute a position, the GPS receiver must know the distance between itself and the Navstar satellites and the exact position of those satellites. Position fixing with GPS then utilises trigonometric theory which states that if an observer knows the distance from three points of known position (the satellites), then that observer must be at one of two points in space. Usually the observer has a rough idea, to within a few degrees of latitude and longitude, of his position and so one of these positions is ridiculous and can be discounted by the computer. Theoretically, if the GPS is being used at sea level, or a known height above it, then only three satellites are needed to calculate position because the receiver is a known distance from a fixed point in space (the centre of the Earth). However, in practice a fourth satellite range is needed in order to correct for timing differences (offsets) between the highly accurate atomic clocks on the satellites, and the less accurate internal clocks of the receiver units (for further details, see Trimble Navigation Ltd, 1993).

☞ **For a wealth of information on GPS, a good starting point is Peter Dana's web pages at the Department of Geography, University of Texas at Austin at the following URL: http://www.utexas.edu/depts/grg/gcraft/notes/gps/gps.html.**

GPS costs and accuracy

Any position fix obtained from a GPS contains a degree of uncertainty. The simplest and cheapest units are typically targeted at the outdoor leisure market (mountaineers, hikers etc.), cost of approximately £200 and give positions with an error of ± 100 m. The most sophisticated GPS systems are used by surveyors who require sub-centimetre accuracy and cost many thousands of pounds. The ultimate accuracy of GPS is determined by the sum of several sources of error, the contribution of any source depending on specific ionospheric, atmospheric and equipment conditions. Sources of error include variations in the speed of radio transmission owing to the Earth's constantly changing ionosphere and atmosphere, drift in the atomic clocks, electrical interference within a receiver and multipath error where the radio signal does not travel directly to the receiver but is refracted *en route*. These errors are all unavoidable physical and engineering facts of life. However, the accuracy of GPS is deliberately degraded by the US Department of Defense, which operates a 'selective availability'. Selective availability is designed to deny hostile military forces the advantages of accurate GPS positional information, can be varied and is by far the largest component of GPS error (Table 4.1).

Table 4.1 Average errors for a good-quality GPS

Typical error source	m
Satellite clock error	1.5
Ephemeris (orbit) error	2.5
Receiver error	0.3
Atmospheric/ionospheric error	5.5
Multipath error	0.6
Total Error	10.4
Typical selective availability error	30.0

There has been some speculation recently that selective availability may be removed altogether. However resistance from the US military is likely to delay this for several years, and in the event of political tension selective availability would be rapidly reinstated.

A further factor in the accuracy of GPS systems is the principle of geometric dilution of precision (GDOP). The errors listed above can vary according to the positions of the satellites in the sky and their angles relative to one another: the wider the angles between the satellites, the lesser the effects of errors and the better the positional measurement. The computer in a good receiver will have routines to analyse the relative positions of all the satellites within the field of view of the receiver and will select those satellites best positioned to reduce error. Thus, GDOP is minimised. Values for position dilution of precision (PDOP) can vary from 4 to 6 under reasonable conditions, under good conditions values < 3 can be obtained. The predicted accuracy of a GPS can be calculated by multiplying the total error in Table 4.1 by the PDOP, giving typical errors of 18–30 m for a good receiver and, in the worst case, about 100 m.

Differential GPS (DGPS)

We have already seen that inaccuracies in GPS signals originate from a variety of sources, vary in magnitude and are difficult to predict. For many uses of GPS, positional errors in the range of 30–100 m are unacceptable. Fortunately, by using a system which measures errors as they occur, and at the same time as positional information is being collected, it is possible to correct much of the inaccuracy. This is achieved by using a second, reference receiver, which remains stationary at a location whose position is known because it has been surveyed to a very high degree of accuracy (if positional accuracy of 2–3 m is required then it would be necessary to survey this point to an accuracy of < 0.5 m). The Navstar satellites' orbit is so high that, if the two receivers on the ground remain only a few hundred kilometres apart, then the signals which reach both will have travelled through virtually the same ionospheric and atmospheric conditions. Therefore, both signals will have virtually the same delays. Selective availability will also delay the signals to both receivers by the same amount. However, receiver noise and multipath refractions will be different for each signal and cannot be corrected.

Instead of using radio signals to calculate its position, the reference receiver uses its position to calculate the actual time taken for the signal to travel between it and the satellite. This is possible because:

- The distance between the reference receiver and the satellite can be calculated at any time. The positions of the satellite can be calculated from orbital details and the position of the reference receiver is, of course, known and stationary.
- The theoretical time taken for the signal to cover this distance is calculated from the speed of transmission of radio waves.
- Any difference (hence the term 'differential') between the two times is the delay or error in the satellite's signal.

The reference receiver continuously monitors the errors and these can be compensated for in one of two ways:

1. The reference receiver transmits the corrections to the 'roving unit' (the other receiver, which is in the field gathering positional data) for real-time correction. One reference receiver can provide corrections for a series of roving units. In real-time correction systems, the instantaneous errors for each satellite are encoded by the receiver and transmitted to all roving units by a VHF radio link. Roving receivers receive the complete list of errors and apply the corrections for the particular satellites they are using. The advantage of a real-time system is that, as the name suggests, positional information of differential accuracy is made available to the operator of the roving unit during the survey. This is especially useful if the roving unit is being used to navigate to a specific location.
2. The reference receiver records and stores corrections for later processing (post-processed GPS). In post-processed systems, the roving units need only to record all their measured positions and the exact time at which they were taken. At some later stage, the corrections are merged with this data in a post-collection differential correction. The advantages of post-processed systems are the ability to operate in areas where radio reception may not be good, greater simplicity and reduced cost. However, navigation to a particular location can only be performed at stand-alone accuracy with a post-processed system.

Accuracy of DGPS

The published accuracy of good quality DGPS systems is less than 2 m (Table 4.2). Operational accuracy achieved in fieldwork in the Turks and Caicos was 2-4 m.

Table 4.2 Published errors for a good quality Trimble DGPS unit (Trimble Navigation Ltd 1993)

Typical error source	m
Satellite clock error	0.0
Ephemeris (orbit) error	0.0
Receiver error	0.3
Typical atmospheric/ionospheric error	0.6
Multipath error	0.6
Total	1.5
Selective availability error	0.0

Cost of DGPS systems

The cost of establishing a DGPS will depend on many variables such as model, computer specifications, the availability of a suitable previously surveyed point to act as the reference base station and location. The costs of a DGPS to the Department of Environment and Coastal Resources of the Turks and Caicos Islands government was £11,935 in 1993 but would cost closer to £2000 today.

Replacement parts can be expensive. For example, the cables connecting the batteries and receiver of the roving unit to the data logger, which typically suffer from high rates of wear and tear in the field, cost US$300 each. A full set of lithium/nickel long-life batteries for the data-logger (necessary if the roving unit is to be operated for several days at a time) costs about US$100. The staff of the Department of Environment and Coastal Resources surveyed their own reference position, which required a high level of technical expertise and took one person twelve days. If such expertise were not available then a technician would have to be imported and paid at commercial rates.

☞ **Useful information on costs of current GPS and DGPS receivers can be found at the following URLs: http://www.navtechgps.com/ and http://www.trimble.com/ among others.**

The need for accuracy assessment

What is accuracy?

Accuracy is referred to in many different contexts throughout this book. The accuracy of a GPS position fix is a measure of the absolute closeness of that fix to the 'correct' coordinates, whereas positional accuracy refers to the accuracy of a geometrically corrected image and is measured with the root mean square (Chapter 6). This section is concerned with thematic accuracy, that is, the non-positional characteristics of spatial data. If data have been subjected to multispectral classification then thematic accuracy is also known as classification accuracy (Stehmen 1997). This accuracy refers to the correspondence between the class label and the 'true' class, which is generally defined as what is observed on the ground during field surveys. In other words, how much of the class labelled as seagrass on a classified image is actually seagrass *in situ*.

☞ **There has been a tendency in remote sensing to accept the accuracy of photointerpretation as correct without confirmation. As a result, digital classifications have often been assessed with reference to aerial photographs. While there is nothing wrong with using aerial photographs as surrogate field data, it is important to realise that the assumption that photo-interpretation is without error is rarely valid and serious misclassifications can arise as a consequence (Biging and Congalton 1989).**

How accurate should a habitat map be?

It might seem surprising that few guidelines exist on the accuracy requirements of habitat maps for particular coastal management applications. The absence of guidelines may be partly attributable to a widespread paucity of accuracy analyses in habitat mapping projects and may also reflect the unsophisticated manner in which remote sensing outputs have been adopted for coastal management. For example, where habitat maps are used to provide a general inventory of resources as background to a management plan, a thematic accuracy of 60% is probably as useful as 80%. However, more sophisticated applications such as estimating the loss of seagrass cover due to development of a marina, would require the highest accuracies possible (currently about 90%).

It is unfortunate that many coastal habitat maps have no accuracy assessment (Green *et al.* 1996), particularly when the accuracies from satellite sensors tend to be low. However, provided that adequate field survey, image processing and accuracy assessments are undertaken, planning activities that depend on coastal habitat maps derived from high-resolution digital airborne scanners such as CASI or Daedalus are likely to be based on more accurate information. The precise advantages of better habitat information are unclear because the biological and economic consequences of making poor management decisions based on misleading information have not been studied. However, managers will have greater confidence in, say, locating representative habitats and nursery habitats for fish and shellfish if more accurate data are available. Examples of thematic accuracy per remote sensing instrument and habitat type are given throughout this book (see Chapters 11–13, 19) but the selection of accuracy requirements remains the user's dilemma. Due consideration must be given to the final use of output maps (e.g. will they have legal ramifications such as prosecution, land cover statistics, etc.) and the consequences of making mistakes in the map.

Calculation of classification accuracy

Imagine an image that has been classified into just two classes, coral reef and seagrass. It would be a serious mistake to accept this image as 100% accurate because it will contain error from a variety of sources. The similarity of reef and seagrass spectra may have caused some reef to be classified as seagrass and *vice versa*. Error in the geometric correction applied to the image and in GPS positioning may have resulted in some correctly classified reef pixels being mapped to locations which are actually seagrass *in situ* (this would be particularly prevalent along the boundaries between the two habitats). A method is needed which quantifies these classification errors by estimating how many reef pixels are in reality seagrass, how many seagrass pixels are reef; hence the reliability, or accuracy, of the classification. There are several complementary methods of conducting this assessment.

Error matrices, user and producer accuracies
An error matrix is a square array of rows and columns in which each row and column represents one habitat category in the classification (in this hypothetical case, reef and seagrass). Each cell contains the number of sampling sites (pixels or groups of pixels) assigned to a particular category. In the example below, two hundred accuracy sites have been collected: one hundred reef and one hundred seagrass. Conventionally the columns represent the reference data and the rows indicate the classification generated from the remotely sensed data.

Table 4.3 Hypothetical error matrix for reef and seagrass

	Reference data			
Classification data	**Reef**	**Seagrass**	**Row Total**	**User accuracy**
Reef	85	25	110	85/110 = 77%
Seagrass	15	75	90	75/90 = 83%
Column Total	100	100	200	
Producer accuracy	85%	75%		

In this simplified example, 85 of the 100 reef sites have been classified as reef and 15 as seagrass. Similarly, 75 of the 100 seagrass sites have been classified as seagrass and 25 as reef. Of the 110 sites which were classified as reef (the sum of the reef row), only 85 were actually reef. Extrapolating this to the whole image, the probability of a pixel labelled as reef on the classified image actually being reef *in situ* is 85/110, or 77%. Likewise, the reliability of the seagrass classification is 75/90, or 83%. The probability that a pixel classified on the image actually represents that category *in situ* is termed the 'user accuracy' for that category. The classification however 'missed' 15 of the reef sites and 25 of the seagrass. The omission errors for the reef and seagrass classes were therefore 15% and 25% respectively.

The 'producer accuracy' is the probability that any pixel in that category has been correctly classified and in this case is 85% for reef pixels and 75% for seagrass. Producer accuracy is of greatest interest to the thematician carrying out the classification, who can claim that 85% of the time an area that was coral reef was classified as such. However, user accuracy is arguably the more pertinent in a management context, because a user of this map will find that, each time an area labelled as coral reef on the map is visited, there is only a 77% probability that it is actually coral reef. In practice, remotely sensed data will usually be classified into more than just two classes; user and producer accuracies may be calculated for each class by using an error matrix.

☞ **Confusion may arise when reading remote sensing literature because different authors have used a wide variety of terms for error matrix and error types. For example, an error matrix has been known as a confusion matrix, a contingency table, an evaluation matrix and a misclassification matrix. Producer accuracy is sometimes called omission error (although omission error is strictly speaking 100 – producer accuracy). Similarly, user accuracy is sometimes called commission error (and again, commission error is strictly speaking 100 – user accuracy). Janssen and van der Wel (1994) provide a useful clarification of these terms in their discussion of accuracy assessment.**

Kappa analysis

It is also desirable to calculate a measure of accuracy for the whole image across all classes, however many there are. The simplest method is to calculate the proportion of pixels correctly classified. This statistic is called the 'overall accuracy' and is computed by dividing the sum of the major diagonal (shaded) by the total number of accuracy sites ((85+75)/200 = 80%). However, the overall accuracy ignores the off-diagonal elements (the omission and commission errors) and different values of overall accuracy cannot be compared easily if a different number of accuracy sites were used to test each classification.

BOX 4.1

Kappa Analysis

Kappa analysis is a discrete multivariate technique used to assess classification accuracy from an error matrix. Kappa analysis generates a Khat statistic or Kappa coefficient that has a possible range from 0 to 1.

$$\text{Kappa coefficient, } K = \frac{N\sum_{i=1}^{r} x_{ii} - \sum_{i=1}^{r}(x_{i+}.x_{+i})}{N^2 - \sum_{i=1}^{r}(x_{i+}.x_{+i})}$$

where r = number of rows in a matrix, x_{ii} is the number of observations in row i and column i, x_{i+} and x_{+i} are the marginal totals of row i and column i respectively, and N is the total number of observations (accuracy sites). For more details see Bishop *et al.* (1975). K expresses the proportionate reduction in error generated by a classification process compared with the error of a completely random classification. For example, a Kappa coefficient of 0.89 implies that the classification process was avoiding 89% of the errors that a completely random classification would generate (Congalton 1991).

Off-diagonal elements are incorporated as a product of the row and column marginal totals in a Kappa analysis (Box 4.1), which can be computed in a standard spreadsheet package. K will be less than the overall accuracy unless the classification is exceptionally good (i.e. the number of off-diagonal elements is very low).

Tau coefficient

Ma and Redmond (1995) recommended use of the Tau coefficient (T – Box 4.2) in preference to the Kappa coefficient. The main advantage of Tau is that the coefficient is readily interpretable. For example, a Tau coefficient of 0.80 indicates that 80% more pixels were classified correctly than would be expected by chance alone. The coefficient's distribution approximates to normality and Z-tests (Box 4.3) can be performed to examine differences between matrices (see Ma and Redmond 1995).

BOX 4.2

Calculation of the Tau coefficient

Tau is calculated from: $T = \frac{P_o - P_r}{1 - P_r}$ where $P_r = \frac{1}{N^2}\sum_{i=1}^{M} n_i.x_i$

P_o is the overall accuracy; M is the number of habitats; i is the ith habitat; N is the total number of sites; n_i is the row total for habitat i and x_i is the diagonal value for habitat i (i.e. number of correct assignments for habitat i).

BOX 4.3

Comparing error matrices with the Tau coefficient

Z-tests between Tau coefficients 1 and 2 (T_1 and T_2) are conducted using the following equations:

$$Z = \frac{T_1 - T_2}{\sqrt{\sigma_1^2 + \sigma_2^2}}$$

where σ^2 is the variance of the Tau coefficient, calculated from:

$$\sigma^2 = \frac{P_o(1-P_o)}{N(1-P_r)}$$

[See **Box 4.2** for definitions of P_o, N and P_r]. For a two-sample comparison, Tau coefficients have a 95% probability of being different if Z = 1.96.

In addition to the error matrix, the quality of an image classification can be quantified via an image-based approach which measures the statistical separability of digital numbers that comprise each habitat class mapped (e.g. the difference between mean digital number (DN) values for forereef, seagrass, mangrove etc.). For example, several authors have examined the separability of classes using canonical variates analyses on the image data (Jupp *et al.* 1986, Kuchler *et al.* 1989, Ahmad and Neil 1994). Others have used analysis of variance on the DN values that comprise each class (Luczkovich *et al.* 1993). There is nothing inherently wrong in this, provided that a high separability of image classes is not assumed to be indicative of an accurate habitat map. The advantage of the error matrix approach is the quantification of the accuracy to which each habitat class produced in a classified image is actually found *in situ*. An accuracy assessment of this type allows the user to make statements such as, 'I can be 70% confident that the area classified as seagrass on the image is actually seagrass in real life'.

Conclusion

Accuracy assessment is an essential component of a habitat mapping exercise and should be planned at the outset of the study. Map accuracy can be determined using several complementary statistical measures:

1. The collective accuracy of the map (i.e. for all habitats) can be described using either overall accuracy, the Kappa coefficient or Tau coefficient. Of these, the Tau coefficient is arguably the most meaningful but many remote sensing studies use overall accuracy and, therefore, since overall accuracy has become a 'common currency' of accuracy assessment, its use is also recommended.
2. User and producer accuracies should be calculated for individual habitats.
3. The Tau coefficient should be used to test for significant differences between error matrices. For example, if habitat maps have been created using several image processing methods, *Z*-tests can be performed on the Tau coefficients to determine which map is most accurate.

References

Ahmad, W., and Neil, D.T., 1994, An evaluation of Landsat Thematic Mapper (TM) digital data for discriminating coral reef zonation: Heron Reef (GBR). *International Journal of Remote Sensing*, **15**, 2583–2597.

Biging, G., and Congalton, R.G., 1989, Advances in forest inventory using advanced digital imagery. *Proceedings of Global Natural Research Monitoring and Assessments: Preparing for the 21st Century*. Venice, Italy, September 1989, **3**, 1241–1249.

Bishop, Y., Fienberg, S., and Holland, P., 1975, *Discrete Multivariate Analysis – Theory and Practice* (Cambridge, Massachusetts: MIT Press).

Congalton, R.G., 1991, A review of assessing the accuracy of classifications of remotely sensed data. *Remote Sensing of Environment*, **37**, 35–46.

Green, E.P., Mumby, P.J., Edwards, A.J., and Clark, C.D., 1996, A review of remote sensing for the assessment and management of tropical coastal resources. *Coastal Management*, **24**, 1–40.

Janssen, L.L.F., and van der Wel, F.J.M., 1994, Accuracy assessment of satellite derived land-cover data: a review. *Photogrammetric Engineering and Remote Sensing*, **60**, 419–426.

Jupp, D.L.B., Mayo, K.K., Kuchler, D.A., Heggen, S.J., Kendall, S.W., Radke, B.M., and Ayling, T., 1986, Landsat based interpretation of the Cairns section of the Great Barrier Reef Marine Park. *Natural Resources Series* No. 4. (Canberra: CSIRO Division of Water and Land Resources).

Kuchler, D., Biña, R.T., and Claasen, D.R., 1989, Status of high-technology remote sensing for mapping and monitoring coral reef environments. *Proceedings of the 6th International Coral Reef Symposium,* Townsville, **1**, 97–101.

Luczkovich, J.J., Wagner, T.W., Michalek, J.L., and Stoffle, R.W., 1993, Discrimination of coral reefs, seagrass meadows, and sand bottom types from space: a Dominican Republic case study. *Photogrammetric Engineering and Remote Sensing*, **59**, 385–389.

Ma, Z., and Redmond, R.L., 1995, Tau coefficients for accuracy assessment of classification of remote sensing data. *Photogrammetric Engineering and Remote Sensing*, **61**, 435–439.

Stehmen, S.V., 1997, Selecting and intepreting measures of thematic classification accuracy. *Remote Sensing of Environment*, **62**, 77–89.

Trimble Navigation Limited, 1993, *Differential GPS Explained*. (Sunnyvale: Trimble Navigation Ltd.).

Part 2

The Acquisition, Correction and Calibration of Remotely Sensed Data

5

Acquisition of Remotely Sensed Imagery

Summary *This chapter describes the major considerations and the costs commonly incurred when acquiring remotely sensed imagery. The chapter begins by evaluating the problem of cloud cover, which interferes with the collection of remotely sensed data and is a major consideration at the beginning of any remote sensing exercise. The chapter then provides advice on acquiring aerial photography, digital airborne multispectral imagery and satellite imagery. Aerial photography and satellite imagery are usually acquired from data archives, whereas airborne multispectral imagery (e.g. CASI) must usually be commissioned for a specific survey. Where archived photography and satellite imagery are not available for a site, new airborne digital imagery or aerial photography can be commissioned or, if appropriate, SPOT or RADARSAT satellites can be specifically programmed to acquire the required data.*

Introduction

Much of this *Handbook* is devoted to deciding which type of imagery is most appropriate for a given objective, cost constraint and accuracy requirement. This chapter is primarily devoted to the acquisition of imagery once a particular method of remote sensing has been chosen. Practical examples are given for imagery commonly used for coastal management applications. A number of additional budgetary and accuracy considerations will be highlighted.

Remotely sensed data are acquired through reference to an image archive or by planning a new remote sensing campaign. The following generalisations may be made:

- *Digital airborne data* are recorded for specific projects resulting in a seemingly haphazard coverage.
- *Satellite data* are obtained from an image archive (although some scope exists for tasking satellites). Data have been archived since 1972 for Landsat MSS, 1984 for Landsat TM, 1986 for SPOT, 1991 for ERS and 1996 for RADARSAT.
- *Aerial photography* is either recorded for a specific project or obtained from an archive.

Digital airborne remote sensing is relatively new and data collection is expensive. Therefore, few archives exist and most digital airborne data must be collected (and paid for!) from scratch. Providing that adequate funds are available, this approach allows the customer to exert control over the instrument set-up and timing of data acquisition. Conversely, satellite-borne data collection is a fairly passive process in which data are recorded during the satellite's orbit of the Earth and images are archived by the agency responsible for disseminating images from that satellite's sensors. In this case, the customer has no direct control over the availability of data. Archives of aerial photographs are widely available for many countries. This is chiefly the product of previous mapping exercises and military reconnaissance.

Dependency on an image archive has several drawbacks. Firstly, data may not be recorded for a given area of interest. The likelihood of this depends on the proximity of that area to centres of development/population – i.e. a lack of images is more likely to be found for isolated oceanic areas than for coastal shelfs. Secondly, the chances of obtaining an image for a desired period may be slim. This is because some satellite sensors may only record a given site at intervals of 16–26 days, weather permitting (see Chapter 3 for further details). Cloud cover may further reduce the availability of images and this subject is discussed in greater detail later in the chapter. It must be pointed out, however, that, while the use of imagery archives may not always be ideal, it is the most common means of acquiring data and is quite adequate for most habitat mapping activities and for areas where cloud cover is not too severe.

The acquisition of satellite imagery is not always straightforward. For example, the customer may be faced with a bewildering choice of pre-processing options ranging from raw data to geo-rectified products. Like most issues in remote sensing, the decision is based on (i) the objective, (ii) the budget and (iii) the amount of time and expertise available for image processing.

Planning an airborne remote sensing campaign is not straightforward either. While digital airborne instruments offer great spectral and spatial versatility, the user must exercise considerable care when selecting an appropriate set up. Logistical advice will be given in this chapter but readers are also directed to chapters concerning sensor specifications for each habitat (Parts 3 and 5).

Having pointed out that imagery acquisition is not always straightforward, the process is generally becoming simpler. This reflects the increasing commercialisation of remote sensing and the move toward better technology. Greater demand for data is accompanied by improved efficiency and customer service on behalf of the suppliers. More sophisticated data storage media (e.g. CD-ROM) also play a key role in streamlining the process.

Cloud cover and seasonal weather patterns

Cloud cover: a critical consideration

Cloud cover is a serious constraint to the use of remote sensing, as it may significantly reduce the number of images available. For practical purposes, an image with more than 25% cloud cover is considered here to be unusable over large areas. Furthermore, clouds may obscure the area of interest within an individual image. Although in a few fortunate cases small areas of interest (within a larger cloudy image) may be cloud-free, finding such occurrences is difficult unless quick looks are used (see hereafter). Even in an image with less than 25% cloud cover, clouds may confound interpretation by obscuring areas which are required for multispectral image classification procedures, such as spectral signatures from training sites (Chapter 10). Misclassification problems may also arise if cloud shadow is not correctly identified. Airborne sensors are not immune from the effects of cloud: even if the sensor is flown below the cloud ceiling, it will still be above the clouds' shadows. However, the flexibility inherent in airborne data collection confers the ability to take advantage of cloud free days.

Clouds are formed when moisture in the air condenses into small, airborne water droplets. Therefore, any factors which influence the temperature and humidity of the air will also affect cloud formation. Two factors are especially significant along coastlines. Firstly, at the start of the day, the land is often cooler than the sea and this situation reverses as the sun warms the ground. Clouds may form in the morning if moist, oceanic air moves onshore or in the afternoon if warm air passes from the land to the cooler sea. This effect is exacerbated in the tropics where large temperature differences can occur between terrestrial and aquatic surfaces. Secondly, the topography of coasts determines the rate at which air moving across the coastline rises or falls. A gently rising coastal plain will cause air to ascend and cool slowly. If the terrain is volcanic or mountainous, rates of cooling will be higher with more cloud developing as a consequence. The extent of the land mass can also be expected to affect cloud formation. Air which is flowing towards a volcanic island may be deflected to one side, as opposed to being uplifted, whereas air may be forced upwards rapidly as it passes across mountainous coastlines of continents.

Considering that cloud cover is such a serious constraint to remote sensing, it is surprising that so little information has been published on the difficulty of successfully completing a remote sensing study because of it. As part of a study of shoreline erosion along the coastline of North Carolina, Welby (1978) obtained 16 Landsat images taken between 1972 and 1977, which constituted 50% of the imagery for that period with 0–10% cloud cover. In those five years, there were approximately 114 viewing opportunities (most of which will have been taken, as the Landsat sensors are rarely turned off while passing over the continental United States). Therefore, the probability of obtaining an image with < 10% cloud was 28%. Allan (1992) estimated that in mid-latitudes there is no more than a 5% chance of obtaining two consecutive images containing less than 30% cloud cover. The situation is usually worse in the tropics and all the more so for the humid areas. Gastellu-Etchegorry (1988) quantified cloud cover in the humid tropics of Indonesia and found that the probability of obtaining an acceptable image depended on both the location and season of interest. During the worst season (November to March), the probability of obtaining a SPOT image in a given month with ≤ 10% cloud cover ranged from 0–25% and during the best season (June to July) the probability was never greater than 50%, but no more than 2% in some areas. Jupp *et al.* (1981) found that for the Great Barrier Reef a suitable Landsat MSS scene may be obtained on average only once a year in optimal months (4.3% of images available), and once every two years in unfavourable months (2.2% of images available).

Probability of a satellite recording a cloud-free or low-cloud cover image

The frequency with which a satellite passes over a coast is determined by its temporal resolution. This section discusses the probability of acquiring imagery with acceptably low levels of cloud cover using meteorological data gathered from 21 recording stations in eight tropical countries (Table 5.1). Four countries are continental and four are island nations which have been categorised as small (< 5000 km^2) or large (> 5000 km^2). Cloud cover data is traditionally measured in oktas whereby an observer divides the sky into eight and estimates how many eighths, or oktas, are covered by cloud. Occasionally, data sets include a rating of 9 to account for the sky being obscured by fog or strong haze.

The probability that any day has 25% cloud cover or less at time t hours ($P_{t25\%}$) was calculated using the following formula:

$$P_{t25\%} = \frac{(N_{t0\%} + N_{t12.5\%} + N_{t25\%})}{N_t}$$

where: $N_{t0\%}$ = total number of records of 0 oktas of cloud (0% cover) at time t, $N_{t12.5\%}$ = total number of records of 1 okta of cloud (12.5% cover) at time t, $N_{t25\%}$ = total number of records of 2 oktas of cloud (25% cover) at time t, N_t = number of records at time t for the recording station.

In addition, the probability of cloud cover < 12.5% and 0% cloud cover at time t can be calculated using $P_{t12.5\%} = (N_{t0\%} + N_{t12.5\%})/N_t$ and $P_{t0\%} = (N_{t0\%})/N_t$ respectively.

Figures 5.1–5.5 plot the probability of three levels of cloud cover (zero, 12.5% and 25%, corresponding to 0, 1 and 2 oktas respectively) for each recording station. Where the data permit the probability of each level of cloud cover is plotted as a function of time of recording. With few exceptions, the highest value of $P_{t25\%}$ was approximately 40%; that for $P_{t12.5\%}$ was about 20%.

Cloud free days were rare. In Tanzania, Belize, and Rabaul, Papua New Guinea $P_{0\%}$ was extremely low, at less than 5% (Figures 5.1–5.3). There was a minute chance of a cloud free day in Port Moresby, Papua New Guinea (Figure 5.3), in the Seychelles and in Trinidad (Figure 5.5). There was not one cloud-free time recorded at any of the five Fijian stations in 15 years (Figure 5.4). The effects of seasonal variation in cloud formation can be seen in many locations. Highest probabilities of low cloud cover occurred in July–February in Tanzania, January–March in Belize and September–December in Port Moresby, Papua New Guinea (Figures 5.1–5.3).

Where observations of cloud cover were made more than once daily, the data show that cloud coverage changes throughout the day and that this daily variation changes with seasons. This was most marked in Zanzibar, Tanzania where from September to April P_{t25}% increased from less than 3% at 0600 hours to over 65% at 1500. In general, probabilities were highest in Tanzania in early morning and late afternoon, and lowest at midday (Figure 5.1). The situation in Belize was similar. From January–March the probability of a day with < 25% cloud was 30–40% at 0600 hours and 40–50% at 1800, but only 5–20% at midday. At other times of year, $P_{t25\%}$ was low (~10%) irrespective of the time of day (Figure 5.2).

Although airborne sensors are theoretically flexible enough so that data can be collected at any time of day, in practice sun glint at high sun angles means that mid-morning and afternoon are often the best times. Fortunately this coincides with the highest probabilities of low cloud cover in places such as Dar es Salaam and Mtwara in Tanzania, for example, but this would certainly not be the case everywhere. The acquisition of remotely sensed data from space is restricted to the time of satellite overpass. Landsat and SPOT satellites pass over the tropics at roughly 10 am local time. The probability of obtaining an image from either sensor with 25% cloud cover or less ($P_{25\%\ at\ 10:00}$) was calculated using:

$$P_{25\%\ at\ 10:00} = \frac{(N_{0\%\ at\ 10:00} + N_{12.5\%\ at\ 10:00} + N_{25\%\ at\ 10:00})}{d}$$

where: $N_{0\%\ at\ 10:00}$ = total number of days with 0 oktas of cloud at about 10 am, $N_{12.5\%\ at\ 1000}$ = total number of days with 1 okta of cloud at about 10 am, $N_{25\%\ at\ 10:00}$ = total number of days with 2 oktas of cloud at about 10 am for all stations for the recording period, d = number of days in the recording period. $P_{12.5\%\ at\ 10:00}$, the probability of obtaining an image from Landsat or SPOT with 12.5% cloud cover or less was calculated using:

$$P_{12.5\%\ at\ 10:00} = (N_{0\%\ at\ 10:00} + N_{12.5\%\ at\ 10:00})/d$$

Comparisons between countries, and with respect to their size and island/continental status cannot easily be made with a data set like this. However, cloud cover is unquestionably a serious consideration in the planning of any exercise using remote sensing in the tropics. As Table 5.2 shows, the probabilities of obtaining usable imagery are low: on average 5.1% for less than 12.5% cloud cover and 12.6% for less than 25%. It must be borne in mind, however, that most of these data on cloud cover were recorded on land and that cloud cover may be considerably lower offshore.

Cloud cover data were available from the Sudan and United Arab Emirates, two arid continental countries, but

Table 5.1 A summary of cloud cover data gathered from meteorological stations in eight tropical countries.

Country	Recording sites	General description	Years	Recording time (local)	Form of data	Figure
Tanzania	Zanzibar	Continental, humid	1981–1990	0600, 1200, 1500	oktas of cloud, daily	Fig. 5.1
	Dar es Salaam					
	Mtwara					
	Tanga					
Belize	Belize City	Continental, humid	1990–1994	0600–1800 hourly	oktas of cloud, daily	Fig. 5.2
United Arab Emirates	Sharjah	Continental, arid	1989–1995	unknown	mean oktas of cloud, monthly	Fig. 5.6
	Ras Al Kaimah					
	Abu Dhabi					
	Dubai					
Sudan	Aqiq	Continental, arid	1980–1987	0600, 1200	mean oktas of cloud, monthly	Fig. 5.6
	Port Sudan					
	Halaib					
Papua New Guinea	Rabaul	Large island, humid	1992	0600, 0900, 1200, 1500, 1800	oktas of cloud, daily	Fig. 5.3
	Port Moresby		1994			
Fiji	Nausouri, Viti Levu	Large island, humid	1974–1988	0900	oktas of cloud, daily	Fig. 5.4
	Nandi, Viti Levu					
	Ono-i-Lau	Small island, humid				
	Rotuma					
	Udu point					
Seychelles	Mahe	Small island, humid	1992–1994	1000	oktas of cloud, daily	Fig. 5.5
Trinidad and Tobago	Piarco	Small island, humid	1980–1995	unknown	mean oktas of cloud, daily	Fig. 5.5

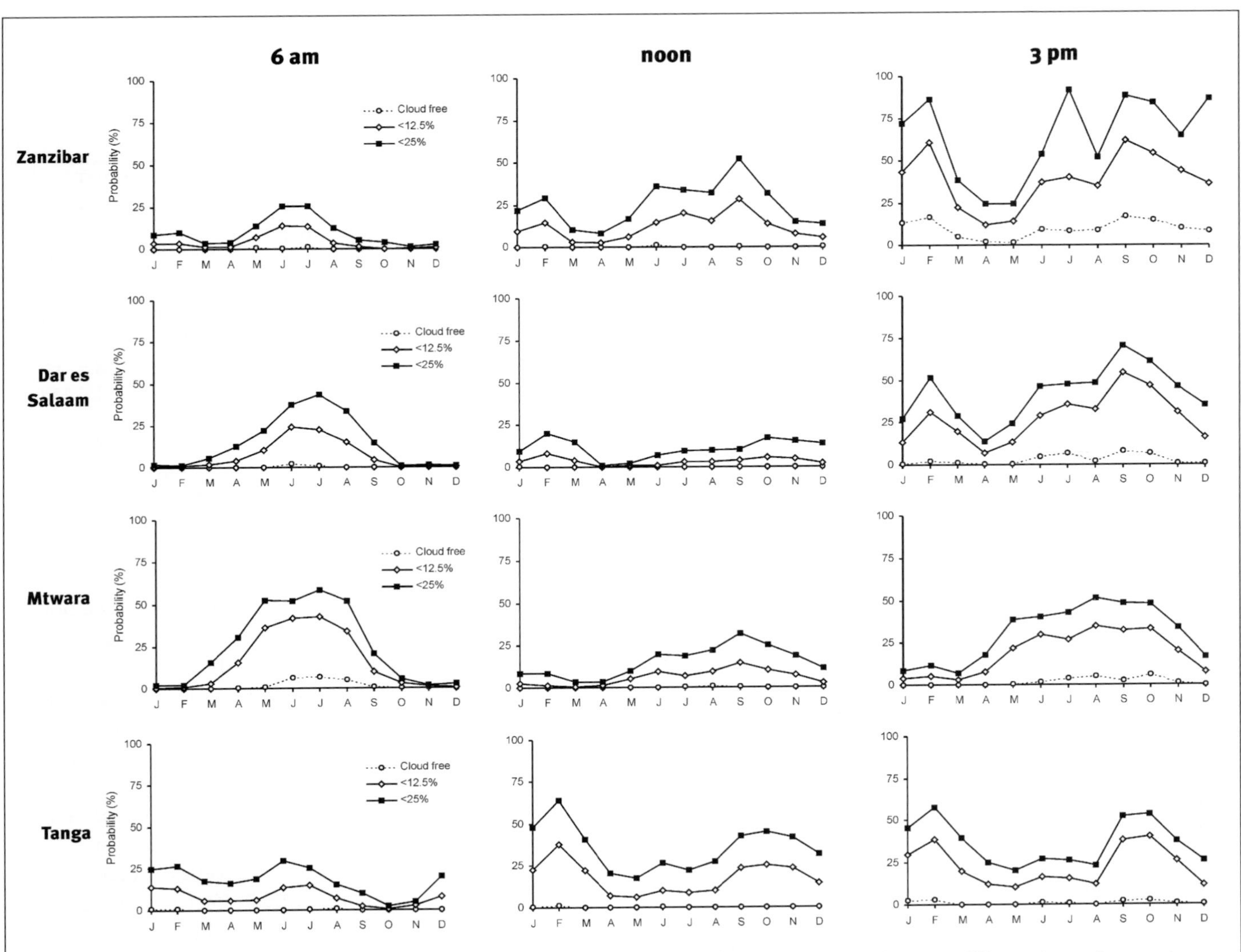

Figure 5.1 The chance of a satellite overpass over Tanzania coinciding with zero, < 12.5% or < 25% cloud cover at different times of year and times of day. Based on weather data from Zanzibar, Dar es Salaam, Mtwara and Tanga recorded at 6 am, noon and 3 pm.

Table 5.2 The probabilities (%) of obtaining a satellite image from either a SPOT or Landsat overpass with acceptable levels of cloud cover at around 10:00 am local time. Tanzanian data generated by interpolation between cloud cover at 6 am and 12 noon; data for Fiji and Papau New Guinea relate to cloud cover at 9 am. Belize and Seychelles cloud cover data specified for 10 am.

Country	Probability of obtaining an image with cloud cover of ≤12.5% or ≤25%	
	$P_{12.5\%}$	$P_{25\%}$
Belize	3.7	7.1
Tanzania	7.5	17.5
Papua New Guinea	5.8	11.0
Fiji	9.9	18.9
Seychelles	3.9	8.6

Note: The time of recording was unknown for the Trinidad data so $P_{12.5\%\ at\ 10:00}$ and $P_{25\%\ at\ 10:00}$ could not be calculated.

only in the form of monthly means, which prevented probabilities from being calculated (Figure 5.6). Although no information on the variation about these means was provided with the data, average cloud cover of less than 20% in both countries during May–September suggests that the probability of an acceptable cloud level on any particular day is good.

Probability of obtaining archived images with acceptable levels of cloud cover

The probability, *P*%, of a satellite sensor acquiring an image with an acceptable level of cloud cover does not mean that *P*% of archived images will be acceptable. Table 5.3 presents the proportion of archived Landsat and SPOT images with various amounts of cloud cover (cloud cover for these images are not given in oktas) for a range of tropical locations.

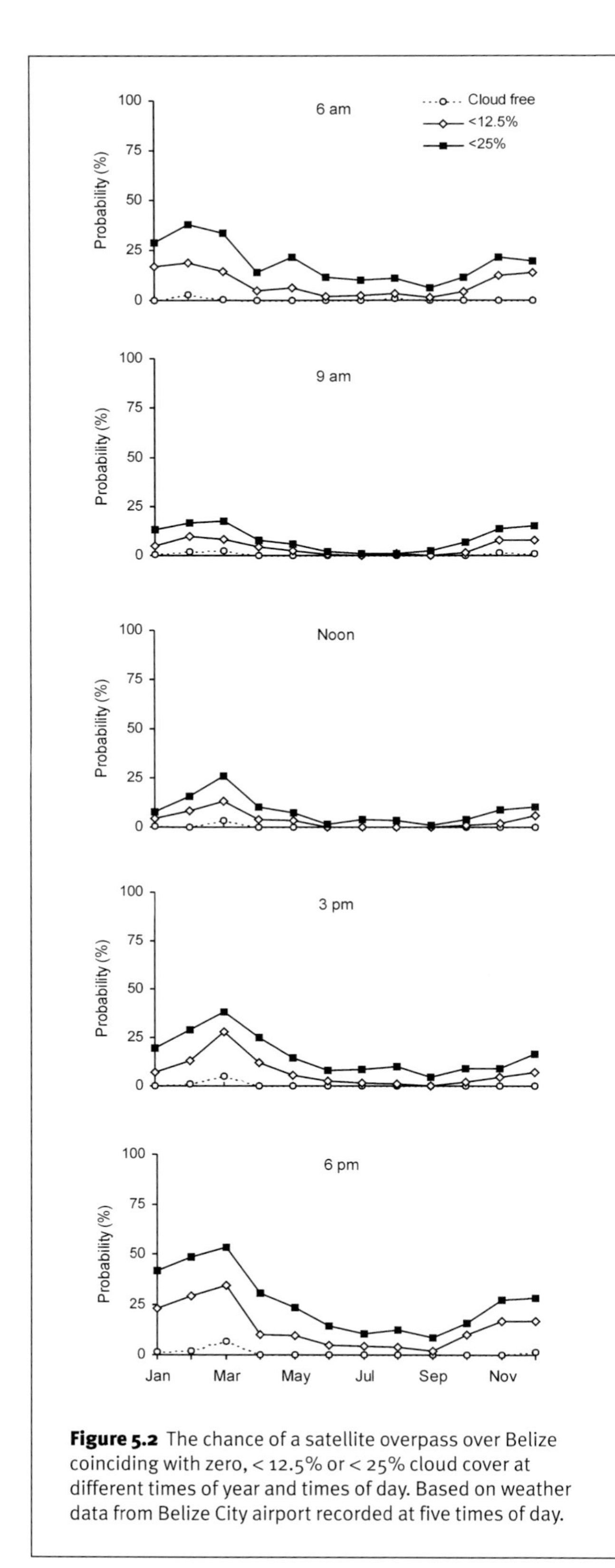

Figure 5.2 The chance of a satellite overpass over Belize coinciding with zero, < 12.5% or < 25% cloud cover at different times of year and times of day. Based on weather data from Belize City airport recorded at five times of day.

Table 5.3 The percentage of archived Landsat and SPOT images which had < 10%, < 20% and < 30% cloud cover. Our sample searches were conducted by the UK National Remote Sensing Centre (NRSC) in 1995 for entire Landsat and SPOT archives, except for Fiji and the Solomon Islands where the search was restricted to 1992–1995. N = number of images archived.

Country	Landsat				SPOT		
	< 10%	< 20%	< 30%	N	< 10%	< 25%	N
Belize	58	87	100	24	37	51	60
Tanzania	25	40	100	96	–	–	–
Fiji	0	100	100	2	48	87	59
Seychelles	16	52	100	30	54	91	34
Sudan	38	52	66	220	59	67	250
Trinidad	20	44	100	25	36	68	60
Turks and Caicos	11	43	52	53	24	51	128
Philippines	18	49	100	396	–	–	–
Solomon Islands	–	–	–	–	65	88	11
Average	21	52	80		36	56	

The archives contain more images with acceptable levels of cloud cover than would be predicted meteorologically, for two reasons. Firstly, sensors are often inactive over the tropics because of perceived low commercial demand for the data and in order to conserve storage space. They may be activated on request, but customers are likely only to do so during periods of predicted clear weather and any images with high cloud content are not stored. Secondly, images with high levels of cloud cover (the majority) are not saved as nobody is likely to buy them. The low image acquisition rate of some sensors whilst over tropical regions can be a major constraint. For example, only two Landsat images were found for Fiji between 1992 and 1995 (Table 5.3).

The implications of low probabilities extend beyond the practicalities of obtaining usable images. The chance of obtaining multiple sets of images becomes exceedingly small because of the multiplicative effect of combining probabilities. For example, if the probability of obtaining a single image with < 12.5% cloud in Papua New Guinea is 5.8% (Table 5.2), then the probability of obtaining two consecutive images with this cloud cover is remote (0.058^2, i.e. 0.33%). A value of 9.9% for $P_{12.5\%}$ means that on average an acceptable image could be acquired about every 10 overpasses of a satellite, in other words every 160 days for Landsat or 260 days for SPOT (if not programmed for off-nadir viewing and with only one platform operational). If a satellite image is required as part of a change detection study in response to a specific event, such as a pollution incident or hurricane, then many months might elapse before an image becomes available (but see the section on programming the SPOT satellite, later in this chapter).

In summary, while it is not possible to predict precisely the probability of acquiring imagery of a given cloud cover for areas beyond those studied here, the results suggest that, broadly speaking within the humid tropics, there is a 5% chance of obtaining an image with < 12.5% cloud cover on a single overpass and a 12% chance of obtaining one with < 25% cloud. Cloud-free images are non-existent in some parts of the tropics and extremely rare in others. The precise probability can vary substantially according to the area of interest, the season and time of day of data acquisition.

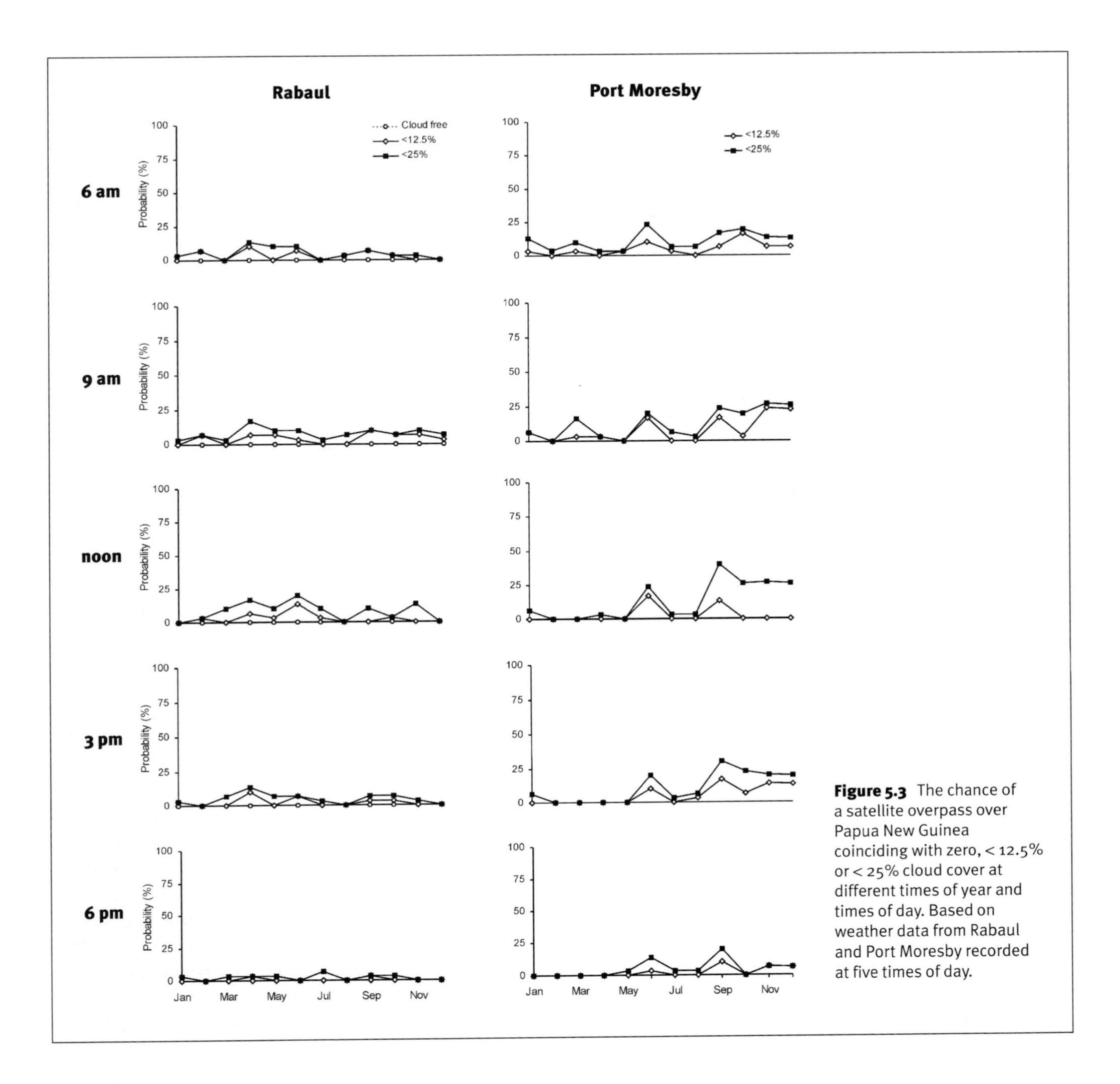

Figure 5.3 The chance of a satellite overpass over Papua New Guinea coinciding with zero, < 12.5% or < 25% cloud cover at different times of year and times of day. Based on weather data from Rabaul and Port Moresby recorded at five times of day.

Aerial photography

Fundamentals of aerial photography

Aerial photography is a diverse field of remote sensing which encompasses a variety of methods, media and considerations. While this chapter cannot describe all of these, a familiarity with the following topics will aid the selection of appropriate archived photography and help plan new photographic campaigns.

Cameras
Most aerial photography is carried out using specialised cameras. Perhaps the most widely used is the continuous strip camera, which exposes film past a narrow aperture that opens at a frequency proportional to the ground speed of the aircraft. Most importantly, the operator does not have to change film frequently (as would be the case with a hand-held camera).

Photographic films
Aerial photography is usually based on films which are sensitive to visible or infra-red portions of the electromagnetic spectrum (Figure 5.7). Normal colour aerial photography is generally the most useful for mapping habitats (Table 5.4). The colour balance of this film, being similar to human eyesight, therefore lends itself to visual interpretation. Wavelengths recorded on normal colour film also have good water-penetration properties. Although most black and white film will provide

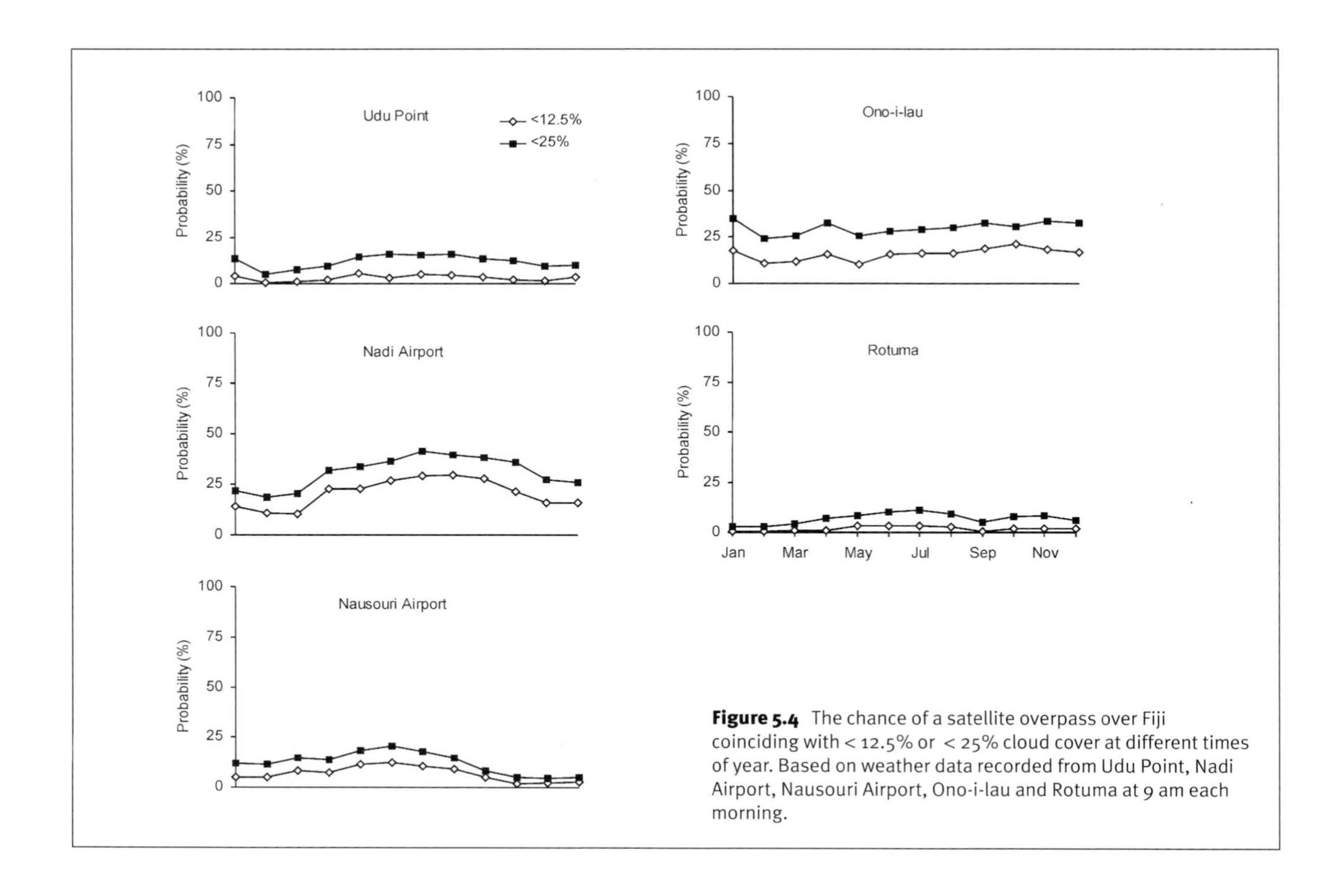

Figure 5.4 The chance of a satellite overpass over Fiji coinciding with < 12.5% or < 25% cloud cover at different times of year. Based on weather data recorded from Udu Point, Nadi Airport, Nausouri Airport, Ono-i-lau and Rotuma at 9 am each morning.

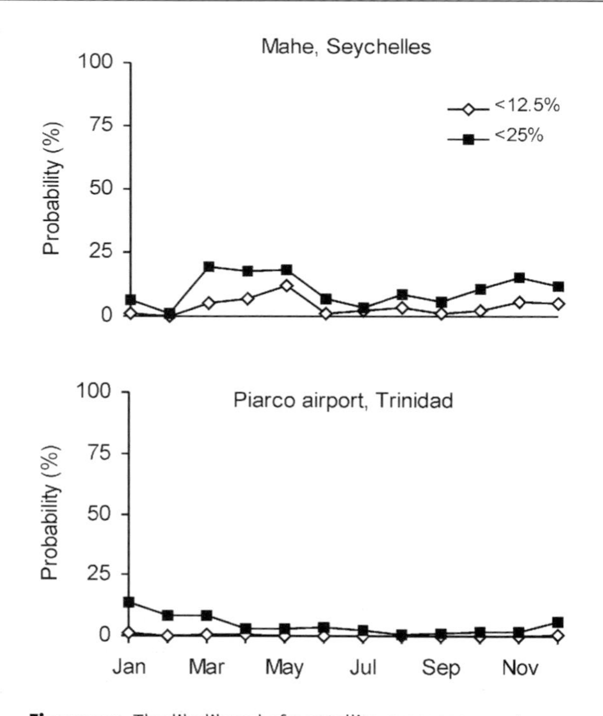

Figure 5.5 The likelihood of a satellite overpass over two humid high island areas coinciding with <12.5% or <25% cloud cover at different times of year. Based on weather data recorded from Mahe airport, Seychelles and from Piarco airport in Trinidad at 10 am.

information on habitats underwater, the interpretation of colour film is superior because the human eye can differentiate many more variations of colour than shades of grey (monochrome photography). Colour infra-red photography records information from beyond the visible spectrum and is especially useful for discriminating between different types of vegetation because near infra-red (NIR) film is sensitive to leaf vigour and health.

Orientation of photographs

Aerial photographs are broadly classified into three types according to their viewing geometry. These are vertical, low oblique and high oblique (Figure 5.8). Vertical photographs are obtained using a camera mount that enables the lens to point directly toward the ground. These photographs are used for photogrammetric purposes or habitat mapping. Low oblique photographs are usually taken from an aircraft passenger window and the Earth's geometry is progressively distorted toward the horizon. The geometry is even worse in high oblique photographs and these are not recommended for mapping (they do, however, provide a useful source of information for interpretation of satellite images (e.g. Figure 5.9, Plate 6).

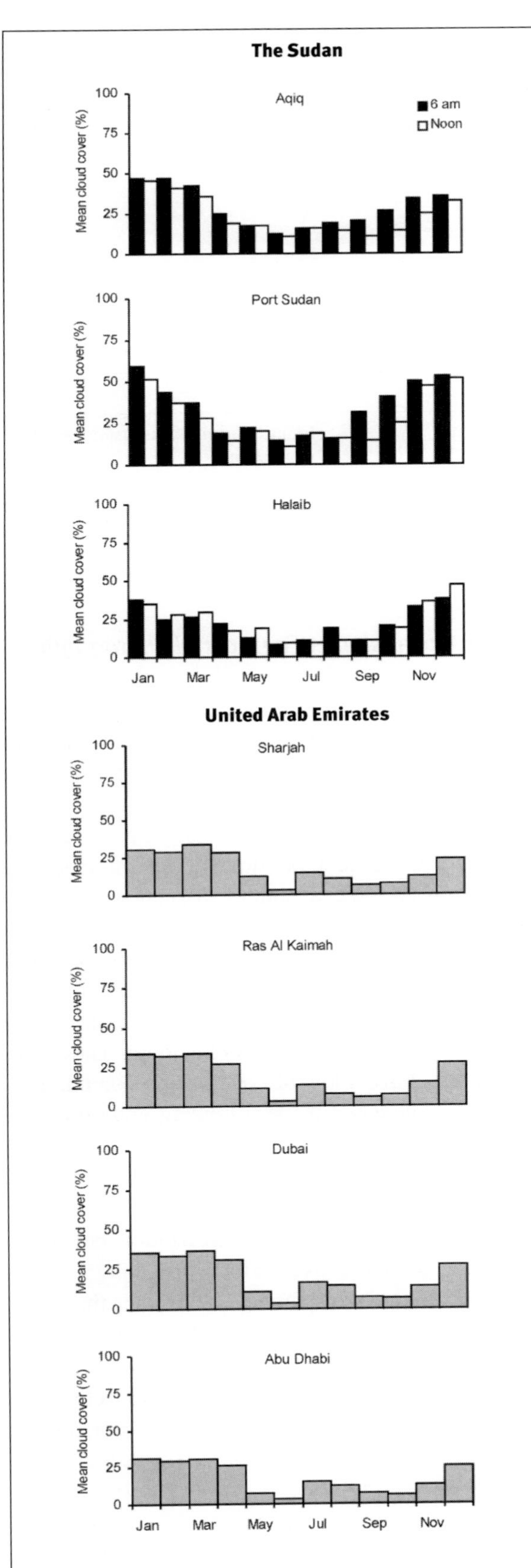

Figure 5.6 Average cloud cover (%) recorded in arid continental countries (Sudan and United Arab Emirates) at different times of year. Based on weather data recorded at three stations in the Sudan (at 6 am and noon) and four stations in the UAE (time unknown).

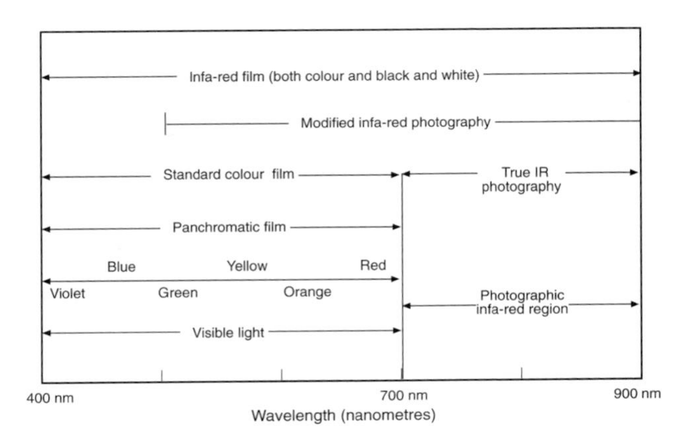

Figure 5.7 Range of sensitivity for various film types within the photographic range of the electromagnetic spectrum. Based on: Paine, D.P., 1981, *Aerial Photography and Image Interpretation for Resource Management* (New York: John Wiley & Sons Inc.).

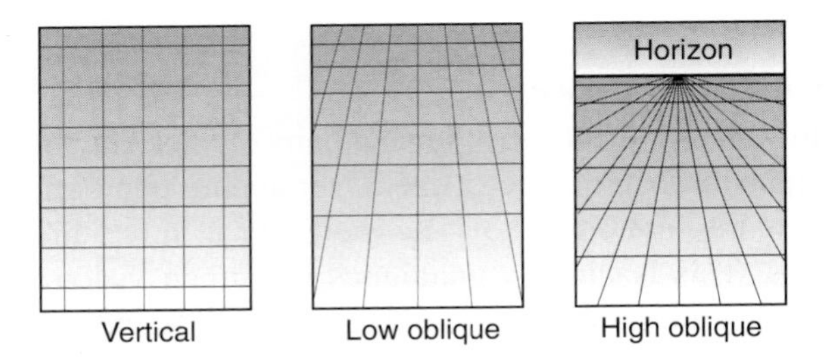

Figure 5.8 Illustration to show the nature of distortions in low and high oblique aerial photographs. Based on: Ciciarelli, J.A., 1991, *A Practical Guide to Aerial Photography with an Introduction to Surveying* (New York: Van Nostrand Reinhold).

Scale

The scale of a photograph defines the ratio between distances apparent on the photograph and those in real life. For example, a distance of 2 cm on a 1:10,000 photograph would represent 200 m on the ground. Most photography for coastal habitat mapping has a scale between 1:10,000 and 1: 50,000. A scale of 1:10,000 is described as larger than (say) 1: 50,000. As scale becomes larger, the degree of spatial detail increases. However, while detail may improve at larger scales, the total area covered by each photograph decreases. If the area covered by each photograph is reduced, a greater number of photographs are required to cover a particular area of interest. Therefore, a trade-off needs to be made between operating at a scale large enough to identify the features of interest and being able to cover the area of interest efficiently. This issue is further complicated by the need to obtain ground control points: smaller areas are less likely to have an adequate number reference points (Chapter 6). Ideally, each photograph should incorporate land features which can be registered to a map at a later stage. A scale of 1:10,000 to 1:20,000 seems optimal for most applications.

Table 5.4 Advantages, disadvantages and applications of main photographic films.

Film	Advantages	Disadvantages	Main application
Normal colour	Easily interpreted as it looks very similar to what we see	More expensive than monochrome	Delineating marine and terrestrial habitats
	Human eye sees 100 times more variations of colour than shades of black and and white i.e. better discrimination of habitats	Colours tend to fade with time	
		Not as sharp as monochrome	
	Good water penetration		
Monochrome (black and white)	Sharp boundaries	Fewer shades of grey than shades of colour i.e. less discriminating power	Photogrammetric mapping
	Good water penetration		
Colour infra-red	Sensitive to vegetation	Poor penetration of water	Identification of diseased terrestrial vegetation
	Good penetration of haze	False colours, so less easy to interpret	

Stereoscopic overlap

Most maps are made by viewing pairs of adjacent photographs through a stereoscope. Overlapping photographs allow the interpreter to view the Earth in three dimensions. This is directly analogous to human vision and reveals information on geomorphology, slope and height. To achieve adequate overlap, it is common to allow for 60% forward overlap along a flight line and 30% side lap with respect to each adjacent flight line (Figure 5.10). The mechanics of stereoscopic photogrammetry will not be explored here but details can be found in Ciciarelli (1991) or Paine (1981) among other books. If stereoscopic viewing is not required, cost savings can be made by purchasing every other photograph rather than the whole series.

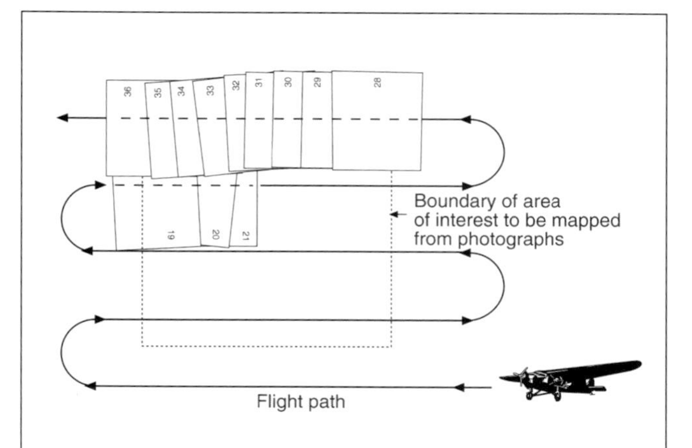

Figure 5.10 A regular flight path followed by an aircraft engaged in aerial photography showing the degree of overlap between photographs. Based on: Ciciarelli, J.A., 1991, *A Practical Guide to Aerial Photography with an Introduction to Surveying* (New York: Van Nostrand Reinhold).

Planning the acquisition of new aerial photography

In the event that new aerial photography must be commissioned, an appreciation of the issues raised earlier in this section should assist planning. In most cases, the decision maker will contract out the survey and will not be concerned with highly specific details – i.e. it is easy to specify a scale of 1:20,000 but quite another to calculate the required altitude, speed, shutter frequency, etc. However, if a knowledge of such details is required, readers are referred to the book by Paine (1981), which also includes some discussion of costs.

On a less rigorous basis, it is quite common to take aerial photographs with a hand-held camera. These will almost certainly have a low oblique orientation but their quality will be enhanced if the following guidelines are adhered to:

- hold the camera as vertically as possible,
- hold the camera near to the aircraft window but do not touch the window (i.e. avoid vibration) ,
- use natural colour transparency film rather than print film (sharper images); select good-quality medium-fast film,
- choose clear skies (> 18 km horizontal visibility) and light winds (to reduce interference from cloud and vibration).

Furthermore, to avoid sunglint:

- fly north–south or east–west in the mid-morning or mid-afternoon,
- point the camera away from sun,
- use a lens with a long focal length (e.g. 200 mm),
- use a polarising filter.

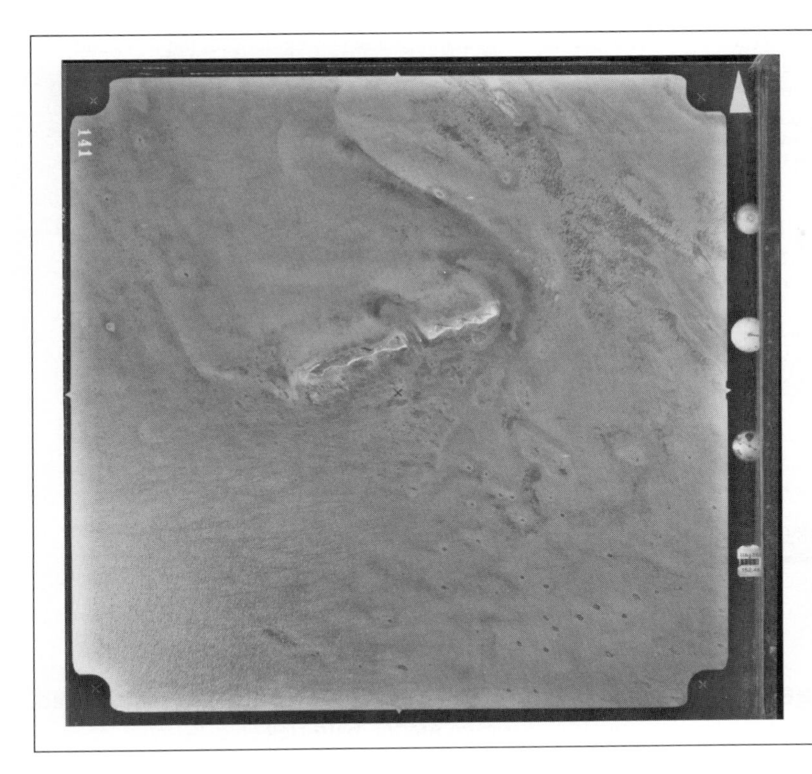

Figure 5.11 Monochrome vertical aerial photograph of Six Hills area of the Caicos Bank. Photographs in this series were used to create 1:10,000 ortho-maps of the islands and can be purchased from the UK Ordnance Survey.

Sources of archived aerial photography

The first places to seek out aerial photography are government ministries and departments. The Department of Lands and Survey (or equivalent) is perhaps the most likely place to archive photographs but relevant coastal bodies may also possess data (e.g. Department of Fisheries and Coastal Management, Forestry Department, Department of Agriculture, Department of Natural Resources).

Military establishments will almost certainly hold archives of aerial photography. However, a word of caution needs be made concerning availability. These photographs are usually taken for reconnaissance mapping and restrictions may be placed on their disclosure. It is quite common to be denied access to photographs and, even if permission is granted, authorisation can take several months. In many countries, aerial photographs cannot be released without permission of the Chief Surveyor (or equivalent).

Mapping agencies are another useful source of photography. The UK Ordnance Survey has conducted mapping and charting exercises for many areas of the world and it is possible to purchase the original photographs (although this can be expensive at £27 per photograph at a scale of 1:10,000, Figure 5.11). Information on the margins of local topographical maps often indicate which organisation to contact.

Most of the institutions described so far are likely to possess monochrome photography, since their mandate is usually photogrammetric mapping for the purpose of topographic map production. While visual interpretation of monochrome photography is inferior to that of colour, if their orientation is orthogonal (i.e. vertical), they are highly desirable for habitat mapping because of their superior geometric qualities over (say) oblique colour aerial photographs.

There are a number of additional sources of aerial photographs, such as research institutions, environmental NGOs, and commercial photographic companies (e.g. photographs for postcards). While photographs from these sources will probably be taken with colour film, their main drawback (if any) is likely to be oblique orientation.

☞ **When contacting an organisation to obtain aerial photography from their archive, you will need to specify the season and/or years of interest, the preferred scale and type of photography and your area of interest in latitude and longitude or local grid coordinates.**

Cost of aerial photography

It is difficult to generalise about the cost of aerial photography because of the number of variables which have to be considered (e.g. area, scale, logistics). Instead, the cost of acquiring aerial photography is illustrated with two case studies from the Caribbean: one for Belize (Table 5.5) and one from the Turks and Caicos Islands (Table 5.6).

Table 5.5 Costs of obtaining aerial photographs for three sites in Belize. Prices based on a quote received by Coral Cay Conservation in 1993. The approximate areas of the three sites are Turneffe Atoll = 330 km², Calabash Cay = 220 km² and Bacalar Chico = 140 km².

Site	Scale	No. of Photographs	Price (US$)
Turneffe Atoll	1:10,000	867	23,100
Calabash Cay	1:10,000	234	9,835
Bacalar Chico	1:10,000	144	8,610
Turneffe Atoll	1:20,000	219	10,210
Calabash Cay	1:20,000	52	5,900
Bacalar Chico	1:20,000	32	5,750
Turneffe Atoll	1:24,000	162	9,700
Calabash Cay	1:24,000	44	5,580
Bacalar Chico	1:24,000	28	8,610
Additional mobilisation and standby charges			18,000

Table 5.6 Cost of acquiring aerial photography for the Turks and Caicos Islands (TCI) for the areas surveyed using CASI (see later). The estimate is based on bringing an aircraft to the TCI for 9 days and using a flying time of 6 hours, while covering an area of 150 km² at a scale of 1:10,000.

	Price (US$)
Mobilisation	9,500
Standby (@ $350 day^{-1})	3,150
Flying (@ $450 hour^{-1})	2,700
Aerial camera rental (@ $600 day^{-1})	5,400
Colour aerial film (1200 foot roll)	1,750
Colour prints (estimated at 100) 350	
Film processing (@ $2.50 foot^{-1}) 500	
Shipping	250
Total	23,600

Digital airborne multispectral imagery

The acquisition of airborne multispectral imagery differs from satellite data in that it is provided on a commercial basis by survey companies or institutions in response to a specific request by the customer. Costs vary hugely depending on the nature of the request, and between different organisations. Costs will usually include the following:

- All the transport costs associated with getting the equipment and operator(s) to and from the site. This can include airfares, freight, insurance, custom duties and import taxes.
- All the costs associated with operating the aircraft. Mobilisation fees are often charged when an aircraft is prepared for a survey job. Charter costs may or may not include pilot, fuel and oil fees. Standby charges will be levied if the plane is being held back from other work in order to be ready to survey (a cost which can mount rapidly if survey is delayed by bad weather). In the event that a local plane is not available, transport costs will be passed on to the customer; if this entails commissioning an aircraft from overseas, these will be large.
- The hire charges for the sensor, auxiliary equipment (e.g. GPS, gyro, field computer) and operator(s) for the surveying period.
- Processing of the data after the field phase. Typically this cannot be done in the field and will take place at the headquarters of the contracted company or institution. These costs will depend upon the level of processing which is desired. Roll/pitch and radiometric correction (see Chapters 6 and 7 respectively) are probably the minimum; geometric correction and mosaicing of images from different flight lines may also be desired.
- All consumable and direct costs. These will include hotel and subsistence costs for the personnel hired, postage, storage media, batteries etc.

Airborne multispectral imagery is not cheap! The purpose of this section is to provide guidelines for those readers considering the acquisition of this type of data and advice on how some costs may be minimised. Costs are discussed for the Compact Airborne Multispectral Imager (CASI) and two Daedalus instruments; the Airborne Multispectral Scanner (AMS) and Airborne Thematic Mapper (ATM). For further details on sensor specifications, refer to Chapter 3.

The first step in the acquisition of airborne multispectral imagery is to contact as many suppliers of imagery as possible and several addresses are listed in Appendix 2. To prepare a quote, a contractor will need to know:

- the purpose of the data,
- the location and size of the target area,
- the topography of the target area (not usually an important issue for coastal habitats),
- the proposed timing of fieldwork and flights (dates and times of day),
- the weather at these times (average humidity, haze, cloud cover),
- the availability of aircraft (some sensors have specific aircraft requirements),
- spectral resolution required (sensor configuration),
- spatial resolution required,
- requirements for the collection of ground control points (Chapter 6),
- requirements for GPS or DGPS processing (Chapter 6),
- processing requirements (e.g. roll/pitch, radiometric, geometric, mosaicing),
- required delivery times.

Table 5.7. Cost of acquiring 3 m pixel size CASI imagery for 150 km² in the Turks and Caicos Islands, exclusive of aeroplane hire costs. A dedicated survey plane is not required for CASI (unlike Daedalus sensors: Table 5.10), which considerably reduces plane hire costs, as a local aircraft (e.g. Cessna 172N) can be used. Hire costs for such an aircraft are in the order of US$200 per hour flying time. (These quotes were obtained in 1995, whereas the quotes discussed in Chapter 19 were obtained in 1997).

Quote 1

Item	Price US$
Preparation – mobilisation, liaison, equipment checks	1,541
Staff time – (two persons) installation of CASI, imaging operations, transit time	4,567
CASI and auxiliary equipment rental	3,400
Airfare and excess baggage to Turks and Caicos (two persons)	1,333
Local travel, subsistence, communications	1,567
Processing (roll/pitch, radiometric correction, DGPS processing)	9,160
Consumable costs – communication, couriers, EXABYTE tapes	133
Total	21,701

Quote 2

Item	Price US$
Airfare (one person)	950
Freight/shipping of CASI and auxiliary equipment	1,400
CASI and auxiliary equipment rental	6,000
Local travel, subsistence, communications (charged at cost)	612
Processing (roll/pitch, radiometric correction, DGPS processing)	2,000
Consumable costs – communication, couriers, EXABYTE tapes	500
Total	11,462

Quote 1 is almost twice as much as quote 2, the difference primarily being in the rates charged for personnel time during imaging and processing. This demonstrates the desirability of requesting quotes from several companies. Moreover, quotes are rarely fixed; this is a competitive field and customers have the ability to negotiate reductions in price.

☞ **Airborne multispectral imagery typically has high spatial and spectral resolution (this is what makes it so attractive for many applications). However, an unavoidable consequence of this is that data volumes are high. For example, a 16-bit CASI file (8 bands, 1m pixel size) of 7 km² is 112 Mbyte in size (i.e. 7000 x 1000 x 1 x 8 x 2 where the 2 constitutes 16-bit data of 2 bytes per pixel). Customers should ascertain the amount of data which will be generated and ensure that they have the ability to handle it.**

The acquisition of CASI imagery

A breakdown of quotes we received from two companies for five days hire of CASI for mapping 150 km² of the Caicos Islands is outlined in Table 5.7 (for reasons of confidentiality, we do not publish their names).

☞ **Specify a time-frame for delivery of processed data, as this can take a while if the operator has many other commitments.**

The customer may chose between various levels of pre-processing when ordering CASI. The quotes in Table 5.7 illustrate that data processing represents a significant portion of the total cost of acquiring CASI – this is because pre-processing takes a considerable amount of time and requires skilled personnel. The levels to which data can be processed are given in Table 5.8. Price will depend on the amount of data and will increase with the level of processing desired. For most applications, level 1b processing would be the minimal acceptable level. Although it is possible to request that data be geometrically corrected to local maps (level 3b), the onus is on the user to obtain ground control points (Chapter 6). Since the user is responsible for acquiring such points, it is probably more cost-effective to conduct the entire geometric correction rather than pay a consultant to effect the final stage.

Table 5.8 Definitions of processing levels.

Processing level	Description
Level 0	Raw 'sensor format' data at original resolution.
Level 1a	Level 0 data reformatted to image files with ancillary files appended.
Level 1b	Level 1a data to which radiometric calibration algorithms have been applied to produce radiance or irradiance and to which location and navigational information has been appended.
Level 2	Geophysical or environmental parameters derived from Level 1a or 1b data; may include atmospheric correction.
Level 3a	Level 1b or 2 data mapped to a geographic co-ordinate system using on-board attitude and positional information only.
Level 3b	Level 1b or 2 data mapped to a geographic co-ordinate system using on-board attitude and positional information with additional ground control points (provided by the customer).
Level 4	Multi-temporal/multi-sensor gridded data products.

Configuring the CASI sensor for specific applications

The wavelengths and bandwidths of CASI's spectral bands are specified by the customer. The configuration of the sensor is vitally important and will depend on the areas being surveyed and the target features which the customer wishes to detect.

The number of CASI bands available is limited by the scanning speed of the instrument and depends on the spatial resolution. At a pixel size of 1 m, a maximum of eight bands may be specified; at 3 m this increases to 15. CASI has 288 channels centred at 1.8 nm intervals (Chapter 3). However, a minimum band width of 10 nm is recommended to increase the signal to noise ratio (Box 5.1), 20 nm is probably more appropriate.

BOX 5.1

Signal-to-noise ratio (S/N)

The ratio of a sensor's signal to noise depends on pixel size, the wavelengths of the spectral band, the width of the spectral band, the type of sensor and the reflectance of the area being imaged. Signal is the useful part of measured radiance - i.e. that originating from the habitat being sampled in a pixel. The noise arises from many sources including electronic limitations of the sensor's detectors and scattered radiance arising from adjacent pixels. The measured signal will decrease if smaller pixel sizes are requested or if the width of spectral bands is reduced. In addition, habitats with lower reflectance (e.g. dense seagrass beds) will contribute less signal than those habitats with greater reflectance (e.g. sand). In short, while high spatial and spectral resolutions are desirable for distinguishing habitat spectra, they do so to the detriment of the S/N ratio. A compromise must be made between achieving a high S/N ratio and selecting high sensor resolutions, particularly when surveying areas which have low reflectance.

When selecting a configuration for CASI, certain decisions can be made easily. For example, there is little point configuring CASI to detect radiation in the infrared portion of the electromagnetic spectrum if the objective is to map submerged habitats (infrared radiation is fully absorbed by only a few centimetres of water). CASI literature provides some direction on which band combinations are appropriate for a variety of applications (Table 5.9), but band configurations are not always published (presumably to protect commercial advantages of some companies).

Configuration of CASI bands for the Turks and Caicos studies

The CASI data discussed throughout this *Handbook* had the 8 band set-up described in Table 5.9. Since a small pixel size was adopted (1 m^2) and the signal to noise ratio was of concern, the minimum bandwidth was set to 20 nm. To distinguish marine habitats, the first five of the bands were placed throughout the visual (water-penetrating) range of the electromagnetic spectrum. The first band had a short wavelength (402.5–421.8 nm) to allow maximum penetration of water. Near infra-red bands 6 to 8 were used to map mangrove habitats. Bands 6 and 7 fell roughly at the reflectance minimum for chlorophyll, whereas band 8 was set to the reflectance maximum of chlorophyll (around 750 nm). The reflectance maxima and minima of chlorophyll were recorded to improve the separability of mangrove spectra (i.e. ratios of these bands offer the greatest discriminating power for chlorophyll-containing surface features). This subject is discussed further in the context of mangrove mapping (Chapters 13 and 17).

The most appropriate configuration of CASI bands will depend on the mapping objectives. This case study from the Turks and Caicos was specific to mapping shallow marine and mangrove habitats. However, since airborne multispectral and hyperspectral remote sensing of marine habitats is in its infancy, few comparable data sets were available and our approach was founded on a 'best educated guess' principle. Had we had an *a priori* expectation

Table 5.9 The configuration of CASI for four different applications. Coastal habitat mapping was carried out in the Turks and Caicos at two spatial resolutions, 1 m (*) and 3 m (**). Pers. comm. = personal communication.

	Application				
Bands	**Coastal habitat mapping***	**Coastal habitat mapping****	**General vegetation**	**Seagrass**	**Sediments/primary productivity**
1	402.5–421.8	402.8–422.2	439.8–447.2	430.4–450.1	429.2–452.5
2	453.4–469.2	430.9–450.3	496.2–502.0	449.7–469.5	479.6–486.1
3	531.1–543.5	452.1–471.5	528.2–533.9	469.1–490.7	496.2–505.5
4	571.9–584.3	473.3–501.6	547.8–551.7	509.8–529.7	514.0–523.3
5	630.7–643.2	505.2–526.4	567.4–571.4	529.3–549.3	523.3–526.4
6	666.5–673.7	530.0–560.2	597.7–601.7	559.3–579.6	526.4–533.7
7	736.6–752.8	561.9–599.4	617.5–621.4	589.9–610.0	546.0–555.3
8	776.3–785.4	629.8–660.2	648.0–652.0	631.1–651.2	563.8–573.1
9		661.9–679.9	667.8–671.8	650.8–669.2	596.0–607.1
10		681.7–692.5	682.2–686.2	668.8–685.4	621.0–630.4
11		694.3–703.3	709.2–715.0	685.0–699.8	646.2–655.6
12		705.1–715.9	779.8–783.8	699.4–714.2	660.6–670.0
13		735.7–759.1	796.1–800.1	780.8–810.2	673.2–684.4
14		766.3–784.4	825.1–829.1	840.8–870.3	707.4–716.8
15			865.1–874.6		745.4–756.6
	Chapters 11–13 and 16–17	Chapters 11–13 and 16–17	D. de Vries (pers. comm.)	D. de Vries (pers. comm.)	Blake *et al.* 1994

of reef spectra, the CASI may have been configured differently. For example, it is now clear that the reflectance maximum of mangrove lies at approximately 710–720 nm (Ramsey and Jensen 1995) and not at the general reflectance maximum of chlorophyll at 750 nm. In light of this development, CASI could have been better optimised for mangrove (this is discussed in detail in Chapters 13 and 17). The following section discusses recent developments in the characterisation of habitat spectra.

Characterisation of habitat spectra

A useful avenue for further research with airborne instruments is the characterisation of habitat spectra. Some airborne sensors can be deployed in spectral mode to record hyperspectral information on spectra (see Chapter 3). Knowledge of the spectra of target habitats will permit a more informed selection of CASI bands to be made. While CASI offers great spectral versatility, this cannot be fully utilised unless an *a priori* expectation of spectra is available. Looking to the future, classification of CASI (and other remotely sensed) data may become better automated or guided through reference to libraries of spectra. In the meantime, customers have to be guided by whatever literature is available, expert opinion and common sense.

☞ **The airborne sensor AVIRIS (Airborne Visible/Infrared Imaging Spectrometer) is also used to record habitat spectra – see Chapter 3.**

Practical considerations

A twin-engined aircraft like the Cessna 402 is recommended for CASI surveys. While these planes and similar models are common in many countries, models which have been adapted for aerial surveys (i.e. with a photo hatch and camera mount) are rare. A major advantage of CASI is that the sensor system, which weighs only about 55 kg, can be mounted in any small aircraft as long as a satisfactory way can be found to mount the sensor head. This was achieved in the Turks and Caicos study by means of a custom door which had been specially designed to replace the usual door of a Cessna 172N (Plate 3). The custom door was rented from Dr J.P. Angers at the Centre Universitaire Saint Louis Mallet in Canada. Hire, freight and insurance costs amounted to US$450. Conversion of a local aircraft negated the need to commission a survey aircraft from overseas. Charter of this Cessna at local rates cost US$150 per hour; aircraft transportation, standby and mobilisation fees (US$11,000) and high flight charges (US$480 per hour) which would have been payable for a foreign plane were avoided for a total saving of US$15,545. Large savings will always be made if suitable modification of a local plane can be achieved without compromising data collection or the relevant aviation safety regulations.

CASI requires a power supply of 17 Amps at 28 Volts DC. If the aircraft cannot supply this power, the equipment can be powered by battery. Ideally, batteries should

Table 5.10 Costs of a Daedalus AMS survey of the Turks and Caicos Islands for two areas: a small area of 10 x 20 km and a much larger area (100 x 150 km) which is approximately that of the entire Caicos Bank. Four spatial and spectral resolution options are listed and all data would be collected as 12-bit. Prices are given in US dollars and are current for 1997.

	10 km x 20 km			100 km x 150 km	
No. of bands	Spatial resolution	No. of survey days	Price (US$)	No. of survey days	Price (US$)
4	1 m	2	62,000	50	300,000
6	2 m	2	62,000	25	165,000
6	3 m	2	62,000	13	126,000
6	5 m	2	62,000	7	96,000

be trickle-charged overnight before use the next morning and a back-up battery carried on all flights.

The CASI operator monitors flight lines and data quality in real-time by means of an on-board image display. The customer can also study CASI images on the day of acquisition and this is strongly recommended to confirm that the correct areas have been surveyed and that the data are not degraded or obscured by clouds or cloud shadow.

Acquisition of Daedalus airborne imagery (Daedalus AMS and ATM)

The acquisition of Daedalus airborne imagery is similar in many aspects to CASI. The main differences are:

- suppliers are usually government departments or large institutions rather than small companies (refer to Appendix 2 for a list of Daedalus operators),
- Daedalus AMS and ATM systems have to be deployed from larger aircraft which have been designed for aerial survey (the weight of a Daedalus AADS 1268 ATM is about 100 kg),
- the channel wavelengths and bandwidths of Daedalus AMS and ATM are fixed.

The first two factors contribute heavily to the final cost of commissioning a survey using the Daedalus. A quote for a Daedalus survey in the Turks and Caicos Islands is given in Table 5.10.

The costs listed above assume no government duties or import taxes. Covering a 10 km x 20 km area requires minimal flight time so the cost is the same for all resolution options. The majority (79%) of this price is to cover the mobilisation of the scanner and aircraft to the TCI and return to the USA. The larger survey makes much better use of the mobilisation expense.

Satellite Imagery

Ordering a search of archived imagery

Satellite data are usually acquired from the nearest remote sensing data centre (a list of which is provided in Appendix 2). If the nearest regional data centre cannot supply the data required, it is worth contacting the suppliers of imagery directly (e.g. SPOT Image, EOSAT). The data centre will ask for the dates and area(s) of interest (see Figure 5.12). The area of interest can be defined either as a rectangle of specified coordinates or as a circle with specified centre coordinate and radius (in km). Some neighbouring images may overlap with the specified area of interest so it is a good idea to specify a large area (e.g. add 50 km to each dimension of the true area of interest).

☞ **When ordering a search of the image archive, request that the list include corner coordinates of each scene.**

The data centre will provide a list of available imagery and include additional details to identify each scene and describe the data quality, cloud cover and so on. For space considerations, the details of image archive searches are not represented here for all imagery types. However, Boxes 5.2 and 5.3 describe the interpretation of SPOT and Landsat archives which are perhaps the most widely used. Selection of suitable imagery clearly depends on the specific objectives in mind. The four most important considerations are:

1. Which images best cover the specified area of interest?
2. What is the degree of cloud cover?
3. What is the quality of the data?
4. What is the date of acquisition?

☞ **Many archiving bodies offer a 'quick look' facility on most images (Figure 5.13, Plate 6).**

Ordering new satellite imagery – programming requests

It is quite possible that a search of the imagery archives will be fruitless. Satellite data are not recorded for every location on the planet and cloud may reduce image availability of those areas that are included (see Figure 5.14).

Unlike Landsat satellites, the SPOT, ERS and RADARSAT satellites can be programmed to take images of specific points on Earth. SPOT images can be acquired from the instrument's usual path or from 'off-nadir' in

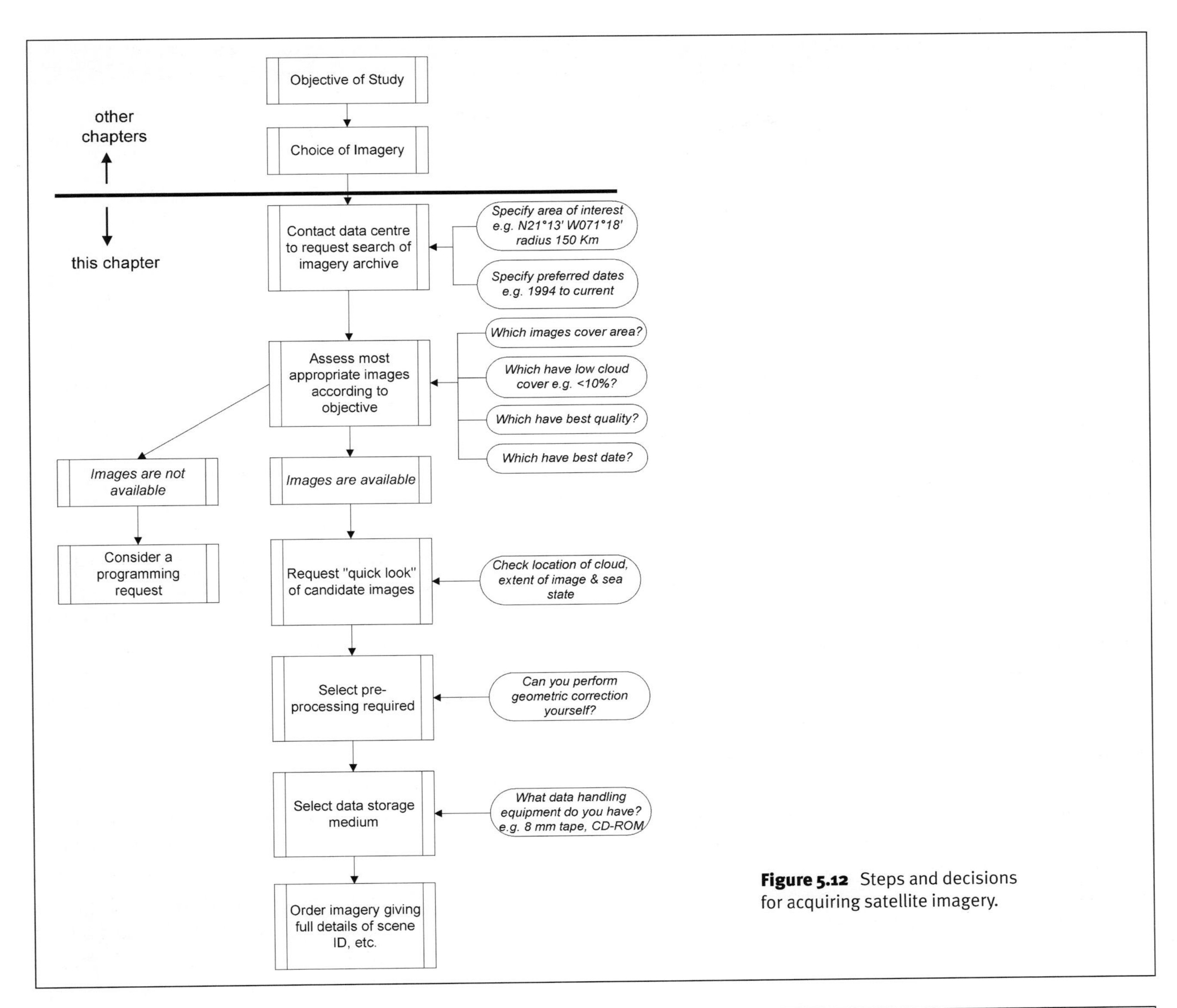

Figure 5.12 Steps and decisions for acquiring satellite imagery.

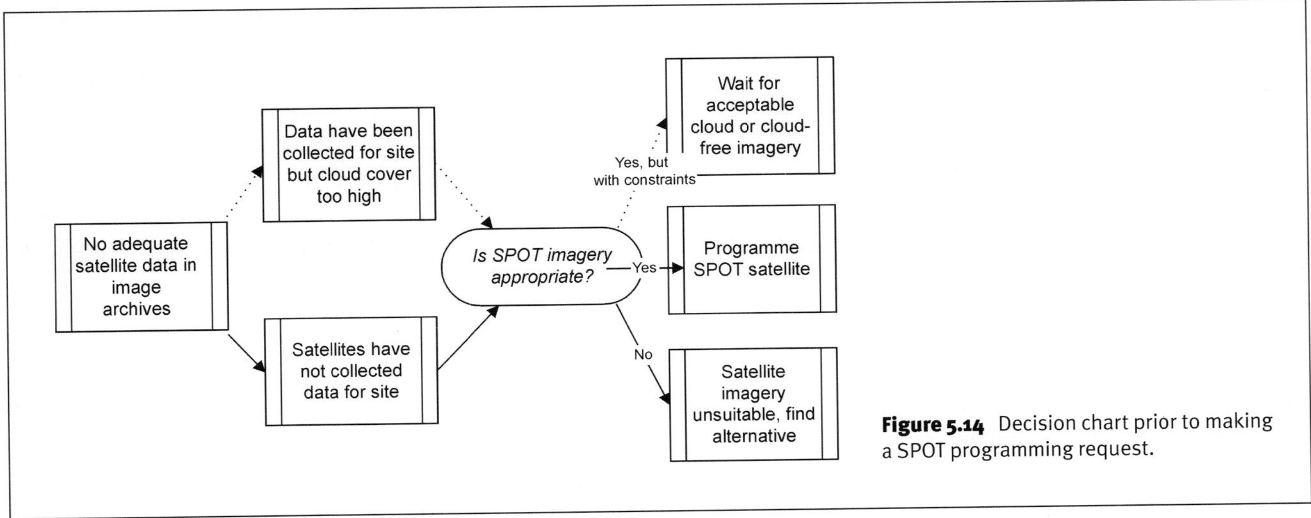

Figure 5.14 Decision chart prior to making a SPOT programming request.

which the sensor is pointed toward a specified area (i.e. to the left or right of the usual viewing location). The capability to view 'off-nadir' affords SPOT great versatility, since the temporal resolution for SPOT scenes can be greatly increased. Off-nadir viewing also permits three-dimensional mapping based on stereo pairs of overlapping images (though this is rarely useful for marine habitats). To make a programming request, the required

BOX 5.2

Interpretation of archived SPOT data

The following extract is a typical archive report from SPOT Image. The terms are described below. Note that an archive would usually contain details of many images. Only one is presented here.

S CLOUD SNOW	KJ TQ CONF	SCENE ID DATE ORB GAIN	TIME INCID V-ANG.	I	MS ORIEN AZIM/SITE	SAT	NW-CORNER SW-CORNER	CENTER	NE-CORNER Dkm SE-CORNER
2 1111	641–307 EEEE T	90/07/07 017 675	15:26:39 -15.1 -13.4	1	X 008.7 082.3/+69.6	(0-0)	N214918/W0752504 N211733/W0725803	N213049/W0723602	N214401/W0721415 + 29 N211217/W0722145

where:
S = SPOT satellite number
DATE = date of image acquisition (month/day/year)
KJ = location of image on SPOT Grid Reference System, GRS (i.e. column 641, row 307). Grid lines are called nodes.
SAT = shift along track. For consecutive scenes along a north–south track of the satellite. A choice of 10 scenes are usually available numbered 0 to 9. The step is 6 km. 0 refers to node of GRS. i.e. specify amount of overlap between scenes.
TQ = technical quality for each quadrant. E, excellent; G, good; P, poor; U, unusable
INCID = angle between look direction at the scene centre and a line perpendicular to the reference ellipsoid (degrees and tenths of a degree).
CONF = viewing configuration given that two instruments are held on satellite. D (dual) only 1 operative; T (twin) both operative and overlapping; I (Independent) each operate independently.
V-ANG = viewing angle between geocentric direction and the vector to the centre of HRV field of view (decimal degrees).
AZIM/SITE = sun azimuth and sun angle at the scene centre (i.e. 82.3 and 69.6)
I = HRV instrument number
TIME = time of image acquisition in Greenwich Mean Time
MS = spectral mode (X, multispectral; P, panchromatic)
CLOUD = cloud cover in each quadrant of the scene (upper left, upper right, second left, etc.) Since 1991, these data have been given for each octant of the scene. Scale is 0 (A) cloud-free; 1 (B) 1–10%; 2 (C) 11–25%; 3 (D) 26–75%; 4 (E) 76–100% cloud cover.
GAIN = gains used per spectral band (order of magnitude, 5–10 nm).
ORIEN = angle through which the image centreline must be rotated in order to align with true east (degrees and tenths of a degree).
ORB = orbit number is reference track passed over during 26-day orbit cycle in which the satellite accomplishes 369 revolutions. By convention track 1 intersects the equator at longitude 330.24°.
(NW)-CORNER = corner coordinates in degrees, minutes and seconds. N refers to north (latitude), W to west (longitude). Also includes centre coordinate.
Dkm = scene centre to KJ node. Distance in km.

date, corner coordinates and allowable cloud cover must be specified.

SPOT Image (the company which supplies SPOT satellite imagery) offers the following programming services (see also Table 5.11):

- Red Service – the system is booked for a period and images are frequently acquired for the specified area of interest. SPOT Image assesses user requirements and conducts a feasibility study prior to programming.
- Blue Service – SPOT Image attempts to cover the area of interest within a given time period (usually a few months). However, priority is given to Red requests. The programming request terminates once a suitable scene is recorded for a given period (e.g. once cloud cover in a scene falls below 10%).
- White Service – equivalent to the Blue Service but cheaper. Only applies if SPOT Image does not already hold a scene for the area of interest in their image archive.

BOX 5.3

Interpretation of archived Landsat data (two example scenes)

The archive for Landsat data does not have as many parameters as that for SPOT. The following information is important:

Entity ID	Date Acquired	Cloud Cover	Quad 1	Quad 2	Quad 3	Quad 4	Path	Row	NW Latitude	NW Longitude	SE Latitude	SE Longitude
LT5104214275X0	07/07/87	2	2	0	2	2	18	048	N21.06.12	W074.45.19	N19.21.36	W073.21.12
LT5102614272X0	22/12/89	0	0	0	0	0	18	048	N21.06.12	W074.46.19	N19.21.36	W073.22.12

Entity ID – A unique 13-digit alpha-numeric code which identifies each Landsat scene.

Date Acquired – dd/mm/yy

Path/Row – Analogous to the KJ Grid Reference System used by SPOT. The Path refers to columns arranged according to longitude. Row refers to latitude. For example, the path/row coordinates for two consecutive images might be 18/048 and 18/049. Although path/row maps are available for all areas covered by the Landsat satellites, prior knowledge of path/row coordinates is not required when requesting an image search.

Cloud Cover – This is usually described for the scene as a whole and then for each individual quadrant (where 1 is top left, 2 is top right, 3 is bottom left and 4 is bottom right). Cloud cover is graded on a 10% incremental basis (i.e. 0 = 0–10%, 1 = 11–20%, 9 = 91–100%). Cloud cover is usually estimated automatically but this process often underestimates the true cloud coverage. Once a list of appropriate scenes has been made, it is wise to request a manual estimate of cloud cover in scenes of interest.

Corner Coordinates – These are given in latitude and longitude (degrees, minutes and seconds). Unlike SPOT, Landsat tends to provide coordinates for the Northwest and Southeast corners.

Many archiving bodies offer a 'quick look' facility on most images (Figure 5.13, Plate 6). A quick look is a condensed composite of the image and is useful for assessing the specific location of cloud and overall boundaries of the satellite scene. Quick looks are either supplied as hard copies or computer graphic files (e.g. sun raster, jpg). Examination of quick looks should complete the selection process.

☞ **Many data centres (e.g. Space Imaging, NRSC) will transfer digital quick looks directly to the user's computer. Electronic mail facilities are required for this process.**

Table 5.11 Cost of programming SPOT satellites based on 1997 prices. Charges are in French Francs. Note: these fees are in addition to cost of each image acquired (detailed in Table 5.12).

Programming service	Details	Cost FF
Red	Access fee, per scene requested and per survey period	12,000
	Surcharge per acquisition attempt and per scene, irrespective of cloud cover in resulting imagery	1,700
Blue	Programming fee per scene acquired with less than 10% cloud cover (irrespective of number of attempts)	7,000
White	As above but only applies to areas not adequately covered in the SPOT image archive	1,000

Table 5.12 Digital image products detailing area of coverage and cost (1997 prices).

Satellite	Image type	Size (km)	Cost
Landsat TM	Full scene (7 bands, system corrected)	185 x 170	US$ 4,400
	Subscene (7 bands, system corrected)	100 x100	US$ 3,100
	Map Sheet (7 bands, system corrected)	½° x 1°*	US$ 2,500
	Micro scene (3 bands)	56 x 56	US$ 2,600
Landsat MSS	Full scene (4 bands, system corrected)	185 x 170	US$ 250
SPOT XS	Full scene (3 bands, level 1A processing)	60 x 60	FF 13,300
	Full scene (3 bands, level 1B processing)	60 x 60	FF 13,300
	Full scene (3 bands, level 2B processing†)	60 x 60	quote
	Stereopair (3 bands, 1A or 1B processing)	60 x 60	FF 20,000
SPOT Pan	Full scene (1 band, level 1A processing)	60 x 60	FF 17,200
	Full scene (1 band, level 1B processing)	60 x 60	FF 17,200
	Full scene (1 band, level 2B processing†)	60 x 60	quote
	Stereopair (1 band, 1A or 1B processing)	60 x 60	FF 26,000
IRS: LISS-III	Full scene (4 bands, path orientated)	141 x 141	US$ 2,970
	Quadrant (4 bands, path orientated)	70 x 70	US$ 2,070
IRS Pan	Full scene (1 band, path orientated)	70 x 70	US$ 2,500
	Subscene (1 band, path orientated)	23 x 23	US$ 900
JERS OPS	Full scene (4 bands, system corrected)	75 x 75	US$ 1,250
JERS SAR	Full scene (1 band, level 2.1)	75 x 75	US$ 1,250
RADARSAT	Full scene (standard beam, map image)**	variable	US$ 3,500
ERS SAR‡	Precision image	100 x 100	Ecu 1,200
	SAR Terrain Geocoded Image	100 x 100	Ecu 2,300

* refer to degrees where 1° of latitude is approximately 111 km
** a wide range of other products are available, refer to data supplier for further information
quote = subject to quotation
†customer must supply maps or ground control points
‡ non-ESA countries must add a royalty fee of US$255 per image. Images also available in photographic format.

☞ **In December 1995, we made a blue programming request to SPOT Image for imagery of the Caicos Bank. Cloud-free images were available by late February 1996 (i.e. 2–3 months later).**

Image products and pre-processing

Most satellite sensors record data over large areas (Table 5.12). Fortunately, if the area of interest is small, it is usually possible to purchase part of an image at a reduced cost (Table 5.12). Furthermore, while it is most useful to acquire digital data, it is not mandatory. SPOT, Landsat and ERS-1 offer photographic products which can be used in a similar fashion to aerial photographs. However, because map production from photographs can be a time-consuming process and digital data provide more information about coastal zones than photographic prints (e.g. the radiance of seagrass standing crop), the purchase of photographic products is only recommended as a last resort (e.g. if image processing software and hardware are not available).

Data suppliers also offer a range of pre-processing options which, for additional cost, will reduce the processing requirements of the customer. Tables 5.13–5.15 provide a general guide although further products are available.

The terms used for pre-processing require some explanation but further details are available from data suppliers (see also Table 5.14).

- *Basic geometric correction* – conducted for Landsat data, this compensates for earth rotation, spacecraft altitude, attitude and sensor variations. A similar process is conducted on SPOT data if level 1B processing (or higher) is requested. Data from the Indian Remote Sensing satellite (Panchromatic and LISS-III Multispectral) are provided with this correction for worldwide coverage (referred to as path orientated).
- *Radiometric normalisation* – equalises the difference in sensitivity for those sensors with CCDs.
- *Map orientated* – data acquired for Europe using the sensors on the Indian Remote Sensing satellite (i.e. Panchromatic and LISS-III Multispectral) can be geometrically corrected to local maps, prior to shipping, although this is approximately £170 more expensive than path-orientated data (for a single multispectral image).

The most commonly requested pre-processing level for SPOT is 1B. Level 2 pre-processing includes further geometric correction but this is often carried out by the customer and is not required (see Chapter 6 for details). The success of the geometric correction depends on the quality of reference information. If the maps are poor, the correction will be poor. Level S processing is only requested for multi-temporal studies and again, the image–image registration is often conducted by the consumer. The prices quoted are from SPOT Image. If geometric correction is required as a pre-processing step (e.g. SPOT level 2), it may be worth obtaining a quote from the nearest satellite data centre.

Table 5.13 Photographic image products detailing scale and cost. Details for ERS SAR have not been included because costs are no different to those for digital products listed in Table 5.12.

Satellite	Photograph type	Coverage (km)	Scale	Cost
Landsat TM	colour composite transparency (+ves or -ves)	185 x 170	1:1,000,000	US$ 2,700
	as above but with print	185 x 170	1:1,000,000	US$ 2,900
	as above but with print	185 x 170	1:500,000	US$ 2,900
	as above but with print	185 x 170	1:250,000	US$ 2,900
Landsat MSS	single band transparency (+ve or -ve)	185 x 170	1:1,000,000	US$ 25
SPOT XS	colour composite transparency 241 x 241 mm*	60 x 60	1:400,000	FF 13,300
	colour composite transparency 241 x 241 mm	30 x 30	1:200,000	FF 13,300
	colour composite transparency 482 x 482 mm	60 x 60	1:200,000	FF 22,200
	colour composite transparency 482 x 482 mm	30 x 30	1:100,000	FF 22,200
	colour composite print	60 x 60	1:200,000	FF 13,300
	colour composite print	30 x 30	1:100,000	FF 13,300
	colour composite print	60 x 60	1:100,000	FF 13,300
	colour composite print	30 x 30	1:50,000	FF 13,300
SPOT Pan	colour composite transparency 241 x 241 mm*	60 x 60	1:400,000	FF 17,200
	colour composite transparency 241 x 241 mm	30 x 30	1:200,000	FF 17,200
	colour composite transparency 482 x 482 mm	60 x 60	1:200,000	FF 26,100
	colour composite transparency 482 x 482 mm	30 x 30	1:100,000	FF 26,100
	colour composite print	60 x 60	1:200,000	FF 17,200
	colour composite print	30 x 30	1:100,000	FF 17,200
	colour composite print	60 x 60	1:100,000	FF 17,200
	colour composite print	30 x 30	1:50,000	FF 17,200
JERS OPS	paper print, VNIR monochrome	75 x 75	1:200,000	US$ 337
	film positive, VNIR natural colour	75 x 75	1:500,000	US$ 1207
JERS SAR	paper print,	75 x 75	1:200,000	US$ 327
	film positive or negative	75 x 75	1:500,000	US$ 1250
RADARSAT	contact print, standard beam mode	variable	1:500,000	US$ 3,290
	film (24 x 22 cm) monochrome -ve or +ve	variable	1:500,000	US$ 3,260

* All SPOT products processed to either 1A or 1B level.

Table 5.14 Summary of pre-processing. Costs are per full scene.

Satellite	Pre-processing	Details	Cost
Landsat TM	system-corrected	data compressed to digital numbers and basic geometric correction	US$ 4,400
Landsat MSS	system-corrected	data compressed to digital numbers and basic geometric correction	US$ 250
SPOT HRV	1A	raw data undergo radiometric normalisation	FF 13,300
	1B	radiometric normalisation and basic geometric correction (locational accuracy approx. 500 m)	FF 13,300
	2A	radiometric normalisation and geometric correction to required map projection (absolute location accuracy is same as 1B but internal geometry is improved)	quote
	2B	radiometric normalisation and geometric correction using ground control points from maps (root mean square error < 50 m provided suitable maps are available)	quote
	S1	radiometric normalisation and registration of image to a reference image which has undergone 1B processing	quote
	S2	radiometric normalisation and registration of image to a reference image which has undergone level 2 processing	quote
ERS SAR	Precision	standard data product without geometric correction	Ecu 1,200
	Terrain Geocoded	precisely geometrically corrected product based on digital elevation models and ground control points	Ecu 2,300

Table 5.15 Discounted costs for older imagery.

Sensor	Supplier	Acquisition criteria	Cost
SPOT XS	SPOT Image: SPOT album	1986 – end 1989	FF 7,000
SPOT Pan	SPOT Image: SPOT album	1986 – end 1989	FF 9,000
Landsat TM	EROS Data Centre	over 10 years	US$ 750

☞ **Although some geometric correction is conducted during pre-processing, the customer will almost always be responsible for registering images to local map coordinates (see Chapter 6 for methods).**

Discounts for 'old' data

In an effort to sell older imagery, SPOT and Landsat TM data are available at discounted rates for some parts of the world. Discounts apply to data acquired before 1990 (SPOT) or data which is over 10 years old (Landsat TM). Refer to Table 5.15 for costs.

Data media and format

Data media

The past few years have seen a tremendous boom in the development of data storage media. Not surprisingly, this has greatly facilitated the transfer and storage of digital imagery.

Data have conventionally been stored on computer compatible tapes (CCT). CCTs contain 2400 feet of tape and are moderately large (30 cm diameter) and, therefore, difficult to transport. Two recording densities are available; 6250 bits per inch (bpi) and 1600 bpi. Since the latter density requires a greater number of tapes for a given image, most data are recorded at 6250 bpi. CCTs were replaced by smaller high-density 8 mm tapes (e.g. EXABYTE) which hold several gigabytes of data. While most types of imagery are still available on CCT (Chapter 4), many modern institutions are equipped with EXABYTE tape drives and cannot read CCTs.

The development of compact disc technology has led to further improvements in data storage and transfer. SPOT data are routinely disseminated on this medium and data importing procedures are generally simpler than for EXABYTE.

Data format
Multispectral digital data may be stored in a variety of formats. Import of data into image processing software will be unsuccessful unless the correct format is specified. The most widely used formats are:

- *Band Sequential (BSQ)* – each band's data are stored sequentially in a separate image file. For example, a Landsat TM scene would be stored as a number of small files containing image information, and 7 large image files (corresponding to bands 1–7). An end-of-file marker separates each spectral band (see Figure 5.15). Landsat data are usually provided in this format.
- *Band Interleaved by Line (BIL)* – all bands of data are stored in one image file. Scanlines are sequenced by interleaving all image bands. For example, a SPOT XS scene would be held as a single file even though it contains three bands. SPOT data are usually provided in this format. Refer to Figure 5.16.
- *Band Interleaved by Pixel (BIP)* – the values for each band are ordered within a given pixel and pixels are ordered sequentially on each tape. Landsat MSS data are occasionally supplied in this format. Refer to Figure 5.17.

Conclusions

Cloud cover is a serious constraint to the remote sensing of coastal zones in humid regions. When planning a remote sensing survey, it is advisable to request an archive search for satellite images with suitably low cloud cover early on or, if planning an airborne campaign, judge which time of year has the most favourable climatic conditions (e.g. the least cloud and humidity).

Aerial photography is often the most convenient archived source of remotely sensed information of an area of interest. If new photography is required, it is advisable

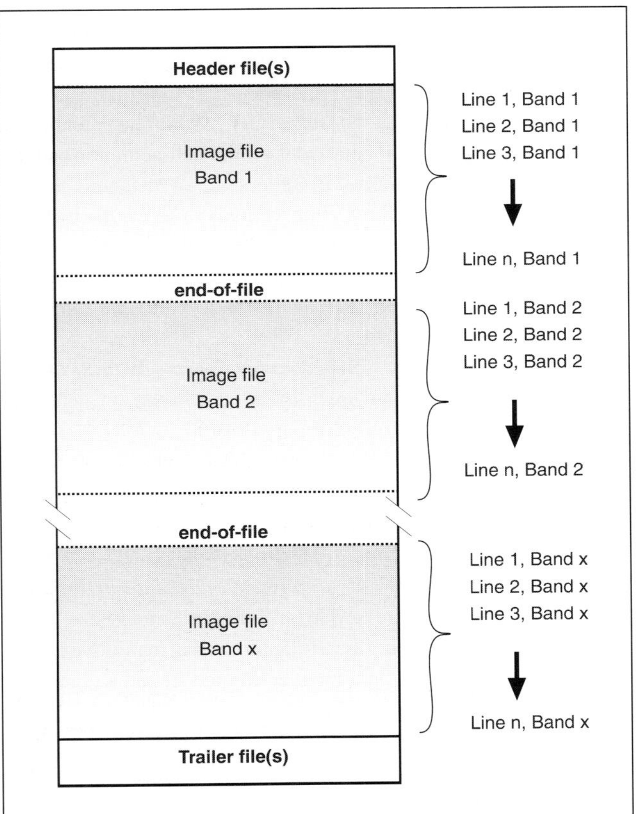

Figure 5.15 Band Sequential (BSQ) storage of digital remotely sensed data showing header file, image files and end-of-file markers for several wavebands (Band 1 to Band x). Data for each band is contained in a separate file. This has the advantage that a single band can be read and viewed easily and that multiple bands can be loaded in any order. Landsat Thematic Mapper data are stored in a type of BSQ format. Based on: Erdas Inc. (1994). *Erdas Field Guide* (Atlanta, Georgia: Erdas Inc.).

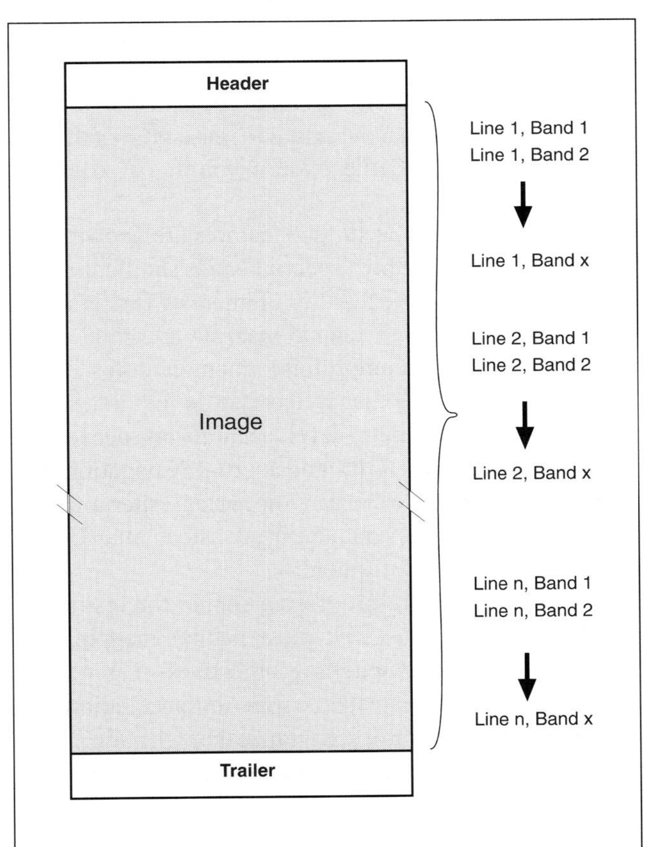

Figure 5.16 Band Interleaved by Line (BIL) storage of digital remotely sensed data. Each record in a BIL file contains one scan line (row) of data for one band. For each scan line, data for all bands (Band 1 to Band x) are stored consecutively within the file. Note: not all BIL format data contain header and trailer files. Based on: Erdas Inc. (1994). *Erdas Field Guide*. (Atlanta, Georgia: Erdas Inc.).

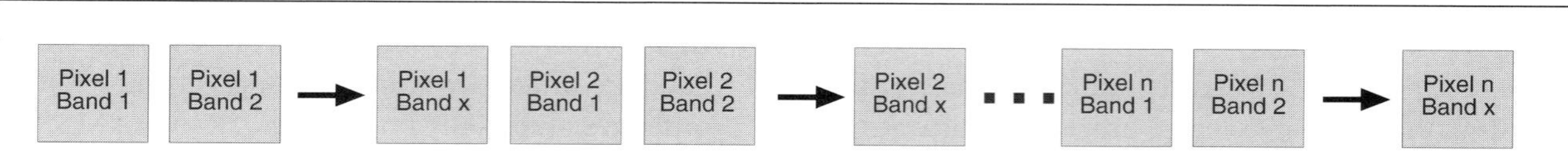

Figure 5.17 Band Interleaved by Pixel (BIP) storage of digital remotely sensed data. For each pixel, data for all bands (Band 1 to Band x) are stored consecutively within the file. Pixels are arranged sequentially in the file or on tape.

to commission a survey aircraft to ensure that the camera points vertically towards the ground, thus reducing oblique viewing and distortion of surface features. The most appropriate combination of photographic scale and film type will differ between survey objectives, but most coastal habitat mapping exercises use true colour film at a scale of 1:10,000 to 1:20,000.

Airborne digital sensors are flexible tools and can present the user with a wealth of instrument configurations (e.g. number of spectral bands, bandwidths, wavelength ranges and pixel size). The most appropriate configuration will depend on the objective of the survey and you should seek the advice of the instrument operator. However, the following criteria should be considered carefully:

1. Although small pixel sizes might be possible, are they necessary given that the signal to noise ratio of the data may decrease and the overall volume of data will increase?
2. If the reflectances of surface features are known (even if only roughly), which spectral bands should theoretically distinguish the features of interest? Bands should be placed at the reflectance maxima and minima for each feature, or a compromise among features.
3. What degree of geometric accuracy is required for the final output? If metre-level accuracy is needed, an aircraft-mounted DGPS and Inertial Navigation System (INS) are advisable. If accuracy criteria can be relaxed, a gyroscope may be used instead of an INS, as this is a cheaper instrument.
4. What degree of pre-processing should the instrument operator undertake? For most habitat mapping and radiative transfer modelling objectives, it is perhaps best to request geometric correction and radiometric correction to at-sensor reflectance (see Chapter 7). Further geometric correction to local maps can usually be undertaken by the consumer.

Satellite images are usually selected from an image archive. However, if suitable data are not available for the area of interest, several sensors can be specifically programmed to acquire data in a limited time frame (e.g. SPOT HRV, ERS and RADARSAT sensors). If the acquisition of recent imagery is not a critical consideration (e.g. for a general purpose map of habitats in a relatively undisturbed area), the cost of image acquisition can be reduced. For example, Landsat TM data which was acquired $\geq$ 10 years ago can be purchased at a discounted rate (roughly 17% of the price of recent imagery).

Photographic prints of satellite imagery are available at similar costs to the original digital data but, since the information content of prints is much less than digital data, their use is only recommended as a 'last resort' where image processing facilities are unavailable.

References

Allan, T.D., 1992, The marine environment. *International Journal of Remote Sensing*, **13**, 1261–1276.

Blake, S.G., Jupp, D.L.B., and Byrne, G.T., 1994, The potential use of high resolution CASI data to determine sedimentation patterns and primary productivity throughout the Whitsunday Islands region, Great Barrier Reef. *Proceedings of the 7th Australasian Remote Sensing Conference*, pp. 43–48.

Ciciarelli, J.A., 1991, *A Practical Guide to Aerial Photography With an Introduction to Surveying* (New York: Van Nostrand Reinhold).

Gastellu-Etchegorry, J-P., 1988, *Remote Sensing With SPOT: An Assessment of SPOT Capability in Indonesia* (Yogyakarta, Indonesia: Gadjah Mada University Press).

Jupp, D.L.B., Mayo, K.K., Kuchler, D.A., Heggen S.J., and Kendall, S.W., 1981, Remote sensing by Landsat as support for management of the Great Barrier Reef. *2nd Australasian Remote Sensing Conference: Papers and Programme* pp. 9.5.1–9.5.6.

Paine, D.P., 1981, *Aerial Photography and Image Interpretation for Resource Management* (New York: John Wiley and Sons Inc.).

Ramsey, E.W., and Jensen, J.R., 1995, Modelling mangrove canopy reflectance by using a light interception model and an optimisation technique. *Wetland and Environmental Applications of GIS* (Boca Raton: CRC Press).

Welby, C.W., 1978, Application of Landsat imagery to shoreline erosion. *Photogrammetric Engineering and Remote Sensing*, **44**, 1173–1177.

6

Geometric Correction of Satellite and Airborne Imagery

Summary *Satellite and airborne images contain a number of geometric distortions that are unavoidable aspects of the data recording procedure and the shape and rotation of the Earth. An uncorrected image will therefore bear a different geometry to that of a map and consequently be hard to use compatibly.*

Geometric correction is the process of correcting these distortions and assigning the properties (and practical value) of a map to an image. The geometry of the image should be warped so that it conforms to that of the required map projection and a coordinate system should be put in place so that the user can either interrogate a point on the image and find its true coordinate, or input a coordinate of a field site and be able find it on the image. Geometric correction is required if:

- *images have to be compared to either field data or maps,*
- *image data have to be compared with other spatial data (e.g. in a GIS),*
- *area or distance estimates are required from the image data,*
- *two images from different dates are to be compared pixel by pixel (e.g. for change analysis).*

For most coastal management applications, geometric correction is essential. The user can pay for various amounts of geometric correction to be performed by the image suppliers, from virtually none to a fully corrected version. As full correction usually involves a final stage of map comparison and/or some GPS fieldwork, it is usual for the user to execute this stage where they have full control over its execution. This also saves money.

Introduction

In the previous chapter the form in which satellite and aerial digital imagery are supplied, their data formats and the pre-processing options available were described. Two important points were made:

Firstly, the raw images are not true and accurate representations of the areas they portray. That is, they are not like a cartographic map where the distances between points on the Earth's surface, relative positions of points, shapes of areas, etc. are accurately represented at a given scale. They are, to a greater or lesser extent, distorted representations. As indicated in the previous chapter (Chapter 5), pre-processing carried out by the image suppliers can remove most of the distortions, but generally some further processing – 'geometric correction' – will be required.

Secondly, the image is composed of pixels whose positions are initially only referenced by their row and column numbers. For the image to be useful in the field, one needs to be able to relate these row and column coordinates to actual positions on the ground or sea surface (and vice versa), be these latitude and longitudes on nautical charts, grid references on maps, or Universal Transverse Mercator (UTM) coordinates read off a Global Positioning System (GPS).

Furthermore, the image contains errors in its geometry, which means that distance, area, direction and shape properties vary across the image. These errors originate during the process of data collection and vary in type and magnitude depending on the sensor used (see sources of geometric error). It is therefore impossible to perform accurately simple calculations, such as the area covered by a particular habitat. The magnitude of these errors will vary from image to image – if two images are to be compared (e.g. for change detection analysis) they must both be geometrically corrected. Equally as important, wider use of remotely sensed data can be made by incorporating it with other information (e.g. political, legal, socio-economic, historical, etc.) in a Geographical Information System (GIS). To do so also requires that the remotely sensed data have the same coordinate system as the other components in the GIS. Geometric correction is the process of correcting the distortions in an image and transforming it so that it has the properties of a map. This allows the coordinates of each point on the image to be assigned coordinates according to a map projection chosen by the operator (see map projections).

☞ **Geometric correction is usually necessary but not always. If the purpose of the study is to estimate relative amounts of land cover, for example, precise positional information may not be required and this lengthy processing step can be omitted.**

It is possible to pay the suppliers of remotely sensed imagery to perform various levels of geometric correction – the different options available are presented in Table 5.14. For example, Landsat images would normally be obtained 'system corrected', whereas SPOT images would normally be obtained with Level 1B correction (see Table 5.14). However, full correction such as SPOT Level 2B is seldom used because many users prefer to avoid both the cost of this service and to maintain control over the geometric correction themselves. If ground control points are needed, suppliers will normally require the user to provide them.

The mathematics of geometric correction are complex and have been the subject of many excellent papers and textbooks (e.g. Mather 1987: Chapter 4). The present chapter deliberately concentrates on the practical aspects of geometric correction. Selected theoretical papers are referred to throughout the text; readers wanting a clear and general explanation of the mathematics are referred to textbooks or the ERDAS Imagine Field Guide.

Sources of geometric error

Remotely sensed images may have distorted geometry for a variety of reasons. These distortions can have one or more causes and vary in magnitude (and hence importance) depending on many factors such as whether the sensor is carried on a satellite or an aircraft, whether the sensor was near-vertical or oblique during data collection, etc.:

- *Panoramic distortion*: the image is a two-dimensional projection of a three-dimensional surface (the Earth's curved surface). Panoramic distortion is the greatest source of geometric error in imagery acquired from sensors with large angular fields of view (satellites rather than airborne sensors).
- *Orientation*: satellite images are unlikely to be oriented towards true north. The orientation of aerial images is determined by the heading of the aircraft and hence the order in which image lines were scanned or photographs taken.
- *Earth's rotation*: Landsat crosses the equator with a heading of about 188° and the earth is rotating eastwards underneath it. As a result the start of each scan line is slightly to the west of its predecessor. The overall effect is to skew the image. The speed of rotation of the Earth's surface varies with latitude being faster at the equator than at higher latitudes, and so the amount of skew is greater. The effect of the Earth's rotation is negligible for airborne sensors.
- *Instrument error*: this is caused by distortions in the optical system, non-linearity of the scanning mechanism and non-uniform sampling rates. It is easier to measure and correct in airborne instruments than for sensors mounted on satellites.
- *Platform instability*: the altitude and attitude of satellites are subject to variation, albeit quite small. By contrast, the roll, pitch and yaw of an aircraft can severely distort aerial images, and are greatly influenced by pilot skill and weather conditions.

☞ **Correcting for roll, pitch and yaw is achieved using specialised software packages specific to the sensor used. This software is not discussed here, though readers are advised to ensure that these corrections are part of the package provided by suppliers of aerial imagery.**

Choosing a map projection, spheroid and datum

Map projection

A map projection is the means by which the three-dimensional surface of the Earth is drawn on a two-dimensional surface (piece of paper or computer screen). At first acquaintance, the topic may appear bewildering to the newcomer. Butler et al. (1987) provide an excellent introduction to marine cartography and is recommended to those unfamiliar with map projections. Not only is there a multitude of different kinds of projections (more than 200 are listed in Bugayevskiy and Snyder (1995) alone, and see Table 6.1) but they are covered by an enormous literature of a predominantly mathematical and technical nature.

Fortunately, in practice, the use of map projections is simpler than the theory behind them. All projections however are similar in that:

Table 6.1 Twenty map projections and their mode of construction, major property and common use

Map projection	Construction	Property	Use
Geographic	n/a	n/a	Data entry, spherical coordinates
Universal Transverse Mercator	Cylinder	Conformal	Data entry, plane coordinates
Polar Stereographic	Plane	Conformal	Polar regions
Oblique Mercator	Cylinder	Conformal	Oblique expanses
Space Oblique Mercator	Cylinder	Conformal	Mapping of Landsat imagery
Modified Transverse Mercator	Cylinder	Conformal	Alaska
Transverse Mercator	Cylinder	Conformal	N–S expanses
Stereographic	Plane	Conformal	Hemispheres, continents
Azimuthal Equidistant	Plane	Equidistant	Polar regions, radio/seismic work
Equidistant Conic	Cone	Equidistant	Middle latitudes, E–W expanses
Gnomonic	Plane	Compromise	Navigation, seismic work
Orthographic	Plane	Compromise	Globes, pictorial
Equirectangular	Cylinder	Compromise	City maps, computer plotting
Miller Cylindrical	Cylinder	Compromise	World maps
Van der Crinten I	n/a	Compromise	World maps
Polyconic	Cone	Compromise	N–S expanses
Lambert Azimuthal Equal Area	Plane	Equivalent	Square or round expanses
Albers Conical Equal Area	Cone	Equivalent	Middle latitudes, E–W expanses
Lambert Conformal Conic	Cone	True direction	Middle latitudes, E–W expanses
Mercator	Cylinder	True direction	Non-polar regions

n/a = not applicable

(i) they transfer features of the Earth's surface to a flat surface which is either a plane, a cone or a cylinder (the latter two may be cut and laid flat without stretching), and
(ii) they inevitably create some error or distortion in transforming a spherical surface to a flat surface.

Ideally, a distortion-free map would have four major properties:

- *Conformality*: the characteristic of true shape. In other words, the shape of any small area is not distorted by the projection.
- *Equivalence*: the characteristic of equal area. In other words, areas on one portion of the map are in scale with areas in any other portion.
- *Equidistance*: the characteristic of true distance measuring. In other words, the scale of distance is constant over the entire map.
- *True direction*: this means that a line connecting two points on the map crosses reference lines (e.g. meridians) at a constant angle.

In practice, no projection is true in all these properties. Each projection is designed to be true in selected properties, at the expense of error in the others or, most commonly, is a compromise among all. Another common feature of map projections is that they require a point of reference on the Earth's surface. Position with respect to this point of reference is expressed by means of coordinates, which in turn are defined in one of two systems.

- *Geographical (spherical) coordinates* are based on a network of lines of latitude (called parallels) and longitude (called meridians). The prime meridian, 0° longitude, passes through Greenwich, England. Latitude and longitude are defined by reference to the point where the prime meridian intersects with the equator.
- *Planar (Cartesian) coordinates* are defined by a column and row position on a planar grid. The origin of a planar system is typically located south and west of the origin of the projection. Coordinates increase going north and east. The origin of the projection is defined by coordinates referred to as the false easting and false northing. UTM is a planar projection.

The desired major property and the region to be mapped (because some are designed only for specific regions) must both be considered when choosing a projection. A comprehensive treatment of the selection of an appropriate map projection would be a very lengthy task; readers requiring detailed information are referred to the work of Lee and Walsh (1984), Maling (1992) and Snyder (1994).

In summary, several factors must be weighed against each other, depending on the uses for your geometrically corrected image:

- *Type of map*: habitat map, navigational chart?
- *Special properties to be preserved*: area, distance, shape?
- *Type of data to be mapped*: road distances, ground cover?
- *Map accuracy*: what is acceptable?
- *Scale*: how big an area?

If you are mapping a relatively small area the decision about map projection is much less important than if you are mapping large areas (countries, continents) because in small areas the amount of distortion in a particular projection may be barely discernible. Opinion may differ but the following act as general guidelines:

- If the country to be mapped is in the tropics, use a cylindrical projection.
- If it is important to maintain proper proportions, data should be displayed using an equal area projection (however, shape may be sacrificed).
- Equal areas projections are well suited to thematic data.
- Where shape is important use a conformal projection.

Further discussion of decision guidelines can be found in Pearson (1990) and Maling (1992).

In practice it is often most convenient to use the same map projection as the most widely used maps of the image area. The projection used in the geometric correction of the Turks and Caicos Islands imagery was Universal Transverse Mercator. South Caicos is in UTM grid zone 19: coordinates are expressed as a six digit 'false' easting followed by a seven digit 'false' northing (e.g. 237118, 2378993). The easting is given with respect to the central meridian of each 6° UTM grid zone (in this zone, 69° west; see *ERDAS Field Guide* for a table of UTM zones), which is given the arbitrary value of 500,000 m east whilst the northing is, in the northern hemisphere, the number of metres north of the Equator. Thus, the coordinates given, represent a point 500,000 - 237,118 m = 262.882 km west of the 69° meridian and 2,378.993 km north of the equator. An important reason for choosing this projection for our imagery was that UTM was the projection used in the maps of the islands (Ordnance Survey maps Series E8112 DOS 309P). Positions on the image could therefore be easily related to map position and vice versa.

Spheroid

Having selected a map projection one needs an estimation of the Earth's shape. Up to now we have assumed that the Earth is a sphere. However the Earth more closely resembles an ellipsoid (also called an oblate spheroid, hence the term map spheroid), with the equator as its semi-major axis. Several ellipsoids/spheroids are available for use (Table 6.2). They differ primarily in that they use dimensions to calculate the ellipsoid which are appropriate for particular regions of the Earth's surface.

The spheroid used in the geometric correction of the Turks and Caicos Islands imagery was Clarke 1866. This was the spheroid used in the maps of the islands (Ordnance Survey maps Series E8112 DOS 309P).

Table 6.2 Examples of some of the spheroids in use around the world

Spheroid	Area of use
Clarke 1866	North and Central America, Caribbean and the Philippines
Clarke 1880	France and Africa
Bessel 1841	Indonesia
Everest 1830	India
Australian National	Australia
Helmert	Egypt

Datum

The Earth is not a smooth uniform shape like an ellipsoid. Its surface bulges in some places and has depressions in others. A geodetic datum is a smooth mathematical surface that closely fits the mean sea-level surface throughout an area of interest, within the area of use of a spheroid. Not surprisingly, there are many datums in use. By qualifying the map coordinates to a particular datum some compensation can be made for irregularities in the Earth's surface. The datum is defined by a spheroid; as a result only certain datums can be used with any particular spheroid. The datums available for use with Clarke 1866 are listed in Table 6.3. Once coordinates have been defined by a particular datum they are unique.

The datum used in the geometric correction of the Turks and Caicos Islands imagery was NAD 27 (Bahamas). This was the datum used in the maps of the islands (Ordnance Survey maps Series E8112 DOS 309P).

☞ **Care must be taken: coordinates shift between different datums. If the area covered by the image lies across two different datums, only one must be used to define the coordinates. Using two different datums in one image can cause errors of many hundreds of metres.**

How to perform a geometric correction

To geometrically correct your original image (called the input or source image) it is necessary to carry out the following steps:

- Collect a set of Ground Control Points (GCPs).
- Use GCPs to solve (calculate the coefficients of) a polynomial equation.
- Use this equation to transform the geometry of image data from the original file system to a selected map projection. This process corrects for the distortion in the input image and is termed 'rectification'.

Table 6.3 Some of the datums that are available for use with the Clarke 1866 spheroid and the areas in which they are used

Datum name	Area applied
Bermuda 1957	Bermuda
Cape Canaveral	Bahamas, Florida
Guam 1963	Guam
L. C. 5 Astro 1961	Cayman Brac Island
Luzon	Philippines (excluding Mindanao)
Luzon (Mindanao)	Philippines (Mindanao)
NAD27 (CONUS) (MRE)	Continental US land areas
NAD27 (East CONUS)	East of Mississippi River including Louisiana, Missouri, Minnesota
NAD27 (West CONUS)	Mean for CONUS (West of Mississippi River)
NAD27 (Bahamas)	Bahamas (Except San Salvador Island, and including the Turks and Caicos Islands)
NAD27 (San Salvador Island)	Bahamas (San Salvador Island)
NAD27 (Canada)	Mean for Canada
NAD27 (Canada) (MRE)	Continental Canada land areas
NAD27 (Canada_AB)	Canada (Alberta, British Columbia)
NAD27 (Canada_MO)	Canada (Manitoba, Ontario)
NAD27 (Canada_NNNQ)	Canada (New Brunswick, Newfoundland, Nova Scotia, Quebec)
NAD27 (Canada_NS)	Canada (Northwest Territories, Saskatchewan)
NAD27 (East Central America)	Antigua and Barbuda, Barbados, Cuba, Dominican Republic, Grand Cayman, Jamaica, Turks and Caicos Islands
NAD27 (West Central America)	Mean for Belize, Costa Rica, El Salvador, Guatemala, Honduras, Nicaragua
NAD27 (Canal Zone)	Canal Zone
NAD27 (Cuba)	Cuba
NAD27 (Greenland)	Greenland (Hayes Peninsula)
NAD27 (Mexico)	Mexico
Old Hawaiian (Hawaii)	Hawaii
Puerto Rico	Puerto Rico, Virgin Islands

- Assess the error that remains in the image.
- Assign data values (DNs) to the pixels on the new grid of the transformed image by extrapolation from the original file. This process is known as 'resampling'.
  The end result is a file called the 'output' or 'reference' image which (i) consists of pixels whose position within the image can either be expressed in terms of row and column, or in the units of a selected map projection, and (ii) which has had the distortion present in the input image corrected to within acceptable levels. Figure 6.1 diagrammatically outlines the process.

1. Collect a set of ground control points (GCPs)

What are GCPs?

GCPs are specific positions which consist of two pairs of (known) coordinates, 'reference' and 'source' coordinates. The reference coordinates of a GCP (x and y) may be UTM eastings and northings for example. The source coordinates of a GCP are data file coordinates in the image being rectified, in other words the column and row numbers (c and r). The nature of the distortion present within the image is calculated and then corrected by comparing the differences between the reference and source coordinates of the GCPs.

Factors to consider when choosing GCPs

GCPs are the locations of ground features which can be unmistakably identified on the image and a map. The suitability of features as GCPs depends largely on the spatial resolution of the imagery. For coarse imagery (e.g. MSS) the tips of headlands or very small offshore islands may be sufficient. For aerial multispectral imagery (e.g. CASI) the ends of jetties or corners of buildings make good GCPs (Figure 6.2, Plate 6). Objects which may change, such as patches of vegetation or small water bodies, are unsuitable as GCPs.

☞ **Due to the nature of coastlines, you often end up with a series of GCPs in a line along a coast, but few GCPs inland and even fewer out to sea. This is likely to result in poor geometric correction of inland and marine parts of an image. Whilst recognising that GCPs are difficult to fix on reefs unless they have mappable features such as small islands, we recommend that, if at all possible, practitioners attempt to get GCPs distributed across the whole image (Plate 5) rather than confined to a belt parallel to the shore. This will significantly improve geometric correction.**

The source coordinates are obtained from the image using the image processing software: this usually involves placing

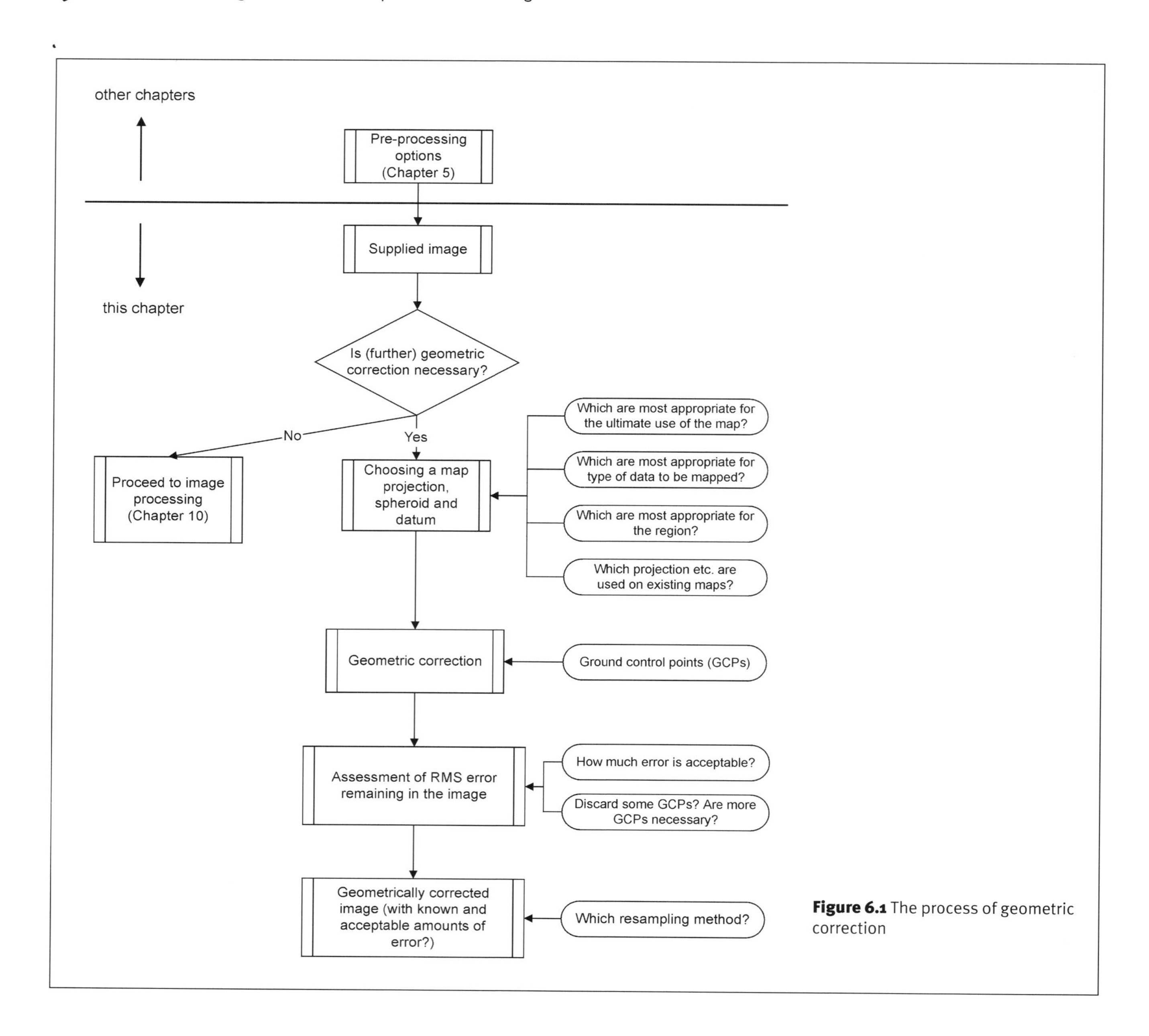

Figure 6.1 The process of geometric correction

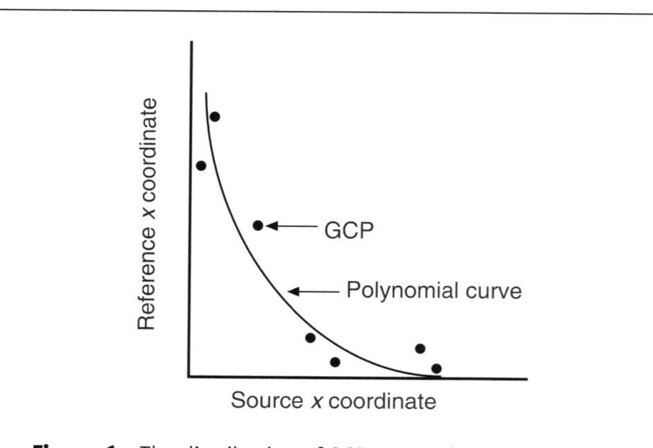

Figure 6.3 The distribution of GCPs around a curve which has been fitted through them and is described by a polynomial equation. For simplicity's sake the reference coordinates have been plotted against the source coordinates in only the *x* direction. Redrawn from: Erdas Inc. (1994). *Erdas Field Guide* (Atlanta, Georgia: Erdas Inc.).

the cursor precisely at the desired location and reading the column and row numbers from the display. The reference coordinates can be obtained in two ways.

(i) They can be read directly from a map, if accurate maps of the area exist at a suitable scale. Obviously the application of remote sensing would be severely restricted if images could only be geometrically corrected in areas which had already been mapped conventionally. Fortunately the second method can be used in unmapped areas.

(ii) Each GCP is visited and its position fixed using a Global Positioning System (GPS). Here, we need only to note that positional information can be obtained using a GPS. For a discussion of the methodological and cost implications of using GPS the reader should refer to the GPS section in Chapter 4.

BOX 6.1

The mathematics of geometric correction — polynomials

Figure 6.3 plots the reference x coordinate against the source *x* coordinate, for a sample set of GCPs. Note: this example is based just on the *x* coordinate for simplicity's sake, though each GCP consists of two coordinates (*x, y*). Geometric correction uses the least squares regression method to calculate the relationship between the reference and source coordinates, which can then be applied to the whole image. This is achieved using polynomial equations, which have the form:

$$A + Bx + Cx^2 + Dx^3 ...... \alpha x^n$$

where A, B, C, D .... α are coefficients and n is the order of the polynomial.

Figure 6.4a shows the reference *x* coordinate regressed against the source *x* coordinate producing a first order polynomial equation. Likewise Figure 6.4b produces a second order polynomial equation and Figure 6.4c a third. Polynomials expressing the column coordinates, *c*, in terms of *x* and *y* would be

First-order polynomial:

$$c = A + (Bx + Cy)^1$$

Second-order polynomial:

$$c = A + (Bx + Cy)^1 + (Dx + Ey)^2$$

Third-order polynomial:

$$c = A + (Bx + Cy)^1 + (Dx + Ey)^2 + (Fx + Gy)^3$$

Similar equations expressing the row coordinates, *r*, in terms of *x* and *y* can be written (also *x* in terms of *c* and *r*, *y* in terms of *c* and *r*). The variables in these equations (*x, y, c, r*) are known for a set of points (the GCPs), so the coefficients can be calculated by the computer, using the least squares regression method to solve the simultaneous equations. Once the coefficients have been calculated they are stored in a transformation matrix and can be used to derive map coordinates from any pair of file coordinates in the source image.

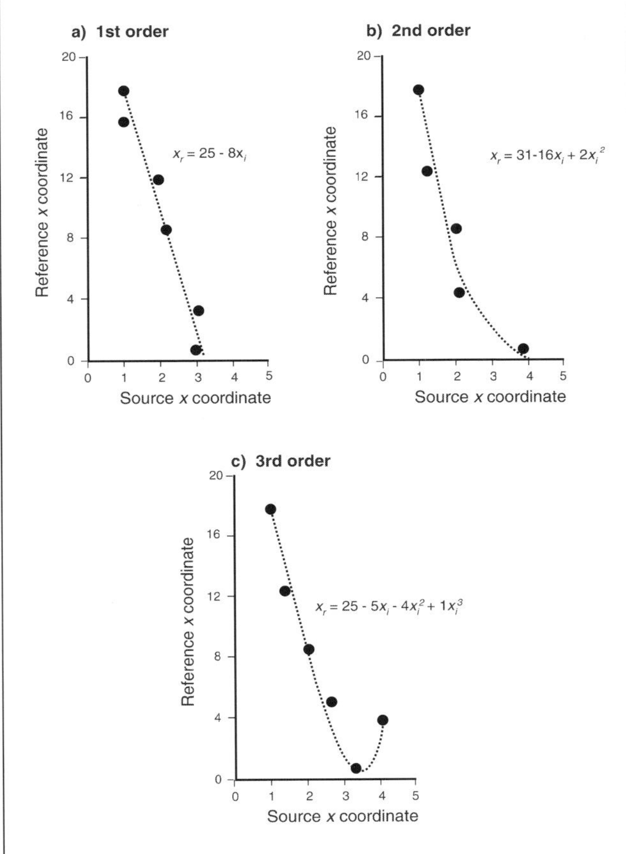

Figure 6.4 Examples of transformations employing first- to third-order polynomials: a) A first-order transformation, b) a second-order transformation, and c) a third-order transformation. Redrawn from: Erdas Inc. (1994). *Erdas Field Guide* (Atlanta, Georgia: Erdas Inc.).

☞ **Standalone GPS cannot be used to collect GCPs for high resolution imagery such as CASI. DGPS is necessary.**

If one image has been geometrically corrected and covers the same area as a second image, then features identifiable in both may be used as GCPs. 'Registration' is the process whereby one image is made to conform to another. If a range of images are to be used it is usual practice to geometrically correct the first one and register the others to it.

2. Use the GCPs to solve a polynomial equation

The row and column coordinates of each pixel have to be mathematically transformed into the coordinates of the chosen map projection. This is done using a transformation matrix which is derived from the GCPs. A transformation matrix contains coefficients calculated from a polynomial equation (the computations will be performed using the image processing software).

The order of a polynomial (Box 6.1) is an expression of the degree of complexity of the equation – generally only first- to third-order polynomial equations are used in geometric correction. The operator should have the ability to select the order of the polynomial to be used.

Factors to consider when choosing a polynomial equation

The correct order polynomial to use depends on:

(i) *The number of GCPs used.* In mathematical terms, 3 GCPs are needed for a first order polynomial, 6 for a second and 10 for a third. This is because the minimum number of points required to perform a transformation of order *t* is $(t+1) \times (t+2)/2$. However, these are just the minimum number of points needed to ensure that it is mathematically possible to compute the coefficients from the equations.

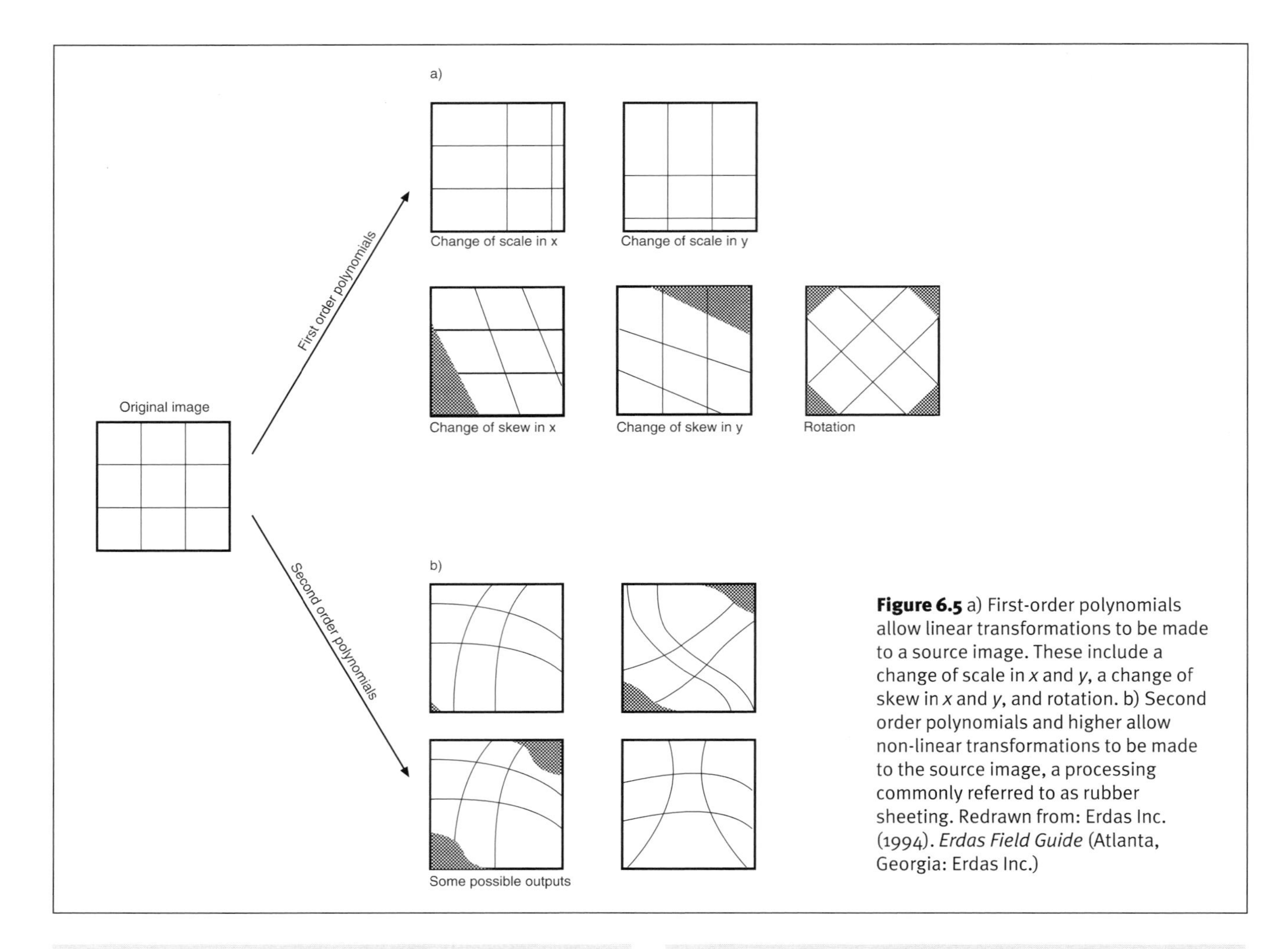

Figure 6.5 a) First-order polynomials allow linear transformations to be made to a source image. These include a change of scale in *x* and *y*, a change of skew in *x* and *y*, and rotation. b) Second order polynomials and higher allow non-linear transformations to be made to the source image, a processing commonly referred to as rubber sheeting. Redrawn from: Erdas Inc. (1994). *Erdas Field Guide* (Atlanta, Georgia: Erdas Inc.)

Statistical texts suggest that 30 GCPs are required but some remote sensing texts suggest that 10–15 give acceptable results. Using these figures as guidelines it is the job of the operator to decide when sufficient points have been used, by studying the x and y residuals of individual points, the total Root Mean Square (RMS) error of all GCPs and their spread (see below).

(ii) *The spatial distribution of GCPs.* It is inappropriate to use a second-order polynomial if a large proportion of GCPs are along a transect (e.g. junctions along a straight road, headlands along a coast) because they only provide information in one dimension, irrespective of how many GCPs are available. A first-order polynomial must be used instead.

(iii) *The distortion in the imagery.* A first-order polynomial will allow a linear transformation of the data. In other words, the transformation allows changes to be made to location in *x* and/or *y*, scale in *x* and/or *y*, skew in *x* and/or *y*, and rotation (Figure 6.5a). This can be used to transform raw imagery to a planar map projection, convert a planar map projection to another planar map projection and to correct small areas (where the curvature of the Earth can be assumed to be small). Second-order and higher polynomials are required to perform more complex, non-linear, image rectifications. This transformation process is commonly known as 'rubber sheeting' (Figure 6.5b). As such, they are generally more widely used. Non-linear changes include the transformation of latitude and longitude data to a planar projection such as UTM and the correction of the effects of the Earth's curvature over large areas. Second-order polynomials are recommended for use with satellite images. Third-order polynomials are used on aerial photographs and with radar imagery.

☞ **Increasing the order of the polynomial used will improve the fit of the GCPs to the polynomial. However caution must be exercised not to use too complex a transformation. It is easy to cause unwanted distortions in other parts of the output image for the sake of obtaining the best possible fit (a minimum RMS error, see below). In practice most transformations are first or second order. The distortions created by higher transformations render the results increasingly less predictable and regular.**

3. Transform the image to the geometry of the chosen map projection

Following the selection of an appropriate map projection, the computer uses the transformation matrix to calculate a new grid system. This is done by inputting the row and column

BOX 6.2

The mathematics of geometric correction — Root Mean Square error.

The inverse of the transformation matrix is used to retransform the reference coordinates of the GCPs back to the source coordinate system. Unless there is a perfect fit of the GCPs to the curve described by the polynomial then some discrepancy will exist between the original source coordinates and the retransformed coordinates. The distance between the GCP original source and retransformed coordinates is called the Root Mean Square (RMS) error.

$$RMS = \sqrt{(x_r - x_o)^2 + (y_r - y_o)^2}$$

RMS error can be calculated using Pythagoras' theorem

where x_0 and y_0 = original GCP source coordinates, x_r and y_r = retransformed coordinates. The units of RMS are the units of the source coordinate system, pixel widths for data file coordinates. Figure 6.6 illustrates that RMS error has a component in the *x* direction (the *x* residual) and one in the *y* (the *y* residual). Residuals are the distance between the source and retransformed coordinates in one direction. The *x* residual is the distance in the *x* direction between original and retransformed *x* coordinate of a GCP, or $x_r - x_0$. Likewise, the *y* residual. If the residuals are consistently out in one direction then more GCPs points should be added in that direction. The residuals can be used to calculate the total RMS and RMS in the *x* and *y* directions for any set of GCPs.

$$R_x = \sqrt{\frac{1}{n}\sum_{i=1}^{n} xr_i^2}$$

$$R_y = \sqrt{\frac{1}{n}\sum_{i=1}^{n} yr_i^2}$$

$$T = \sqrt{R_x^2 + R_y^2}$$

$$T = \sqrt{\frac{1}{n}\sum_{i=1}^{n} xr_i^2 + yr_i^2}$$

or where R_x = *x* RMS, R_y = *y* RMS, *T* = total RMS, *n* = number of GCPs, *i* = GCP number, xr_i = *x* residual for GCP *i*, yr_i = *y* residual for GCP *i*.

coordinates of each pixel of the input image into the polynomial equation (whose coefficients are now known). The result is a set of output (map) coordinates for each pixel.

☞ **The output image will have the same map projection as the GCP reference coordinates. For example, the GPS which we used in the Turks and Caicos produced coordinates in UTM (zone 19), using spheroid Clark 1886 and datum NAD 27 (Bahamas). Images geometrically corrected with GCPs collected with this GPS must therefore be set to the same map projection. If another map projection is required this must be changed afterwards. This is quite a straightforward operation in a good image processing package.**

4. Assess the error that remains in the image

Much of the geometric error in the original image is corrected by rectification, but not all of it. Polynomial equations are selected to minimise error during the coordinate conversion but an error free conversion is practically impossible.

Each GCP influences the value of the coefficients in the transformation matrix. A 'bad' GCP will introduce large amounts of error. It is possible to calculate the total error being introduced using a transformation matrix derived from a particular set of GCPs, and the amount of error each GCP is responsible for (in both *x* and *y* directions). The set of GCPs can therefore be edited and it is the responsibility of the operator to decide how much error is acceptable.

Root Mean Square (RMS) error

The accuracy of the polynomial transformation is assessed by calculating the Root Mean Square (RMS) error for each GCP. RMS is evaluated both in the *x* and *y* directions (Box 6.2). A high RMS error in both directions indicates an unreliable GCP. This can be a result of choosing an inappropriate GCP (see *Factors to consider when choosing GCPs*, above), error in position fixing on the ground (perhaps GPS error), errors in reading coordinates from a map, or errors in fixing the position on the image using a cursor. It may be possible to check and correct for these errors. If not, then the GCP in question should be discarded. If GCPs have consistently high residuals in the same direction, either x or y, more GCPs should be added in that direction. This is particularly common in imagery taken off-nadir, an option that is available with SPOT and IRS data (see Chapter 3).

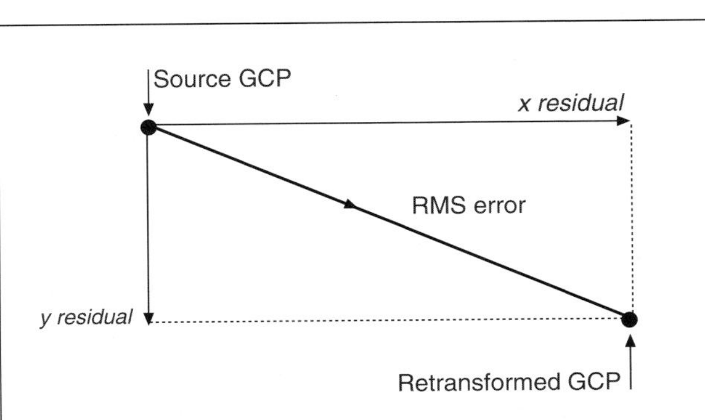

Figure 6.6 The relationship between residuals and the RMS error of a single GCP. Redrawn from: Erdas Inc. (1994). *Erdas Field Guide* (Atlanta, Georgia: Erdas Inc.)

Factors to consider when assessing the accuracy of the geometric correction

GCPs should be checked, unreliable ones discarded, and more added until the total RMS error reaches an acceptable level. In deciding the acceptable level of RMS error the following must be considered carefully:

(i) The end use of the imagery.
(ii) The type of data being used. Is it high or low spatial resolution?
(iii) The accuracy of the GCP's reference coordinates. Is it greater or less than the RMS?
(iv) The positional accuracy of ancillary (field) data to be used. For example, if ground-truthing is to be conducted with a non-differential GPS then there is little point in expending lots of effort to achieve a RMS error equivalent to less than 40–100m.

When assessing RMS error remember that its units are the units of the source coordinate system, i.e. pixel widths. Therefore if an accuracy of 40 m is required from a rectification of Landsat MSS imagery then total RMS error must not exceed 0.5. Phrased the other way round, with a total RMS of 0.5, all points within a radius of 0.5 pixels are considered to be correct (that is, close enough to use).

☞ **As a rule of thumb a RMS error of less than a single pixel is desirable for satellite imagery. Note that the RMS error is merely a measure of the goodness of fit of GCPs to the polynomial used, and should therefore only be used as a general indication of the accuracy of the geometric correction. For a rigorous assessment of the accuracy, further GCPs, not used in the transformation, are required. This step is frequently omitted for practical reasons.**

5. Resample the image

Once the GCPs have been edited, they are used to derive a transformation matrix which will geometrically correct the image to a suitable level of accuracy. The first step is to assign a grid to the image with map coordinates, and blank (zero) data values for all pixels. The next step is to assign a data value (brightness level, digital number) to each pixel – but this is not a simple operation (because the geometry of the input image has now been changed: it is not therefore possible to overlay images so that each pixel in the output image is covered by a single pixel in the input image). In other words, once geometric correction has been performed, there is no direct relationship between input and output pixels (see Figure 6.7). Data values cannot therefore be transferred directly from the source to the rectified image and a method of interpolation is needed.

Three resampling methods are used to achieve this (see Atkinson 1985 for the mathematical theory behind these resampling techniques).

- *Nearest Neighbour*: The digital number of the pixel that is closest to the retransformed coordinates becomes the new data value of the pixel in the output image. This is the least computationally intensive method.
- *Bilinear Interpolation*: The data value of the pixel in the output image is interpolated from the four nearest pixels to the retransformed coordinates, using a linear function.
- *Cubic Convolution*: The data value of the pixel in the output image is interpolated from a 4 x 4 grid compromising the 16 surrounding pixels, using a linear function. This is the most computationally intensive method.

Factors to consider when choosing resampling methods

The nearest neighbour method transfers original data values without averaging them, and as such can be performed before radiometric correction and classification. Changes in data values across boundaries are preserved and not smoothed out (as in bilinear interpolation and cubic convolution). This is an important consideration if the data is to be used for discriminating boundaries of habitats.

Bilinear interpolation should be used when changing the cell size of the data, e.g. on merging SPOT and TM scenes, and produces smoother, less 'blocky' images.

Cubic convolution is recommended when the cell size of the data is being dramatically reduced such as in merges between aerial photographs and TM data.

Special considerations for the geometric correction of airborne digital data

The theory behind the geometric correction of airborne imagery will be discussed in relation to CASI and is basically similar to the processes applied to satellite imagery which are described above. However, there are two differences:

1. CASI is an airborne sensor, and so the data is affected by the roll, pitch and yaw of the aircraft which occurs during normal data collection (such movement is much greater for an aircraft than a satellite). The amount of roll, pitch and yaw must be measured concurrently with the collection of CASI data, to correct for these effects during post-processing.
2. The position of the aircraft can be measured in-flight. This positional information is synchronised with the CASI scan lines and recorded directly onto tape. The need to collect GCPs is therefore, in theory, negated.

☞ **This second point is an important feature of CASI for mapping submerged habitats away from the shore. In such**

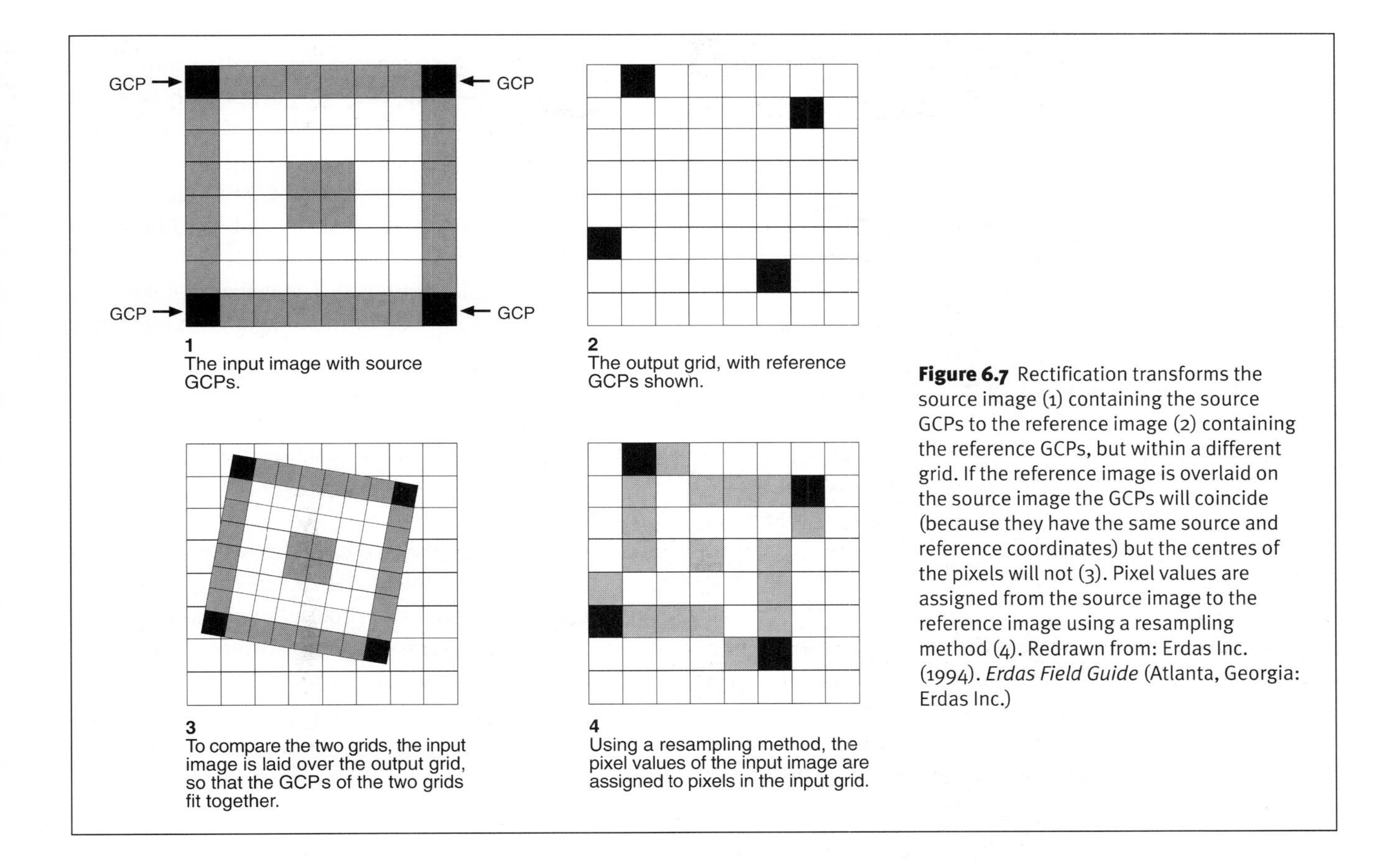

Figure 6.7 Rectification transforms the source image (1) containing the source GCPs to the reference image (2) containing the reference GCPs, but within a different grid. If the reference image is overlaid on the source image the GCPs will coincide (because they have the same source and reference coordinates) but the centres of the pixels will not (3). Pixel values are assigned from the source image to the reference image using a resampling method (4). Redrawn from: Erdas Inc. (1994). *Erdas Field Guide* (Atlanta, Georgia: Erdas Inc.)

cases, features which can be identified on a map are frequently scarce and the placement of markers to coincide with an overflight is generally impractical.

Table 6.4 lists the instruments and software options that are available for the geometric correction of CASI data, and the level of accuracy it is possible to obtain with them. The method chosen not only determines the accuracy of the geometric correction process but also the equipment to be purchased or hired, the type of aircraft to be used and hence the cost of the project.

Roll and pitch measurement

CASI gyro system

The CASI system can be configured with a standard aircraft gyro which measures the amount of roll and pitch. The gyro is integrated into the CASI system and the information from it is recorded directly on to a data tape so that it is synchronised with specific scan lines.

Advantages: A gyro is lightweight, easily transportable and relatively inexpensive. There is no separate cost for the gyro because it is included as part of the standard CASI package.

Disadvantages: A gyro only records approximately nine measurements of roll and pitch per second. CASI on the other hand records as many as 40 scan lines per second. Roll and pitch data must therefore be interpolated between scan lines. Gyro measurements are comparatively coarse, in the order of 0.1 degrees of arc.

Inertial Navigation System (INS) and Inertial Measurement Unit (IMU)

Both INSs and IMUs measure aircraft roll and pitch. IMUs are devices more commonly used on marine survey vessels to compensate for vessel motion. Information from an INS can be recorded directly onto the data tape and synchronised with specific scan lines. This is not possible with an IMU.

Advantages: High recording rate – both INSs and IMUs make approximately 200 measurements per second. High accuracy – both INSs and IMUs measure roll and pitch to an accuracy of about 0.03 degrees of arc. IMUs are readily available for rental.

Disadvantages: It is difficult to locate INSs for rental purposes. They are normally installed on larger passenger aircraft and used for navigational purposes (the reason for which they were developed). They are sometimes found on smaller aircraft that are equipped for very precise survey work – generally twin turbo-propeller aircraft. Both are expensive to either rent or purchase, and therefore may only be suitable for large surveys. For example, the rental rate for an IMU will normally be based on a percentage of the purchase price, approximately 15% for the first month and 10% for subsequent months. Although there are different versions of IMUs, typical rental charges could be in the range £7,500–£10,000 for the first month and £5,150–£6,500 for subsequent months.

Roll, pitch and yaw correction

This is achieved using specialised software packages specific to CASI and should be carried out by the suppliers of the CASI data. The software may be developed in-house by suppliers, in which case details of the process will probably not be provided, though readers are advised to ask for details of the accuracy of this correction.

Position measurement

A gyro system does not measure the position of the aircraft, whereas INS and IMU do.

INS and IMU
Higher accuracies of geometric correction can be obtained by employing an INS or IMU. In addition to roll and pitch, they also record aircraft heading and azimuth, and do so very accurately. Just as roll and pitch data, the heading and azimuth data are recorded many more times per second than the actual CASI data, so extrapolation between scan lines is unnecessary. *Advantages* and *Disadvantages*: as above.

Airborne GPS
The CASI system is designed to record GPS data during image collection. A survey quality GPS receiver is mounted in the aircraft and the GPS data is recorded directly onto the tape and synchronised with the scan line data.

☞ **The base station data must be recorded at the same time interval (normally 1 per second) as the airborne GPS receiver.**

Advantage: Relatively easy to make this installation and to record the data, and GPS receivers are relatively cheap to hire or buy (in the region of £50–£70 per day to hire).

Disadvantages: GPS receivers are usually set to record at a rate of once per second. They are not normally set faster because the amount of data to be collected and written to the tape drive would be prohibitive. A recording rate of once per second contrasts severely with the CASI imager which records as many as 40 scan lines per second. In addition, the aircraft travels at typical operational speeds of 45–60 $m.s^{-1}$ (90–120 nautical miles per hour). The velocity of the aeroplane and the recording rate of the imager both mean that GPS data have to be interpolated to individual scan lines, which introduces error. When combined with the error inherent in the GPS system, this can cause a large error overall (Table 6.4), which may be unacceptable.

Differential GPS
Differential GPS processing will improve the accuracy of the positional data (Table 6.4). A survey quality GPS receiver can be mounted at a precisely surveyed point, if available, and set to record continuously during the period of airborne survey. If a precisely surveyed ground location is not available it may be possible to set a survey quality receiver at a given point for 24 hours and use commercially available software to produce a base station accuracy of about 1–5m (see Chapter 4).

Ground Control Points (GCPs)

Prior to an aerial survey a number of GCPs may, if necessary, be surveyed and marked so as to be visible on the imagery (e.g. reflective crosses on a dark background). The positions of these GCPs are then used in geometric correction. Alternatively, or additionally, GCPs may be located after data collection using a differential GPS to fix the position of prominent features on the imagery (e.g. road intersections, ends of jetties).

Advantages: high accuracy can be obtained without the need to hire expensive INS or IMU units.

Disadvantages: large numbers of markers are required for each CASI strip. Moreover, it is difficult to lay markers

Table 6.4 The accuracy which is obtainable from a range of instruments capable of measuring the roll and pitch and/or the position of an aircraft carrying CASI. The accuracy of CASI geometric correction using three software options is also listed. These options are discussed more fully in the text. n/a = not applicable. CCRS = Canada Centre for Remote Sensing.

Positional accuracy	Data collected	Gyro	INS/IMU	GPS alone	Differential GPS	GPS with DEM and GCPs	Standard ITRES software	GCPs and commercial software	GCPs and CCRS software
	Roll + pitch	✓	✓	✕	✕	✕	n/a	n/a	n/a
	Position	✕	✓	✓	✓	✓	n/a	n/a	n/a
100 m		✓	✓	✓	✓	✓	✓	✓	✓
10–20 m		✓	✓	✓	✓	✓	✓	✓	✓
5–10 m		✓	✓	✕	✓	✓	✓	✓	✓
3–5 m		✓	✓	✕	✓	✓	✕	✓	✓
0.5–1 m		✕	✓	✕	✕	✓	✕	✕	✓

on the seabed or fix the position of submerged features. Even in terrestrial or littoral habitats above high water (e.g. mangrove forests) suitable positions from which to take fixes are few and far between.

Geometric correction using a digital elevation model (DEM)

Another source of geometric error arises from the topography of a region. For relatively low-lying coastal areas these errors are not likely to be significant for satellite data, but they are relevant to high resolution airborne imaging, and particularly if significant relief is present.

Correction of topographic errors (orthocorrection) requires the use of a digital elevation model (DEM). These are 3-dimensional spatial datasets (x, y, and z, where z is elevation) which record surface topography. Essentially they are the digital equivalent of a contour map. It is unfortunate that DEMs at comparable resolution to that of images are extremely rare for tropical coasts. Enquiries to the local mapping agency will ascertain their availability. In the absence of a commercially available DEM it is possible to construct one by digitising from a contour map and using appropriate spatial interpolation algorithms. This however, is labour intensive and fraught with technical problems (e.g. Weibel and Heller 1991, Wise 1997). For these reasons topographic correction of tropical coastal regions is rarely executed.

Alternative ways of carrying out geometric correction

Three options are available to incorporate the positional data with the CASI data (Table 6.4). Cost varies depending on the software selected, and amount of supporting work which is necessary (e.g. collection of GCPs in the field, production of DEM).

1. *Commercially available software (e.g. ERDAS Imagine or PCI)*: The CASI data (once roll, pitch and yaw corrected by the supplier) can be geometrically corrected using GCPs and routines in commercially available software. This process can be labour intensive – 12 days in the case of the worked example from the Turks and Caicos (see following section). Using this method and ERDAS Imagine, an RMS error of 3.1 m was obtained.

☞ **Currently available commercial software cannot incorporate the airborne and/or base station GPS data.**

2. *Standard ITRES software*: this is designed to utilise the airborne and/or base station GPS data. Costs for a project should normally include a cost for geometric correction using the ITRES software. By way of example, we received a quote of £450 for the correction of 500 MB of data (i.e. approximately what one would get from flying an area of 140 km^2 with a 3 m pixel and 15 bands of data). There is a small set-up fee on any data processing but after that costs are roughly proportional to the size of the dataset. For further information, see ITRES homepage on http://www.itres.com/.

☞ **Standard ITRES software lacks the capability to employ GCP and/or DEM data.**

3. *Canada Centre for Remote Sensing (CCRS) software*: this software package is designed for use with airborne digital data such as CASI which can incorporate GPS, GCP and DEM data. CCRS can be hired on a subcontract basis to geometrically correct data. The accuracies (0.5–1 m) given in Table 6.4 are based on processing using this software. Again, the cost will depend on the size of the dataset to be processed and, once fixed set-up costs are covered, this will be proportional to the size of the dataset. As an example, a dataset which consisted of 30 flight lines, each approximately 10,000 scan lines long (with 8 bands of data), 40 MB per line and a total size of 1.2 GB, would cost £2000 to geocorrect. Set-up costs are fixed, but actual processing costs are linear.

   For further information see CCRS homepage on: http://www.ccrs.nrcan.gc.ca/.

The example given here does not include costs for the GCP and DEM data, which need to be provided by the customer.

☞ **Readers who require extremely good geometric correction (< 1 m) are advised to investigate the CCRS service further.**

Case study: geometric correction of CASI data

Our aim was to acquire high resolution (1–3 m) CASI images for the assessment of mangrove and offshore submerged habitats in the Turks and Caicos Islands, and to use an onboard gyro and DGPS to make the necessary corrections. This procedure was important, particularly for the offshore habitats, due to the absence of potential GCPs.

Practical problems encountered

Unfortunately, geometric correction by this route yielded errors of approximately 200 m east–west and 1200 m north–south. We believe the problem was that the Cessna 172N, which was used to carry the CASI, had an old engine and was under-powered for the task. As a result it may have been

Table 6.5 The source and destination coordinates for a series of ground control points (GCPs). The *x* and *y* residuals and RMS error have been calculated for a second-order polynomial for each GCP. GCP#2, 14 and 16 (shaded) have unacceptably high RMS error (RMS error significantly more than 6m, which is approximately 150% of the accuracy of the differential GPS in the Turks and Caicos, was defined as acceptable).

GCP Identifier	*x* Source Coordinates	*y* Source Coordinates	*x* Destination Coordinates	*y* Destination Coordinates	*x* Residual	*y* Residual	RMS Error
GCP #1	249.04	957.29	236699.2	2379927	1.565	1.851	2.424
GCP #2	117.00	984.65	236824.0	2379989	-2.538	-10.02	10.33
GCP #3	113.55	864.43	236860.6	2379861	-0.451	-0.666	0.804
GCP #4	113.33	840.30	236869.2	2379835	-1.598	0.705	1.747
GCP #5	50.873	602.01	237005.7	2379575	-0.458	0.665	0.807
GCP #6	33.856	586.93	237028.8	2379563	-2.521	0.761	2.633
GCP #7	64.658	359.62	237075.3	2379290	-1.902	5.180	5.518
GCP #8	75.428	189.85	237120.5	2379074	4.683	-2.358	5.243
GCP #9	79.736	128.01	237136.1	2379003	5.516	1.258	5.657
GCP #10	182.05	75.662	237062.7	2378911	-4.289	-4.541	6.247
GCP #11	148.66	28.262	237109.1	2378865	-0.679	-1.525	1.669
GCP#12	281.13	2054.3	236359.4	2381148	-1.438	-0.031	1.439
GCP #13	280.05	1219.9	236588.9	2380215	3.494	1.753	3.909
GCP #14	123.24	116.38	237111.7	2378980	-9.007	-0.327	9.012
GCP #15	58.250	321.76	237091.9	2379232	4.878	-5.617	7.439
GCP #16	95.676	420.60	237015.0	2379361	4.747	12.91	13.75

Table 6.6 The effect of editing a series of GCPs is illustrated. GCPs #2, 14 and 16 have been deleted and the RMS error of each GCP is now acceptable (approximately 6m or less).

GCP Identifier	*x* Source Coordinates	*y* Source Coordinates	*x* Destination Coordinates	*y* Destination Coordinates	*x* Residual	*y* Residual	RMS Error
GCP #1	249.04	957.29	236699.2	2379927	-0.101	0.508	0.518
GCP #3	113.55	864.43	236860.6	2379861	2.799	-0.678	2.880
GCP #4	113.33	840.30	236869.2	2379835	1.437	0.660	1.581
GCP #5	50.873	602.01	237005.7	2379575	-0.241	0.386	0.455
GCP #6	33.856	586.93	237028.8	2379563	-2.852	0.449	2.887
GCP #7	64.658	359.62	237075.3	2379290	-4.453	4.717	6.487
GCP #8	75.428	189.85	237120.5	2379074	1.390	-2.188	2.593
GCP #9	79.736	128.01	237136.1	2379003	2.159	1.762	2.787
GCP #10	182.05	75.662	237062.7	2378911	-1.302	-0.603	1.435
GCP #11	148.66	28.262	237109.1	2378865	0.376	1.679	1.720
GCP#12	281.13	2054.3	236359.4	2381149	0.180	0.256	0.313
GCP #13	280.05	1219.9	236588.9	2380215	-1.205	-0.905	1.507
GCP #15	58.250	321.76	237091.9	2379232	1.810	-6.042	6.308

flying slightly nose-up during data collection. The maximum altitude attained by the aeroplane carrying CASI and two people was 5200 m. If the aircraft was nose-up in flight, the CASI sensor would have been pointing forward and imaging areas slightly ahead of the actual aircraft position (the position which was recorded by the GPS). An aircraft pitch angle of 10° would create a discrepancy of 917 m between the area being imaged and the position of the aircraft. As a consequence of these errors we were unable to use any of our image lines of submerged habitats except for those that included appreciable amounts of coastline within them. For these latter cases, it was possible to use GCPs to effect a rea-

Table 6.7 The combined effect of editing a series of GCPs and increasing the order of the polynomial used in calculating the transformation matrix. If all GCPs are used with a first-order polynomial the RMS error are clearly too high. Excluding those GCPs which have been identified as unacceptable in Table 6.5 still does not produce an acceptable RMS error with a first order polynomial. A second order polynomial is therefore required, and is appropriate given the number of GCPs available and the nature of the imagery (CASI is an airborne sensor and requires rubber-sheeting). However, a second-order polynomial will not produce acceptable RMS errors if GCPs #2, 14 and 16 are included. If those GCPs are excluded RMS errors are comfortably within acceptable limits, and the transformation matrix calculated from this polynomial can be used in geometric correction. A third-order polynomial is not appropriate in this case.

RMS error	All GCPs – 1st order polynomial	Excluding 2, 14 and 16 – 1st order polynomial	All GCPs – 2nd order ploynomial	Excluding 2, 14 and 16 – 2nd order polynomial
x	22.37	3.83	6.89	1.98
y	34.99	4.77	10.84	2.36
Total	45.18	6.12	12.84	3.08

Table 6.8 The transformation matrix calculated from a second-order polynomial using the GCPs in Table 6.6 (Note: e-05 means x 10^{-5}, e-06 means x 10^{-6})

	x'	y'
Constant	25293085	11761466
x coefficient	237.35	121.93
y coefficient	-44.53	-22.79
x^2 coefficient	-0.0002	-8.6e-06
xy coefficient	-6.3e-05	-4.9e-05
y^2 coefficient	1.2e-05	7.4e-06

The computer calculates the map coordinates (x', y') of any source coordinate (x, y) by using the above coefficients in the following equations:

$x' = 237.35\,x - 44.53\,y - 0.0002\,x^2 - 6.3\text{e-}05\,xy + 1.2\text{e-}05\,y^2 + 25293085$

$y' = 121.93\,x - 22.79\,y - 8.6\text{e-}06\,x^2 - 4.9\text{e-}05\,xy + 7.4\text{e-}06\,y^2 + 11761466$

sonable geometric correction. We were fortunate that the adjacent land included intersections of roads, walls of salt ponds and the corners of buildings which were used as GCPs. If the images had lacked these features, geometric correction would not have been possible.

Solutions and the correction process

The image supplied by the CASI operator was corrected for roll, pitch and yaw. Further geometric correction was performed using GCPs and following the procedure outlined in Figure 6.1.

1. *Is geometric correction necessary?* Yes. In this case, we wished to use the CASI data to map the different mangrove habitats and measure Leaf Area Index (LAI) across a large area of South Caicos. Field data had been collected from within the mangroves at positions determined using the differential GPS. In order to calibrate the image with these field data it was necessary to relate spectral information to field data at these points.
2. *How many GCPs are to be used?* Sixteen features were identified both on the image and *in situ*. They included the corner of the concrete base of some fuel tanks (Figure 6.2, GCP #9) and the ends of piers (Figure 6.2, GCP #10 and 11).
3. *How are the positions of GCPs to be determined?* By differential GPS *in situ*.
4. *What order polynomial is appropriate?* First- or second-order polynomials are appropriate for digital aerial imagery, though the latter is more likely to be necessary in order to carry out non-linear transformations.
5. *Which map projection?* UTM, zone 19, spheroid Clark 1866, datum NAD 27 (Bahamas). This is the output projection for the GPS coordinates (and hence the map projection for our GCPs). Therefore, it is necessary to resample to this projection.
6. *Is the RMS error of individual GCPs (x and y residuals) acceptable?* No. See Table 6.5. Among 16 GCPs examined, the *x* residual of GCP #14 and the *y* residuals of GCPs #2 and 16 gave rise to unacceptable RMS error. These three GCPs were therefore discarded.
7. *Is the total RMS error acceptable?* Yes. With a second-order polynomial individual *x* and *y* residuals, RMS error and total RMS error is now acceptable (Table 6.6). The effect of choosing this combination of GCPs and a second-order polynomial is displayed in Table 6.7.
8. *Which resampling technique is appropriate?* Nearest neighbour, because we wish to relate actual reflectance values to mangrove LAI. Therefore we want to preserve data values without averaging.

The computer now resamples the image using the computed transformation matrix (Table 6.8) to convert file coordinates to map coordinates.

Cost considerations

Satellite imagery
Our satellite imagery of the Turks and Caicos Islands was geometrically corrected using GCPs taken from 1:25,000 Ordnance Survey Maps. Between 25–30 GCPs were used for each image. Preparation time (reading and familiarisation with the software) took two days. The time taken to perform geometric correction was one day per image. The complete set of Ordnance Survey maps cost £126.25.

CASI imagery
Seven days were wasted in discovering that the synchronisation of GPS data to the CASI had not worked, in communication with the suppliers and in trying to rescue the coordinates from the data. Field collection of GCPs took four days; geometric correction of the images another eight. Total 19 days.

Conclusions

Failure to remove geometric distortions from remotely sensed imagery may have four consequences:

(i) an inability to relate spectral features of the image to field data,
(ii) an inability to make a comparison, pixel by pixel (change detection), with another image obtained at a different time,
(iii) an inability to obtain accurate estimates of the area of different regions of the image, and
(iv) an inability to integrate the image with other geographical data.

Since the overwhelming majority of applications of remote sensing involve one or more of these activities, geometric correction is a vitally important step in the analysis of image data and great care should be taken over performing it.

References

Atkinson, P., 1985, Preliminary results of the effect of resampling on thematic mapper imagery. 1985 ACSM–ASPRS Fall Convention Technical Papers. Falls Church, Virginia: American Society for Photogrammetry and Remote Sensing and American Congress on Surveying and Mapping.

Bugayevskiy, L.M., and Snyder, J.P., 1995, *Map Projections: a Reference Manual.* (London: Taylor and Francis).

Butler, M.J.A., LeBlanc, C., Belbin, J.A., and MacNeill, J.L., 1987, Marine resource mapping: an introductory manual. *FAO Fisheries Technical Paper*, **274**, (Rome: Food and Agriculture Organization of the United Nations).

Lee, J.E., and Walsh, J.M., 1984, Map projections for use with the geographic information system. *U.S. Fish and Wildlife Service*, FWS/OBS–84/17.

Maling, D.H., 1992, *Coordinate Systems and Map Projections.* (Oxford: Pergamon Press).

Mather, P.J., 1987, *Computer Processing of Remotely-Sensed Images: An Introduction.* (Chichester: John Wiley & Sons).

Pearson, F., 1990. *Map Projections: Theory and Applications.* (Boca Raton, Florida: CRC Press, Inc).

Snyder, J.P., 1994, Map projections – a working manual. *U.S. Geological Survey Professional Paper 1395.* (Washington: U.S. Government Printing Office). 383 pp.

Weibel, R., and Heller, M., 1991, Digital terrain modelling. In *Geographical Information Systems,* Volume 1, edited by D.J. Maguire, M.F. Goodchild and D.W. Rhind (Harlow: Longman), pp. 269–297.

Wise, S.M., 1997, The effect of GIS interpolation errors on the use of DEMs in geomorphology. In *Landform Monitoring, Modelling and Analysis*, edited by S.N. Lane, K.S. Richards and J.H. Chandler (Chichester: Wiley), pp. 139–164.

7

Radiometric Correction of Satellite and Airborne Images

Summary *Sensors record the intensity of electromagnetic radiation as digital numbers (DN), typically with values ranging from 0–255. These DN are specific to the sensor and the conditions under which they were measured and are therefore arbitrary units. To convert DN to physical units we must use calibration information provided by the designers of the sensor. A simple analogy is that if we were to pace out a distance in strides, which is a measure unique to ourselves and thus of limited use, we could use a calibration (1 pace = 0.8 m) to convert our distance measure to metres. This is the first step in radiometric calibration. The second and third steps seek to remove the influence of solar and atmospheric conditions at the time the image was acquired.*

Radiometric correction involves the following three steps:

- *conversion of DN values to spectral radiance at the sensor (using sensor calibration),*
- *conversion of spectral radiance to apparent reflectance (using intensity of incoming solar illumination),*
- *removal of atmospheric effects due to the absorption and scattering of light (atmospheric correction).*

These corrections can be time-consuming and problematic, particularly when it is hard to ascertain the appropriate calibration values or where atmospheric conditions are unknown. They are, however, essential if you are intending (i) to compare images in time or space, (ii) to model interactions between electromagnetic radiation and a surface feature, or (iii) to use band ratios for imagery analysis. Radiometric corrections are not essential for visual interpretation of imagery or for the production of one-off habitat maps by multispectral classification.

Introduction

Digital sensors record the intensity of electromagnetic radiation (EMR) from each spot viewed on the Earth's surface as a digital number (DN) for each spectral band (see Chapter 1). The exact range of DN that a sensor utilises depends on its radiometric resolution. For example, a sensor such as Landsat MSS measures radiation on a 0–63 DN scale whilst Landsat TM measures it on a 0–255 scale, and NOAA AVHRR on a 0–1023 scale (see Chapter 3 for sensor details). Although the DN values recorded by a sensor are proportional to upwelling EMR (radiance), the true units are W m^{-2} $ster^{-1}$ μm^{-1} (Box 7.1).

The majority of pre-1990 remote sensing research was based on raw DN values in which actual spectral radiances were not used. However, there are problems with this approach. The spectrum of a habitat (say seagrass) is not transferable if measured in digital numbers. The values are image specific – i.e. they are dependent on the viewing geometry of the satellite at the moment the image was taken, the location of the sun, specific weather conditions, and so on. It is generally far more useful to convert the DN values to spectral units. This has two great advantages. Firstly, spectra with meaningful units can be compared from one image to another. This would be required where the area of study is larger than a single scene or if monitoring change at a single site where several scenes taken over a period of years are being compared. It is also vital for radiative transfer modelling (see Chapter 16) where the interaction between EMR and the object is examined. Secondly, there is growing recognition that remote sensing could make effective use of 'spectral

BOX 7.1

Units of electromagnetic radiation

The unit of electromagnetic radiation is W m^{-2} $ster^{-1}$ μm^{-1}. That is, the rate of transfer of energy (Watt, W) recorded at a sensor, per square metre on the ground, for one steradian (three dimensional angle from a point on Earth's surface to the sensor), per unit wavelength being measured. While being quite a mouthful, this measure is referred to as the spectral radiance (i.e. the amount of radiance at a particular wavelength). Prior to the launch of a sensor, the relationship between measured spectral radiance and DN is determined. This is known as the sensor calibration. It is worth making a brief note about terminology at this point. The term radiance refers to any radiation leaving the Earth (i.e. upwelling, toward the sensor). A different term, irradiance, is used to describe downwelling radiation reaching the Earth from the sun (Figure 7.1). The ratio of upwelling to downwelling radiation is known as reflectance. Reflectance does not have units and is measured on a scale from 0 to 1 (or 0–100%). DN is simply the radiance measured at-satellite expressed in uncalibrated arbitrary units.

libraries' – i.e. libraries of spectra containing lists of habitats and their reflectance characteristics (Box 7.1).

While spectral radiances can be obtained from the sensor calibration, several factors still complicate the quality of remotely sensed measurements. The spectral radiances obtained from the calibration only account for the spectral radiance measured at the satellite sensor. By the time EMR is recorded by a satellite or airborne sensor, it has already passed through the Earth's atmosphere twice (sun to target and target to sensor). During this passage (Figure 7.1), the radiation encounters two processes which (i) reduce its intensity (by absorption) and (ii) alter its direction (by scattering). Absorption occurs when electromagnetic radiation interacts with gases such as water vapour, carbon dioxide and ozone. The degree of absorption varies across the spectrum and most remote sensing is carried out in those wavebands where absorption is at its lowest (termed 'transmission windows'). Transmission windows (regions with high transmission) cover the 300–1300 nm (visible/near infra-red), 1500–1800 nm, 2000–2500 nm and 3500–4100 nm (middle infra-red) and 7000–15000 nm (thermal infra-red) wavebands (Figure 1.1). Scattering results from interactions between EMR and both gas molecules and airborne particulate matter (aerosols). These molecules and particles range in size from the raindrop (> 100 μm) to the microscopic (< 1 μm). Scattering will redirect incident electromagnetic radiation and deflect reflected EMR from its path (Figure 7.1).

Absorption and scattering (Box 7.2) create an overall effect of 'haziness' which reduces the contrast in the image. Scattering also creates the 'adjacency effect' in which the radiance recorded for a given pixel partly incorporates the scattered radiance from neighbouring pixels. It is also worth pointing out another important piece of ter-

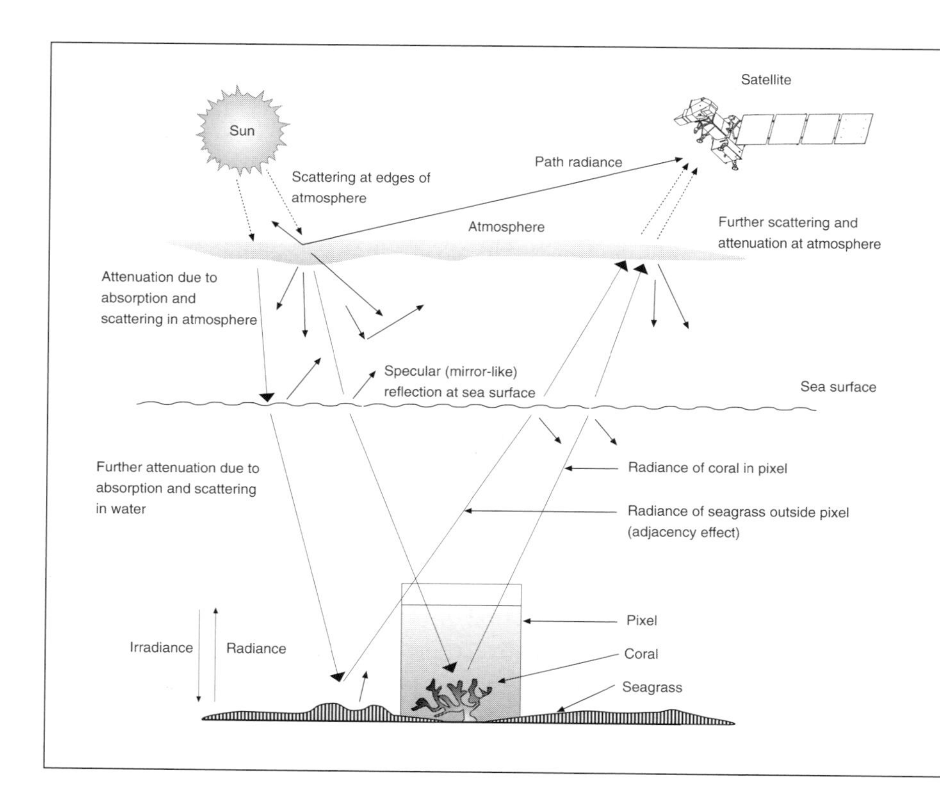

Figure 7.1 Simplified schematic of atmospheric interference and the passage of electromagnetic radiation from the Sun to the satellite sensor

BOX 7.2

Types of scattering

A *Selective scattering (wavelength dependent)*

- *Rayleigh scattering*: caused by very small particles and molecules with radii far less than the electromagnetic radiation of interest. The extent of scattering is inversely proportional to the fourth power of the wavelength (exponent = -4). Therefore shorter wavelengths, such as blue light in the visible spectrum, are affected the greatest.
- *Mie scattering*: caused by particles with radii between 0.1 and 10 μm such as dust, smoke and salt (aerosols). Like Rayleigh scattering, Mie scattering affects shorter wavelengths greater than longer ones. However, with an exponent ranging from -0.7 to -2, the disparity is not as great as that for Rayleigh scattering.

B *Nonselective scattering (wavelength independent)*

This is created by particles whose radii exceed 10 μm (e.g. water droplets and ice fragments in cloud). All visible wavelengths are affected by these particles.

minology at this stage. Radiation that has been scattered in the Earth's atmosphere and has reached the sensor without contacting the Earth's surface is known as the 'path radiance' (Figure 7.1).

In order to make a meaningful measure of reflectance at the Earth's surface, the atmospheric interferences must be removed from the data. This process is called 'atmospheric correction'. The entire process of radiometric correction involves three steps (Figure 7.2).

The spectral radiance of features on the ground are usually converted to reflectance. This is because spectral radiance will depend on the degree of illumination of the object (i.e. the irradiance). Thus spectral radiances will depend on such factors as time of day, season, latitude, etc. Since reflectance represents the ratio of radiance to irradiance, it provides a standardised measure that is directly comparable between images.

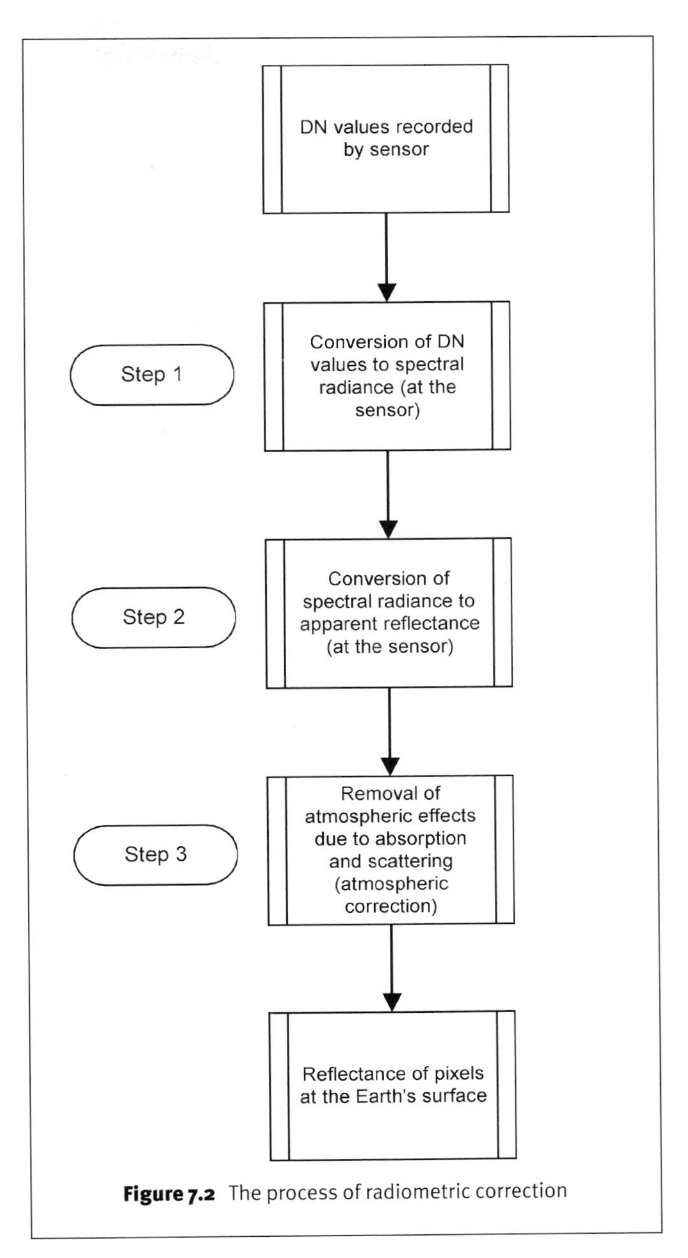

Figure 7.2 The process of radiometric correction

Steps needed to convert DN values to reflectance

1. Conversion of DN to spectral radiance

This is a fairly straightforward process which requires information on the 'gain' and 'bias' of the sensor in each band (Figure 7.3). The transformation is based on a calibration curve of DN to radiance which has been calculated by the operators of the satellite system. The calibration is carried out before the sensor is launched and the accuracy declines as the sensitivity of the sensor changes over time. Periodically attempts are made to re-calibrate the sensor. The methods of calibration set out below are the best advice we can offer at the time of going to press.

The calibration is given by the following expression for at satellite spectral radiance, L_λ:

$$L_\lambda = Bias + (Gain \times DN)$$

☞ **Make sure you know what units your gains and biases are in. For our EOSAT-supplied Landsat images, they were in mW cm^{-2} ster^{-1} μm^{-1} but for our SPOT images they were in W m^{-2} ster^{-1} μm^{-1}. However, in National Landsat Archive Production System (NLAPS) format Landsat images, gains and biases in the header can be in W m^{-2} ster^{-1} μm^{-1}.**

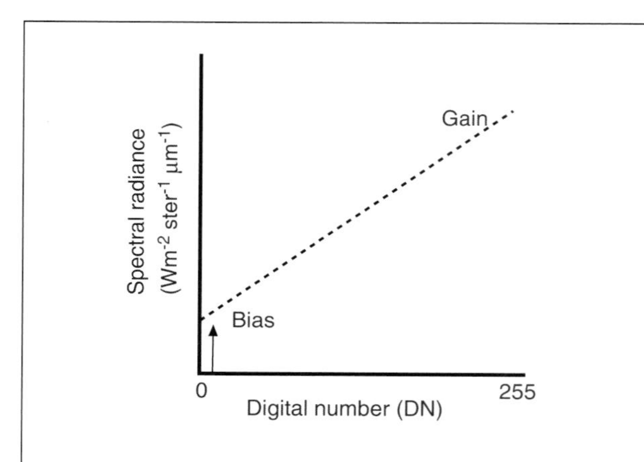

Figure 7.3 Calibration of 8-bit satellite data. Gain represents the gradient of the calibration. Bias defines the spectral radiance of the sensor for a DN of zero. The conversion of DN to spectral radiance follows the standard $y = a + b.x$ equation. I.e. spectral radiance = bias + (gain x DN).

Calibration of Landsat TM data

Each waveband has its own gain and bias values, which must be ascertained from the image header files or ancillary information often supplied with the imagery. Using the simple linear equation above, DN are converted into spectral radiance. The method for calculating gain and bias varies according to when the imagery was processed (in the case of imagery obtained from EOSAT). Gains and biases are calculated from the lower ($Lmin_\lambda$) and upper ($Lmax_\lambda$) limits of the post-calibration spectral radiance range for each band (Table 7.1 and Figure 7.3). If the imagery was processed before October 1st 1991, use Equation 7.1a below. If it was processed after October 1st 1991 use Equation 7.1b.

☞ **Date of processing is not the date when the image was acquired. For example, a TM scene is recorded in 1989 then purchased in 1995. The processing is carried out in 1995 and therefore, equation 1b should be used.**

The values of *Lmax* and *Lmin* for each waveband of Landsat TM can be obtained from published tables (Table 7.1) or, if the data were processed after October 1st 1991, in the header file which accompanies the data (see Chapter 5). The header file is in ASCII (American Standard Code for Information Interchange) format and the gains/biases are stored in fields 21–33. Unfortunately, the band numbers are not given. However, the sequence is band 1 to band 7. Also note that the values in the header file need converting. They are given as in-band radiance values (mW cm^{-2} ster^{-1}) which need to be converted to spectral radiances across each band (mW cm^{-2} ster^{-1} μm^{-1}). This is done by dividing each value by the spectral band width. Band widths for the TM sensors carried on Landsats 4 and 5 are listed in Table 7.2.

$$Gain = \frac{L\max_\lambda - L\min_\lambda}{Total\ scale\ of\ DN\ values}$$

Equation 7.1a

$$Bias = L\min_\lambda$$

where the total scale of DN values for TM data is 255. For MSS data (originally on a scale from 0–63 but normally resampled) use 127 or 255 (depending on the data format).

$$Gain = \frac{L\max_\lambda}{254} - \frac{L\min_\lambda}{255}$$

Equation 7.1b

$$Bias = L\min_\lambda$$

Calibration of Landsat MSS data

The procedure for converting DN to radiance for Landsat MSS is similar to that required for Landsat TM. Although the total scale of DN values for MSS data are usually 127 or 255, it should be borne in mind that the MSS sensor has a 6-bit resolution and the original data were recorded on a scale of 0–63. For convenience, the data archiving agency usually re-samples this to 7-bit or 8-bit data. When calculating gains and biases, Equation 7.1a should be used regardless of the processing date (i.e. do not use Equation 7.1b). Unlike Landsat TM, the $Lmax_\lambda$ and $Lmin_\lambda$ are not given in the header files. Refer to Table 7.3 for guidance as to which values should be used for imagery of a given date.

Calibration of SPOT XS and SPOT Pan

SPOT Image have made the conversion of DN to radiance much easier for the user by supplying hardcopy of the calibration information with the imagery. *SPOT Image* calculate the appropriate gains and biases for each scene and provide the Absolute Calibration Gain (A) in W m^{-2} ster^{-1} μm^{-1} for each band. To obtain the radiance of each band, simply divide each DN value by A.

☞ **These units can be different to those used by Landsat. 10 W m^{-2} ster^{-1} μm^{-1} are equivalent to 1 mW cm^{-2} ster^{-1} μm^{-1}.**

Calibration of airborne data

Airborne data are calibrated in a similar manner to satellite imagery. However, this process is usually carried out by the instrument operator and is not elaborated further here.

2. Spectral radiance to apparent reflectance

Satellite data

This step ratios the upwelling radiance recorded at the sen-

Table 7.1 Landsat Thematic Mapper (TM) post-calibration dynamic ranges for US processed data. Spectral radiances, $Lmin_\lambda$ and $Lmax_\lambda$ in mW cm^{-2} ster^{-1} µm^{-1}. *Source*: EOSAT.

	Prior to August 1993 (Scrounge-ERA processing)		Prior to 15 Jan 1984 (TIPS-ERA processing)		After 15 Jan 1984 and before 1 October 1991 (TIPS-ERA processing)	
Band	$Lmin_\lambda$	$Lmax_\lambda$	$Lmin_\lambda$	$Lmax_\lambda$	$Lmin_\lambda$	$Lmax_\lambda$
TM1	-0.152	15.842	0.000	14.286	-0.15	15.21
TM2	-0.284	30.817	0.000	29.125	-0.28	29.68
TM3	-0.117	23.463	0.000	22.500	-0.12	20.43
TM4	-0.151	22.432	0.000	21.429	-0.15	20.62
TM5	-0.037	3.242	0.000	3.000	-0.037	2.719
TM6	0.20	1.564	0.484	1.240	0.1238	1.560
TM7	-0.015	1.700	0.000	1.593	-0.015	1.438

Table 7.2 Bandwidths for Landsat 4 and 5 Thematic Mappers (µm).

	TM1	TM2	TM3	TM4	TM5	TM6	TM7
Landsat 4	0.066	0.081	0.069	0.129	0.216	1.000	0.250
Landsat 5	0.066	0.082	0.067	0.128	0.217	1.000	0.252

Table 7.3 Landsat Multispectral Scanner (MSS) post-calibration dynamic ranges for US processed data. Spectral radiances, $Lmin_\lambda$ and $Lmax_\lambda$ in mW cm^{-2} ster^{-1} µm^{-1}. *Source*: Markham and Barker (1987).

Satellite dates	MSS 1 (Band 4 on Landsat 1-3)		MSS 2 (Band 5 on Landsat 1-3)		MSS 3 (Band 6 on Landsat 1-3)		MSS 4 (Band 7 on Landsat 1-3)	
	$Lmin_\lambda$	$Lmax_\lambda$	$Lmin_\lambda$	$Lmax_\lambda$	$Lmin_\lambda$	$Lmax_\lambda$	$Lmin_\lambda$	$Lmax_\lambda$
Landsat-1 all	0.0	24.8	0.0	20.0	0.0	17.6	0.0	15.3
Landsat-2								
Before 16/7/75	1.0	21.0	0.7	15.6	0.7	14.0	0.5	13.8
After 16/7/75	0.8	26.3	0.6	17.6	0.6	15.2	0.4	13.0
Landsat-3								
Pre-launch	0.4	25.0	0.3	20.0	0.3	16.5	0.1	15.0
Before 1/6/78	0.4	22.0	0.3	17.5	0.3	14.5	0.1	14.7
After 1/6/78	0.4	25.0	0.3	17.9	0.3	14.9	0.1	12.8
Landsat-4								
Before 26/8/82	0.2	25.0	0.4	18.0	0.4	15.0	0.3	13.3
26/8/82–31/3/83	0.2	23.0	0.4	18.0	0.4	13.0	0.3	13.3
After 1/4/83	0.4	23.8	0.4	16.4	0.5	14.2	0.4	11.6
Landsat-5								
Before 6/4/84	0.4	24.0	0.3	17.0	0.4	15.0	0.2	12.7
6/4/84–8/11/84	0.3	26.8	0.3	17.9	0.4	15.9	0.3	12.3
After 9/11/84	0.3	26.8	0.3	17.9	0.5	14.8	0.3	12.3

sor (L_λ) against the irradiance from the sun, taking account of the solar elevation at the time of imagery acquisition. It converts pixel values from radiance units to reflectance, or more precisely, to apparent reflectance which is reflectance at the top of the atmosphere (i.e. not allowing for the effects of atmospheric attenuation). In the context of satellite imagery, apparent reflectance is often termed exoatmospheric reflectance, ρ, and is expressed as:

$$\rho = \frac{\pi . L . d^2}{ESUN . \cos(SZ)}$$

ρ = unitless planetary reflectance at the satellite (on a scale of 0–1)

π = 3.141593

L = Spectral radiance at sensor aperture in mW cm^{-2} ster^{-1} µm^{-1} (or W m^{-2} ster^{-1} µm^{-1})

d^2 = Earth-Sun distance in astronomical units = $(1 - (0.01674 \cos(0.9856 (JD-4))))^2$ where JD is the Julian Day (day number in the year) of the image acquisition

$ESUN$ = Mean solar exoatmospheric spectral irradiance in mW cm^{-2} µm^{-1} (or W m^{-2} µm^{-1})

SZ = sun zenith angle in degrees when the scene was recorded.

Table 7.4 Landsat Thematic Mapper (TM) Solar Exoatmospheric Spectral Irradiances (*ESUN*) in mW cm^{-2}µm^{-1}. *Source*: EOSAT.

Band	Landsat-4	Landsat-5
TM1	195.8	195.7
TM2	182.8	182.9
TM3	155.9	155.7
TM4	104.5	104.7
TM5	21.91	21.93
TM7	7.457	7.452

Table 7.5 Landsat Multispectral Scanner (MSS) Solar Exoatmospheric Spectral Irradiances (*ESUN*) ± standard deviation. (mW cm^{-2} µm^{-1}). *Source*: EOSAT.

Satellite	MSS 1	MSS 2	MSS 3	MSS 4
Landsat-1	185.2 ± 0.8	158.4 ± 0.6	127.6 ± 0.4	90.4 ± 0.5
Landsat-2	185.6 ± 0.1	155.9 ± 0.4	126.9 ± 0.6	90.6 ± 0.3
Landsat-3	186.0 ± 0.3	157.1 ± 0.6	128.9 ± 0.5	91.0 ± 0.6
Landsat-4	185.1 ± 0.2	159.3 ± 0.7	126.0 ± 0.3	87.8 ± 1.0
Landsat-5	184.9 ± 0.2	159.5 ± 0.3	125.3 ± 0.1	87.0 ± 0.7

Table 7.6 SPOT Solar Exoatmospheric Spectral Irradiances (ESUN) in W m^{-2} µm^{-1}. *Source*: SPOT Image.

	HRV1				HRV2			
	Pan	XS1	XS2	XS3	Pan	XS1	XS2	XS3
SPOT 1	1680	1855	1615	1090	1690	1845	1575	1040
SPOT 2	1705	1865	1620	1085	1670	1865	1615	1090
SPOT 3	1668	1854	1580	1065	1667	1855	1597	1067

☞ **Both Landsat and SPOT products provide the sun elevation angle. The zenith angle (SZ) is calculated by subtracting the sun elevation from 90°.**

Since ρ has no units it does not matter whether mW and cm^{-2} or W and m^{-2} are used to describe *L* and *ESUN*. However, it is important that once a unit is chosen for *L*, the same unit is used for *ESUN*. *ESUN* can be obtained from Tables 7.4–7.6.

☞ **Values of ESUN are occasionally given in the units of W m^{-2} ster^{-1} µm^{-1}. To convert these to W m^{-2} µm^{-1} simply multiply by π (the number of steradians in a hemisphere i.e. 180°).**

Airborne data

The conversion of spectral radiance to apparent reflectance is more complex for airborne imagery. Since the instrument is usually held in the lower portion of the atmosphere, it is not possible to use a simple predictive equation to estimate solar irradiance (as is done with satellite measurements). The collection of airborne imagery is a far more dynamic process and an individual flight line (image) may take many minutes to complete. Not surprisingly, the solar irradiance is likely to vary markedly during this time (e.g. due to periodic obstruction of clouds overhead). To account for fluctuations in solar irradiance during imagery acquisition, an incident light sensor (ILS) is mounted on the aircraft. The sensor points skyward and records a solar irradiance value for every scan made by the airborne imager. Both datasets are recorded simultaneously, which allows each scan line to be corrected to apparent reflectance. In this case then, 'apparent reflectance' refers to reflectance at the aircraft (rather than at the edge of the atmosphere as in satellite measurements).

Like the calibration of airborne data to spectral radiance, apparent reflectance conversion is usually carried out by the instrument operator.

A practical word of caution needs to be said at this point. It is important that the ILS only records irradiance when the instrument is absolutely vertical. If irradiance is recorded while the aircraft rolls, the measurements will be biased and asymmetric i.e. the sensor will point toward the sun on one roll (high irradiance value recorded); it will then point away from the sun during the opposite roll (low irradiance value recorded). We encountered this problem in the Turks and Caicos Islands. The solution was to average solar irradiance over the entire scene and apply the same correction to all scan lines in the image.

3. *Removal of atmospheric effects due to absorption and scattering*

As outlined earlier, radiation passing through the atmosphere can undergo considerable attenuation resulting in a measurement of reflectance being different from the true reflectance of the surface under investigation. For example, up to 80% of the signal recorded by the Coastal Zone Color Scanner may be attributable to scatter. The removal of atmospheric effects is important and considerable research effort is directed at the problem. A detailed discussion of the methods available for atmospheric correction is beyond the scope of this handbook (see Kaufman, 1989 for further reading) but the key methods will be outlined. Although the description that follows is targeted toward satellite imagery, the same principles apply to airborne imagery.

Atmospheric correction techniques can be broadly split into three groups (Figure 7.4) and these will be described separately. A quick-look summary is given in Table 7.7. We recommend the atmospheric modelling approach (c) for most applications.

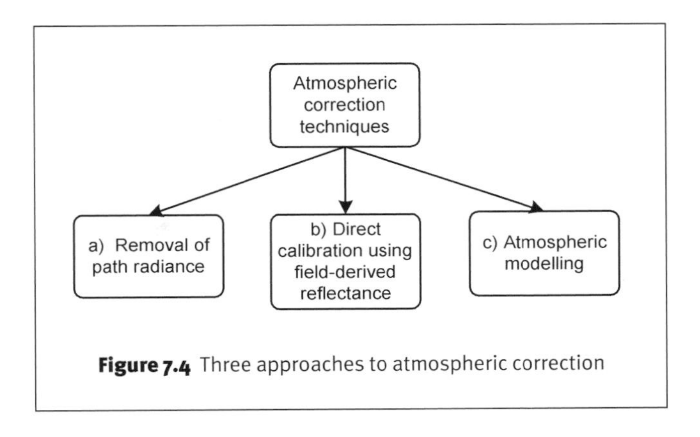

Figure 7.4 Three approaches to atmospheric correction

a) Removal of path radiance (scatter)

These techniques attempt to remove scattering effects by estimating the path radiance (Figure 7.1). The methods are generally used prior to band-ratioing a single image and are not usually employed for image to image comparisons. Band ratios are widely used in vegetation studies (e.g. the Normalised Difference Vegetation Index: Chapter 13). The underlying premise for band-ratioing is that recorded radiances can be adjusted so that their ratio is approximately proportional to the ratio of surface reflectances. In short, the ratio cancels out atmospheric transmittance and topographic effects. However, this is only valid if path radiance can be removed.

These techniques were originally proposed for use with Landsat MSS imagery and are based on the wavelength dependent nature of scattering. There are four principal methods of removing path radiance.

- *Dark pixel subtraction* (Chavez *et al.* 1977) – (also known as the histogram minimum method). The method is based on an assumption that somewhere in the image is a pixel with zero reflectance, such that the radiance recorded by the sensor is solely attributable to path radiance. To remove path radiance, the minimum pixel value in each band is subtracted from all other pixels. The technique can be refined by identifying a region on the image that supposedly has zero reflectance (e.g. deep clear water, deep shadows, fresh tarmac). In MSS images it is assumed that scattering is zero in the infra-red band 4 but present in the bands with shorter wavelengths. Any reflectance recorded in bands 1–3 over these dark areas is assumed to result from atmospheric path radiance rather than real reflectance from the Earth's surface. Bi-plots of band 4 against bands 1–3 are generated for pixels of the dark regions. Regression techniques are then used to calculate the *y*-intercept which represents the path radiance in bands 1–3 (see Mather 1987). This is then subtracted from all pixels in the imagery.

BOX 7.3

Practical notes on calculating radiance and apparent reflectance using image processing software

Use of modelling software: The easiest method of implementing the equations listed above is through simple modelling procedures in image processing software.

Process one band at a time: It is advisable to run the model on one spectral band at a time. This reduces the total data volume because processed input DN files can be deleted before moving on to the next band.

Checking for errors in the model: It is advisable to test each stage of the model manually by calculating expected radiance/reflectance and comparing the values to those in the output. To do this, view both the input and output images simultaneously and compare the values for a series of points.

File size: Since DN are typically recorded on a scale of 0– 255 (or greater scales) and conversion to reflectance will yield values between 0–1, it is necessary to specify an appropriate data scale for the output. The data could be output as 'single floating point' 32-bit as this will allow for the decimal point and 8 decimal places. Unfortunately, storing data in this format (32 bits = 4 bytes per pixel) makes for large file sizes and it is often useful to examine the actual data range and choose a more efficient data type for storage. For example, if the maximum reflectance is 0.61 and the majority of values lie between 0.28550 and 0.59152, the data could be multiplied by 10,000 and stored as 16-bit integers that offer a range of 65,536 reflectance values. On a cautionary note, however, care should be taken not to truncate the dynamic range of the data.

- *Regression method* (Potter and Mendlowitz 1975) – an homogeneous region of the image is chosen and a series of bi-plots drawn between each band pair.

☞ **The area needs to exhibit homogeneity but does not have to have low reflectance as above.**

The slope of the plot reflects the relative reflectance of the homogeneous region in the two bands. In the absence of noise and atmospheric effects, the plot should pass through the origin. The offsets from the origin therefore represent the path radiance in each band, which can be used to make a correction (Figure 7.5).

- *Covariance matrix method* (Switzer *et al.* 1981) – this method extends the regression method to treat all bands simultaneously (rather than on a per-band basis). It is mathematically complex and readers are directed to the original paper if further details are required.
- *Regression intersection method* (Crippen 1987) – the most mathematically complex of the techniques. It is based on a similar regression principle to those above but incorporates regions of homogeneous pixels with different characteristics. Theoretically, the absolute values for path radiance are more realistic than those obtained using the covariance matrix method. However, the relative values of path radiance between bands are more accurate in the covariance matrix method. If this methodology is used, it is best to use it in conjunction with the covariance matrix method (see Crippen 1987).

b) Direct calibration using field-derived reflectance

This method requires some field work to measure the reflectance of two (or preferably more) regions of the area covered by the image. Reflectance measurements are made using a spectral radiometer. This generates a calibration curve that relates satellite-recorded radiance to field-recorded reflectance (i.e. it attempts an entire radiometric correction rather than just correcting for atmospheric effects alone). The rest of the imagery is then converted to reflectance using linear extrapolation. If such field work is not feasible, it is possible to use reflectance values from comparable studies. However, this makes the tenuous assumption that the reflectance recorded in one region (say, deep water in the Bahamas) is directly comparable to that in other regions (e.g. deep water in Tahiti). There are a number of reasons to question such an assumption (see Table 7.7).

c) Atmospheric modelling

This is perhaps the most sophisticated method used to compensate for atmospheric absorption and scattering. So much is known about atmospheric structure that a number of radiative transfer models have been developed to predict the effects of the atmosphere on recorded EMR. These can be used to correct images. Ideally, modelling approaches are best used when scene-specific atmospheric data are available (e.g. aerosol content, atmospheric visibility). However, such information is rarely available and while a range of models exist, the 5S radiative transfer code (Simulation of the Sensor Signal in the Solar Spectrum) is described here because its widespread use and because it includes a variety of standard options which allow this approach to be used with limited ancillary data (Tanre *et al.* 1990).

Using the 5S Radiative Transfer Code: the physics behind the model will not be described here but readers may refer to the paper by Tanre *et al.* (1990) for further details. The model predicts the apparent reflectance at the top of the atmosphere using information about the surface reflectance and atmospheric conditions (i.e. it works in the opposite way that one might expect). Since the true apparent reflectance has been calculated from the sensor calibration and exoatmospheric irradiance (above), the model can be inverted to predict the true surface reflectance (i.e. the desired output). In practice, some of the model outputs are used to create coefficients that may then be applied to the image file.

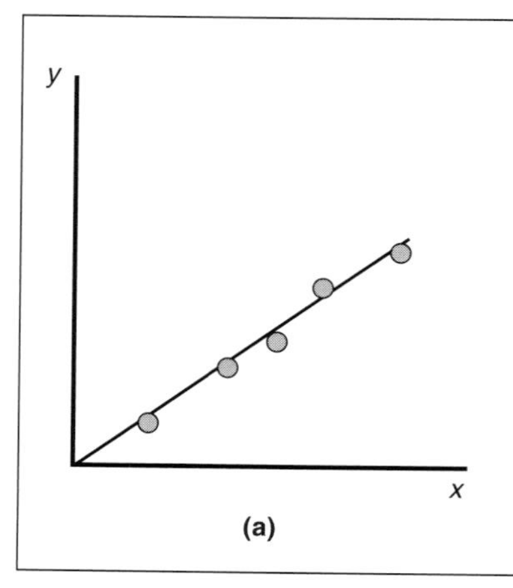

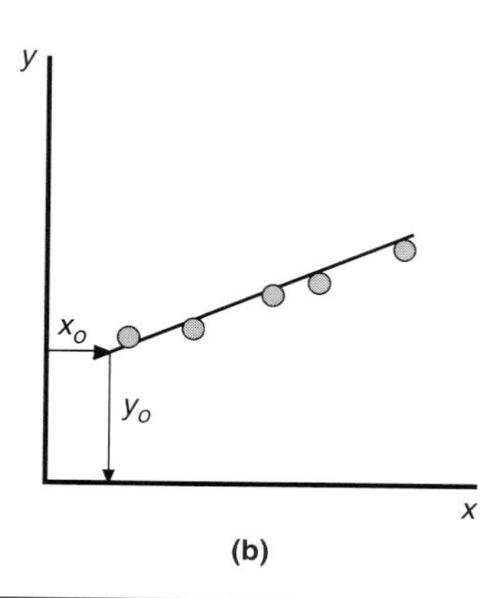

Figure 7.5 Plots of measured radiance (y) on ground radiance (x) of a homogenous material for a range of illumination intensities. (a) For accurate measurements (e.g. with a hand-held spectrometer) which pass through the origin, and (b) for **satellite** measured radiance which is offset (x_o, y_o) from the origin due to calibration offsets of the sensor and the addition of atmospheric path radiance.

Table 7.7 Quick guide to atmospheric correction techniques.

Nature of technique	Name	Maths complexity	Assumptions	Advantages	Disadvantages	Use
Removal of Path Radiance	Dark pixel subtraction (DPS)	Simple	‡ Presence of a pixel with zero reflectance ‡ Selected 'dark pixels' are representative of whole scene ‡ Only considers path radiance	‡ Easily implemented ‡ Susceptible to sensor noise (lowest value may be representative of sensor detector with lowest calibration offset)	‡ Assumption rarely true	‡ Rarely used ‡ May actually degrade quality of imagery
	Regression (R)	Simple	‡ Path radiance of one band is known accurately (preferably by an independent method) ‡ Selected pixels are representative	‡ Easily implemented, better than DPS	‡ Assumptions may be false ‡ If the determination of path radiance for the reference band is poor, the absolute values of path radiance for other bands are poor ‡ Only considers path radiance	‡ Mainly used prior to ratioing bands (not for image–image comparison)
	Covariance matrix (CM)	Difficult	‡ Path radiance of one band is known accurately (preferably by an independent method) ‡ Selected pixels are representative	‡ Possibly better than R ‡ Relative estimates of path radiance between bands may be accurate	‡ More complex and the superiority of the technique relatives to R is questionable ‡ Assumptions may be false ‡ If the determination of path radiance for the reference band is poor, the absolute values of path radiance for the other bands are poor ‡ Should be used in conjunction with RIM ‡ Only considers path radiance	‡ Mainly used prior to ratioing bands (not for image–image comparison)
	Regression intersection (RI)	Difficult	‡ None (Pixels are selected from several areas of imagery and are more likely to be representative)	‡ Provides good absolute values of path radiance ‡ Requires no ancillary information	‡ Mathematically complex ‡ Should be used in conjunction with CM ‡ Only considers path radiance	‡ Mainly used prior to ratioing bands (not for image–image comparison)
Radiance-Reflectance Conversion		Moderate	‡ Ground-based reflectance measurements are comparable to satellite measurements ‡ Calibration sites on image are representative		‡ Assumptions may be false because pixels may not represent whole scene ‡ Calibrations of satellite (± 6.8%) and ground-based (±10%) not reliable ‡ Solar angles may differ ‡ Surficial reflectance may change between image acquisition and field work ‡ Geo-location of samples on imagery ‡ Requires sampling-intensive field work	‡ Rarely used
Atmospheric Modelling	5S Radiative Transfer Code	Simple to moderate	‡ Models are fair representations of the system ‡ Incorporate viewing geometry and adjacent effect	‡ Take into account absorption and scatter ‡ If standard model profiles are fitted to imagery, it is difficult to estimate accuracy	‡ Ideally require ancillary metereological data which are often difficult to acquire	‡ Suitable for image–image comparisons and ratioing

BOX 7.4

Example of output of 5S radiative transfer code for TM band 1 of a Landsat image

GEOMETRICAL CONDITIONS IDENTITY

T.M. OBSERVATION; MONTH: 11 DAY : 22 UNIVERSAL TIME: 14.55 (HH.DD); LATITUDE:21.68; LONGITUDE: -72.29; SOLAR ZENITH ANGLE: 51.16; SOLAR AZIMUTH ANGLE: 142.10; OBSERVATION ZENITH ANGLE: 0.00; OBSERVATION AZIMUTH ANGLE: 0.00; SCATTERING ANGLE:128.84; AZIMUTH ANGLE DIFFERENCE: 142.10

ATMOSPHERIC MODEL DESCRIPTION

ATMOSPHERIC MODEL IDENTITY: TROPICAL (UH2O=4.12 G/CM2 ,UO3=.247 CM)
AEROSOLS TYPE IDENTITY : MARITIME AEROSOLS MODEL
OPTICAL CONDITION IDENTITY : VISIBILITY 35.00 KM OPT. THICK. 550NM 0.1823

SPECTRAL CONDITION

TM 1 VALUE OF FILTER FUNCTION WLINF = 0.430 MICRON / WLSUP = 0.550 MICRON

TARGET TYPE

HOMOGENEOUS GROUND; SPECTRAL CLEAR WATER REFLECTANCE 0.042

INTEGRATED VALUES

APPARENT REFLECTANCE 0.108; APPAR. RADIANCE (W/M2/SR) 2.642
TOTAL GASEOUS TRANSMITTANCE 0.987

INT. NORMALIZED VALUES

% OF IRRADIANCE AT GROUND LEVEL			REFLECTANCE AT SATELLITE LEVEL		
% OF DIR. IRR.	% OF DIFF. IRR.	% OF ENV. IRR	ATM. INTRINS	BACKG.	PIXEL
0.668	0.325	0.006	0.076	0.007	0.025

INT. ABSOLUTE VALUES

IRR. AT GROUND LEVEL (W/M2)			RAD. AT SATEL. LEVEL (W/M2/SR)		
DIR. SOLAR	ATM.	DIFF. ENV.	ATM. INTRIN	BACKG.	PIXEL
43.582	21.169	0.421	1.858	0.180	0.604

INTEGRATED FUNCTION FILTER 0.061 (MICRONS)
INTEGRATED SOLAR SPECTRUM 122.586 (W/M2)

INTEGRATED VALUES

	DOWNWARD	UPWARD	TOTAL
GLOBAL GAS TRANS.	0.992	0.995	**0.987**
WATER GAS TRANS.	1.000	1.000	1.000
OZONE GAS TRANS.	0.992	0.995	0.987
CARBON DIOXIDE	1.000	1.000	1.000
OXYGEN	1.000	1.000	1.000
RAYLEIGH SCA.TRANS.	0.882	0.922	0.813
AEROSOL SCA. TRANS.	0.959	0.983	0.943
TOTAL SCA. TRANS.	0.849	0.915	**0.776**

	RAYLEIGH	AEROSOLS	TOTAL
SPHERICAL ALBEDO	0.129	0.044	**0.156**
OPTICAL DEPTH	0.164	0.188	0.352
REFLECTANCE	0.068	0.009	**0.077**
PHASE FUNCTION	1.043	0.102	0.540
SINGLE SCAT. ALBEDO	1.000	0.990	0.994

☞ Although the 5S code is described here, an updated version known as the 6S code is also available. The main advantage of the 6S code is that it is better adapted for airborne imagery.

Stage 1 Run the 5S code for each band in the imagery
The inputs of the model are summarised in Table 7.8 but the exact sequence and format of the code is described in the software. Note that it is possible to input either a general model

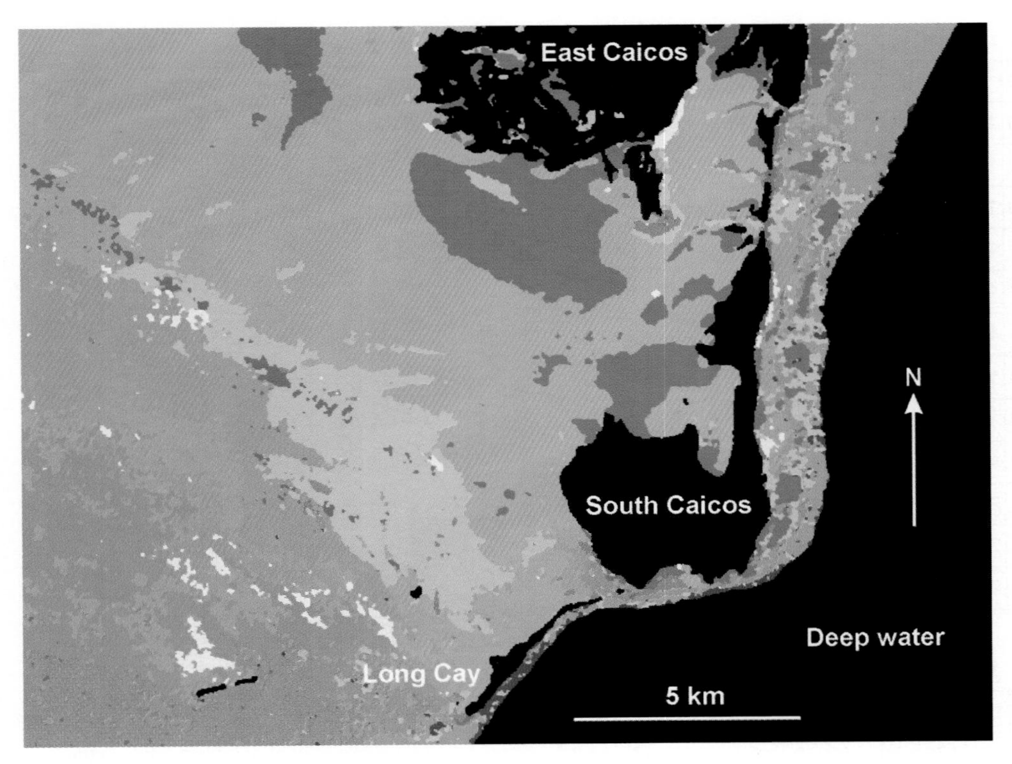

Finescale habitat map (13 classes) of east part of Caicos Bank based on a supervised classification of Landsat TM data subjected to water column correction and contextual editing. Such maps appear convincing. However, the overall accuracy of the map is only 31% and as such it is of little use as a basis for planning and management. An unsupervised classification of the same area, with classes assigned to habitats using extensive local knowledge, looked similarly plausible but had an overall accuracy of only 16%.

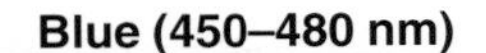

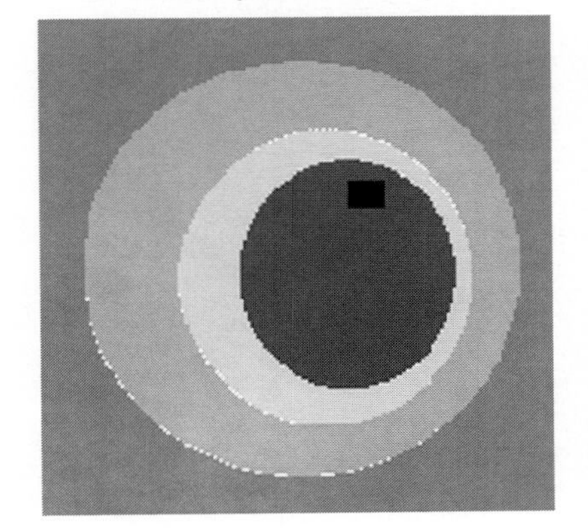

Green (530–560 nm)

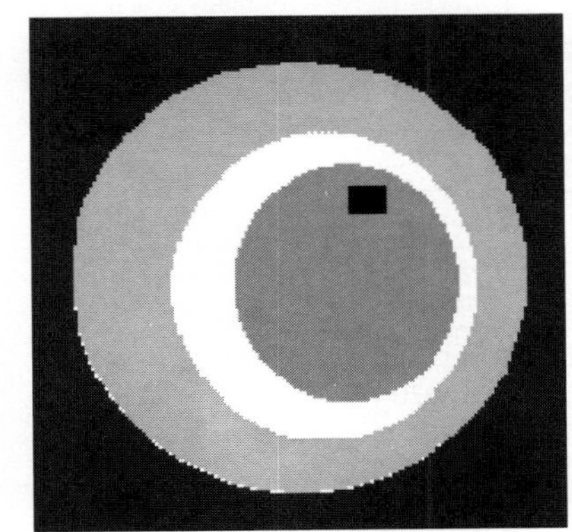

Red (625–675 nm)

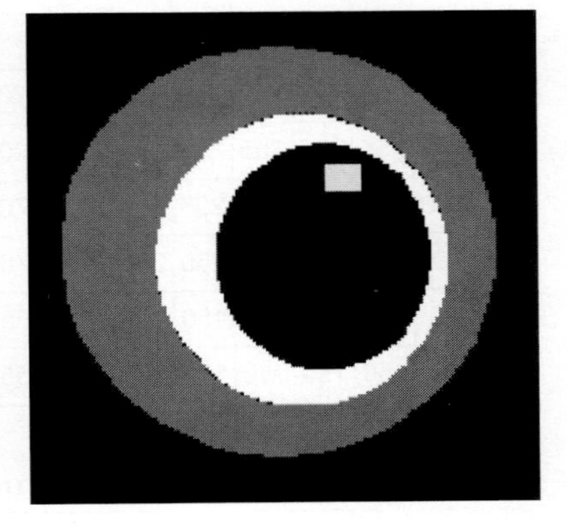

Colour composite image

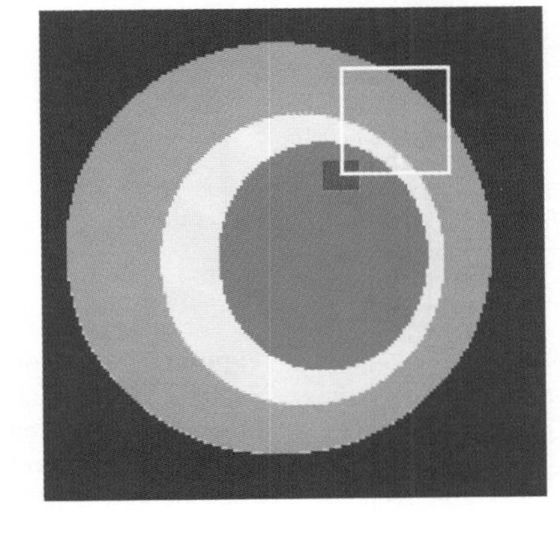

Key Habitat/object type

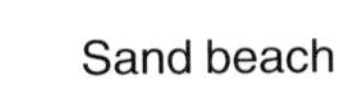

Figure 1.2 Hypothetical images of a small imaginary atoll in the Indian Ocean recorded by a sensor measuring the amount of sunlight reflected in the blue, green and red parts of the visible spectrum. The upper three images show how each sensor waveband records the different habitats/objects. The colour composite image at the bottom is created by displaying the blue band image on the computer monitor's blue gun, the green band image on the green gun, and red band image on the red gun. Because the sensor is recording at wavelengths similar to those used in human vision, the colour image is similar to what we would see if we flew over the atoll. The inset area is discussed further in Figure 1.3 (Plate 2).

Blue (450–480 nm)

180	180	180	180	180	180	180	180	180	130	130	130	130
180	180	180	180	180	180	180	180	180	180	130	130	130
180	180	180	180	180	180	180	180	180	180	180	130	130
180	180	180	180	180	180	180	180	180	180	180	180	130
180	180	180	180	180	180	180	180	180	180	180	180	180
200	190	180	180	180	180	180	180	180	180	180	180	180
200	200	200	190	180	180	180	180	180	180	180	180	180
200	200	200	200	200	180	180	180	180	180	180	180	180
90	200	200	200	200	200	180	180	180	180	180	180	180
90	90	90	145	200	200	200	187	180	180	180	180	180
90	90	90	90	90	200	200	200	180	180	180	180	180
0	0	90	90	90	90	200	200	200	180	180	180	180
0	0	90	90	90	90	90	200	200	200	180	180	180

Green (530–560 nm)

170	170	170	170	170	170	170	170	170	40	40	40	40
170	170	170	170	170	170	170	170	170	170	40	40	40
170	170	170	170	170	170	170	170	170	170	170	40	40
170	170	170	170	170	170	170	170	170	170	170	170	40
170	170	170	170	170	170	170	170	170	170	170	170	170
255	212	170	170	170	170	170	170	170	170	170	170	170
255	255	255	212	170	170	170	170	170	170	170	170	170
255	255	255	255	255	170	170	170	170	170	170	170	170
150	255	255	255	255	255	170	170	170	170	170	170	170
150	150	150	202	255	255	255	202	170	170	170	170	170
150	150	150	150	150	255	255	255	170	170	170	170	170
10	10	150	150	150	150	255	255	255	170	170	170	170
10	10	150	150	150	150	150	255	255	255	170	170	170

Table 1.3 Spectra of the five different habitat/object types in the inset area. Note the four mixels, that is pixels which overlie two different habitats.

Surface category	Blue DN	Green DN	Red DN
Deeper water	130	40	10
Shallow lagoon	180	170	70
Sandy beach	200	255	255
Forest	90	150	20
Red-roofed buildings	0	10	200

Red (625–675 nm)

70	70	70	70	70	70	70	70	70	10	10	10	10
70	70	70	70	70	70	70	70	70	70	10	10	10
70	70	70	70	70	70	70	70	70	70	70	10	10
70	70	70	70	70	70	70	70	70	70	70	70	10
70	70	70	70	70	70	70	70	70	70	70	70	70
255	162	70	70	70	70	70	70	70	70	70	70	70
255	255	255	162	70	70	70	70	70	70	70	70	70
255	255	255	255	255	70	70	70	70	70	70	70	70
20	255	255	255	255	255	70	70	70	70	70	70	70
20	20	20	137	255	255	255	139	70	70	70	70	70
20	20	20	20	20	255	255	255	70	70	70	70	70
200	200	20	20	20	20	255	255	255	70	70	70	70
200	200	20	20	20	20	20	255	255	255	70	70	70

Figure 1.3 Details of the inset area of Figure 1.2 (Plate 1) showing the digital numbers (DNs) recorded for each waveband. Groups of 5 x 5 pixels have been merged together to allow the digital numbers recorded for each pixel to be displayed more easily. Each merged pixel is thus 10 x 10 m and the inset is 130 x 130 m.

PLATE 3

Coast of Negros Island, Philippines showing aquaculture development in a mangrove area.
[*Photograph:* A. Edwards]

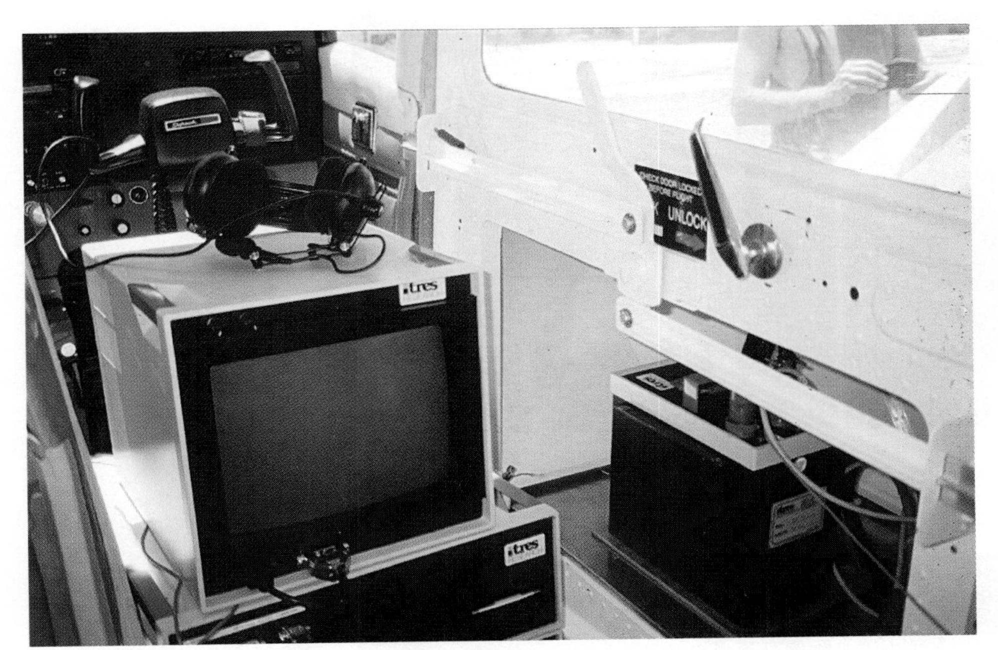

The Compact Airborne Spectrographic Imager (CASI) sensor (to the right, mounted inside specially modified door) and its computer control equipment set up in a Cessna 172N on South Caicos. The operator sat behind the computer monitor with the pilot in front. A heavy-duty battery provided power for the CASI instrument as the aeroplane's power supply was insufficient.
[*Photograph:* A. Edwards]

The specially modified Cessna door used to house the CASI instrument. The lens of the sensor can be seen projecting through the underside of the door.
[*Photograph:* A. Edwards]

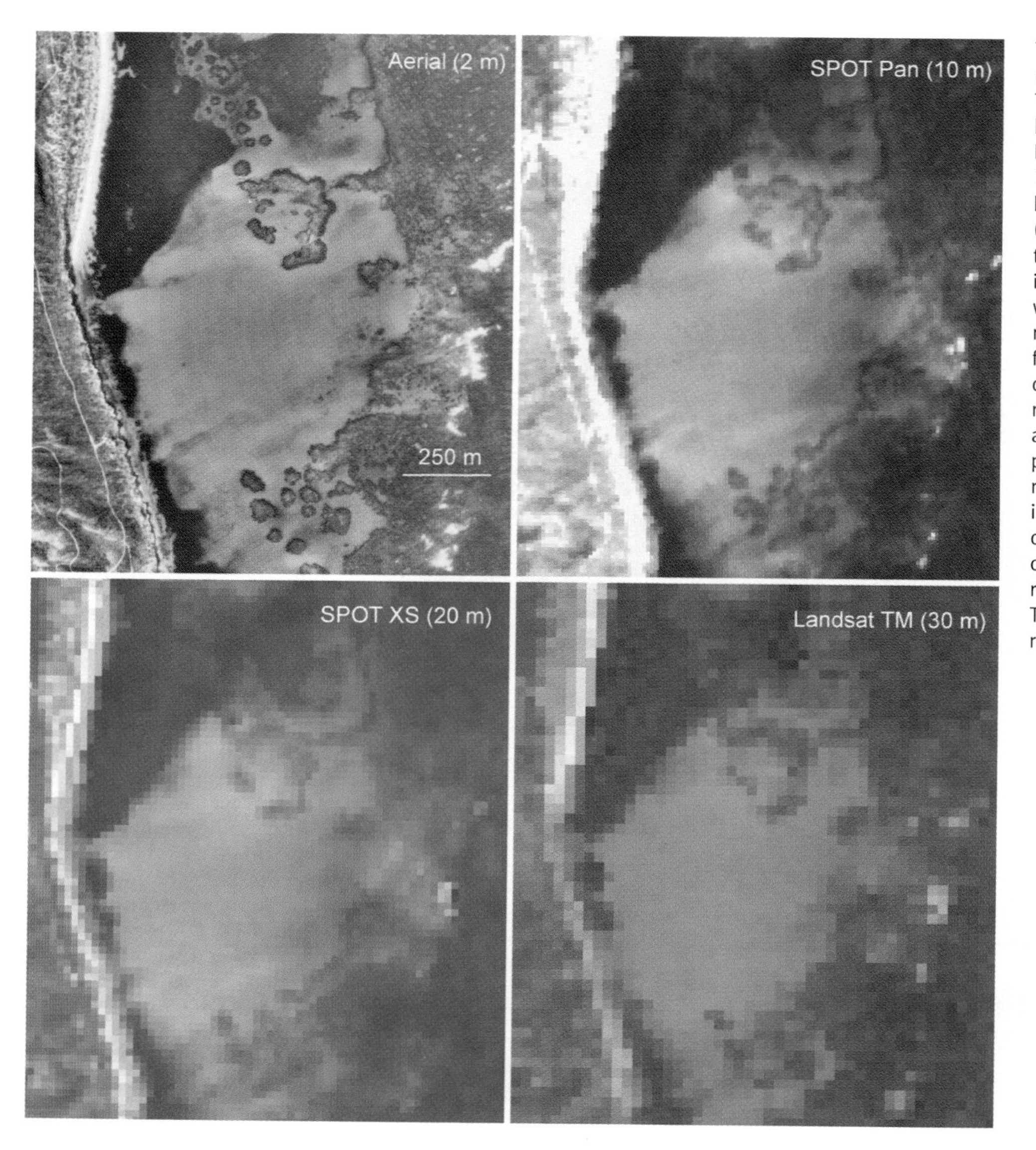

The effect of differing spatial resolution on image quality. The area shown includes a sand beach with some seagrass leaf litter on the strand line with *Thalassia* dominated seagrass beds immediately offshore (dark areas of fairly uniform texture). On the right of the area is a fringing coral reef with waves visible breaking on the reef crest. In between the fringing reef and the seagrass is calcareous sand (pale) with coral reef patches on it. The images are from a black and white aerial photograph scanned at 2 m resolution, a SPOT Panchromatic image (10 m resolution), and colour composite images derived from SPOT XS (20 m resolution) and Landsat Thematic Mapper (30 m resolution).

Oblique aerial photograph of the reef flat at Moorea, French Polynesia showing major geomorphological zones (see Figure 9.1). Mapping such zones using remote sensing is relatively straightforward due to the shallow water depth (< 3 m) and sharp boundaries. [*Photograph:* P. Mumby]

PLATE 5 FIELD SURVEY TECHNIQUES

Preparing to survey a deeper reef site (> 5 m depth) near Long Cay on the Caicos Bank using a one-metre quadrat to record percentage cover of different biota and substrate types. [*Photograph:* P. Mumby]

Underwater survey of reef at about 10 m depth. These surveys were used to construct a classification scheme appropriate for the area and guide supervised classification. [*Photograph:* E. Green]

Carrying out accuracy assessment from a boat in shallow water using a glass-bottom viewer to determine habitat categories for a stratified random sample of sites. [*Photograph:* P. Mumby]

A white tarpaulin (2 x 2 m) is secured to the sea bed at a known point (fixed with a differential GPS) prior to a survey flight to provide a ground control point (GCP) for the geometric correction of airborne imagery. [*Photograph:* P. Mumby]

Measuring attenuation of irradiance of a given wavelength at different depths using a light sensor lowered through the water column. [*Photograph:* E. Green]

Measuring the spectrum of a coral (*Porites* sp.) using a spectroradiometer. An identical instrument is placed over a uniform reflecting surface (white card) above the water surface. The ratio of the measured spectra gives the reflectance spectrum of the coral. [*Photograph:* P. Mumby]

Figure 5.9 High oblique colour aerial photograph of Cockburn Harbour, Turks and Caicos Islands. Although the geometry of this photograph is not suited to mapping, it would aid the interpretation of other imagery. Seagrass beds and some reef habitats are clearly visible. See Figure 6 in **Background to the Research behind the *Handbook*** for location of this photograph. [*Photograph:* P. Mumby]

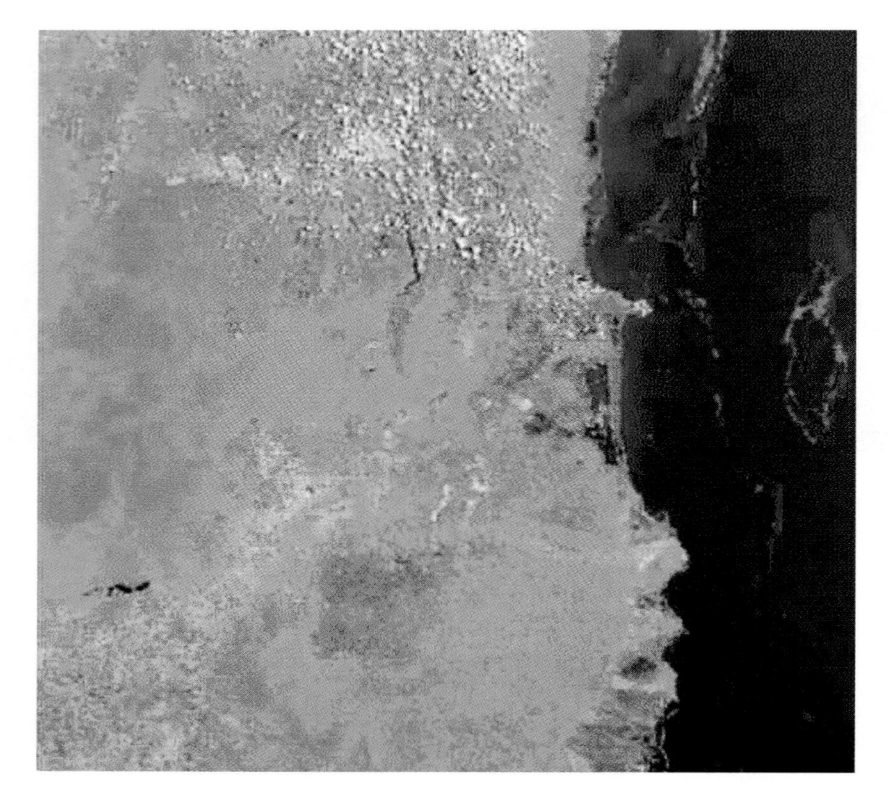

Figure 5.13 Quick-look of a Landsat TM scene of Belize, Central America showing the mainland, barrier reef and an offshore atoll. Coastal areas are cloud free although there is some cloud inland. (*Source:* EOSAT).

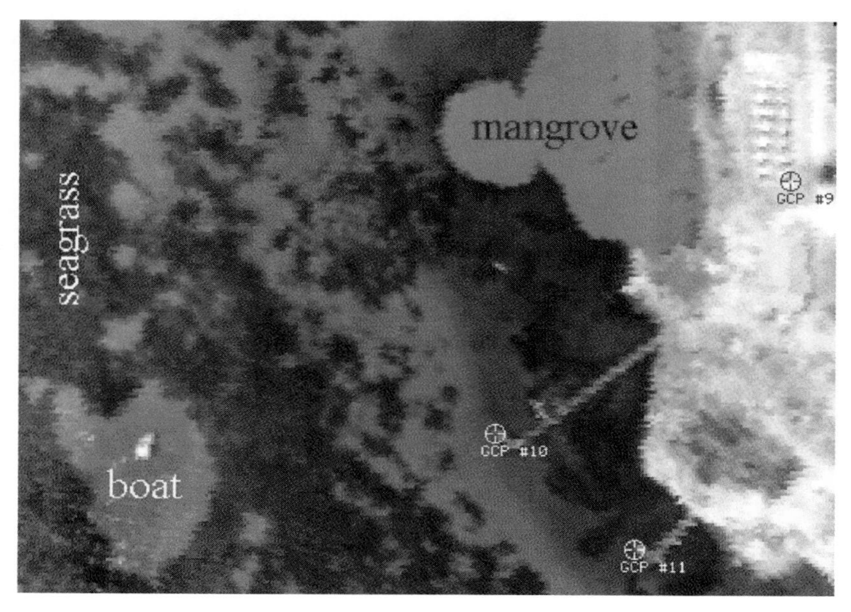

Figure 6.2 The selection of suitable Ground Control Points (GCPs). This figure shows three of the GCPs which we used to geometrically correct a CASI image from South Caicos. GCP #9 is the corner of a set of oil tanks, GCP #10 and #11 are the outer corners of the ends of boat jetties. All three could be located unambiguously both on the image and in the field. The boat, although moored, would not have been a suitable GCP because its position could have changed over a few metres as it moved with the wind.

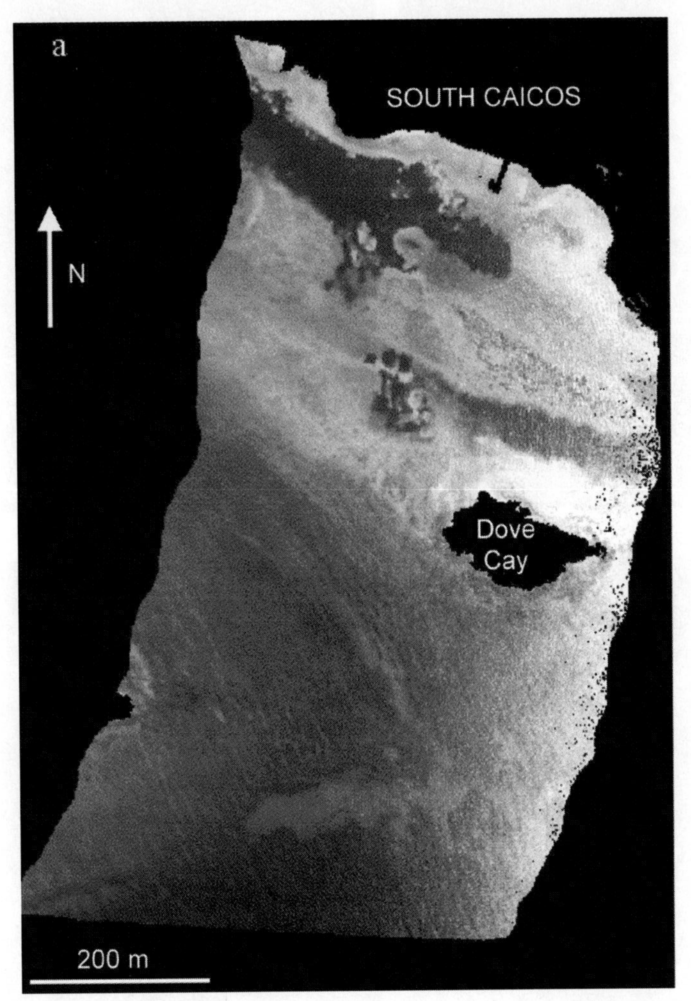

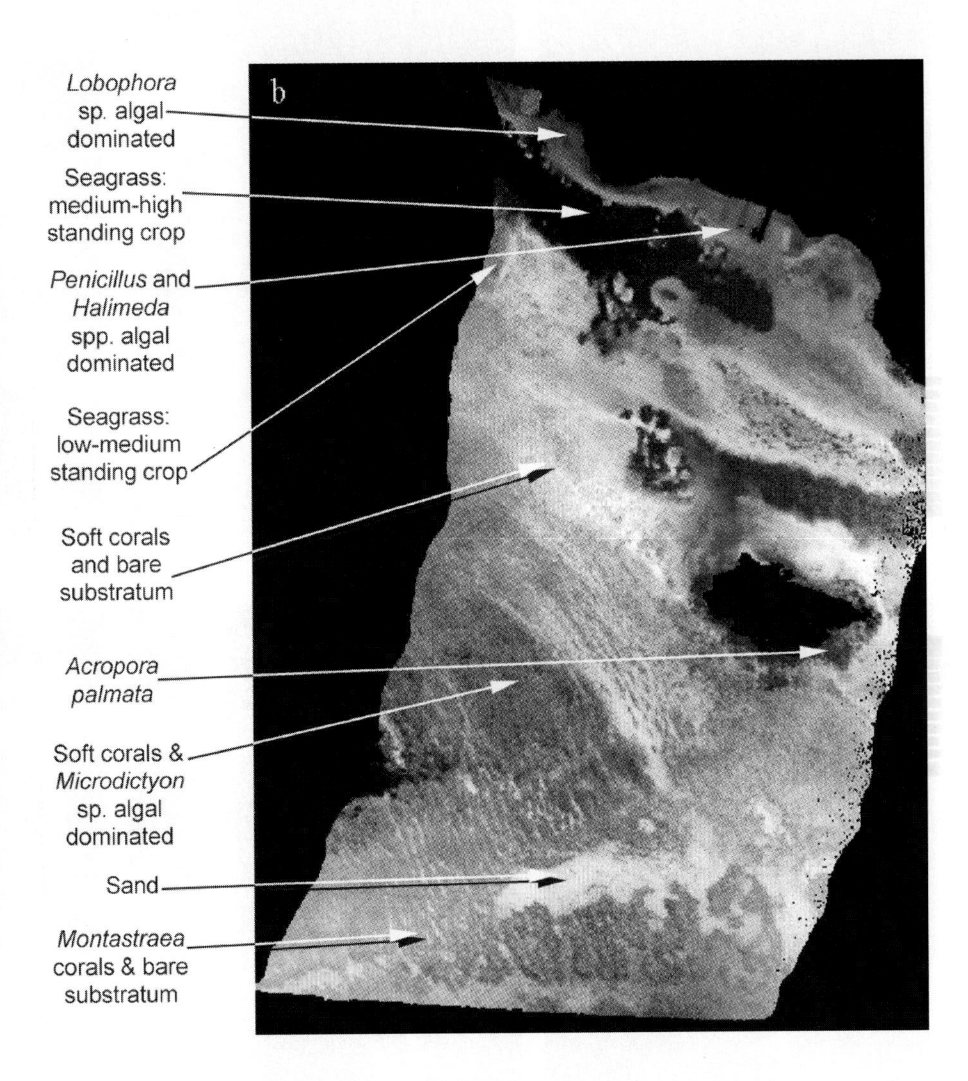

Figure 8.6 CASI imagery from Cockburn Harbour (TCI) showing comparison between raw data (a) and depth invariant bands (b). The depth invariant data reveal far more detail on the buttress and valley structures to the deeper (bottom) end of the scene. The processing has also reduced the effect of sunglint. Major habitat types are labelled.

PLATE 8 CORAL AND MACROALGAL HABITATS OF THE CAICOS BANK

1. Elkhorn coral, *Acropora palmata*, habitat in high energy shallow water on exposed coasts. [*Photograph:* J. Ridley, Coral Cay Conservation]
2. Gorgonian plain habitat dominated by soft corals. [*Photograph:* A. Edwards]
3. An unusual shallow water macroalgal habitat of the central Caicos Bank dominated by the green alga *Acetabularia*. [*Photograph:* A. Edwards]
4. *Montastraea* coral dominated habitat. [*Photograph:* A. Edwards]
5. Soft bottom community dominated by the green calcareous alga *Halimeda*. [*Photograph:* A. Edwards]
6. An unusual habitat dominated by the coralline red alga *Amphiroa* and a brown sponge. [*Photograph:* E. Green]
7. Macroalgal habitat dominated by the brown alga *Lobophora* and red alga *Laurencia*. [*Photograph:* E. Green]
8. Macroalgal habitat dominated by the green alga *Microdictyon*. [*Photograph:* M. Weedon]

Table 7.8 Inputs to the 5S radiative transfer code for atmospheric correction. With the exception of inputs highlighted in **bold**, general inputs can be used where specific information is not available. See text for further information.

Parameter	Specific Inputs	General Inputs
Viewing and illumination geometry*+	‡ **Type of sensor (e.g. SPOT)** ‡ **Date and time of image acquisition** ‡ **Latitude and longitude of scene centre**	‡ None
Atmospheric profile	‡ Temperature (K) ‡ Pressure (mB) ‡ Water vapour density ($g.m^{-3}$) ‡ Ozone density ($g.m^{-3}$)	‡ Tropical ‡ Mid latitude summer ‡ Mid latitude winter ‡ Subarctic summer ‡ Subarctic winter
Aerosol components	‡ Dust-like component (%) ‡ Oceanic component (%) ‡ Water soluble component (%) ‡ Soot component (%)	‡ Continental aerosol model ‡ Maritime aerosol model ‡ Urban aerosol model
Aerosol concentration	‡ Aerosol optical depth at 550 nm	‡ **Meteorological visibility (km)****
Spectral band	‡ Lower and upper range of band (µm)	‡ Band name (e.g. SPOT XS1)
Ground reflectance	(a) Choose homo- or heterogeneous surface (b) If heterogeneous, enter reflectance of target surface, surrounding surface and target radius (km)	As specific inputs, except 5S supplies mean spectral value for green vegetation, clear water, sand, lake water

* this information is available in image header file and/or accompanying literature

** a value of 35 km is appropriate for the humid tropics in clear weather (e.g. the Bahamas)

+ it is occasionally difficult to find the time of image acquisition (e.g. for Landsat MSS data). To solve this problem, input an approximate time into the 5S code. The 5S model output will display the calculated values of solar zenith and azimuth (based on the information supplied by the user). Since the actual solar zenith and azimuth are found in the header files / accompanying literature, it is possible to assess the "correctness" of the estimated acquisition time. The correct acquisition time is reached when the output solar angles agree with those stated in the header file.

for the type of atmospheric conditions or, if known, specific values for atmospheric properties at the time the image was taken. In the event that no atmospheric information is available, the only parameter that needs to be estimated is the horizontal visibility in kilometres (meteorological range).

Stage 2 Calculate the inversion coefficients and spherical albedo from the 5S code output

The 5S code provides a complex output but only some of the information is required by the user. An example of 5S output (Box 7.4) highlights the important information in bold.

The following inversion coefficients, A_l and B_l are calculated as follows (see Box 7.4 for input values):

$$A_l = \frac{1}{\text{Global gas trans.} \times \text{Total scattering trans.}}$$

$$= \frac{1}{0.987 \times 0.776} = 1.305$$

$$B_l = \frac{-\text{Reflectance}}{\text{Total scattering transmittance}}$$

$$= \frac{-0.077}{0.776} = -0.099$$

coefficients A_l and B_l are then combined to give Y.

$$Y = (A_l \times \text{Apparent reflectance}\,[\rho]) + B_l$$

☞ **The imagery has already been converted to apparent reflectance (section 2). p therefore represents the value of each pixel in the imagery. The final parameter needed is the spherical albedo (S):**

$$S = 0.156$$

which varies from band to band.

Stage 3 Implementation of model inversion to imagery

(i) Both A_l and B_l are calculated from the 5S model (as above).

(ii) A simple model has to be made in the image processing software to generate a file of Y values using apparent reflectance (pixel values) and the constants A_l and B_l.

(iii) The following equation is then used to obtain surface reflectance p_s:

$$p_s = \frac{Y}{1 + SY}$$

This is best implemented as a second stage to the Y-generating model above.

☞ **If erroneous reflectance values of greater than 1 are obtained, it is necessary to alter the parameters in the 5S code until the range of reflectance is appropriate. For example, the atmospheric visibility may have been incorrect and should be changed.**

Obtaining the 6S Radiative Transfer Code: The code can be obtained over the internet at:

ftp://kratmos.gsfc.nasa.gov/6S/.

Look at the README file for further instructions.

Time considerations

A summary of the learning and processing times for radiometric correction is given for guidance (Table 7.9). Although the reader would also need to spend time learning the techniques, we hope that the duration will be less than we encountered. For example, we spent 4 days reading fairly disparate literature about radiometric correction. If the relevant information had been available in a simple summary as detailed here, this would probably have reduced the reading time to one day.

Table 7.9 Guidelines to learning and processing times for radiometric correction.

Activity	Time
Background reading	4 days
Learning software to implement equations for DN–radiance–apparent reflectance conversion	1 day
Implementation of DN–radiance–apparent reflectance equation	4 hours/scene
Learning 5S Radiative Transfer Code and creating processing algorithm	1 day
Running the 5S code and interpretation of output	15 minutes/band
Implementation of the inversion coefficients to imagery	2 hours/scene

Conclusions

Remotely sensed satellite data should be radiometrically corrected if you are intending (i) to compare images in time or space, (ii) to model interactions between electromagnetic radiation and a surface feature, or (iii) to use band ratios for imagery analysis. One-off habitat mapping with multispectral classification does not require radiometric correction. However, the radiometric properties of airborne data are much more variable than satellite data because they change during imagery acquisition (i.e. during flight). It follows that radiometric correction is required for airborne data even when conducting a one-off habitat map. Failure to do so may complicate mosaicing of images between flight lines and constrain multispectral classification, because habitat spectra will vary along each flight line according to instantaneous atmospheric conditions.

Radiometric correction is a three stage process. For satellite data, steps 1 and 2 are standard and the user only faces a choice of methods for atmospheric correction (step 3). Wherever possible, use of a radiative transfer code such as 5S or 6S is recommended for this step. Although the radiometric correction of airborne data is usually the instrument operator's responsibility, the consumer should ensure that each scan-line of data has been corrected to at-sensor reflectance using an incident light sensor.

References

Chavez, P.S., Berlin, G.L., and Mitchell, W.B., 1977, Computer enhancement techniques of Landsat MSS digital images for land use/land cover assessments. *Remote Sensing of Earth Resources*, **6**, 259.

Crippen, R.E., 1987, The regression intersection method of adjusting image data for band ratioing. *International Journal of Remote Sensing*, **8**, 137–155.

Kaufman, Y.J., 1989, The atmospheric effect on remote sensing and its correction. In *Theory and Applications of Optical Remote Sensing*, edited by G. Asrar (Chichester: John Wiley & Sons), pp. 336–428.

Markham, B.L., and Barker, J.L., 1987, Radiometric properties of U.S. Processed Landsat MSS data. *Remote Sensing of Environment*, **22**, 39–71.

Mather, P.M., 1987, *Computer Processing of Remotely Sensed Data* (Chichester: John Wiley & Sons).

Moran, M.S., Jackson, R.D., Clarke, T.R., Qi, J., Cabot, F., Thome, K.J., and Markham, B.L., 1995, Reflectance factor retrieval from Landsat TM and SPOT HRV Data for Bright and Dark Targets. *Remote Sensing of Environment*, **52**, 218–230.

Potter, J.F., and Mendlowitz, M.A., 1975, On the determination of haze levels form Landsat data. *Proceedings of the 10th International Symposium on Remote Sensing of the Environment* (Ann Arbor: Environmental Research Institute of Michigan). p. 695.

Price, J.C., 1987, Calibration of satellite radiometers and the comparison of vegetation indices. *Remote Sensing of Environment*, **21**, 15–27.

Switzer, P., Kowalik, W.S., and Lyon, R.J.P., 1981, Estimation of atmospheric path-radiance by the covariance matrix method. *Photogrammetric Engineering and Remote Sensing*, **47**, 1469–1476.

Tanre, D., Deroo, C., Dahaut, P., Herman, M., and Morcrette, J.J., 1990, Description of a computer code to simulate the satellite signal in the solar spectrum: the 5S code. *International Journal of Remote Sensing*, **11**, 659–668.

8

Water Column Correction Techniques

Summary *When trying to map or derive quantitative information about underwater habitats, the depth of water significantly affects the remotely sensed measurement. It does so to such an extent that it can cause us to confuse the spectra of (say) bare sand and seagrass. This chapter describes a method for compensating for the effects of variable depth. Implementation of the method was found to be cost-effective for improving the thematic accuracy of coastal habitat maps and essential for deriving empirical relationships between remotely sensed data and features of interest in the marine environment.*

Introduction

The previous two chapters dealt with the geometric and radiometric correction of digital imagery. These processes are required for almost all remote sensing applications whether land-orientated or marine. This chapter describes an additional pre-processing step which is only required when assessing underwater habitats.

When light penetrates water its intensity decreases exponentially with increasing depth. This process is known as attenuation and it exerts a profound effect on remotely sensed data of aquatic environments. The severity of attenuation differs with the wavelength of electromagnetic radiation (EMR). In the region of visible light, the red part of the spectrum attenuates more rapidly than the shorter-wavelength blue part (see Figure 8.1). As depth increases, the separability of habitat spectra declines (Figure 8.1). In practice, the spectra of sand at a depth of 2 m will be very different to that at 20 m – yet the substratum is the same. In fact, the spectral signature of sand at 20 m may be similar to that of seagrass at (say) 3 m. The spectral radiances recorded by a sensor are therefore dependent both on the reflectance of the substrata and on depth. These two influences on the signal will create considerable confusion when attempting to use visual inspection or multispectral classification to map habitats. Since most marine habitat mapping exercises are only concerned with mapping benthic features, it is useful to remove the confounding influence of variable water depth. This chapter describes a fairly straightforward means of compensating for variable depth but it is only truly applicable to clear waters such as those in coral reef environments. For more turbid waters, see the comments at the end of the chapter.

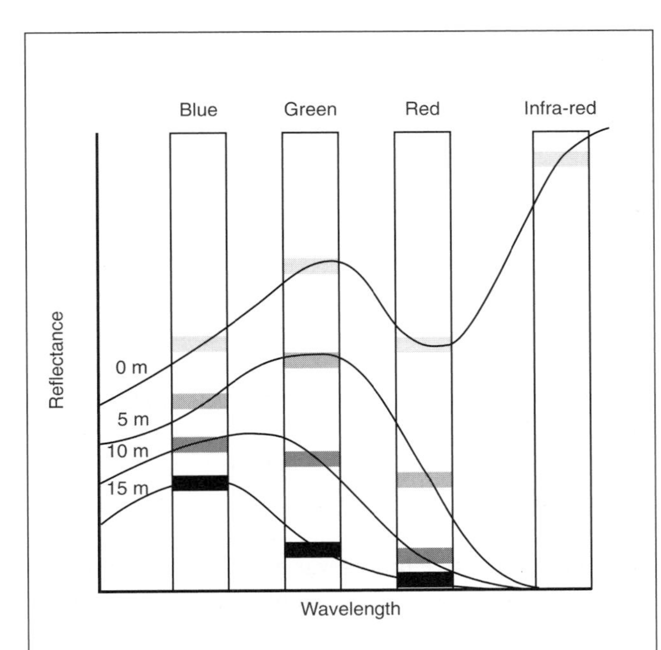

Figure 8.1 Diagram to show how the spectra for a habitat (such as macroalgae or seagrass) might change with increasing depth for a four waveband sensor measuring radiance in the blue, green, red and near infra-red parts of the electromagnetic spectrum. Differential attenuation of the four wavebands in the water column results in both a decreased ability to discriminate between different habitats with increasing depth and different spectra being recorded for the same habitat at different depths.

Light attenuation in water

The exponential decay of light intensity with increasing depth results from two processes, absorption and scattering. These processes have been discussed in an atmospheric context (Chapter 7) but it is worth pointing out the nature of attenuation in water.

Absorption

Absorption involves the conversion of electromagnetic energy into other forms such as heat or chemical energy (e.g. photosynthesis in phytoplankton). The main absorbers in seawater are:

- algae (phytoplankton),
- inorganic and organic particulate matter in suspension (excluding algae),
- dissolved organic compounds (yellow substances) which result from the breakdown of plant tissue,
- water itself, which strongly absorbs red light and has a smaller effect on shorter wavelength blue light (hence the blue colour of clear water).

Absorption is wavelength-dependent (Figure 8.1). The chlorophyll in algae appears ‘green’ because it reflects in the central portion of the visible spectrum (i.e. the green) and absorbs strongly at either end. Dissolved organic compounds absorb strongly at the short wavelength (blue) part of the spectrum and reflect strongly in the yellow-red end (hence they impart a yellow colour to the water).

Scattering

EMR may interact with suspended particles in the water column and change direction. This process of scattering is largely caused by inorganic and organic particulate matter and increases with the suspended sediment load (turbidity) of the water. For further details see Open University (1989).

Classification of water bodies

The clarity of water bodies varies on many scales. For example, many coastal areas exhibit a seaward gradient of turbid to clear waters created by increases in depth (i.e. less resuspension of sediments) and reduced input from terrestrial sources such as sediment-laden rivers. On a large scale, oceans also vary in their overall turbidity.

Jerlov (1951) formally classified oceanic water types according to their optical attenuation properties. Type I waters were represented by extremely clear oceanic waters. Most clear coastal waters were classified as Type II because attenuation tends to be greater than that for oceanic waters of low productivity. However, many water bodies were found to lie between Types I and II and the former was subsequently split into Types IA and IB (Jerlov 1964). Most coral reef waters fall into categories I or II. Type III waters are fairly turbid and some regions of coastal upwelling are so turbid that they are unclassified. The subject will not be explored further here; it was outlined because these terms are often mentioned in the literature. For further information, readers are referred to Jerlov (1976).

Compensating for the influence of variable depth on spectral data

Removal of the influence of depth on bottom reflectance would require: (i) a measurement of depth for every pixel in the image, and (ii) a knowledge of the attenuation characteristics of the water column (e.g. concentrations of dissolved organic matter). Good digital elevation models of depth are rare, particularly for coral reef systems where charts are often inaccurate (but see Zainal 1994). As a compromise, Lyzenga (1978, 1981) put forward a simple image-based approach to compensate for the effect of variable depth when mapping bottom features (hereafter referred to as water column correction). Rather than predicting the reflectance of the seabed, which is prohibitively difficult, the method produces a ‘depth-invariant bottom index’ from each pair of spectral bands. The technique was tested for water in the Bahamas and is only appropriate where water clarity is good (i.e. most reef and seagrass areas; Jerlov water Types I or II). The procedure is divided into several steps.

☞ **Do not be put off by the equations described below. They are easily implemented as shown in the worked example which follows.**

Step 1 – removal of scattering in the atmosphere and external reflection from water surface

Most published accounts of the water-correction method suggest prior application of a crude atmospheric correction (Lyzenga 1978, 1981, Spitzer and Dirks 1987, Armstrong 1993, Maritorena 1996). This is based on the ‘dark pixel subtraction’ method described in Chapter 7. A large number of pixels are sampled from ‘deep water’ and their average radiance (or DN) is then subtracted from all other pixels in each band respectively:

$$\text{Atmospherically corrected radiance} = L_i - L_{si}$$

where L_i is the pixel radiance in band i and L_{si} is the average radiance for deep water in band i. However, if a full atmospheric correction has already taken place so that

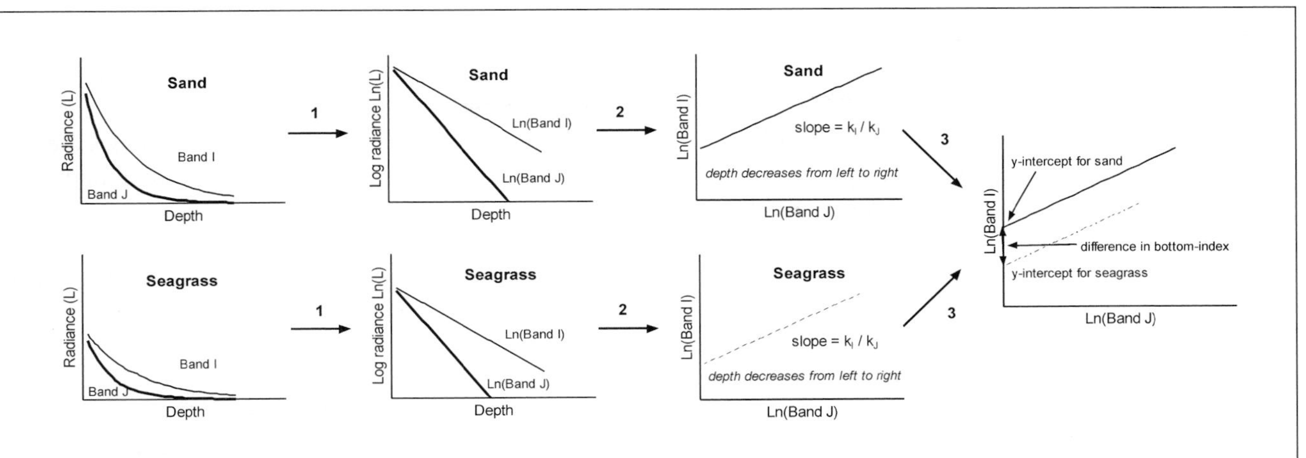

Figure 8.2 Processes of water column correction, showing the steps involved in creating depth-variant indices of bottom type for sand and seagrass

Step 1. Exponential attenuation of radiance with depth linearised for bands *i* and *j* using natural logarithms. (Band *i* has a shorter wavelength, and therefore attenuates less rapidly, than band *j*).

Step 2. Plot of (transformed) band *i* against (transformed) band *j* for a unique substratum at various depths. Gradient of line represents the ratio of attenuation coefficients, k_i/k_j. The ratio is the same irrespective of bottom type.

Step 3. Plotting of multiple bottom types. Each bottom type has a unique *y*-intercept (regardless of its depth). The *y*-intercept therefore becomes a depth-invariant index of bottom type.

pixel values have been converted to surface reflectance (Chapter 7), this process is unnecessary and values of surface reflectance can be used directly. In short, L_{si} can be ignored if a full atmospheric correction has been undertaken and this is preferred to using the cruder, dark pixel subtraction method.

Step 2 – linearise relationship between depth and radiance

In relatively clear water, the intensity of light will decay exponentially with increasing depth (Figure 8.2). If values of light intensity (radiance) are transformed using natural logarithms (ln), this relationship with depth becomes linear (Figure 8.2, step 1). Transformed radiance values will therefore decrease linearly with increasing depth. If X_i is the transformed radiance of a pixel in band i, this step is written as:

$$X_i = ln\ (L_i - L_{si})$$

for data which have not been atmospherically corrected;

$$X_i = ln\ (L_i)$$

for data which have been atmospherically corrected.

Step 3 – calculate the ratio of attenuation coefficients for band pairs

The irradiance diffuse attenuation coefficient (hereafter referred to as attenuation coefficient, *k)* describes the severity of light attenuation in water for that spectral band. It is related to radiance and depth by the following equation where a is a constant, r is the reflectance of the bottom and z is depth:

$$L_i = L_{si} + a.r.e^{(-2k_i z)}$$

Theoretically, it would be possible to rearrange the equation and generate an image of bottom type, *r* (reflectance) which is the measurement we seek. However, this approach is not feasible because there are too many unknown quantities – i.e. the value of the constant a, the attenuation coefficient for each band and the depth of water at each pixel. The method developed by Lyzenga does not require the actual calculation of these parameters but gets around the problem by using information from more than one band. All that is required is the **ratio** of attenuation coefficients between pairs of spectral bands. Use of ratios cancels out many of these unknowns and the ratios can be determined from the imagery itself. Two bands are selected and a bi-plot made of (log transformed) radiances (or reflectances) for the same substratum at differing depths (Figure 8.2, step 2). Since the effect of depth on measured radiance has been linearised and the substratum is constant, pixel values for each band will vary linearly according to their depth (i.e. points will fall on this straight line). The slope of the bi-plot represents the relative amounts of attenuation in each band. In fact, the slope represents the ratio of attenuation coefficients between bands. Conceptually, the line represents an axis of radiance (reflectance) values for a unique bottom type. As one moves along the line, the only change is depth.

Step 4 – generation of a depth-invariant index of bottom type

If radiance (reflectance) values for another bottom type were added to the bi-plot (Figure 8.2, step 3), a similar line would be obtained – once again, the only change between data points would be depth. However, since the second bottom type will not have the same reflectance as the first, the new line will be displaced either above or below the existing line (e.g. if line 1 was derived from sand which generally has a high reflectance, and line 2 was generated from seagrass with lower reflectance, the latter line would lie below that for sand). The gradient of each line should be identical because the ratio of attenuation coefficients k_i/k_j is only dependent on the wavelength of the bands and clarity of the water.

An index of bottom type can be obtained by noting the y-intercept for each bottom type (Figure 8.2, step 3). For example, while pixel values lying on the line for sand show considerable variation in radiance, they all represent the same bottom type and have the same y-intercept. The y-intercept for pixels of seagrass is considerably different. The y-axis therefore becomes an axis (or index) of bottom type.

Of course, not all pixel values for a given bottom type lie along a perfectly straight line (Figure 8.3). This is because of natural variation in bottom reflectance, patches of turbid water and sensor noise. Nevertheless, each pixel can be assigned an index of bottom type once the ratio of attenuation coefficients has been estimated (k_i/k_j). This is accomplished by 'connecting' each pixel on the bi-plot to the y-axis using an imaginary line of gradient k_i/k_j. Pixel values on the bi-plot are then converted to their corresponding positions on the y-axis (index of bottom type). Using this method, each pixel value is converted to an index of bottom type, which is independent of depth.

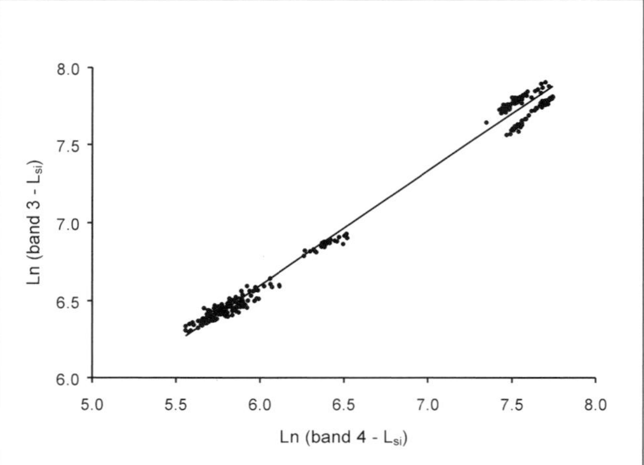

Figure 8.3 Bi-plot of log-transformed CASI bands 3 and 4. Data obtained from 348 pixels of sand with variable depth from 2–15 m

These depth-invariant indices of bottom type lie along a continuum but pixels from similar habitats will have similar indices.

The mathematics of the depth-invariant index are simple and are based on the equation of a straight line:

$$y = p + q \,.\, x$$

where p is the y-intercept, q is the gradient of the regression of y on x. The equation can be rearranged to give the y-intercept:

$$p = y - q \,.\, x$$

which in the case of Figure 8.2 (step 3) becomes,

$$depth\text{-}invariant\ index_{ij} = \ln(L_i - L_{si}) - \left[\left(\frac{k_i}{k_j}\right).\ln(L_j - L_{sj})\right]$$

or, if a full atmospheric correction has been undertaken already,

$$depth\text{-}invariant\ index_{ij} = \ln(L_i) - \left[\left(\frac{k_i}{k_j}\right).\ln(L_j)\right]$$

Each pair of spectral bands will produce a single depth-invariant band of bottom type. If the imagery has several bands with good water penetration properties (e.g. Landsat TM, CASI), multiple depth-invariant bands can be created. The depth-invariant bands may then be used for image processing or visual inspection instead of the original bands.

The theory described here scales the depth-invariant index to the y-intercept. Lyzenga (1978, 1981) provided a further modification which re-scaled the values to an axis orthogonal to the bi-plot slope. However, such a refinement does not alter the functionality of the process and it has not been included here for simplicity's sake.

When to implement water column correction

The methods described in this chapter have three main applications for mapping coastal habitats:

1. Multispectral classification of marine habitats. The accuracies of some habitat maps of the Turks and Caicos Islands showed significant improvement once water column correction had taken place. Improvements to accuracy were greatest for imagery which had at least three water-penetrating spectral bands, from which three (or more) depth-invariant bottom indices could be created and input to the spectral classification (for a more detailed discussion, refer to Chapter 11).
2. Establishing quantitative empirical relationships between image data and marine features. Water column correction was found to be essential for relating digital image data to seagrass standing crop (see Chapter 16).

3. Visual interpretation of digital data. By removing much of the depth-induced variation in spectral data, water column correction makes for a better visual assessment of habitat types.

Worked example of depth-invariant processing

Band selection

Pairs of spectral bands are selected which have different bottom reflectances but good penetration of water (i.e. visible wavebands) e.g. Landsat TM 1/2, 1/3, 2/3; Landsat MSS 1/2; SPOT XS 1/2. Two CASI bands are presented in this example (although the process was carried for several band pairs).

Calculation of deep water radiance (reflectance) in the absence of a full atmospheric correction

Calculation of the parameter L_{si}, deep water radiance (reflectance) does not take long. A group of pixels are selected which represent deep water (i.e. greater than 40 m). The pixel data are analysed either in the image processing software or transferred to a spreadsheet (e.g. Microsoft Excel). The mean and standard deviation of radiance (reflectance) are calculated for each band. Armstrong (1993) recommended subtracting two standard deviations from the mean to account for sensor noise. This (lower) value is then used as L_{si} in subsequent calculations.

Selection of pixels of uniform substratum and variable depth

A series of groups of pixels are selected from the imagery that have the same bottom type but different depths (totalling approximately 100 pixels). Areas of submerged sand are good because they are fairly recognisable to an interpreter without much field experience (Figure 8.4). Both bands should exhibit attenuation so avoid choosing areas of the imagery in which one of the bands is saturated. For example, shallow water (< 1 m) reflectances for Landsat TM band 1 will show little variation whereas reflectance for band 3 may vary considerably. The resulting bi-plot will have a gradient close to zero which cannot be used to determine the ratio of attenuation coefficients. Similarly, if the area chosen is too deep, one band may not penetrate and will only contain one value (L_{si} – the deep water radiance for that band). Pixel data for the selected areas in both bands are transferred to a spreadsheet and converted to natural logarithms.

Calculation of ratio of attenuation coefficients

Bi-plots of the transformed data are made and examined (Figure 8.3). Regions of saturation (vertical or horizontal lines of pixels on the bi-plot) are removed. The gradient of the line is not calculated using conventional least squares regression analysis (which is the standard equation given by most statistical packages). This is because the result depends on which band is chosen to be the dependent variable. Therefore, rather

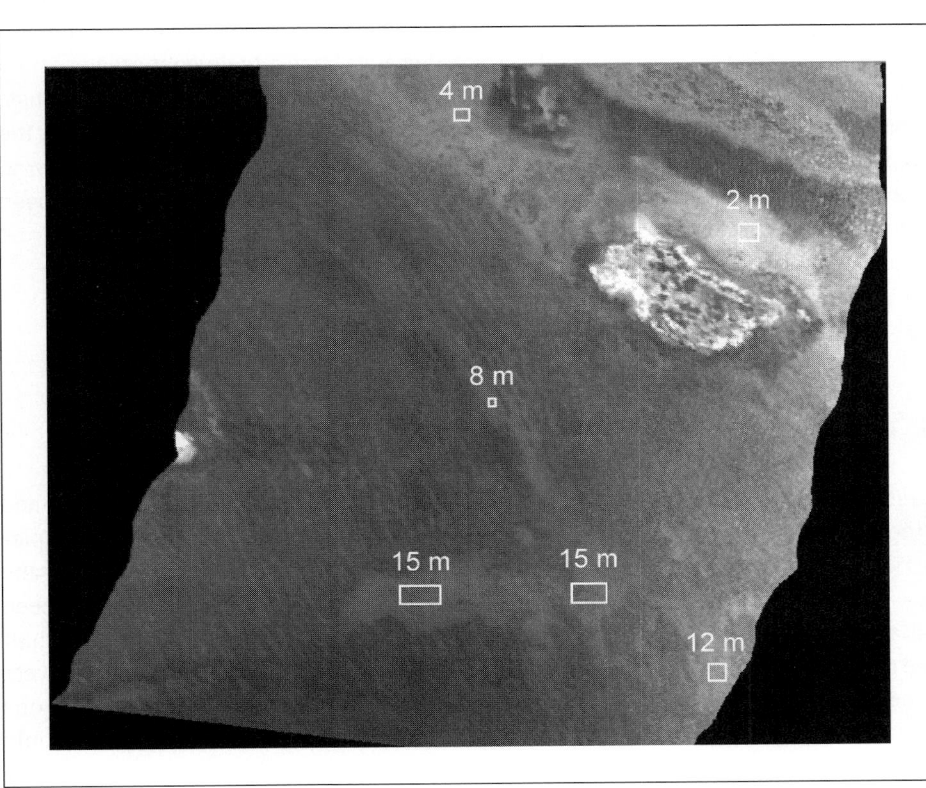

Figure 8.4 CASI image of Cockburn Harbour (Turks and Caicos Islands) showing the selection of pixels of sand at variable depth

than calculating the mean square deviation from the regression line in the direction of the dependent variable, the regression line is placed where the mean square deviation (measured perpendicular to the line) is minimised. The following equations are used. Most spreadsheets have functions allowing the calculation of variance and covariance (σ_{ii} is the variance of band i, σ_{ij} is the covariance between bands i and j).

$$k_i / k_j = a + \sqrt{(a^2 + 1)}$$

where

$$a = \frac{\sigma_{ii} - \sigma_{jj}}{2\sigma_{ij}}$$

and

$$\sigma_{ij} = \overline{X_i X_j} - (\overline{X_i} \times \overline{X_j})$$

(i.e. you will need a new column in the spreadsheet which multiplies X_i by X_j. From left to right, the inputs to the above equation are the average of this column (the mean of the products of X_i and X_j), the average of X_i and the average of X_j).

In practice, this method of calculating the bi-plot gradient (k_i/k_j) does not differ largely from the standard least squares method. For example, the standard calculation of gradient from CASI data in Figure 8.3 gave a value of 0.73. Calculation of the gradient by minimising the mean square deviation perpendicular to the regression line resulted in a value of 0.74. The latter value was used for the processing algorithm.

Implementation of depth-invariant algorithm to whole image (band pairs)

Prior to execution of depth-invariant processing all areas of land and cloud should be masked out. It is best to set pixels for these areas to zero. The depth-invariant algorithm is implemented in image processing software once the ratios of attenuation coefficients have been calculated for band pairs. In keeping with the theory provided above, the following equation is used: e.g. for bands i and j in the absence of a full atmospheric correction;

$$\textit{depth-invariant index}_{ij} = \ln(L_i - L_{si}) - \left[\left(\frac{k_i}{k_j}\right).\ln(L_j - L_{sj})\right]$$

This can be implemented in one process. For example, a flow model of the process would have the following structure (Figure 8.5). The results are presented in Figure 8.6 (Plate 7).

☞ **If some values of depth-invariant bottom-index are negative, an offset can be incorporated to make all data positive. If the minimum value was -8.3 and the maximum value was 10.5, all values could be offset by adding 8.4. The new scale would be 0.1 – 18.9.**

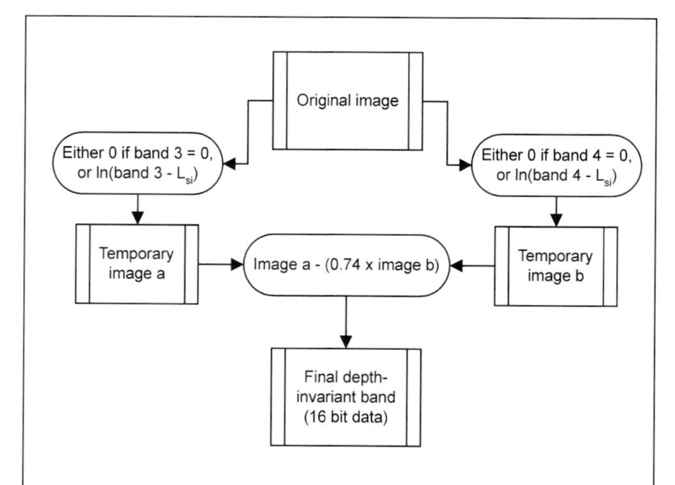

Figure 8.5 Model representation for implementation of depth invariant processing algorithm. The 'either, or' arguments are included to prevent zeros creating a problem in the calculation. To encompass a wide range of output pixel values (depth invariant indices), the final output and temporary image files should be stored as 16-bit data. Values of L_{si} have not been included but the ratio of attenuation coefficients from Figure 8.3 (0.74) has been included for clarification.

To demonstrate the effectiveness of the depth-invariant transformation, the statistical precision of radiance values can be calculated before and after the transformation. For example, 115 pixels were sampled from sand areas of variable depth to determine the ratio of attenuation coefficients of Landsat MSS bands 1 and 2. The precision (standard error/mean) of these values was 0.008 for band 1 and 0.019 for band 2. This is because band 2 undergoes greater attenuation than band 1 and therefore has more variable values. The precision of depth-invariant bottom-index values for the same pixels was one to two orders of magnitude greater at 0.0006. The variability in radiance due to depth has been greatly reduced for a given substratum type.

Time needed to undertake water column correction

The subject took approximately five days to research and the first water correction process took two days to implement (for a pair of SPOT bands). However, once the process had been carried out and the technique had been implemented in software, further bands took less than half a day to analyse. If the process outlined above is followed, we estimate that a satellite scene can be corrected in one and a half days and that subsequent band pairs will only take several hours of analysis each.

Table 8.1 Some published values of attenuation coefficients and their ratios for guidance purposes
For further details refer to original papers

Location	Sensor and band	Ratio of attenuation coefficients	Attenuation coefficient (m^{-1})	Reference
Bahamas	Landsat MSS 1/2	0.24		Lyzenga 1981
TCI	Landsat MSS 1/2	0.35		pers. obs.
Bahamas	Landsat TM 1/2	0.74 (0.81)†		Armstrong 1993
Bahamas	Landsat TM 2/3	0.49 (0.49)†		Armstrong 1993
Bahamas	Landsat TM 1/3	0.36 (0.38)†		Armstrong 1993
TCI	Landsat TM 1/2	0.93		pers. obs.
TCI	Landsat TM 2/3	0.39		pers. obs.
TCI	Landsat TM 1/3	0.31		pers. obs.
TCI	SPOT XS 1/2	0.36		pers. obs.
Pacific Ocean	Landsat TM 1		0.054	Maritorena 1996
Pacific Ocean	SPOT XS 1		0.089	Maritorena 1996
Pacific Ocean	SPOT XS 2		0.391	Maritorena 1996
Pacific Ocean	SPOT XS 1/2	0.23		Maritorena 1996
Moorea, S. Pacific	Landsat TM 1		0.105	Maritorena 1996
Moorea, S. Pacific	SPOT XS 1		0.124	Maritorena 1996
Moorea, S. Pacific	SPOT XS 2		0.429	Maritorena 1996
Moorea, S. Pacific	SPOT XS 1/2	0.29		Maritorena 1996

TCI: Turks and Caicos Islands
† Values in parentheses denote in situ measurements in the field.

Published values of attenuation coefficients and their ratios

The attenuation coefficients of different spectral bands will vary according to the specific water types being sampled in the imagery. However, several authors have published attenuation coefficients and their ratios. While these cannot be used directly because they are image specific, they may provide a general guideline for those experimenting with the method for the first time (Table 8.1).

Extensions of the Lyzenga method

The major limitation of depth-invariant processing is that turbid patches of water will create spectral confusion. However, where water properties are moderately constant across an image, the method strongly improves the visual interpretation of imagery and should improve classification accuracies (see relevant chapters later). Depth-invariant processing was found to be beneficial in the Turks and Caicos Islands (Mumby *et al.* 1998) where the horizontal Secchi distance at a depth of 0.5 m is of the order of 30–50 m. At present, it is not possible to give a threshold turbidity with which depth-invariant processing will be ineffective. In the absence of such information, we point out that Tassan (1996) has described a theoretical depth-invariant model for water of greater turbidity than specified by Lyzenga (1981). This method is mathematically complex and still requires field validation. For these reasons, it is not described further here. Those readers attempting to correct for depth effects in turbid waters are urged to consult Tassan's paper (and possible follow-ups) directly.

An alternative 'quick-fix' approach may be possible where several water bodies can be clearly distinguished. In this case, the attenuation coefficients for each band will differ in each water type (e.g. seawater verses plumes of riverine water with a high content of yellow substance). If these water bodies are separated by subsetting the imagery (digitising outlines), depth-invariant processing may be applied to each segment individually (i.e. for each set of attenuation coefficients). We have not attempted this process but it might prove feasible under some circumstances.

Conclusion

Depth-invariant processing of digital spectral data is essential for establishing quantitative empirical relationships with marine features. It can also improve the accuracy of marine habitat maps and makes for improved visual interpretation of imagery.

If the study objective involves habitat mapping using multispectral classification (Chapter 10), water column correction is most appropriate for imagery with several water-penetrating spectral bands (e.g. Landsat TM, CASI).

References

Armstrong, R.A., 1993, Remote sensing of submerged vegetation canopies for biomass estimation. *International Journal of Remote Sensing*, **14**, 621–627.

Jerlov, N.G., 1951, Optical Studies of Ocean Water. *Report of Swedish Deep-Sea Expeditions*, **3**, 73–97.

Jerlov, N.G., 1964, Optical Classification of Ocean Water. In *Physical Aspects of Light in the Sea*. (Honolulu: University of Hawaii Press), pp. 45–49.

Jerlov, N.G., 1976, *Applied Optics*. (Amsterdam: Elsevier Scientific Publishing Company).

Lyzenga, D.R., 1978, Passive remote sensing techniques for mapping water depth and bottom features. *Applied Optics*, **17**, 379–383.

Lyzenga, D.R., 1981, Remote sensing of bottom reflectance and water attenuation parameters in shallow water using aircraft and Landsat data. *International Journal of Remote Sensing*, **2**, 71–82.

Maritorena, S., 1996, Remote sensing of the water attenuation in coral reefs: a case study in French Polynesia. *International Journal of Remote Sensing*, **17**, 155–166.

Mumby, P.J., Clark, C.D., Green, E.P., and Edwards, A.J., 1998, Benefits of water column correction and contextual editing for mapping coral reefs. *International Journal of Remote Sensing*, **19**, 203–210.

Open University, 1989, *Seawater: Its Composition, Properties and Behaviour* (Exeter: A. Wheaton and Co. Ltd).

Spitzer, D., and Dirks, R.W.J., 1987, Bottom influence on the reflectance of the sea. *International Journal of Remote Sensing*, **8**, 279–290.

Tassan, S., 1996, Modified Lyzenga's method for macroalgae detection in water with non-uniform composition. *International Journal of Remote Sensing*, **17**, 1601–1607.

Zainal, A.J.M., 1994, New technique for enhancing the detection and classification of shallow marine habitats. *Marine Technology Journal*, **28**, 68–74.

Part 3

Habitat Classification and Mapping

9

Methodologies for Defining Habitats

Summary *What is understood by 'habitat mapping' may vary from person to person so it is important to make a clear definition of habitats prior to undertaking the study. The objectives of most habitat mapping exercises can be classified into five groups: (i)* ad hoc *definition of habitats without field data, (ii) application-specific studies concerning only a few habitats, (iii) geomorphological studies, (iv) ecological studies, and (v) studies which combine more than one type of information (e.g. geomorphology and biotic assemblages).*

An ad hoc *approach to defining habitats is relatively cheap but is only recommended in cases where the accuracy of habitat maps may not be important. Habitat-specific studies should include the habitat of interest and those additional habitats which are most likely to be confused with it, thus permitting the accuracy of mapping to be evaluated. Geomorphological classifications can be assigned to imagery with little or no field work and some examples of such classifications are given. The establishment of ecological habitat classifications (e.g. assemblages of bottom-dwelling species and substrata) requires multivariate analysis of field data, such as hierarchical cluster analysis. These methods simplify the dataset and provide the classification scheme with an objective basis. The characteristic and discriminating features of each habitat can be identified and used to make a clear and unambiguous description of the classification scheme, thus facilitating its use and interpretation. Combined geomorphological and ecological classifications may be the most appropriate for remote sensing of tropical coastal areas.*

Introduction

It is vital to have a clear idea of the study objective(s) before conducting field work or selecting imagery. For example, stating that the study aims to map 'coral reef habitats' is not necessarily adequate because the definition of 'coral reef habitat' is likely to differ from person to person. In this example, the definition could embody reef geomorphology, assemblages of reef-dwelling organisms, the reef's physical environments, or a combination of each.

The mapping objectives of most remote sensing exercises fall into one of five categories:

1. studies using an *ad hoc* definition of habitats,
2. studies which focus on a particular habitat type for a specific application,
3. studies which are principally concerned with mapping geomorphology,
4. ecological studies which define habitats through quantification of biotic assemblages,
5. studies which combine more than one type of information (e.g. geomorphology and biotic assemblages).

This chapter outlines the methods and pitfalls of these five approaches.

1. *Ad hoc* definition of habitats

Habitats can be defined in an *ad hoc* fashion if the analyst is familiar with the area concerned or if a comparable habitat classification scheme is available. This approach does not usually include the collection of new field data and is often favoured because of its relatively low cost. However, there are several important drawbacks to making habitat maps without reference to field data.

a) The fundamental definition of classes may be incorrect and not applicable to the area concerned.
b) Even if the habitat classification scheme is appropriate, habitats may be identified incorrectly on the imagery (i.e. inaccuracies during image interpretation).

c) Maps tend to have a vague definition of habitats.
d) Accuracy assessment is not possible without independent field data so the reliability of the map is uncertain.

The resulting maps can be difficult to interpret meaningfully, particularly by individuals who are not familiar with the interpreter's concept of a habitat. By way of example, Table 9.1 outlines a marine habitat classification scheme that we created prior to conducting marine surveys of the Turks and Caicos Islands explicitly for habitat mapping. Categories were based on one person's existing familiarity with the area (two years teaching marine science on the Caicos Bank) and another person's experience from surveying similar ecosystems elsewhere in the Caribbean.

Field data were collected soon after the scheme was created and in the light of these data, the preliminary scheme was found to be unsatisfactory. Firstly, the scheme presented in Table 9.1 was inconsistent and combined geomorphological categories such as 'spur and groove' with ecological categories such as '*Thalassia* dominated (high density)'. The revised scheme (Mumby *et al.* 1998) was confined to ecological categories. Secondly, the scope of the scheme was fairly limited and failed to include many of the ecological classes that were only discovered during field survey (e.g. assemblages of encrusting sponge, seagrass and calcareous red algae). Thirdly, and perhaps most importantly, some of the predicted classes were not actually found at the study site. For example, whilst an 'algal rubble/*Porites* spp. zone' is common in Belize, it was not found as a distinct habitat on the Caicos Bank.

In conclusion, unless the accuracy of habitat maps is not deemed to be particularly important, an *ad hoc* approach to defining habitats is not recommended.

Table 9.1 Preliminary marine classification for habitats of the Turks and Caicos Islands. *Note*: the scheme was created without field data and was later found to be unsuitable for habitat mapping.

General habitat type	Specific habitat classes
Reef	Forereef
	Spur and groove
	Gorgonian plain
	Montastraea reef
	Acropora palmata zone (i.e. branching corals)
	Reef crest
	Back reef
	Mixed back reef community (seagrass / corals)
	Algal rubble, *Porites* spp. zone
	Carbonate pavement
	Patch reef
Bare Substratum	Sand
	Mud
	Hard substratum
Seagrass	*Thalassia* dominated (high density)
	Thalassia dominated (low density)
	Syringodium dominated (high density)
	Syringodium dominated (low density)
	Mixed seagrasses (high density)
	Mixed seagrasses (low density)
Algal dominated	Calcareous green algae
	Fleshy brown algae

2. Application-specific studies

Some remote sensing studies may be highly focused on specific surface features and, therefore, not concerned with mapping all habitats in an area. For example, a manager may be interested to know the extent of fringing red mangrove (*Rhizophora* spp.) along a particular area of coast. Provided that the manager's definition of fringing red mangrove is clear (e.g. areas where the mangrove species composition exceeds 80% for *Rhizophora* spp.), a comprehensive habitat classification scheme is not required. However, it is sensible to extend the scope of the mapping exercise to include those habitats which most strongly resemble the habitat of interest (e.g. stands of black mangrove, *Avicennia* spp.). By incorporating those habitats which are most likely to create misclassifications and errors in the habitat maps, the accuracy of the mapping objective can be determined. In this example, an accuracy assessment may show that fringing red mangrove and black mangrove are often confused and, therefore, the manager must attribute less confidence to the estimate of red mangrove cover.

3. Geomorphological classifications

Most remote sensing studies of coral reefs have focused on mapping geomorphological classes (Green *et al.* 1996). Labelling such classes is relatively straightforward because several geomorphological classification schemes exist (Hopley 1982, Kuchler 1986, Holthus and Maragos 1995). Examples of some geomorphological classes are given in this chapter to provide guidance (Plate 4) and encourage standardisation.

Geomorphological classes for coral reefs

Holthus and Maragos (1995) provide a comprehensive and nicely illustrated guide to reef geomorphology; readers are directed to this publication for detailed information. In addition, several appropriate geomorphological terms for reef flat, forereef and lagoon habitats are illustrated in Figures 9.1 to 9.3.

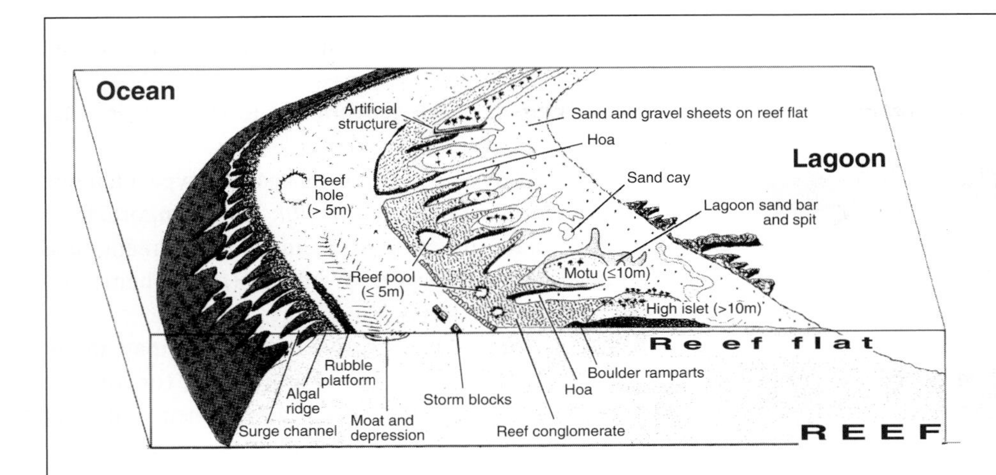

Figure 9.1 Geomorphological terms for reef flats and forereefs. Redrawn from: Holthus, P.F., and Maragos, J.E., 1995, Marine ecosystem classification for the tropical island Pacific. In *Marine and Coastal Biodiversity in the Tropical Island Pacific Region*, edited by J.E. Maragos, M.N.A. Peterson, L.G. Eldredge, J.E. Bardach and H.F. Takeuchi (Honolulu: East-West Center), pp. 239–278.

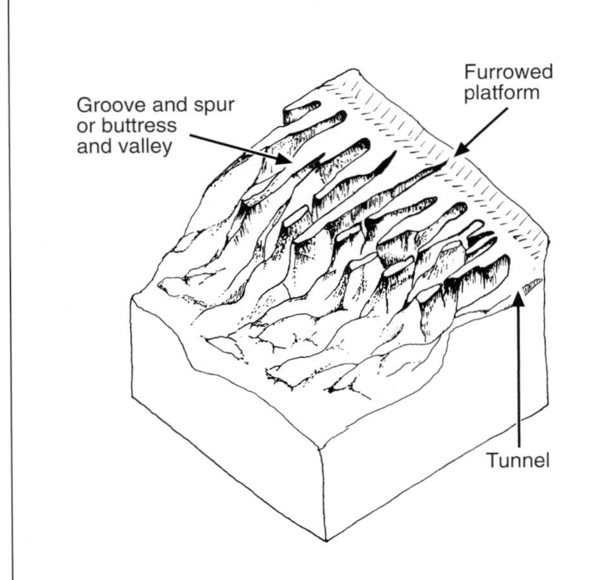

Figure 9.2 Geomorphological terms for forereef zones. Redrawn from: Holthus, P.F., and Maragos, J.E., 1995, Marine ecosystem classification for the tropical island Pacific. In *Marine and Coastal Biodiversity in the Tropical Island Pacific Region*, edited by J.E. Maragos, M.N.A. Peterson, L.G. Eldredge, J.E. Bardach and H.F. Takeuchi (Honolulu: East-West Center), pp. 239–278.

Mangrove community types

Odum *et al.* (1982) defined several mangrove communities based on their morphology and environment (Figure 9.4). This classification is useful for field studies and can be related to mangrove function (Odum *et al.* 1982).

4. Ecological classification of habitats

Unlike geomorphological classifications, ecological assemblages do not lend themselves easily to standard classifications. 'Ecological' definitions of habitat may be limited to assemblages (communities) of plant and animal species or widened to include species (or higher taxonomic or functional descriptors) and the substrata which collectively comprise the upper layer of the seabed (benthos). Assemblages of species and/or substrata often exhibit considerable variability and several distinct assemblages may inhabit each geomorphological zone (see Fagerstrom 1987). As such, it is often more difficult to distinguish ecological assemblages whereas geomorphology can usually be interpreted straight from an image (i.e. in the absence of field survey). As we have seen

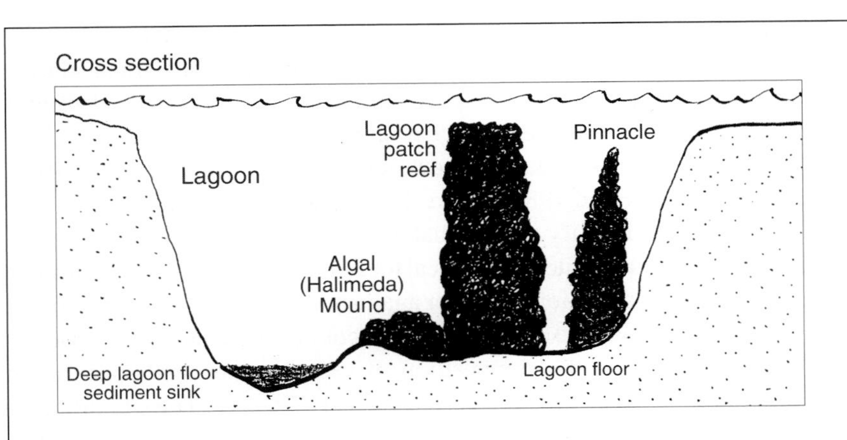

Figure 9.3 Geomorphological terms for lagoon habitats. Redrawn from: Holthus, P.F., and Maragos, J.E., 1995, Marine ecosystem classification for the tropical island Pacific. In *Marine and Coastal Biodiversity in the Tropical Island Pacific Region*, edited by J.E. Maragos, M.N.A. Peterson, L.G. Eldredge, J.E. Bardach and H.F. Takeuchi (Honolulu: East-West Center), pp. 239–278.

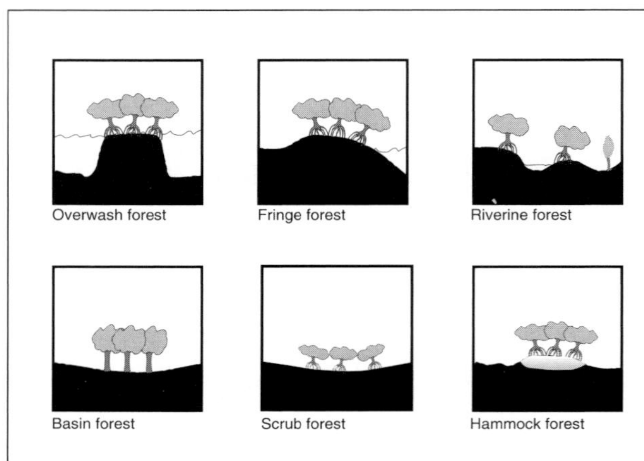

Figure 9.4 Mangrove community types based on morphology and hydrological conditions. Based on: Odum, W.E., McIvor, C.C., and Smith III, T.J., 1982, *The Ecology of the Mangroves of South Florida: A Community Profile* (Washington DC: US Fish and Wildlife Service, Office of Biological Services FWS/OBS – 81/24).

already, to infer ecological assemblages without field survey is potentially misleading.

It must also be borne in mind that geomorphological zones tend to have more distinct boundaries than ecological habitats (assemblages) which tend to exhibit change along gradients (e.g. progressive changes in species composition with changing depth). This makes the classification of ecological habitats somewhat inexact and one might ask how different two habitats must be before they are considered separate. In fact, many classification schemes have a hierarchical structure to reflect this uncertainty. At one end of the hierarchy, habitats are clearly distinct with little in common (e.g. coral reefs and seagrasses). Whereas at the other end of the hierarchy, habitat types might share a similar complement of species and are only separated because their dominant species differ (e.g. reefs dominated by the macroalgae *Lobophora* spp. or *Microdictyon* spp. – Plate 8).

There is no absolutely correct method of categorising (classifying) ecological habitats so the choice of methods depends on the objective(s) of the study. Generally, the classification aims to reflect the major habitat types found in the area of interest as faithfully as possible. Regardless of which method is eventually selected, the definition of habitats will always be slightly arbitrary, particularly where gradients of assemblages are involved. It follows that the imposition of habitat boundaries on a map will also be somewhat arbitrary which, practically speaking, means that habitat maps can never be 100% accurate.

A good habitat classification scheme must be interpreted easily and be unambiguous. To fulfil these criteria, habitat types should be determined objectively and have semi-quantitative, or preferably quantitative, descriptors that characterise habitats and discriminate between them. Linking habitat maps to reality is important for several reasons:

1. Maps can be related explicitly to the species/life-forms/substrata in each habitat.
2. Quantitative descriptors of the habitat type facilitate the recognition of habitats *in situ*. This is important for further field survey (e.g. accuracy assessment) and enables other surveyors to interpret the scheme and adopt it in other areas.
3. Each habitat has an objective basis and it may, therefore, be thought of as a unit of assemblage (or community) diversity. Planners may then identify representative habitats (McNeill 1994) or assess patterns and hotspots of habitat diversity.

Ideally, the scheme should have a hierarchical structure to encompass a broad range of user needs, technical ability and image types. For example, it is easy to visualise a coastal mapping strategy which uses Landsat TM to make a national marine habitat map with coarse descriptive resolution (i.e. few classes; see Chapter 11), and that this would be augmented using airborne digital imagery with finer descriptive resolution (a greater number of habitats), at specific sites of interest.

An ecological dataset may include; species cover (abundance), substratum cover, tree height, canopy cover, biomass, and so on. Each of these variables describe part of the ecological and physical characteristics of a habitat. After fieldwork, the data analyst may face a complex data set which includes measurements for multiple variables at each site. To extract the characteristics of each habitat (i.e. natural groupings of data), some form of multivariate analysis is required to simplify the data. Multivariate statistics have been developed for over a century and aim to simplify and describe complex data sets. The rest of this chapter focuses on the use of multivariate statistics for defining habitats. A broader discussion of the classification of marine habitats is given in Mumby and Harborne (1999).

Multivariate classification of field data

The definition of habitat types from field data is not difficult but several important decisions must be made during the analysis (Figure 9.5). An overview of these decisions is provided here, but readers are referred elsewhere for a more detailed discussion (e.g. Digby and Kempton 1987, Hand 1981, Clarke 1993, Clarke and Warwick 1994). To identify groups (habitats) in the data, the most appropriate suite of statistical tools are known as multivariate classification or cluster analysis. A biological example will be used to explain the concept of cluster analysis. Imagine that the composition of seagrass and algal species had been measured at each of (say) 70 sites. The similarity (or dissimilarity) between each pair of sites can be deter-

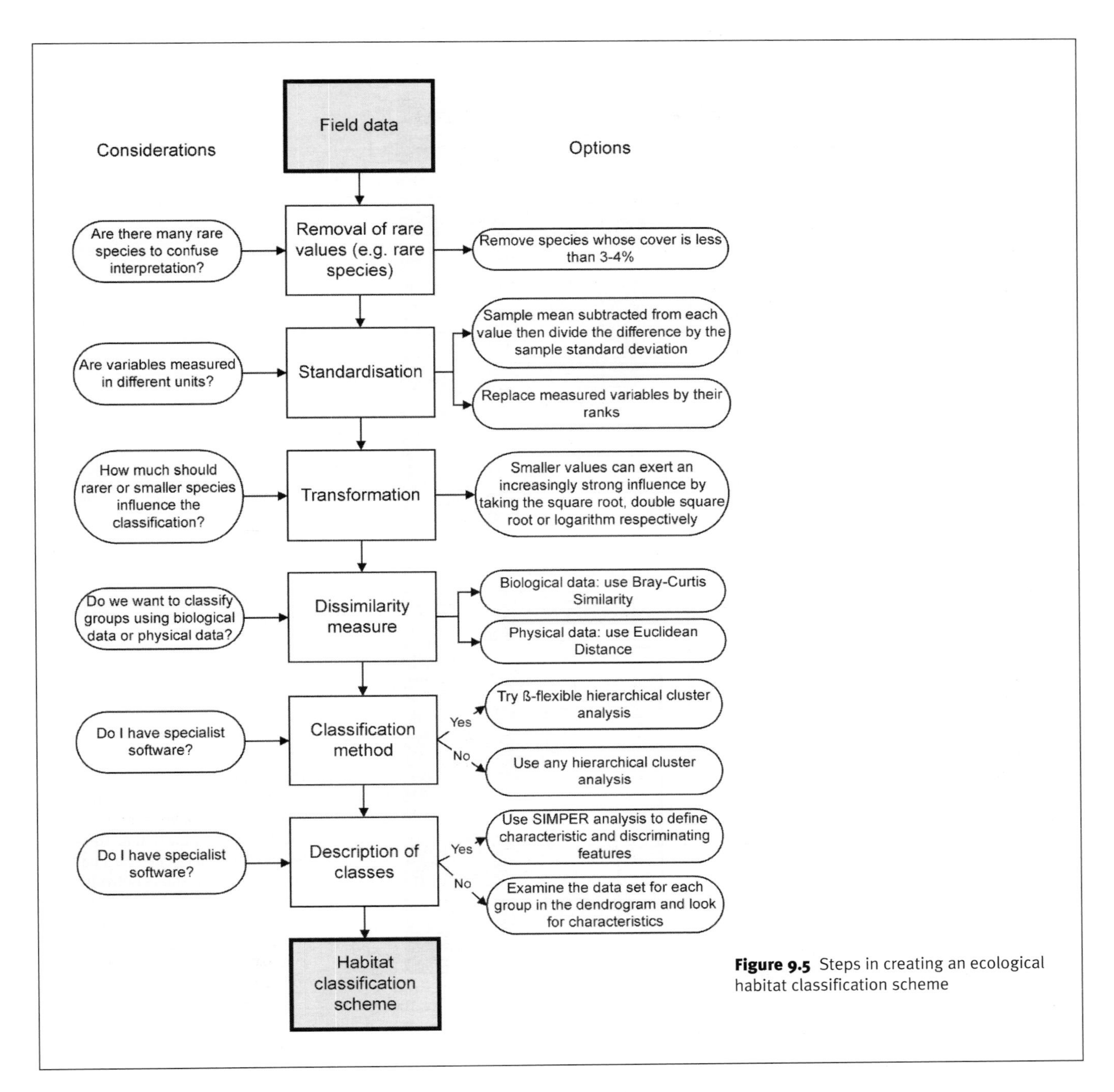

Figure 9.5 Steps in creating an ecological habitat classification scheme

mined using a similarity coefficient. For every pair of sites, the coefficient evaluates the similarity in abundance of each species. The result is usually the algebraic sum of similarities for each species and ranges from 0–1. Two sites with a similarity of 0, have no species in common whereas a similarity of 1 constitutes an identical complement of species with identical abundances at each site. Simulation experiments with various similarity measures have found the Bray-Curtis similarity coefficient (Bray and Curtis 1957) to be a particularly robust measure of ecological distance (Faith *et al.* 1987). The similarity coefficient essentially exchanges the multivariate data set (e.g. abundances of, say, 20 species) for a single measure of distance/similarity between each pair of sites. A classification algorithm is then used to group sites according to their relative similarities. Those sites that are most alike will cluster (group) together whereas those sites that are more dissimilar are unlikely to join the same cluster. The clusters are represented in a tree-like diagram called a dendrogram (see Figure 9.6).

There are dozens of methods available for generating clusters and a discussion of their relative merits is beyond the scope of this *Handbook*. The key point to bear in mind is that no method is perfect: there is no absolute way to describe the grouping of sites. This is because the dissimilarity of sites varies semi-continuously (e.g. along gradients) and the cluster analysis frequently has to make quite arbitrary decisions over which sites should cluster with which others. If specialist software is available (e.g. PATN: Belbin 1995), a flexible cluster analysis can be used: in this

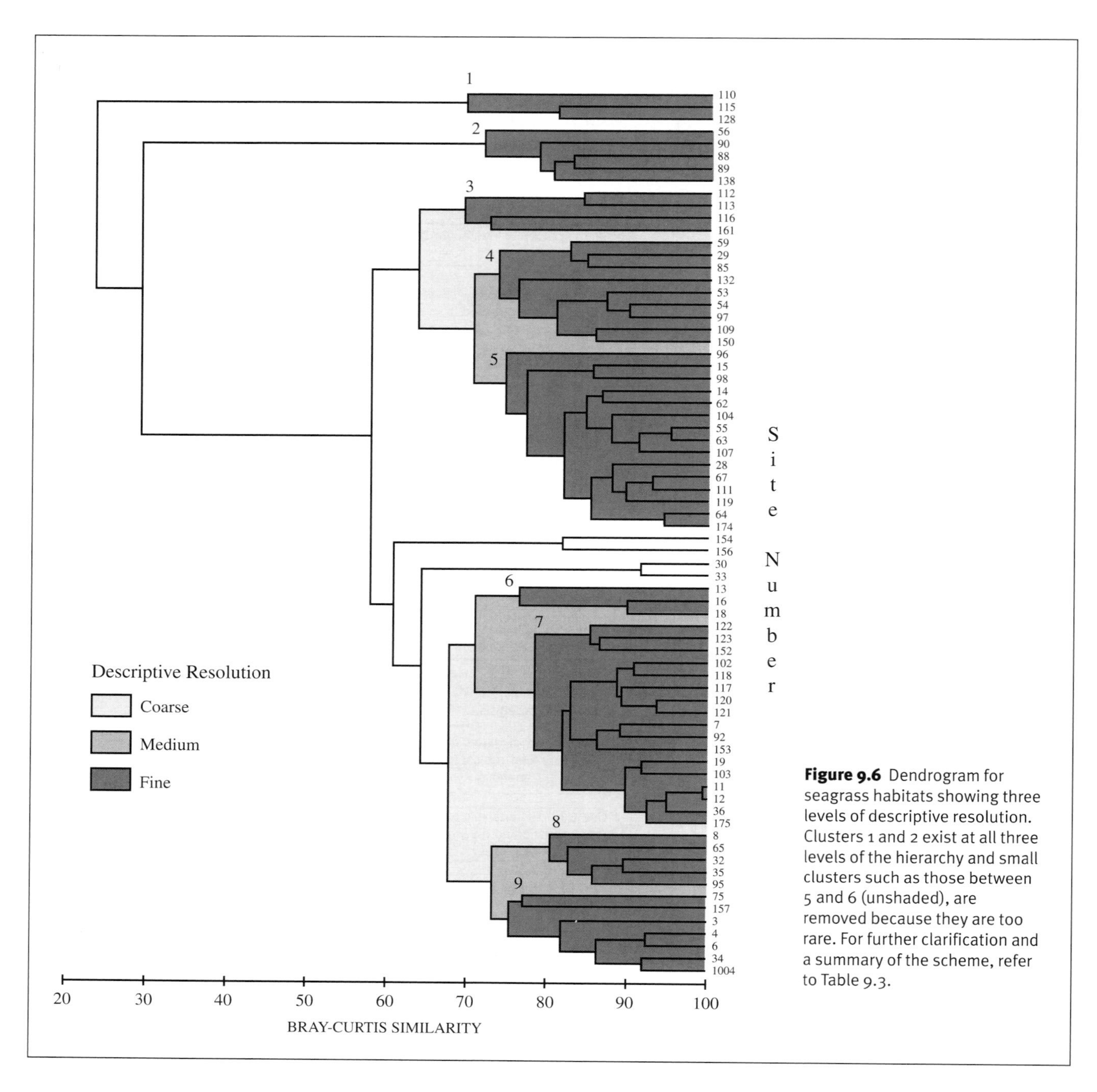

Figure 9.6 Dendrogram for seagrass habitats showing three levels of descriptive resolution. Clusters 1 and 2 exist at all three levels of the hierarchy and small clusters such as those between 5 and 6 (unshaded), are removed because they are too rare. For further clarification and a summary of the scheme, refer to Table 9.3.

case, classes are 'believed' if they remain stable using several different approaches. Alternatively, one of the most popular and widely available algorithms is hierarchical classification with group-average sorting (Clarke and Warwick 1994).

Hierarchical methods will result in subgroups forming from groups and so on. A hierarchical breakdown of habitat types is useful for defining different descriptive resolutions. Large dissimilar groupings (at the left of Figure 9.6) are more different from one another than the smaller groups toward the right of the dendrogram. Three hierarchical levels of descriptive resolution are illustrated in Figure 9.6. Since the similarity between groups will be partly reflected in different reflectance characteristics, it follows that remote sensing will distinguish the larger groups more easily than the smaller (more similar) groups.

Pre-processing options

Before data are analysed, pre-processing options should be considered (Figure 9.5). It might be necessary to remove rare species as inclusion may create unnecessary 'noise' and obscure the groupings of sites. If different types of data are being combined such as seagrass standing crop and percent algal cover, each variable can be standardised using the methods in Figure 9.5. If data are not transformed, larger values will exert a greater influence on the cluster analysis. The extent of this effect can be altered using transformations of varying severity. For example, when clustering percentage cover data from coral reefs, we elected not to transform the data, thus allowing the dominant species or substrata to exert an appropriately large influence on the habitat grouping. This was deemed necessary because remote sensing is more likely to discriminate habitats on the basis of their major benthic components rather than more cryptic species or substrata. Finally, Mumby *et al.* (1996) present a method of weighting individual species during the calculation of similarity matrices. This might be appropriate if some species are considered to be more important from a conservation perspective but could equally well apply to different substrata, canopy cover values and so on.

Describing characteristics of habitats

The characteristics of each habitat are determined by examining the data in each cluster. In the absence of specialised software such as PRIMER (Clarke and Warwick 1994), the easiest approach is to calculate the mean and standard deviation of each variable in the cluster. For example, if 30 sites had formed a single cluster (habitat A), the average algal cover, seagrass species composition (etc.) could be calculated for this habitat. Those species and substrata with the highest mean covers/densities are the dominant features of the habitat. It is also worth calculating the coefficient of variation (COV) which is the ratio of the mean to standard deviation. Those species with the highest COVs have the most consistent cover/density in the habitat, although they may not necessarily be dominant. Dominant and consistent features (species, substrata) may be considered characteristic of the habitat. However, different habitats can share the same characteristic features. To identify discriminating features, comparisons must be drawn between pairs of habitats.

A more objective method of identifying characteristic and discriminating features, is Similarity Percentage analysis (SIMPER), described by Clarke (1993) and available in the software, PRIMER (Plymouth Routines in Multivariate Ecological Research). To identify discriminating features, SIMPER calculates the average Bray-Curtis dissimilarity between all pairs of inter-group samples (i.e. all sites of habitat 1 against all sites of habitat 2). Because the Bray-Curtis dissimilarity measure incorporates the contribution of each feature (e.g. each species), the average dissimilarity between sites of habitat 1 and 2 can be expressed in terms of the average contribution from each species. The standard deviation provides a measure of how consistently a given species will contribute to the dissimilarity between habitats 1 and 2. A good discriminating species contributes heavily to interhabitat dissimilarity and has a small standard deviation.

Characteristic features can be identified using the same principle; the main difference is that average similarity is calculated between all sites of each habitat. The species which consistently contribute greatly to the average similarity between sites are considered characteristic of the habitat. The following example illustrates this procedure.

The following SIMPER result (Table 9.2) was obtained from the classification of seagrass habitats (Figure 9.6). Clusters 8 and 9 were compared to identify discriminating features. *Penicillus* spp. is clearly the best discriminating genus, accounting for over 40% of the dissimilarity between habitats. *Halimeda* spp. also offer a fair means of discrimination. Standing crop (of seagrass) would be a better discriminator than *Laurencia* spp. because its ratio of average contribution/standard deviation is higher.

An example of a habitat classification scheme

A summary of each habitat type can be presented which broadly follows the structure of the dendrogram. Table 9.3 gives such a summary for habitats categorised by the dendrogram in Figure 9.2.

Ignoring the first three clusters, which were discarded as representing rare habitats, descriptive resolution was partitioned to give 6 fine-level habitats (clusters 4–9) which were arranged into 4 medium-level habitats. Medium descriptive resolution included sites of low seagrass standing crop (clusters 6 and 7), low to medium standing crop (clusters 8 and 9) and medium to high standing crop (clusters 4 and 5; although two of the sites had low

Table 9.2 SIMPER analysis of dissimilarity between seagrass clusters (habitats) 8 and 9 (Figure 9.6 and Table 9.3). The term 'average abundance' represents the average abundance, biomass, density (etc.) of each feature. 'Average contribution' represents the average contribution of feature *i* to the average dissimilarity between habitats (overall average = 33.7%). Ratio = contribution average/standard deviation. Percentage contribution = average contribution/average dissimilarity between habitats (33.7). The list of features is not complete so percent values do not sum to 100%.

Feature	Average abundance		Average contribution	Ratio	Percentage contribution
	Habitat 9	Habitat 8			
Penicillus spp.	56.0	7.1	11.5	2.1	42.9
Halimeda spp.	25.1	18.8	5.6	1.6	21.7
Laurencia spp.	0.3	3.0	2.1	0.7	7.8
Standing crop	2.2	1.7	1.7	1.9	6.2

Table 9.3 Summary of seagrass habitat classes showing cluster numbers from dendrogram (Figure 9.2). *Note*: the first three clusters (1, 2 and 3) were excluded from the scheme below because it became apparent, through further field survey, that these habitats were rare and thus could not feasibly be mapped.

Descriptive Resolution			
Coarse	Medium	Fine	
Rare habitat classes (removed)		*Halodule wrightii* of low standing crop (5 g.m^{-2})	1
		Syringodium filiforme of low standing crop (5 g.m^{-2})	2
		Thalassia testudinum, *Syringodium filiforme*, and *Halodule wrightii* of low to medium standing crop (< 10 g.m^{-2})	3
Seagrass habitats 4, 5	*Thalassia* and *Syringodium* of medium to high standing crop	*Thalassia testudinum* and *Syringodium filiforme* of standing crop 5-80 g.m^{-2}	4
		Thalassia testudinum and *Syringodium filiforme* of standing crop 80-280 g.m^{-2}	5
Sand habitats 6, 7, 8	low standing crop *Thalassia* and sparse algae	*Thalassia testudinum* of low standing crop (5 g.m^{-2}) and *Batophora* sp. (33%).	6
		Thalassia testudinum of low standing crop (5 g.m^{-2}) and sand	7
	low to medium standing crop	medium dense colonies of calcareous algae – principally *Halimeda* spp. (25 m^{-2}). *Thalassia testudinum* of low standing crop (< 10 g.m^{-2})	8
Algal habitats 9	*Thalassia* and dense calcareous algae	dense colonies of calcareous algae – principally *Penicillus* spp. (55 m^{-2}) and *Halimeda* spp. (100 m^{-2}). *Thalassia testudinum* of medium standing crop (~80 g.m^{-2})	9

standing crop). At coarse descriptive resolution, the low standing crop classes were merged with sand-dominated habitats of the reef classification. This decision was taken because the lower end of the visual assessment scale for seagrass standing crop includes extremely low biomass values (< 5 g.m^{-2}) and the distinction between seagrass of low standing crop and sand-dominated habitats is somewhat arbitrary. Medium to high standing crop classes were merged to give a coarse descriptive resolution labelled seagrass.

Software

Most general purpose statistical software offers hierarchical cluster analysis (e.g. SPSS, Minitab). The following also offer more specialised routines:

- PRIMER contains objective methods for describing the characteristics of each cluster (group). It costs approximately £330 and is available from: M. Carr, Plymouth Marine Laboratory, Prospect Place, West Hoe, Plymouth PL1 3DH, UK;
- PATN contains a comprehensive range of multivariate tools. It costs approx. £500 and is available from Fiona Vogt, CSIRO Division of Wildlife and Ecology, PO Box 84, Lyneham, ACT 2602, Canberra, Australia.

5. Combined hierarchical geomorphological and ecological classifications

Given that benthic geomorphology and biotic/substratum cover strongly influence the radiance recorded by a remote sensing instrument, habitat mapping may subsume both geomorphological and ecological habitat classifications (sections 3 and 4, above) into a single scheme (Mumby and Harborne 1999). A merged habitat classification scheme provides a geomorphological and ecological component to each polygon on a habitat map (e.g. use of 'shallow lagoon floor (< 12 m) + *Thalassia testudinum* and *Syringodium filiforme* of standing crop 5–80 g.m^{-2}, in a legend). Both components can have a hierarchical structure (e.g. patch reef versus dense patch reef or diffuse patch reef; coral versus branching corals or sheet corals). The structure of a combined habitat classification is systematic in that geomorphological and ecological classes are not mixed or used interchangeably, and it also provides flexibility. For example, the geomorphological class 'shallow lagoon floor (<12 m)' might also be coupled with the ecological class 'medium dense colonies of calcareous algae'.

Providing that supporting documentation is clear, use of a hierarchical classification scheme allows some areas to be mapped in greater detail than others without confusing interpretation. Where assignment of a label is uncertain, the designation should reflect this. For example, if the depth of the lagoon is unknown, the geomorphological component should be labelled at a coarser level of the hierarchy (i.e. lagoon). Similarly, the ecological component '*Thalassia testudinum* and *Syringodium filiforme* of standing crop 5–80 g.m^{-2}' may be used in areas which are data rich but 'seagrass' might be more appropriate elsewhere.

In practice, a coastal mapping strategy is envisaged which uses Landsat TM data to make a regional marine habitat map of coarse descriptive resolution, and that this would be augmented using airborne digital imagery (e.g. CASI) or possibly colour aerial photography, with finer descriptive resolution, at specific sites of interest. A hierarchical habitat classification scheme helps integrate such hierarchical mapping activities.

Cost considerations

A total of 170 survey sites were visited in the Turks and Caicos Islands mapping campaign. Percentage cover/density data were collected from six 1 m^2 quadrats at each site, in addition to date, time, water depth, GPS position and a visual estimate of the size of the habitat patch. These data took 4 person-days to enter into a database and a further two days were necessary to produce the ecological classification used throughout the *Handbook*.

Conclusion

The meaning of 'habitat' should be made explicit at the outset of any habitat mapping study. *Ad hoc* habitat definitions are not recommended because they are prone to being vague and the accuracy of habitat maps cannot be determined in the absence of field survey. Highly focused studies should embody the habitat of interest and those additional habitats which are most likely to be confused with it, thus permitting the accuracy of mapping to be evaluated. Geomorphological classifications can be assigned to imagery with little or no fieldwork. The derivation of ecological habitat classifications should usually involve objective multivariate cluster analyses of field data. Cluster analysis can be tailored to study objectives by using appropriate pre-processing of field data (i.e. to place emphasis on specific elements of the dataset). Habitat classes should be described quantitatively to facilitate use of the classification scheme by other surveyors and improve its interpretation. A hierarchical approach to habitat classification is useful where coastal areas are mapped with varying detail (e.g. where more than one sensor is used) and to satisfy a broad range of user needs and expertise.

References

Belbin, L., 1995, *PATN Pattern Analysis Package* (Canberra: CSIRO Division of Wildlife and Ecology).

Bray, J.R., and Curtis, J.T., 1957, An ordination of the upland forest communities of Southern Wisconsin. *Ecological Monographs*, **27**, 325–349.

Clarke, K.R., 1993, Non-parametric multivariate analyses of changes in community structure. *Australian Journal of Ecology*, **18**, 117–143.

Clarke, K.R., and Warwick, R.M., 1994, *Change in Marine Communities: an Approach to Statistical Analysis and Interpretation* (Plymouth: Natural Environment Research Council).

Digby, P.G.N., and Kempton, R.A., 1987, *Multivariate Analysis of Ecological Communities* (London: Chapman and Hall).

Fagerstrom, J.A., 1987, *The Evolution of Reef Communities* (New York: John Wiley & Sons).

Faith, D.P., Minchin, P.R., and Belbin, L., 1987, Compositional dissimilarity as a robust measure of ecological distance. *Vegetatio*, **69**, 57–68.

Green, E.P., Mumby, P.J., Edwards, A.J., and Clark, C.D., 1996, A review of remote sensing for the assessment and management of tropical coastal resources. *Coastal Management*, **24**, 1–40.

Holthus, P.F., and Maragos, J.E., 1995, Marine ecosystem classification for the tropical island Pacific. In *Marine and Coastal Biodiversity in the Tropical Island Pacific Region*, edited by J.E. Maragos, M.N.A. Peterson, L.G. Eldredge, J.E. Bardach and H.F. Takeuchi (Honolulu: East-West Center), pp. 239–278.

Hand, D.J., 1981, *Discrimination and Classification* (Chichester: John Wiley & Sons).

Kuchler, D., 1986, *Geomorphological Nomenclature: Reef Cover and Zonation on the Great Barrier Reef* (Townsville: Great Barrier Reef Marine Park Authority Technical Memorandum 8).

McNeill, S.E., 1994, The selection and design of marine protected areas: Australia as a case study. *Biodiversity and Conservation*, **3**, 586–605.

Mumby, P.J., Clarke, K.R., and Harborne, A.R., 1996, Weighting species abundance estimates for marine resource assessment. *Aquatic Conservation: Marine and Freshwater Ecosystems*, **6**, 115–120.

Mumby, P.J., Green, E.P., Clark, C.D., and Edwards, A.J., 1998, Digital analysis of multispectral airborne imagery of coral reefs. *Coral Reefs*, **17**, 59–69.

Mumby, P.J., and Harborne, A.R., 1999, Development of a systematic classification scheme of marine habitats to facilitate regional management and mapping of Caribbean coral reefs. *Biological Conservation*, **88**, 155–163.

Odum, W.E., McIvor, C.C., and Smith III, T.J., 1982, *The Ecology of the Mangroves of South Florida: A Community Profile* (Washington DC: US Fish and Wildlife Service, Office of Biological Services FWS/OBS – 81/24).

10

Image Classification and Habitat Mapping

Summary *Classification is the process of identifying image pixels with similar properties, organising them into groups and assigning labels (e.g. habitat names) to those groups. The end product of classification is a map of habitats or other features of interest.*

Many methods exist to achieve this, ranging in complexity, cost and time. The simplest involves an analyst tracing the boundaries of habitat patches on an aerial photograph with a pencil. A more sophisticated approach involves the use of digitising tablets to draw polygons around different habitats but, like any visual interpretation or photo-interpretation method, still suffers from the subjectivity of the operator. For digital imagery by contrast, a computer can be instructed to classify an image objectively using clearly defined rules. There are two main approaches to classifying multispectral digital images; unsupervised and supervised classification. Both techniques are described and their uses and limitations discussed. Unsupervised classification, where the computer automatically classifies pixels in an image into a number of classes (set by the operator) on the basis spectral similarity, was found useful in planning field survey. Supervised classification, where field survey data are used to identify pixels of particular habitat types and guide the derivation of spectral signatures for those habitats from the image data, was usually found to generate the most accurate classifications. These techniques are implemented in standard remote sensing software packages. The more sophisticated classification algorithms may require quite powerful computer hardware (or take a long time to carry out), particularly for large images. A combination of visual interpretation, unsupervised classification, field survey inputs and supervised classification techniques ultimately provide the most accurate habitat maps.

Introduction

Inspection of a satellite image reveals information on the colour, tone, texture and pattern of different areas which are in some way related to the underlying habitats of the region. Any attempt to use this information, for example to produce habitat maps, is based on the principle that different types of habitat reflect electromagnetic radiation in distinctive ways across the wavelengths being measured creating their own spectral signatures (spectra). Unfortunately this is quite often not the case and provides the main physical limitation to remotely sensed habitat mapping.

No matter how biologically or commercially significant two habitats may be, they will not be distinguishable on a remotely sensed image unless they have different spectra. Most conventional methods of classification separate out areas using only the spectral information of individual pixels and do not deal with parameters of texture and pattern. These techniques vary in complexity, and range from visual interpretation and direct mapping from an image (photo-interpretation) through multispectral image classification techniques such as the automated statistical clustering of image data ('unsupervised classification') to statistical clustering directed by the user ('supervised classification'). They are all used routinely for habitat mapping. New techniques are being developed which use alternative approaches (e.g. classification by neural networks) to include classification of pixels according to pattern and texture. However these methods have not yet been demonstrated to be superior to standard multispectral classification and have not been incorporated in widely available image processing software; so we do not describe them here.

The aim of this chapter is to guide the reader as to appropriate methods for converting image data to habitat maps. Subsequent chapters deal in more detail with the specifics of mapping reefs, seagrass, mangroves and bathymetry.

Image classification methods

Photo-interpretation is a method by which the user visually inspects an image and identifies habitats. Multispectral classification is a computer-based method by which a number of wavebands are input to a statistical clustering algorithm which organises individual pixels into distinctive groups (classes) which are assumed to represent specific habitat types. Each method will be discussed in turn.

Photo-interpretation (visual interpretation)

Visual interpretation is the process whereby an image is studied and habitats identified by eye. The image may be a photographic print or colour hard copy of a satellite image bought from a data supplier or digital data viewed on a computer monitor. Digital imagery is advantageous in that it may be processed (e.g. contrast stretched) to emphasise the habitats of interest. For example, land can be masked out of an image and the full dynamic range of the display device applied only to the aquatic habitats. Image interpretation should be carried out using either a range of individual wavebands or an optimally-combined colour composite. For the latter case it will be necessary to choose the most appropriate wavebands for the habitats in question, e.g. Landsat Thematic Mapper bands 3, 2, and 1 as an RGB colour composite for submerged habitats.

Habitats are identified visually by an analyst based on experience, then delineated and labelled manually. Decisions on the boundaries between different habitats are made with reference to local knowledge, common sense, field data, aerial photographs or maps and a sound understanding of remote sensing theory. In essence the analyst edits and classifies the image in context (see Box 10.1). These decisions are reached without any computational effort or statistical processing. Delineation of boundaries is achieved using a tracing overlay and pencil or a mouse-driven cursor on a computer screen.

Whether using a pencil or mouse-driven cursor, photo-interpretation requires a skilled and experienced operator to make sound decisions. Figure 10.1 (Plate 9) illustrates the challenge that visual interpretation presents: it is a Landsat TM image (bands 1, 2 and 3) of a 12 x 10 km area of uniform depth with patch reefs to the west, and areas of dense algae to the east. Some of the boundaries between habitats are clear, because the contrast with adjacent habitats is high, and should be relatively straightforward to delineate (e.g. boundaries around A and B). Area C is a seagrass bed – its eastern boundary (white line) is clear but to the west the seagrass grades into sand. There is no clear boundary and the analyst must decide where the boundary lies (red line). Ideally, supplementary data will be available to guide this decision but typically this is not the case and the analyst is obliged to rely on common sense and local knowledge. Similarly it would be difficult to draw a boundary around areas D and E although their colour and texture would seem to indicate different habitats. The south-eastern boundary of F is clearly defined but closing the polygon around the north-western boundary would be problematical. However once all these decisions have been made and polygons drawn to define habitats a map can be created (Figure 10.1c).

A skilled interpreter can distinguish about 30 grey tones (brightness levels) in a panchromatic image, and many more different hues in a colour image. However each pixel

BOX 10.1

Contextual editing

Contextual editing is perhaps best thought of as 'the appli- cation of common sense to habitat mapping'. Contextual rules are adopted intuitively by photo-interpreters. For example, whilst a photo-interpreter might be unable to distinguish the colour of (say) coral and seagrass on an aerial photograph, a seagrass bed may be correctly labelled on the basis of its context, i.e. in (say) the centre of a lagoon. Likewise small circular structures surrounded by a halo can be labelled as patch reefs and not small patches of seagrass (which are rarely circular). When thematic maps are created from digital data using image classification, contextual information has to be added as a series of explicit decision rules. To ensure that contextual editing does not create bias or misleading improvements to map accuracy, the decision rules must be applicable throughout an image and not confined to the regions most familiar to the analyst. The simplest of these for coral reefs is the presence and absence of coral and seagrass habitats on the outer fringing reef. Surprisingly, contextual editing has not been widely adopted for digital remote sensing of coral reefs despite terrestrial studies such as Groom *et al.* (1996) which have demonstrated that contextual editing can significantly improve map accuracies. Since coral reefs often exhibit predictable geomorphological and ecological zonation with gradients of depth and wave energy, contextual editing should substantially improve the accuracy of coral reef habitat maps.

Contextual editing can be applied to an image after classification in which case it improves accuracy (Chapter 11) or simultaneously in the case of photo-interpretretation.

may represent 2^{8+8+8} or approximately 1.7 x 10^7 unique spectral signatures (three bands at 8-bit resolution). The human eye is incapable of distinguishing this number of tones and hues. Furthermore a colour composite image is displayed on a screen by projecting three bands through the red, green and blue guns of the monitor. If other bands are available (i.e. in imagery which has more than three bands), they cannot be displayed and are not analysed by the interpreter whose decisions are dependent upon the bands selected. Visual interpretation therefore does not come close to utilising the full information content of an image. This can only be achieved using statistically based digital processing, and is usually accomplished by multispectral classification.

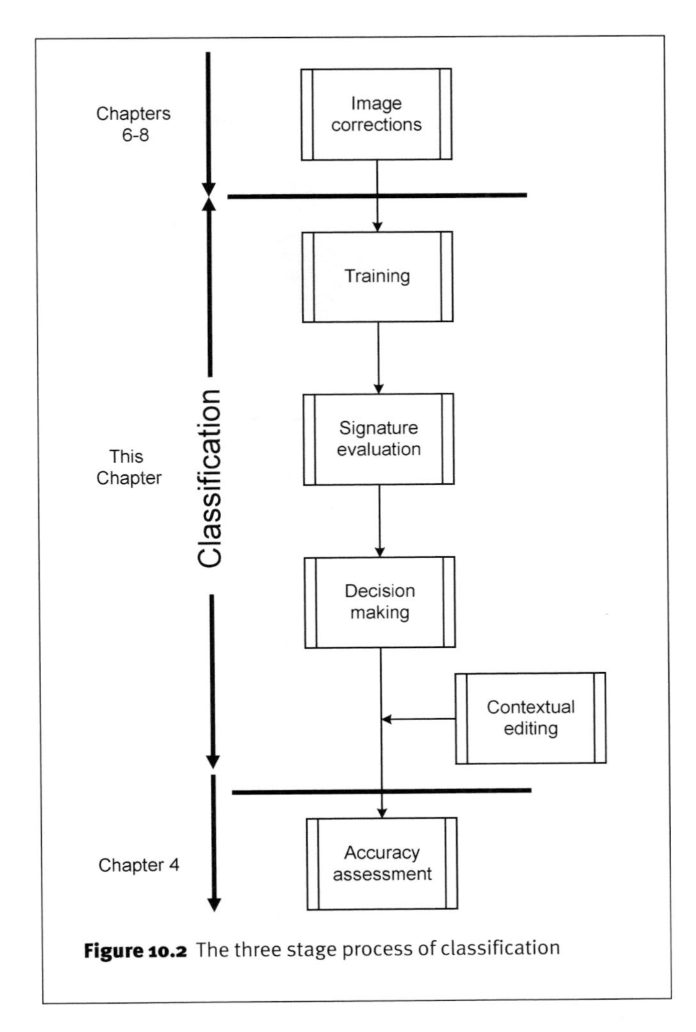

Figure 10.2 The three stage process of classification

Multispectral classification of image data

Multispectral classification is a three stage process (Figure 10.2) which identifies statistically-based clusters to group pixels together into different classes:

- *Training*: the process of defining the spectral envelope of each class. A seagrass class, for example, may be defined by minimum and maximum pixel values in three wavebands, thus defining a 3-dimensional spectral envelope that is cubic in shape. This simple statistical description of the spectral envelope is termed the 'signature'.
- *Signature evaluation*: signatures derived from the above are checked for representativeness of the habitats they attempt to describe and to ensure a minimum of spectral overlap between signatures of different habitat. Signatures may need to be deleted, merged or manipulated
- *Decision making*: the process of actually sorting all image pixels into classes (defined by signatures) using mathematical algorithms called 'decision rules'. This stage is sometimes referred to as classification though, in practice, training and signature evaluation are integral parts of the whole process.

Classification is usually followed by an assessment of accuracy (Chapter 4).

☞ **The mathematics and theory behind multispectral classification algorithms are well covered by a series of remote sensing textbooks and are not dealt with here. Only a very brief overview is provided here. Readers requiring further information are directed to the very helpful books by Mather (1987) and Wilkie and Finn (1996). The *ERDAS Field Guide* (1994) is also extremely instructive, regardless of the software you are using. This section of Chapter 10 concentrates on presenting practical, rather than theoretical, advice on performing unsupervised and supervised classifications.**

Training

There are two approaches. Unsupervised training uses statistical clustering techniques to derive the dominant spectral clusters within an image. This requires little user input, hence the term unsupervised. It is hoped that these clusters, defined by spectral envelopes (or signatures), correspond to the main habitat types of interest. Supervised training utilises knowledge of the user to inform and direct the definition of spectral envelopes. This is achieved by the user delineating some known areas of a particular habitat on the image, whose spectral properties are then taken to be statistically representative of that habitat. Readers requiring information on the mathematics of the two training methods are referred to Hord (1982).

Unsupervised training

N-dimensional data (where n = the number of bands being classified) can be plotted abstractly in feature space, as an n-dimensional scatter-plot (Figure 10.3 shows pixels in two dimensions). Unsupervised training is sometimes called 'clustering' because the computer is instructed to find the natural concentrations of pixels (or clusters) on

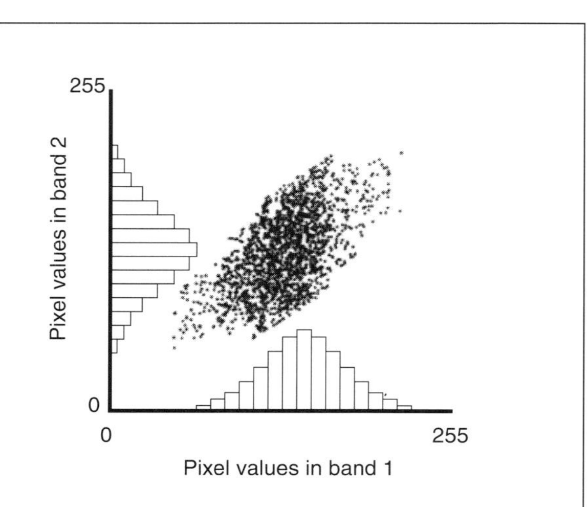

Figure 10.3 Data for two bands plotted as a scatterplot in feature space and defining an ellipse in feature space for a particular habitat. Histograms of the normally distributed data values in each band are also shown. Redrawn from: Erdas Inc. (1994). *Erdas Field Guide* (Atlanta, Georgia: Erdas Inc.)

this plot. This is more or less an automated process though the analyst is required to define the number of classes and set statistical parameters for the computer to work within and to interpret the meaning of the classes produced in the light of field data.

Classes are produced using a clustering algorithm. A widely used algorithm is ISODATA (Iterative Self-Organising Data Analysis Technique) which iteratively classifies the image and is implemented on commercial image processing packages such as ERDAS Imagine. Tou and Gonzalez (1974) explain the mathematics of the calculations. The process is as follows:

1. The clustering algorithm calculates a number of arbitrary cluster means in feature space. The number of clusters is selected by the analyst.
2. The computer then calculates the spectral distance between each pixel in the image and each of these arbitrary cluster means. Each pixel is assigned to the cluster whose mean is the nearest in feature space. The way in which spectral distance is calculated and the criteria used to select the nearest pixel is achieved using a decision rule, chosen by the analyst (see *Decision making*, this chapter).
3. The second iteration begins. The mean of all the pixels in each cluster is then recalculated: a new mean is produced, which now occupies a slightly different position in feature space.
4. The computer then calculates the spectral distance between each pixel in the image and each of these new means. Each pixel is assigned to the cluster whose mean is the nearest in feature space. Some pixels will now be closer to a different mean and assigned to this new cluster accordingly.
5. The iterations are repeated. At the end of each iteration the normalised percentage of pixels which remain assigned to the same cluster as in the previous iteration is calculated. When this number reaches a level set by the analyst (the convergence threshold) the computer terminates the calculations. A convergence threshold of 95% therefore means that 95% of the image pixels were not assigned to new classes in the last iteration of the algorithm.

Minimal input is required from the analyst who has only to decide the maximum number of classes to be produced, the convergence threshold and the maximum number of iterations. If the convergence threshold is never reached, the program terminates once the maximum number of iterations has been performed. ISODATA is a computationally intense program to run. To avoid wasting large amounts of time therefore, it is inadvisable to strive for too high a convergence: 95% is satisfactory for most applications.

☞ **As a general rule of thumb it is a good idea to direct the algorithm to generate twice as many classes as desired. After the classification is complete classes can then be merged together by the analyst. This gives the analyst greater control over outputs.**

The product of an unsupervised classification is a set of classes, each with a defined spectral envelope. These must now be related to field data, labelled and where appropriate merged. The maximum number of classes specified at the start of an unsupervised classification is therefore an important consideration. Too few, and the classification may fail to discriminate all the habitats possible with the data. Too many, and the analyst is faced with the time consuming task of evaluating and merging many, very similar classes together. Also, there is a risk that if too many classes are specified then the iterative programme may not be able to reach its convergence threshold.

☞ **If radiometric and atmospheric correction has not been carried out, habitat spectra (spectral signatures) will only apply to the image from which they were derived. If radiometric and atmospheric correction has been carried out, then the spectra should theoretically be transferable to other radiometrically and atmospherically corrected images.**

Case study: unsupervised classification of Landsat TM data

For a case study, the unsupervised classification of Landsat Thematic Mapper data of the Caicos Bank, Turks and Caicos Islands will be used. Figure 10.4 (Plate 10) illustrates the unsupervised classification of Figure 10.1a, using the ISODATA clustering algorithm, into reef, algae, seagrass and sand classes.

Classification into just two classes (Figure 10.4a) does not discriminate all the habitats possible with these data. Classification into four classes (Figure 10.4b) produces a thematic map of similar topology to that produced by visual interpretation (Figure 10.1c). While there may seem little point in classifying the image into more classes than is required we have found it useful to produce about twice as many classes (Figure 10.4c) and merge/delete the extra classes with reference to field data. Classification into more than twice the desired number (12 in this case) produces more detail than is easily interpretable (Figure 10.4d).

Figure 10.4c can be interpreted with reference to field data. These field data are either collected before classification or, preferably afterwards in which case the classified image is used to plan the field campaign. Table 10.1 presents hypothetical data collected from 20 sites in each class in Figure 10.4 c. Classes 4 and 5 appear to be seagrass and classes 2, 3 and 8 sand. They can therefore be merged and recoded with confidence. However class 6 could be either seagrass or algae, and class 1 could be either algae or reef. More field data, another classification method, or contextual editing (Box 10.1) must be used to sort out this confusion.

Table 10.1 Hypothetical field data from 20 sites in each of the eight classes generated by unsupervised classification (Figure 10.4c) of the area shown in Figure 10.1a. This allows the unsupervised classification to be interpreted. For example, of the twenty class 1 sites visited, 10 were found to be reef, 9 algae, and 1 seagrass. Thus reef and algae are confused by the unsupervised clustering. By contrast, classes 2, 3 and 8 are sand in 80–85% of sites visited and can be assigned to this habitat with reasonable confidence. See text for further discussion.

	Habitat				
Class	Reef	Algae	Seagrass	Sand	Total
1	10	9	1	0	20
2	0	2	2	16	20
3	1	1	1	17	20
4	0	7	11	2	20
5	1	2	17	0	20
6	1	8	9	2	20
7	0	15	4	1	20
8	0	1	2	17	20

Supervised training

Supervised training is 'analyst-led' in the sense that the analyst controls the process closely by selecting pixels that represent known features or habitats. In essence the analyst instructs the computer by saying 'these pixels are typical of habitat A. Group all pixels in the image most similar to these into one class and call it class A. These pixels are typical of class B. Group all pixels in the image most similar to these into another class and call it class B, and so on'. For example, an analyst may wish to separate an image into two classes, high and low density seagrass. Using field data, areas of low and high-density seagrass are identified on the image. Signatures are derived from the data values of the pixels in these areas and used to sort all image pixels into either low-density seagrass, or high-density seagrass.

☞ **In practice there will always be a third class consisting of unclassified pixels or those pixels which cannot be grouped into either seagrass class. These pixels could be labelled as 'other bottom types'. Some pixels will however always be misclassified: signature evaluation (see this chapter) attempts to reduce misclassification as much as possible.**

Supervised training therefore requires *a priori* information about the image data and habitats to be mapped. This information is usually obtained from field surveys (Chapter 4) and can be organised in a hierarchical habitat classification scheme (Chapter 9). Aerial photography and existing maps are sometimes used as surrogate field data but caution must be applied when using such sources (e.g. maps may contain errors through misinterpretation of aerial photographs on which they were based, or they may be out of date).

The groups of pixels which represent known features or habitats on the image are known as 'training samples'. It is important that all pixels within the training samples are representative of the (potential) habitat class that is to be mapped – if not, the analyst will introduce misclassification errors. Training samples may be identified in a number of ways (though different image processing packages will differ in the range of options available). Some typical methods include the use of:

- *A vector layer*: for example, it is known that the habitat within 0.5 km of a known point (perhaps a GPS coordinate) is gorgonian plain. A polygon, in this case a circle, is drawn over the image and used later to extract statistical information about the pixels within it (Figure 10.5, Plate 10).
- *A seed pixel*: field data are available from a known point. To continue the previous example, it is known that there are extensive gorgonian areas around the GPS coordinate. The pixel containing this position is located within the image. It is used as a seed pixel – contiguous pixels are compared to it and accepted as being similar or rejected as being different under criteria specified by the analyst. For example, a maximum

permissible spectral difference between contiguous pixels and the seed pixel may be set, above which pixels are rejected as being too dissimilar. If a pixel is accepted it is grouped with the seed and the process repeated until no contiguous pixels satisfy the selection criteria – in essence the training sample grows outward from the seed pixel, hence the name. The analyst may also wish to impose a geographical constraint on the growth of the training sample, so that no pixels more than, say, 2 km from the seed will be accepted (Figure 10.5, Plate 10).

- *One or more classes produced by an unsupervised classification*: for example, if there are *a priori* reasons for accepting some of the classes produced by a preliminary unsupervised classification then those signatures may be used in supervised training in conjunction with signatures from training samples identified by other methods. Thus supervised and unsupervised training may be used in combination.

☞ **Several training sites should be acquired for each class to be mapped but there are no hard and fast rules as to the optimum number of training sites per habitat. Too few and it may be difficult to obtain a spectral signature which is truly representative of that class. Too many and the analyst wastes time in collecting and evaluating signatures which do not significantly improve the final signature. In classifications of the Turks and Caicos Islands data, 10–30 training sites per class seemed to be sufficient (Chapter 19).**

Signature evaluation

Once training has produced a set of training samples these can be evaluated before the decision making stage. Imagine a situation where training has produced ten training samples for each of three classes. Three single signatures are required, one for each class, which are (i) truly representative of that habitat, and (ii) different from the other two. However there will be variation within each group of ten training samples, and some may differ markedly from others in the same class. This variation could be the result of:

- Spectral variation in the habitat of interest
- Position error in the GPS: this could cause mismatch between recorded field positions and pixels in the image.
- Misinterpretation of field data or aerial photographs/maps: this is not restricted to human error. There may be errors in the photographs or maps themselves.
- Variation in water quality or colour: if the water was turbid at one survey site at the time of satellite overpass then it is possible that the signature taken from it would be erroneously 'dark'.

It is important to identify aberrant training samples as their inclusion in the decision making process can have far-reaching effects. Some will have to be deleted, others merged and renamed before classification can proceed to decision making.

Signatures are commonly evaluated in three ways:

1. A classification is performed of the pixels within the training samples for each habitat and the proportion classifying as the habitat recorded by the field survey is calculated. Theoretically 100% of the pixels in a training sample should classify correctly but in reality training samples are never homogeneous enough to achieve that. However, high percentages of correctly classified pixels should be expected. This makes intuitive common sense. If the pixels in the areas deemed to be representative of a particular habitat do not classify as that habitat then what else will? The results are frequently presented in a contingency matrix (Table 10.2).

Table 10.2 Hypothetical contingency matrix based on the 'supervised' coarse habitat classification of the area in Figure 10.1a. Eighty-five percent of the pixels in the training sample for class 'Reef' classify as Reef, 10% as Algae, etc. In this example the signatures from training sample 'Sand' need to be evaluated further because most of those pixels actually classify as class Algae.

	Reef	Algae	Seagrass	Sand
Reef	85	10	0	5
Algae	7	79	5	9
Seagrass	0	7	91	2
Sand	2	58	2	36

2. The 'spectral distance' between signatures is measured. Statistically distant signatures are said to have a high separability. Separability is usually calculated in one of three ways, by computing divergence, transformed divergence or the Jeffries-Matusita distance. The mathematics behind the computation of these will not be discussed here, but further details are available in Swain and Davis (1978). What is important is that there should be high separability between signatures from different types of training samples and low separability amongst signatures from the training samples of a particular habitat.
3. The mean and standard deviation of each signature are used to plot ellipse diagrams in two or more dimensions (Figure 10.6). If the spectral characteristics of the signatures are completely distinct from each other then ellipses (or ellipsoids) will not overlap, but in practice some overlap usually occurs. If the amount of overlap between a pair of signatures is large, then those habitats cannot be separated using the spectral characteris-

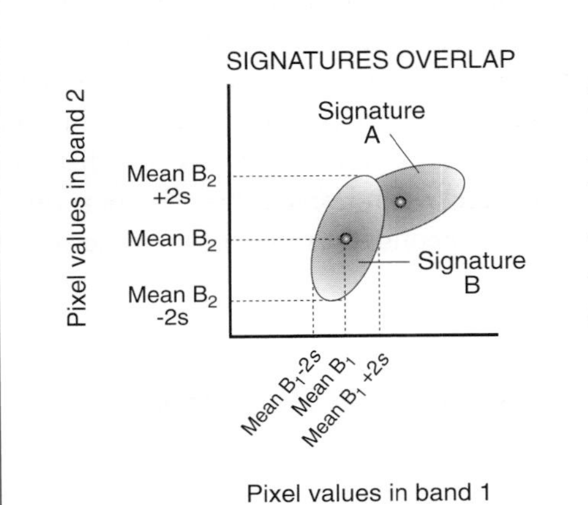

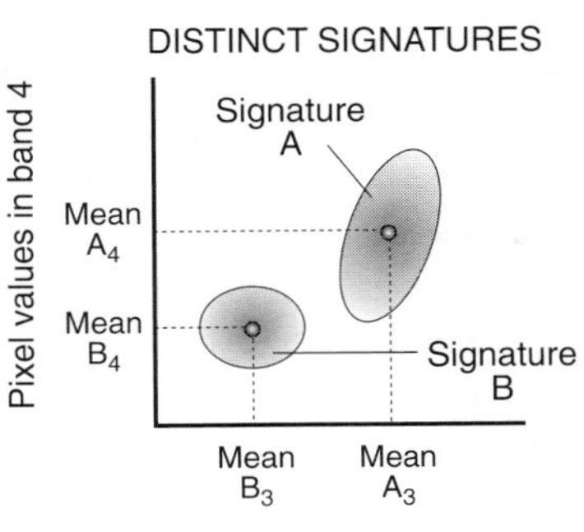

Figure 10.6 Evaluation of spectral signatures (envelopes). In the first case the ellipses, defined by the means ± two standard deviations, overlap considerably and do not allow the spectral signatures to be distinguished in the two bands and thus do not allow the two habitats to be classified with any degree of confidence using these two bands. In the second case there is no overlap in feature space, allowing accurate classification using the two bands. In reality, some overlap is expected. Redrawn from: Erdas Inc. (1994). *Erdas Field Guide* (Atlanta, Georgia: Erdas Inc.)

tics of the bands which have been plotted. Plotting ellipses allows the analyst to identify which signatures are most similar (and which classes will therefore suffer most from misclassification).

Decision making

The third and final step of classification uses the evaluated signatures to assign every pixel within the image to a particular class. This is achieved using a decision rule. Decision rules set certain criteria: a candidate pixel is assigned to a class if its signature 'passes' the criteria which have been set for that class. Pixels which fail to satisfy any criteria remain unclassified.

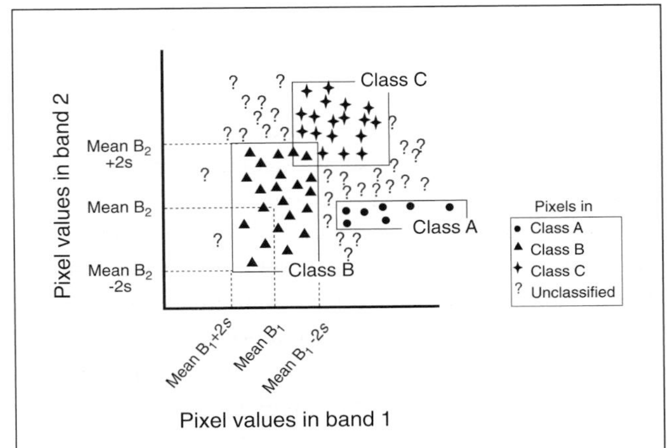

Figure 10.7 Parallelpiped classification using the means ± two standard deviations in two bands as limits to habitat classes. Class A is clearly distinguished from classes B and C but these two classes have a small area of overlap in feature space. Various decision rules can be applied to classify pixels in the area of overlap, or they can be left unclassified. Redrawn from: Erdas Inc. (1994). *Erdas Field Guide* (Atlanta, Georgia: Erdas Inc.)

1. Non-parametric rules. There is one common non-parametric decision rule:

Parallelpiped decision rule: the data values (in each band) of candidate pixels are compared to upper and lower limits (Figure 10.7). The limits may be maximum and minimum values, means plus and minus a number of standard deviations or any other limits set by the analyst based on the knowledge of the data acquired during signature evaluation.

Imagine an example based for simplicity's sake on data values in just one band where the minimum and maximum values for seagrass pixels are 75 and 150 respectively, whereas for sand, pixel values range from 130–220. All pixels with data values between 75–129 in this band will be classified as seagrass, those with values between 151–220 as sand. However, pixels with values of 130–150 lie in a region of overlap. The analyst must set up further decision rules to decide how such pixels will be classified.

There are three options. Firstly, the signatures are ranked and then the pixel will be classified to the class with the highest rank. Rank is determined by the analyst based upon knowledge of which is the most likely class. Secondly, the pixel can be defined according to a selected parametric decision rule. Thirdly, the pixel may be left unclassified. The parallelpiped decision rule is fast and simple, and is often good for an initial first classification that can help the analyst to narrow down the number of possible classes. However, pixels that are quite distant spectrally from the mean of a class, are nonetheless classified in it if within the specified boundaries. This may lead to a less accurate classification.

2. Parametric decision rules. There are three common parametric decision rules:

- *Minimum distance decision rule*: the spectral distance between a candidate pixel and the mean signature of each class is calculated (Swain and Davis 1978). The pixel is assigned to the spectrally closest class. Every pixel is thus classified. However, this means that pixels which should be unclassified because they are spectrally distant from all classes are nonetheless assigned to the closest. Thresholding is a method of reducing this effect, by setting limits and assigning pixels outside the limits to a separate, 'unclassified' class. The main disadvantage of the minimum distance decision rule is that it does not consider class variance. Classes with high variance are typically under-classified: distant pixels belonging to these classes may actually be spectrally closer to the mean of another class than they are to the mean of their own. Classes with low variance classify more accurately because a greater proportion of those pixels are spectrally closer to their own mean than they are to means of other classes. This decision rule is not computationally intensive and so can be rapidly carried out.

- *Mahalanobis distance decision rule*: this rule is similar to the minimum distance decision rule except that the covariance and variance of each class is taken into account in calculating 'spectral distance' (Swain and Davis 1978). As such more computing time is necessary to compute the Mahalanobis distance. Another potential disadvantage is that a normal distribution of the data in each band is assumed – if this is not the case, then the mathematics are invalidated and a better classification may be obtained with a parallelpiped or minimum distance classification.
- *Maximum likelihood decision rule*: this is the most sophisticated of the common parametric decision rules because it takes into account the most variables. Both the variability of classes and the probability of a pixel belonging to each class are taken into account in calculating the distance between a candidate pixel and the mean of all classes. The basic equation for this decision rule assumes that the probabilities of a pixel being in each class are equal. However if the analyst has *a priori* reasons to believe that these are not equal then the probabilities can be weighted – in other words if there are good reasons to believe that a pixel is twice as likely to be sand as seagrass then the sand probability can be weighted to twice that of seagrass. A maximum decision rule with weighted probabilities is known as a Bayesian decision rule – readers interested in the mathematics of the Bayesian rule are referred to Hord (1982).

☞ **Decision making is an integral stage in unsupervised classification, but the analyst is still required to select a decision rule which is appropriate to the image data and type of classes desired.**

In the context used here, the terms parametric and non-parametric refer to decision rules that assume that the data is normally distributed and those that do not. Strictly speaking, parametric decision rules should only be applied if the data within wavebands is normally distributed. For pragmatic reasons however, this criteria is often relaxed but with a concomitant price to pay in terms of final accuracy.

Case study: supervised classification of CASI data for mangrove mapping

Figure 10.8 is a greyscale representation of a CASI (Compact Airborne Spectrographic Imager) false colour composite of mangroves on South Caicos using bands in the red (630.7–643.2 nm) and near infra-red (736.6–752.8 nm and 776.3–785.4 nm). Submerged areas have been masked out of the image.

Training

1. An ecologically-based habitat classification scheme was developed for the mangrove areas of the Turks and Caicos Islands using hierarchical agglomerative clustering with group-average sorting applied to the calibration data (Chapter 9). The calibration data were 4th root transformed in order to weight the contribution of tree height and density more evenly with species composition (the range of data was an order of magnitude higher for density and height). This identified seven classes which separated at a Bray-Curtis similarity level of 85% (Figure 10.10). Categories were described in terms of average species composition (percent species), mean tree height and mean tree density (Table 10.3). One category, *Laguncularia* dominated mangrove, was discarded because white mangrove was rare in this area – in both the calibration and accuracy field phases it was observed at only two locations. Three other ground cover types were recorded, though no quantitative information was obtained beyond the location of 'typical' sites: (i) sand, (ii) saline mud crust, and (iii) mats of the halophytic succulents *Salicornia perennis* and *S. portulacastrum*. These nine habitat categories (six mangrove, three other) were used to direct the image classification of the CASI data, and the collection of accuracy data.
2. For each habitat class signatures were collected from the CASI image using seed pixels centred at the position of each field site. Each field site was 5 x 5 m square (chosen to reflect the accuracy of the differential GPS) so a geographical constraint of that size was used to limit the growth of the polygons. However, in a few instances larger areas were uniformly covered in one type of mangrove habitat in which case polygons could be drawn by hand to obtain signatures.

Signature evaluation

3. Of the 46 field sites used to create training samples we wished to be sure of two factors:

   i) that training samples for each mangrove class (e.g. black mangrove) were sufficiently similar and representative, in which case we would merge them to produce a single signature file for black mangrove

   ii) and that the signature for each mangrove class was sufficiently dissimilar to warrant classification, e.g. that the signature of red tall mangrove was sufficiently dissimilar to that of black mangrove.

Tables 10.4, 10.5, and 10.6 illustrate how we used measures of Euclidean 'spectral distance' and transformed divergence to inform and direct task i) above. The same approach was used to ascertain dissimilarity for task ii). For black mangroves the signatures separate into two quite distinct groups (Table 10.5). Black mangrove sites 35, 71, 50, 63, 59, 60 and 78 are reasonably similar to each other (they have low Euclidean distances and are therefore close in feature space). But, black mangrove sites 61 and 52 are clearly different – the Euclidean distances are consistently high when compared to the other *Avicennia* sites (and are thus far apart from them in feature space) and are also very different from each other. The same is true when similarity is measured using transformed divergence (Table 10.6). Thus, the signatures from sites 61 and 52 appear aberrant. We assume this to be due to uncertainty in GPS/geometric correction resulting in us collecting spectral information a short distance away from our true field location. Their deletion from the signature file created a 'purer' signature more representative of black mangrove. Signatures from sites 35, 71, 50, 63, 59, 60 and 78 were merged to form a single black mangrove signature and the process repeated for all other habitat types. This naturally causes a reduction in the dataset used to classify the image (Table 10.7). We disposed of ~ 10% of the field sites during signature evaluation. This

Figure 10.8 Greyscale representation of a false colour composite CASI image of mangroves on the west coast of South Caicos Island. CASI bands in the red: 630.7–643.2 nm, and near infra-red: 736.6–752.8 nm and 776.3–785.4 nm were used to make the colour composite and in supervised classification of this image (Figure 10.9, Plate 11). The image is approximately 1 km across.

Table 10.3 Descriptions for each of the mangrove habitat categories identified in Figure 10.10. n = number of calibration sites in each category. Species composition: Rhz = *Rhizophora*, Avn = *Avicennia*, Lag = *Laguncularia*, Con = *Conocarpus*.

		Species composition %				Tree height (m)	Tree density (m^{-2})
Habitat category description	n	Rhz	Avn	Lag	Con	mean (range)	mean (range)
Conocarpus erectus (Buttonwood)	6	0	0	0	100	2.4 (1.8-4.5)	0.6 (0.5-1.0)
Avicennia germinans (Black)	11	0	100	0	0	2.6 (0.8-6.0)	0.6 (0.2-1.0)
Short, high density, *Rhizophora mangle* (Red)	10	100	0	0	0	1.1 (0.5-2.0)	8.0 (6.0-10.0)
Tall, low density, *Rhizophora mangle* (Red)	25	100	0	0	0	3.7 (2.0-7.0)	0.3 (0.2-0.5)
Short mixed mangrove, high density	10	62	38	0	0	1.7 (0.8-2.5)	8.1 (5.0-15.0)
Tall mixed mangrove, low density	14	56	43	0	1	3.5 (2.0-5.0)	0.6 (0.2-1.2)
Laguncularia dominated mangrove (White)	2	35	5	45	0	3.8 (3.5-4.0)	2.2 (0.5-4.0)
Unclassified	3						

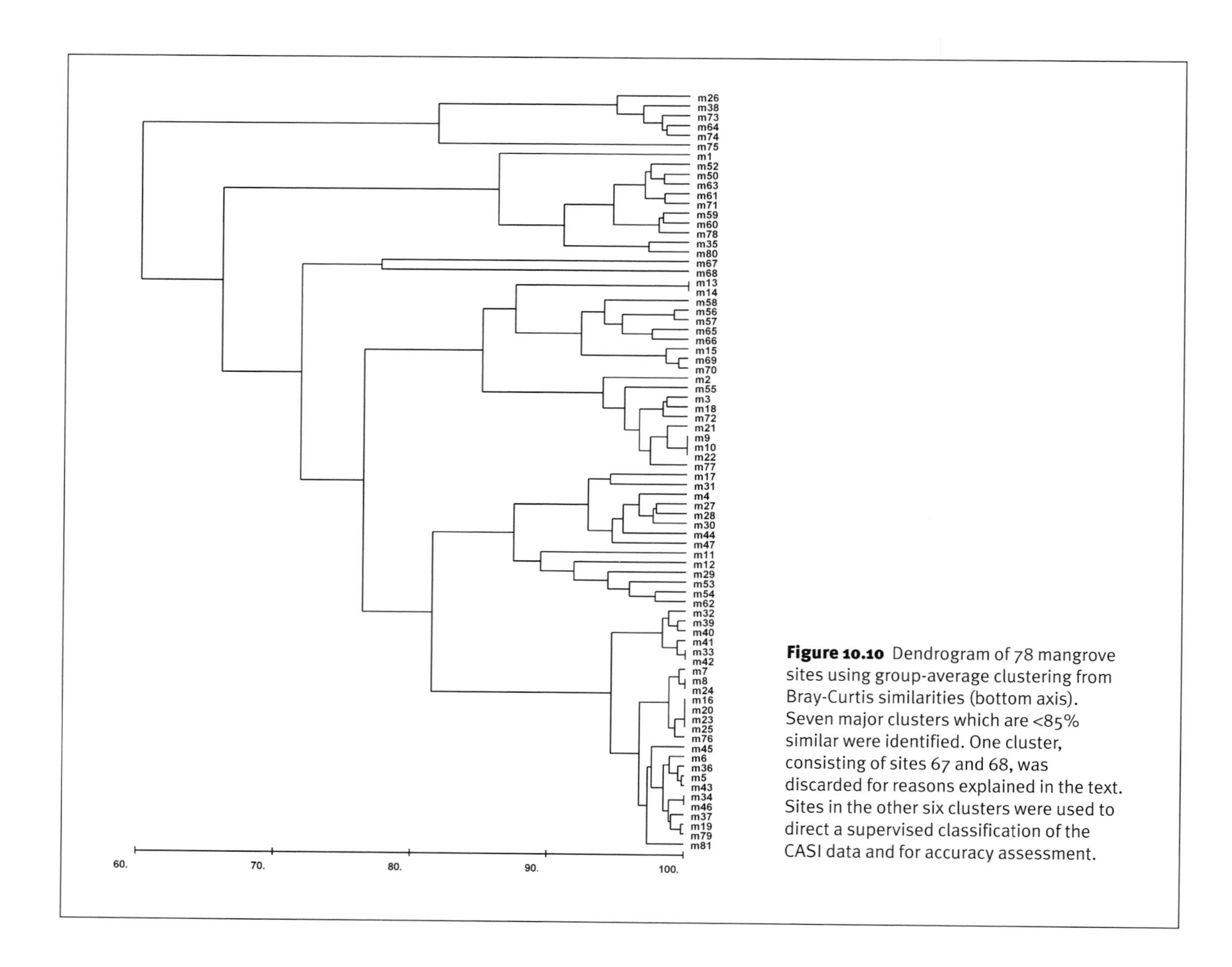

Figure 10.10 Dendrogram of 78 mangrove sites using group-average clustering from Bray-Curtis similarities (bottom axis). Seven major clusters which are <85% similar were identified. One cluster, consisting of sites 67 and 68, was discarded for reasons explained in the text. Sites in the other six clusters were used to direct a supervised classification of the CASI data and for accuracy assessment.

Table 10.4 Training sample evaluation to produce signatures for each mangrove class. A total of 46 sites were seeded on the image and the individual signatures from each assessed for similarity using Euclidean distance and transformed divergence (Table 10.5 and 10.6). Training samples from some sites (shaded) were discarded as being unrepresentative of that habitat due to their proven dissimilarity. n = number of pixels derived from the seed at each site. The bottom row shows the total number of sites and seeded pixels for each habitat category.

Habitat type											
Red tall		Black		Mixed short		Mixed tall		Red short		Buttonwood	
Site #	*n*	Site #	*n*	Site #	*n*	Site #	*n*	Site #	*n*	Site #	*n*
33	16	35	28	58	12	31	15	1	36	73	30
40	12	61	15	65	16	68	16	2	36	38	21
39	16	71	16	56	9	67	16	3	42	64	20
41	15	52	16	75	20	53	20	4	72	74	19
46	16	50	24	57	20	62	12	5	72		
42	9	63	15	66	18	54	16				
37	16	59	20	69	18	44	9				
34	12	60	24	70	16						
79	16	78	28								
36	12										
43	12										
45	16										
81	16										
13	184	9	218	8	129	7	104	5	258	4	90

Table 10.5 A matrix recording the Euclidean distance between signatures from pairs of black mangrove field sites. Sites 61 and 52 have consistently high Euclidean distances indicating that they are quite separate from the typical black mangrove class and were eventually deleted from the file (see text).

Site	black 35	black 61	black 71	black 52	black 50	black 63	black 59	black 60	black 78
black 35									
black 61	1222								
black 71	812	1119							
black 52	1597	1302	2079						
black 50	992	920	331	2030					
black 63	323	1489	1055	1675	1275				
black 59	286	1366	696	856	944	389			
black 60	695	1716	1445	1647	697	427	789		
black 78	633	883	542	1001	583	770	871	881	
mean	763	1256	878	1327	837	799	758	981	664

Table 10.6 A matrix which measures the transformed divergence between signatures from pairs of black mangrove field sites. Transformed divergence is scaled between 0 and 2000. A transformed divergence of 2000 means the signatures are totally separate, 0 that they are identical in the bands being studied. Sites 61 and 52 have consistently high transformed divergence and were eventually deleted from the file.

Site	black 35	black 61	black 71	black 52	black 50	black 63	black 59	black 60	black 78
black 35									
black 61	256								
black 71	418	1999							
black 52	127	1750	139						
black 50	684	1664	487	2000					
black 63	726	2000	634	2000	649				
black 59	245	989	490	1804	185	532			
black 60	489	1224	259	1692	549	681	496		
black 78	364	1653	788	2000	693	493	996	590	
mean	367	1281	579	1279	767	857	637	664	831

Table 10.7 The proportion of pixels in Table 10.4 that were used to derive signatures for the six different mangrove habitat types

Habitat	Number of pixels initially collected	Number of pixels discarded	Number of pixels used
Red short	258	0	258
Red tall	184	21	163
Black	186	31	15
Buttonwood	90	0	90
Mixed tall	104	9	95
Mixed short	129	39	90
Total	951	100	851

should be taken into account when planning field surveys (i.e. sufficient sites should be visited in the field to allow editing of the signature file later).

4. The final signatures for each mangrove habitat were tested for contingency (Table 10.8). Most signatures were satisfactory – high proportions of the pixels used to define each habitat type would classify as that habitat type using that signature file. However, there was considerable confusion between mixed tall mangrove and both black and red tall, and about 80% of mixed short mangrove classified as other habitats! (Table 10.8). The latter was probably a result of a small sample size (only six mixed short mangrove sites were recorded) – ideally more field data should have been collected to try and rectify this situation but that was not possible. A contingency matrix forewarns the analyst of possible misclassification difficulties. In this case the overall user accuracy of mixed short mangrove was very low (16.7%).

Table 10.8 A useful preliminary test prior to classifying the whole image is to test to what extent the signatures accurately classify known training samples. A contingency matrix for the classification of a CASI mangrove image using the edited signature file (after signature evaluation). High proportions of the training sample pixels used to define the signature for the first four mangrove habitat types classify correctly. Therefore the signature file is suitable for mapping these habitats. The exceptions are mixed short mangrove: most of whose training sample pixels classify as mixed tall mangrove; and mixed tall mangrove where over 40% of pixels classify as red tall or black mangrove (perhaps not surprisingly as the class is primarily a mixture of these). A poor user's accuracy could be expected for these classes if this file was used.

	Proportion of pixels in habitat *x* training sample which classify as that habitat					
Training sample habitat	Red short	Red tall	Black	Buttonwood	Mixed tall	Mixed short
Red short	93.4	0.0	0.0	0.0	1.8	4.8
Red tall	0.0	79.9	6.9	0.0	12.6	0.0
Black	0.0	0.0	77.8	1.6	3.3	16.8
Buttonwood	0.0	0.0	8.1	63.5	2.7	25.7
Mixed tall	1.9	20.3	24.8	4.5	39.5	8.9
Mixed short	2.7	0.0	17.8	20.5	84.9	19.2

Decision rule

5. A maximum likelihood decision rule was performed using these signature files to produce a mangrove habitat map (Figure 10.9, Plate 11).

This is a particularly clear case of signature evaluation, presented as an example. It is unusual that the decision is so straightforward. Readers should be prepared to spend considerable amounts of time (days) evaluating their signature files, perhaps several times over. The process becomes more difficult as the number of image bands, number of habitats and similarity of habitats increases.

☞ **Supervised classification should be based on the most detailed descriptive resolution that your field survey data will allow, even if your anticipated outputs are of lower descriptive resolution. Individual spectral envelopes for each habitat can then be progressively merged to provide characteristic habitat spectra. Merging is conducted with reference to the spectral distance between signatures in Euclidean space. For example, if habitat A is represented by 35 signatures (sites) which form two distinct groups in Euclidean space (e.g. sites 1–9 and 10–35), the two groups should be treated separately. In this example, two spectra would be retained for habitat A. However, if these two spectral envelopes had been merged into a single spectrum, the result would lie somewhere between the two component spectra and might not represent the spectrum of that habitat at all, leading to confusion with distinct habitats whose spectra may actually lie between the two habitat A spectra. An example of where this may occur is where habitat A may be some seagrass habitat at two depths or in areas of differing water quality. Thus, in our experience it is better to merge at the thematic class level than to combine spectral envelopes.**

Choosing a classification method

How does one choose whether to use visual interpretation or multispectral image classification? Photo-interpretation is easy to understand and can be performed using relatively inexpensive equipment but it is both very time-consuming (due to the digitisation process) and subjective. The subjectivity means that the outputs of visual interpretation will be as good or bad as the interpreter, and will vary from interpreter to interpreter. Good results are only likely to be obtained if well-trained, highly experienced personnel are used. Also it is difficult to compare visually interpreted products (e.g. for change detection) because differences may be attributable to inconsistencies in interpretation rather than real changes on the ground. Multispectral classification requires sophisticated computing equipment and a good comprehension of the statistics being used, but rapidly and objectively produces a digital map. As long as radiometric and atmospheric correction has been carried out carefully, habitat maps made at different times should be comparable.

If the capacity to perform multispectral classification does not exist then we recommend photo-interpretation for (i) small areas < 100 km^2 to be mapped at a fine level of habitat discrimination or (ii) medium size areas < 1000 km^2 to be mapped at a coarse level of habitat discrimination. In cases where habitat maps have been made from visually interpreted aerial photographs (e.g. Sheppard *et al.* 1996) habitat patches less than about 10 m across are not usually digitised, presumably because of the amount of time it would take to draw around every patch that size. Visual interpretation is therefore unlikely to make full use of high resolution data such as aerial photographs or CASI. Circumstances vary widely making it hard to generalise accurately, but the main advantages and disadvantages of photo-interpretation and multispectral classification are summarised in Table 10.9.

Table 10.9 The advantages and disadvantages of visual and digital image processing techniques

Photo-interpretation	Multispectral classification
ADVANTAGES	ADVANTAGES
The human visual system is good at making qualitative judgements on colour, tone, texture, patterns and relative size and orientation.	Multidimensional spectral information is analysed quantitatively and simultaneously.
Computer analysis is not necessary and can therefore be performed by more cheaply and with minimally trained personnel (though a knowledge of the habitats to be mapped is necessary).	All radiometric brightness levels are quantitatively analysed, and so full spectral range is utilised.
	Objective approach; once the statistical parameters for classification have been set the process is objective.
DISADVANTAGES	DISADVANTAGES
The human eye is sensitive to only a small number of levels of brightness. e.g. to only 30 different grey tones. Full spectral range of image data is not utilised.	Only uses per-pixel quantitative information, i.e. accounts for 'colour' and 'tone' of individual pixels but not texture or patterns because these require information over wider areas.
The ability of the human visual system to distinguish colour is greatly reduced by variations in the background colour.	Computationally intensive.
Subjective approach; accuracy of interpretation likely to vary between individuals and with experience (Drake 1996).	Requires a good understanding of the statistics being used.
Problems arise in converting the visual (mental) interpretation of an analyst into an image file on the computer or hard copy map. Polygons may be drawn manually around the habitats which the interpreter has identified visually and labelled accordingly. This is extremely time consuming for large areas such as those covered by satellite images or very heterogeneous areas such as patch reef/seagrass complexes. Furthermore while a polygon may be obvious to the human eye, the action of physically drawing and closing a boundary around its edges is difficult.	

Photo-interpretation and multispectral classification should be thought of as complementing each other. A good classification can be achieved using a combination of approaches. For example, the patch reefs and algal areas (A and B in Figure 10.1b, Plate 9) are not distinguished by unsupervised classification (Figure 10.4c, Plate 10) but appear different in Figure 10.1a to the naked eye. This is because the differences in tone and pattern can be separated by the human brain but not an unsupervised classification. In this case a combination of the two techniques in which visual interpretation was used to guide a field survey for supervised classification would reduce confusion between the two classes and would probably result in a more accurate classification.

Conclusions

There is no 'best' classification method; different methods should be judged in terms of their cost and accuracy for specific applications (see Chapter 19). In the interest of clarity parametric/non-parametric and supervised/unsupervised approaches to classification have been referred to separately in this section. However they should not be considered mutually exclusive – the best classifications are probably achieved using a combination of these approaches (Figure 10.11). In practice visual interpretation is the first step in any classification with the image being studied in the light of existing knowledge and data. Differences between major habitats can usually be identified at a glance and closer inspection familiarises personnel with the more subtle patterns in the digital data (without the need to physically draw boundaries between). An unsupervised classification can then be used to direct and plan a field campaign – the results of this unsupervised classification should always be studied in the light of the initial visual interpretation of the image to confirm that the classes appear sensible. Once field data have been collected these steps should serve to facilitate informed decision making during the more interactive supervised supervisions. Most classifications can be improved by careful contextual editing. Examples of this approach are given in Chapter 11. If a habitat map produced by multispectral classification is to be used for management or monitoring it is important that it is accompanied by an assessment of its accuracy (see Chapter 4).

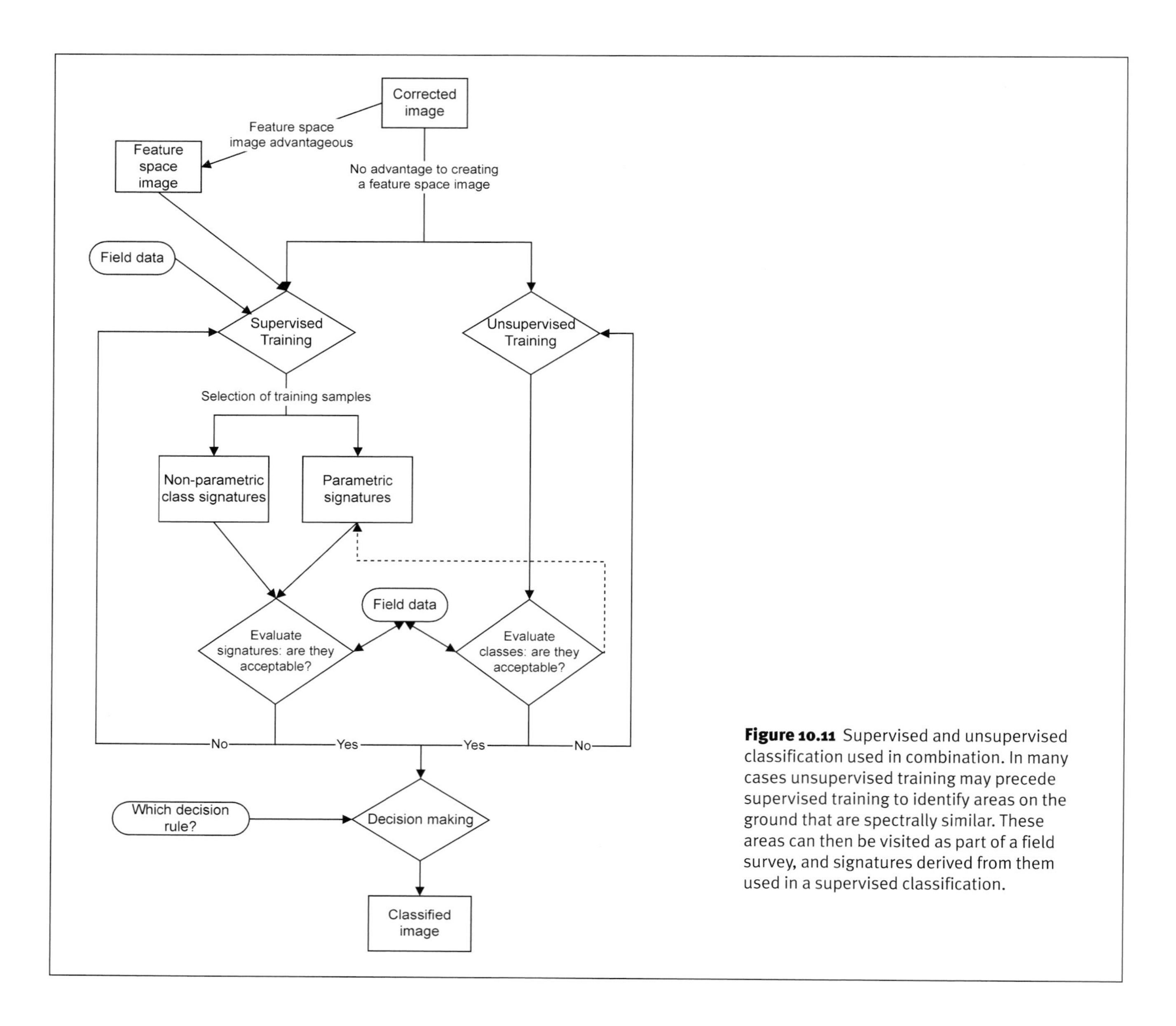

Figure 10.11 Supervised and unsupervised classification used in combination. In many cases unsupervised training may precede supervised training to identify areas on the ground that are spectrally similar. These areas can then be visited as part of a field survey, and signatures derived from them used in a supervised classification.

References

Drake, S., 1996, Visual interpretation of vegetation classes from airborne videography: an evaluation of observer proficiency with minimal training. *Photogrammetric Engineering and Remote Sensing*, **62** (8), 969–978.

Erdas Incorporated, 1994, *Erdas Field Guide* (Atlanta, USA: Erdas Incorporated).

Groom, G.B., Fuller, R.M., and Jones, A.R, 1996, Contextual correction: techniques for improving land cover mapping from remotely sensed data. *International Journal of Remote Sensing*, **17**, 69–89.

Hord, R.M., 1982, *Digital Image Processing of Remotely Sensed Data.* (New York: Academic Press).

Mather, P.M., 1987, *Computer Processing of Remotely-Sensed Images: an Introduction*. (Chichester: John Wiley & Sons).

Sheppard, C.R.C., Matheson, K., Bythell, J.C., Blair Myers, C., and Blake, B., 1995, Habitat mapping in the Caribbean for management and conservation: use and assessment of aerial photography. *Aquatic Conservation of Marine and Freshwater Ecosystems*, **5**, 277-298.

Swain, P.H., and Davis, S.M., 1978, *Remote Sensing: the Quantitative Approach*. (New York, USA : McGraw Hill Book Company).

Tou, J.T., and Gonzalez, R.C., 1974, *Pattern Recognition Principles*. (Reading, Massachusetts, USA: Addison-Wesley Publishing Company).

Wilke, D.S., and Finn, J.T., 1996, *Remote Sensing Imagery for Natural Resources Monitoring*. (New York, USA: Columbia University Press).

11

Mapping Coral Reefs and Macroalgae

Summary *The applications of remote sensing for coral reef management include baseline cartographic mapping, mapping reef geomorphology and mapping reef habitats. Generally, more detailed mapping objectives require more sophisticated remote sensing techniques. Most satellite-mounted sensors are able to provide information on reef geomorphology and limited ecological information such as the location of corals, sand, algal and seagrass habitats with an accuracy ranging from 50–70%. The most cost-effective satellite sensors for habitat mapping are Landsat TM (for areas greater than one SPOT scene) and SPOT XS for areas within a single SPOT scene (i.e. < 60 km in any direction). Colour aerial photography can resolve slightly more detailed ecological information on reef habitats but, for general purpose mapping, satellite imagery is more effective (i.e. has slightly greater accuracy and uses up much less staff time). At low altitude, infra-red aerial photography can be used to estimate live-coral cover over shallow (< 1 m deep) reef flats. However, since the low altitude restricts the areal coverage of each photograph, this method is only appropriate for small areas.*

The most accurate means of making detailed reef habitat maps appears to be use of airborne multispectral instruments such as CASI (Compact Airborne Spectrographic Imager). In the Caribbean, CASI can map assemblages of benthic species and substrata with an accuracy of > 80% (cf. < 37% for satellite sensors and 57% for colour aerial photography).

A wide variety of field survey techniques are available for assessing coral and macroalgal habitats. For broadscale habitat mapping (e.g. with satellite imagery), manta tow and visual belt transects are recommended and described. Rapid quantitative methods such as quadrats, line intercept transects and underwater video are recommended for imagery with greater resolution like CASI.

Introduction

Coral reefs

Coral reefs generally exist in clear waters, which are highly suited to optical remote sensing. Their nutritional biology dictates that they usually live in regions of relatively clear water which transmit sufficient solar irradiance to allow their endosymbiotic algae (zooxanthellae) to photosynthesize (for overview of coral biology see Dubinsky 1990). This requirement has been capitalised upon by the remote sensing community and satellite investigations of coral reef structure date back to 1975, shortly after the launch of Landsat MSS (Smith *et al.* 1975).

Remote sensing has been touted to provide information on several parameters that are of importance to reef management. From a remote sensing point of view, coral reef boundaries are probably the easiest of these to map. This information may be used for routine planning requirements and locating the boundaries of management zoning schemes (e.g. Kenchington and Claasen 1988). The next level of sophistication is to distinguish principal geomorphological zones of the reef (e.g. reef flat, reef crest, spur and groove zone). For management purposes, such information may be used in a similar manner to that of 'coral reef maps'. These maps may also have more sophisticated uses, which include the stratification of field sampling regimes and relating reef geomorphology to reef function and productivity (discussed by Fagerstrom 1987).

Maps of reef geomorphology have been made with considerable success (see forthcoming sections). However, mapping the ecological components of coral reefs appears to be considerably more difficult using conventional satellite-borne imagery (Bainbridge and Reichelt 1989). Ecological components may be defined in several ways, such as assemblages of coral species (e.g. Done 1983), assemblages of major reef-dwelling organisms, or assemblages of species and substrata (e.g. Sheppard *et al.* 1995). The choice depends on the ecological objective and type of reef. For example, coral species assemblages would be appropriate in places where coral cover was high (see Plate 8), but perhaps less appropriate where coral cover rarely exceeded (say) 20%. Irrespective of their foundation, maps with an ecological basis will be referred to here as 'habitat maps'. Habitat maps have considerable potential in the management of reef biodiversity at ecosystem, community and (indirectly) at species scales. This has often been sought by protecting representative areas of habitat (McNeill 1994).

Perhaps the 'holy grail' of all coral reef remote sensing is the determination of live-coral cover. Whilst the issue of coral reef health is complex (see UNESCO 1995), one of the most widely accepted parameters to consider is live coral cover with a view to relating changes in cover to the processes responsible. As we shall see, however, optical remote sensing is severely limited at measuring this parameter and has only been successful in shallow reef flat environments.

Macroalgae

Technically speaking, the term 'macroalgae' describes algae that are large enough to see by eye rather than a taxonomic grouping. Most macroalgae can be categorised by lifeform into: (i) fleshy algae, (ii) calcareous algae (which have a calcium carbonate skeleton), (iii) turf algae (which form a mat < 1 cm high on the substratum), and (iv) crustose coralline algae (which form flat plates on the substratum and may cement coral fragments). Macroalgal species are divided among three large groups (green, brown and red algae), which are named according to the colour of their dominant photosynthetic and accessory pigments (Plate 8).

Green algae (Chlorophyta) contain chlorophyll and are well represented in the tropics (Littler *et al.* 1989). Some of these algae (e.g. *Enteromorpha* spp.) favour stressful environments where nutrients are high and herbivory low. Others are calcified (e.g. *Halimeda* spp.) and contribute heavily to the sandy sediments of reef areas. Brown algae (Phaeophyta) predominantly contain the brown pigment, fucoxanthin, but colonies may range in colour from beige to almost black. In temperate seas, brown algae can form vast kelp forests (e.g. *Macrocystis* sp. in California) but their abundance and diversity in tropical seas is reduced. Some of the most common tropical genera include *Sargassum* and *Turbinaria* which are often associated with reef flats, and *Lobophora* which is fairly ubiquitous. The red algae (Rhodophyta) are the largest and most diverse group but arguably the least well understood (Littler *et al.* 1989). They contain large quantities of the the pigment phycoerythrin, which can often resemble the pigmentation of brown algae (fucoxanthin). Red algae are extremely important reef-building organisms, which may form reef crests (e.g. *Lithophyllum* spp.) and large calcareous plates (*Sporolithon* spp.).

From a coastal management perspective, it would be useful to map algal-dominated habitats for biodiversity management and locating mariculture sites for harvesting commercial seaweeds such as *Porphyra* spp. and *Euchema* spp. Maps of algal biomass may also be useful in understanding coastal productivity and in managing algal mariculture. Algal habitats may also be correlated to the abundance of commercially significant gastropods such as the queen conch, *Strombus gigas*. While this remains to be proved, the topic of correlating marine resources to habitats is discussed in Chapter 18.

Structure of this chapter

This chapter sets out methods and anticipated outcomes of mapping coral reefs and macroalgae. It begins by evaluating the likely outcomes of using various satellite and airborne methods for reef and macroalgal assessment. This section is split into a general overview based on the literature and then a more detailed discussion of habitat mapping which draws mainly on results from the Turks and Caicos Islands. It also discusses the usefulness of various image processing methods. In several places, the accuracies of habitat maps are discussed with reference to other types of habitat such as seagrass and sand. This is because mapping of coral reefs and macroalgae is not an isolated process and is affected by the presence of neighbouring habitats.

The latter part of the chapter describes the field survey techniques used to describe coral and macroalgal systems. This is a vast subject area which is covered in several comprehensive texts. As such, the scope of this chapter is to give some general guidance and an overview of relevant literature.

Overview of descriptive resolution and sensors

Coral reefs

The majority of studies in Table 11.1 provided geographic information on reef zones. These zones are primarily of a geomorphological, rather than an ecological nature and include forereef, reef crest, algal rim, spur and groove, and so on. Any attempt to make detailed comparisons between the ability of different sensors is difficult because of the differences in reef terminology, study sites and objectives employed in each study. Generally speaking however, sensors with higher resolution will offer greater detail. A recent study of the reef zonation at Heron Island was conducted using Landsat TM (Ahmad and Neil 1994). The authors compared their results to those obtained by Jupp *et al.* (1986) who originally used Landsat MSS to map the island. TM revealed a similar number of principal reef zones to the MSS, but allowed the principal zones to be divided into a greater number of subzones. Reef zones were mapped more precisely with the TM.

A few studies have described coral cover in terms of density (e.g. Zainal *et al.* 1993, Ahmad and Neil 1994). Measurements of density of this type are usually qualitative and are generally not supported by field data. For example, Ahmad and Neil (1994) include reef subzones described as, 'medium coral head density', 'low coral head density' and 'high coral head density'. The vagueness of these terms reflects practical difficulties experienced in assigning numeric quantitative ranges to coral head density (which may require considerable fieldwork). This is not a criticism of the authors; it merely reflects a general limitation of remote sensing to provide information on reef structure and composition beyond a geomorphological level. This issue will be returned to later when discussing the overall usefulness of remote sensing products in an ecological context. A notable exception to this generalisation is seen in Bour *et al.* (1996). Coral density in New Caledonian reefs was categorised into low density (10–29% cover), medium density (30–70% cover) and high density (71–100% cover). These density bands were mapped using SPOT XS and the authors point out that these results may be useful for long-term monitoring of coral reefs.

Sheppard *et al.* (1995) achieved an overall map accuracy of 57% when using 1:10,000 colour aerial photography to map several coral reef, algal and seagrass habitats of Anguilla. In a study of similar habitats in the Turks and Caicos Islands, Mumby *et al.* (1998b) achieved greater accuracy (81%) using CASI. Satellite imagery of the same study site was found to be generally unsuccessful for mapping individual reefal habitats (accuracy < 37%). This synopsis of habitat mapping will be discussed in greater detail in the following section.

To date, the only successful remote sensing method for providing synoptic data on coral colonies and live coral cover is the use of low altitude aerial photography (see Catt and Hopley 1988, Thamrongnawasawat and Catt 1994). An altitude of 3000 feet (914 m) is generally considered optimal (Hopley and Catt 1989). Healthy corals strongly reflect near infra-red electromagnetic radiation. Therefore, these wavelengths are able to discriminate coral growth from coralline, turf and macro- algae. However, the major limitation of the technique is the poor depth of penetration (< 1 m) and therefore its use is largely confined to assessing the coral cover of emergent and extremely shallow reef flats at low tide (Hopley and Catt 1989). True colour photography will penetrate to approximately 15–20 m but does not offer such good discriminating power. The greatest resolution reported for low altitude aerial photography of reef flats discerned colonies of live massive and branching corals, dead corals and sand (Thamrongnawasawat and Hopley 1995). CASI is capable of sub-metre pixel sizes but the signal to noise ratio is likely to be low, especially for deeper reefs or habitats with low reflectance. The use of CASI for measuring shallow-water live coral cover is yet to be evaluated.

Coral species discrimination does not appear to be possible (Table 11.1) although this is barely surprising considering the spatial resolution of most sensors and the arguments made above.

Table 11.1 The descriptive resolution of various remote sensing methods for mapping coral reefs. Numbers denote referenced papers (pp. 172–174). Updated from Green *et al.* (1996).

Type of class	Landsat MSS	Landsat TM	SPOT XS	SPOT Pan	Merged TM/Pan	Airborne MSS	API
Coral reef (general)	30	22, 25, 30, 37,	37, 30	30	28, 30	29	14, 41
Reef zones (geomorphology)	2, 3, 4, 5, 13, 16, 17, 18, 19, 32	12, 37	2, 6, 11, 21, 37, 42	26			10, 15, 23, 24, 27, 33, 34
Coral density		1, 39, 40, 43, 44	7, 8, 9				
Coral reef habitats		43, 44				29	31
Coral colonies							10, 20, 36
Live-coral cover (only shallow reef flats)							35, 36, 38

Macroalgae

Relatively little has been written explicitly about macroalgal remote sensing. In temperate waters, large phaeophytes have been mapped using a variety of sensors including Landsat TM for drifting *Sargassum* spp. (Tiefang *et al.* 1990) and CASI for regions of *Ecklonia* spp. in Perth, Australia (Jernakoff and Hick 1994). For tropical reef flats, aerial photography has been used to map *Sargassum* spp., *Turbinaria* spp. and other shallow-water algal assemblages (Maniere and Jaubert 1984, Catt and Hopley 1988, Mumby *et al.* 1995a). Zainal (1994) successfully used Landsat TM and a digital bathymetry model to map algal beds around Bahrain. Using this approach, Zainal achieved user accuracies of 86% for *Sargassum* spp. and 75% for algal turf. These accuracies were considerably higher than those reported by Sheppard *et al.* (1995) for mapping phaeophytes in Anguilla with aerial photography (38%). They were also greater than satellite-based results from Mumby *et al.* (1997) who mapped beds of the brown alga, *Lobophora variegata*, with a user accuracy of 0% (SPOT Pan) to 41% (Landsat TM and SPOT XS). The reasons for this disparity are unclear but may be a consequence of the detail to which habitats were defined in each study (studies by Sheppard *et al.* and Mumby *et al.* included a greater number of habitat classes and consequently tended to be less accurate). CASI was found to map this habitat with a user accuracy of 82% (Mumby *et al.* 1998b). Again, these results will be described further later.

With regard to assessing macroalgal biomass or the coverage of commercially important species, the literature seems extremely sparse. Meulstee *et al.* (1986) found an empirical relationship between the biomass of temperate, inter-tidal macrophytes and the density of colour in aerial photographs (error 10%). However, we are unaware of any further examples.

Descriptive resolution for mapping coral and macroalgal habitats

This section compares the ability of different satellite and airborne remote sensing methods for mapping reefal habitats. Image processing methods are assessed first using a case-study from the Turks and Caicos Islands. The descriptive resolution of each sensor is then assessed by extending the case-study from the TCI and describing a second case-study from the Caribbean which used 1:10,000 colour aerial photography to map the coastal zone of Anguilla (Sheppard *et al.* 1995).

Image processing methods

Creation of habitat maps from digital remote sensing usually involves either supervised or unsupervised classification of spectral bands (see Chapter 10). The addition of water column correction and contextual editing should improve map accuracies for submerged habitats (see Chapters 8 and 10 for implementation details). Surprisingly, water column correction has not been widely adopted as a pre-processing step for the remote sensing of tropical coastal waters which often satisfy the exponential attenuation requirements of the model. Reviewing relevant papers, Green *et al.* (1996) found only four studies out of forty-five (9%) had attempted water column correction. From a practical perspective, this prompted the question: What benefits, in terms of thematic map accuracy, accrue from water column correction and is the extra processing time cost-effective? A similar argument may be made for contextual editing. These issues are discussed below.

The following evaluation of these image processing techniques is derived from our case study in the Turks and Caicos Islands (Mumby *et al.* 1998a). Three hierarchical levels of habitat discrimination are given which include coral reef, algal and seagrass classes. For satellite comparisons, the fine, intermediate and coarse levels include 13, 8 and 4 habitats respectively. CASI imagery was confined to a smaller area so the fine, intermediate and coarse descriptive resolutions include 9, 6 and 4 habitats (see Table 11.2). Four levels of image processing were assessed and these are defined in Table 11.3.

☞ **This section uses two measures of accuracy – overall accuracy and the Tau coefficient (see Chapter 4).**

Supervised multispectral classification of reflectance data

Habitat map accuracies were lowest for this level of image processing (Figure 11.1a).

Water column correction

Water column correction of CASI imagery made a significant ($P < 0.01$) improvement to maps with fine habitat discrimination (Figure 11.1b). Inter-habitat similarity was high (Bray-Curtis Similarity, 60–80%) at this descriptive resolution and variable depth exerted a strong effect on accuracy, thus requiring water column correction. At coarse descriptive resolutions (coral, algae, sand, seagrass), the habitats were sufficiently dissimilar to one another (Bray-Curtis Similarity, 10–15%) that depth invariant processing was not essential for habitat mapping (although it did make a minor improvement and may be more advantageous in other areas).

Map accuracy was significantly improved for Landsat TM at coarse and intermediate descriptive resolutions.

Table 11.2 Description and characteristics of benthic habitats determined from hierarchical classification of field data (except for *Acropora palmata* which was added later), showing mean percentage cover, densities and standing crop where appropriate. Class assignment is described for three levels of habitat discrimination: coarse (C), intermediate (I) and fine (F). Habitats present in CASI imagery are marked with an asterisk.

Description and characteristic features	Class assignment number		
	C	I	F
Living and dead stands of *Acropora palmata*			*
Microdictyon marinum (77%), *Sargassum* spp (4%), medium soft coral density (5 m^{-2}) and rubble (10%)	1	1	1
Bare substratum (40%), low soft coral density (3 m^{-2}), *Microdictyon marinum* (30%), *Lobophora variegata* (12%)	1	2	2*
Bare substratum (80%), medium soft coral density (5 m^{-2})	1	2	3*
Bare substratum (60%), high soft coral density (8 m^{-2}), *Lobophora variegata* (14%), high live coral cover (18%) of which ~ 9% is *Montastraea* spp.	1	2	4*
Lobophora variegata (76%) and branching red/brown algae (9%)	2	3	5*
Sand and occasional branching red algae (< 6%)	3	4	6*
Amphiroa spp. (40%), sand (30%), encrusting sponge (17%), sparse *Thalassia testudinum* and calcareous green algae	2	5	7
Thalassia testudinum of low standing crop (5 $g.m^{-2}$) and *Batophora* sp. (33%)	3	6	8
Thalassia testudinum of low standing crop (5 $g.m^{-2}$) and sand	3	6	9
Medium–dense colonies of calcareous algae – principally *Halimeda* spp. (25 m^{-2}) *Thalassia testudinum* of low standing crop (~10 $g.m^{-2}$)	3	7	10
Dense colonies of calcareous algae – principally *Penicillus* spp. (55 m^{-2}) and *Halimeda* spp. (100 m^{-2}) *Thalassia testudinum* of medium standing crop (~80 $g.m^{-2}$)	2	7	11*
Thalassia testudinum and *Syringodium filiforme* of 5–80 $g.m^{-2}$ standing crop	4	8	12*
Thalassia testudinum and *Syringodium filiforme* of 80–280 $g.m^{-2}$ standing crop	4	8	13*

Table 11.3 Definition of image processing steps for the Turks and Caicos Islands case-study using satellite and CASI data.

Processing method	Description
Supervised multispectral classification of reflectance data	multispectral classification of at-surface reflectance data (satellites) and at-sensor reflectance data (CASI)
Water column correction	use of depth invariant bands (Chapter 8) for multispectral classification
Contextual editing	multispectral classification of at-surface reflectance data (satellites) and at-sensor reflectance data (CASI) *followed* by contextual editing
Water column correction and contextual editing	use of depth invariant bands for multispectral classification *followed* by contextual editing

However, water column correction was not significantly beneficial for sensors which produced a single depth invariant band (SPOT XS and Landsat MSS). Supervised classification of a single band is limited because the statistical separation of habitat spectra is confined to one dimension. Therefore, whilst depth-invariant processing of SPOT XS and Landsat MSS data may have reduced the effects of variable depth, it did so at the expense of the number of spectral bands available to the classifier (i.e. dimensions of the discriminant function). Landsat TM and CASI were amenable to depth-invariant processing because three and six depth-invariant bands (respectively) were available for supervised classification. We suggest that where only a single depth-invariant bottom-index can be created, the benefit of accounting for variable depth is out-weighed by the need to input several bands to the classifier – even if the component bands exhibit depth effects. This conclusion is undoubtedly site specific and may not hold in areas where variation in bathymetry is much greater.

Contextual editing

To ensure that contextual editing does not create bias or misleading improvements to map accuracy, the decision rules must be applicable throughout an image and not confined to the regions most familiar to the interpreter. The simplest of these for coral reefs is the presence or absence of coral and seagrass habitats on the forereef (Figure 11.2).

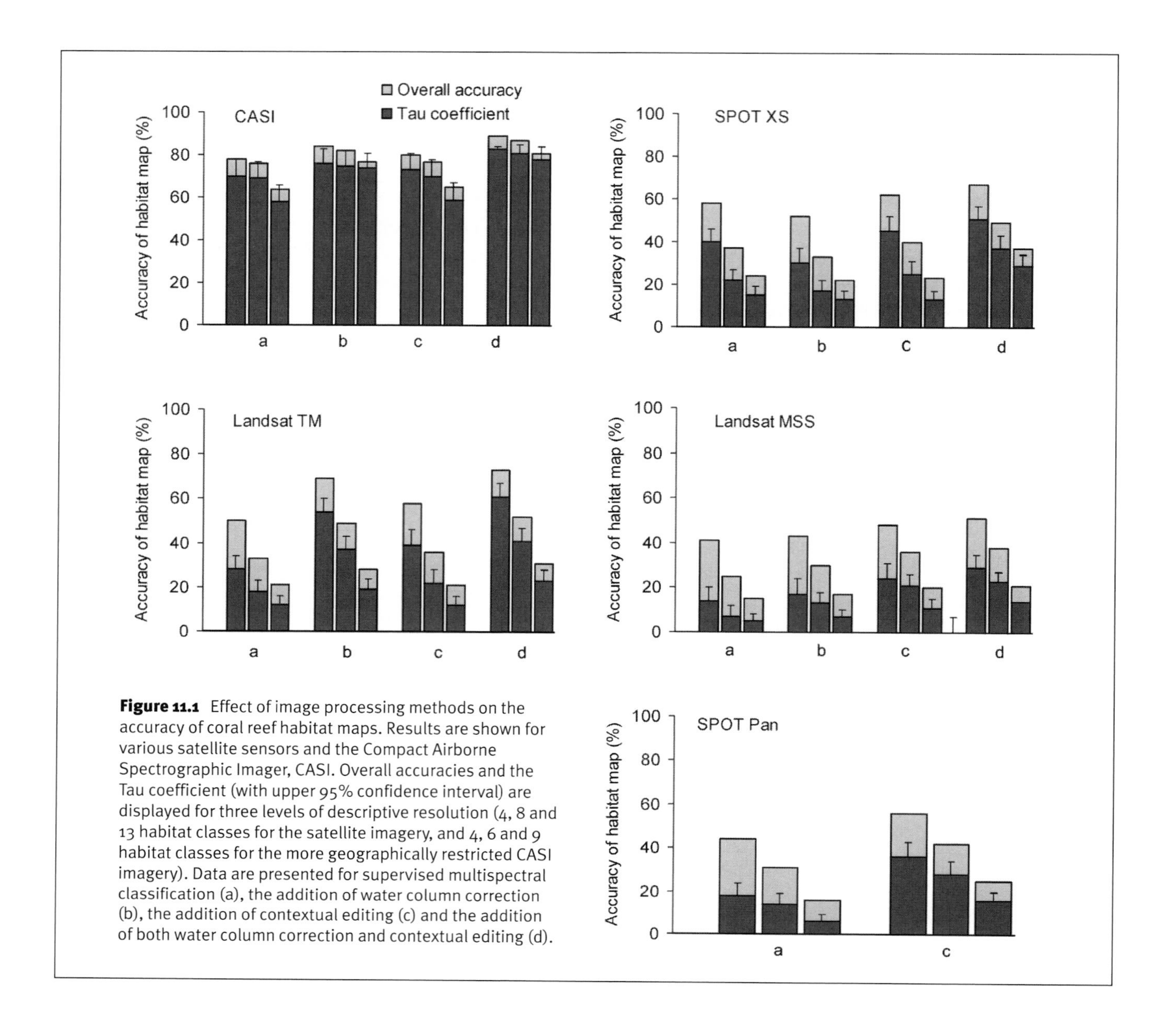

Figure 11.1 Effect of image processing methods on the accuracy of coral reef habitat maps. Results are shown for various satellite sensors and the Compact Airborne Spectrographic Imager, CASI. Overall accuracies and the Tau coefficient (with upper 95% confidence interval) are displayed for three levels of descriptive resolution (4, 8 and 13 habitat classes for the satellite imagery, and 4, 6 and 9 habitat classes for the more geographically restricted CASI imagery). Data are presented for supervised multispectral classification (a), the addition of water column correction (b), the addition of contextual editing (c) and the addition of both water column correction and contextual editing (d).

A full list of contextual rules are given in Table 11.4. with reference to Figure 11.3 (Plate 12). Spectral confusion between these habitats was greatest in Landsat MSS and SPOT Pan, which had the poorest spatial and spectral resolutions respectively. Consequently, thematic maps from these sensors showed the greatest improvements in accuracy (Figure 11.1c). The capability of Landsat TM was improved for coarse-level habitat mapping but more detailed maps were unaffected. Accuracies derived from SPOT XS were not significantly affected by contextual editing although small gains were found (3–4%). Similarly, overall accuracies of CASI images were not significantly affected by contextual editing.

Water column correction and contextual editing

When combined, depth compensation and contextual editing made a significant improvement upon basic image processing (Figure 11.1d and 11.1a). The combined approach was collectively more accurate than the singular implementation of water column correction or contextual editing (although the improvement was not always significant). The relative importance of depth compensation and contextual editing varied between sensors. Habitat maps from Landsat TM and CASI benefited from depth-invariant processing whereas Landsat MSS was more amenable to contextual editing. Accuracies from SPOT XS showed the greatest improvement when depth compensation and contextual editing were used together. This resulted from a change in inter-habitat misclassification, which was brought about by water column correction. Prior to this processing step, sand was a major source of confusion between habitats. Water column correction, which was optimised for sand habitats, improved the mapping of sand but confusion between coral and seagrass spectra increased (presumably because of the dependency on a single depth-invariant band). The

Table 11.4 Contextual rules used during post-classification editing. Misclassified habitats were re-coded to the habitats shown. Locations are illustrated in Figure 11.3 (Plate 12).

Location	Misclassified habitat	Re-coded habitat
Seaward margin of fringing reef	seagrass (high standing crop)	*Montastraea* reef
Non-seaward area of fringing reef	seagrass (low to medium standing crop)	bare substratum and *Microdictyon* spp.
Shallow bays landward of reef crest	bare substratum and *Microdictyon* spp.	seagrass (high standing crop)
Shallow bays landward of reef crest	*Microdictyon* spp. dominated	seagrass (high standing crop)
Shallow bays landward of reef crest	*Amphiroa* spp. and encrusting sponge	seagrass (high standing crop)
Deep areas of bank	*Amphiroa* spp. and encrusting sponge	bare substratum and soft corals
Seaward margin of fringing reef	*Amphiroa* spp. and encrusting sponge	*Montastraea* reef

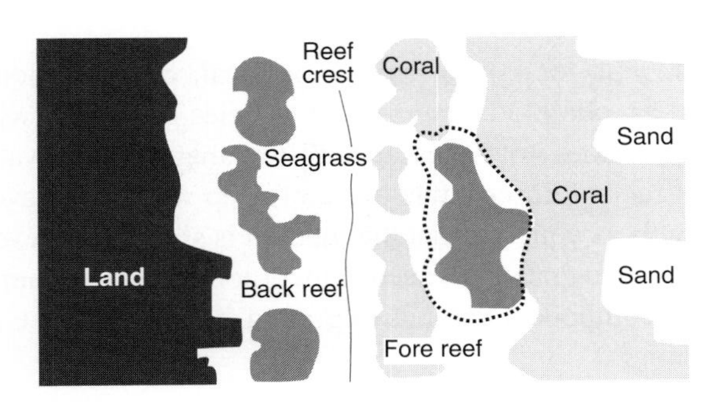

Problem:
Seagrass is occasionally confused spectrally with coral reef patches particularly where the latter include significant levels of macroalgae.
Context:
However, seagrass is not the found on the forereef.
Decision rule:
Seagrass patches on the forereef should be recoded as coral.
Action:
Firstly, visually delimit forereef area (seaward of reef crest where waves are breaking) using cursor on the screen. Secondly, change any seagrass polygons on the forereef to coral.

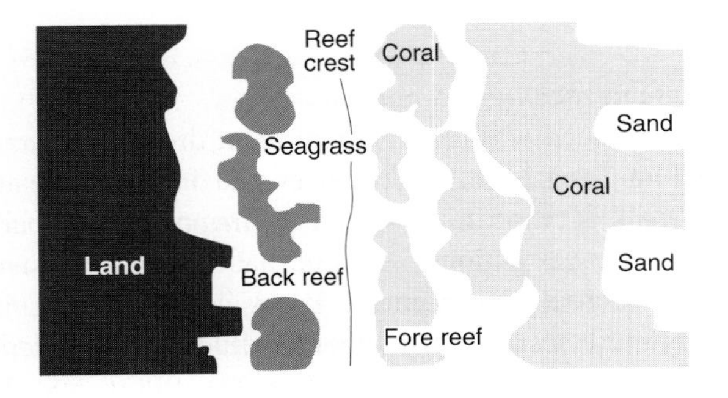

Figure 11.2 Contextual editing of a schematic reef profile. A patch of coral had been incorrectly classified as seagrass in the upper image (denoted with broken line). Seagrass pixels in this polygon were re-coded to a coral category resulting in the corrected image below.

change of classification error to corals/seagrass permitted subsequent contextual editing to make a significant improvement to accuracy.

Cost-effectiveness

Refinements to basic image processing may be considered cost-effective if the additional investments in time are justified by improved map accuracy. Since thematic map accuracy is finite, continued processing will inevitably reach a stage where the 'accuracy pay-off' declines and processing becomes progressively ineffective. The daily accruement of accuracy from water column correction and contextual editing remains high (Table 11.5) and on the premise set out above, it seems that both processing steps are cost-effective.

Other image processing methods

We have discussed the use of water column correction and contextual editing, both of which improve the classification accuracy of mapping coral and macroalgal habitats. There are at least two other processing methods which have been employed in conjunction with reef mapping.

Bour and co-authors have described two alternative methods for mapping reefs in New Caledonia with SPOT XS imagery. Both involve plotting bands 1 and 2 against one another to examine the spread of data. Ordinarily, the bands are quite highly correlated which limits their potential for spectrally discriminating habitats. The first method de-correlates the bands by rotating their axes until they align with the main variability in the data (de Vel and Bour 1990). A texture layer was also created and entered into the supervised classifier with the rotated bands. The authors found that the approach improved on standard supervised classification of SPOT data but the accuracies were not compared.

The second method resulted in the creation of 'coral index' for mapping coral density (Bour *et al.* 1996). Regions of the original bi-plot of SPOT bands are related to features on the reef by reference to known points on the imagery. Conceptually, this process is the reverse of running a supervised classification on a 2-band image and

Table 11.5 Summary of processing steps for habitat mapping showing the total implementation time, the % gain in accuracy per day's processing effort and the methods which significantly improve upon basic image processing (denoted ✔). % accuracy gains are averaged for all satellite images and expressed for coarse (C), intermediate (I) and fine (F) descriptive resolution. *Note*: the 3 days required to conduct multispectral classification has been added to each processing step – i.e. water column correction took an additional 1/2 day to implement so the total processing time would be 3.5 days. Figures in the table assume that the user is familiar with the processing methods and does not have to learn them from scratch (if that is the case, refer to processing times in Chapter 19).

Processing method	Total time taken (days)	% gain in accuracy per day			Imagery					
		C	I	F	CASI	Landsat TM	Landsat TM/Pan	SPOT XS	SPOT Pan	Landsat MSS
Supervised multispectral classification of reflectance	3	16	11	6						
Water column correction	3+½	16	11	7	✔	✔	✔	✘	n/a	✘
Contextual editing	3+¼	17	12	7	✘	✘	✘	✘	✔	✔
Water c.c. & contextual editing	3+¾	17	2	8	✔	✔	✔	✔	n/a	✔

then looking at the signatures plotted in 2-band feature space (which is usually done statistically by the classification algorithm, see Chapter 10). Bour *et al.* (1996) found that an ellipse may be drawn on this bi-plot which defines coral reefs of varying density. A mathematical transformation is then carried out to align band 2 with the direction of the ellipse. This stretches the range of values for coral density along a single axis which becomes a coral index. The range of values are then divided into three sections which constitute sparse, medium and dense corals.

Conceptually, this process is similar to water column correction. Rather than creating a depth-invariant index, the transformation aligns axes to a gradient in coral density. However, Bour and co-authors' method does not take account of variable depth and has been found to be of limited use in anything but shallow water (Peddle *et al.* 1995). In addition, the degree to which a 'coral index' improves thematic map accuracy is yet to be demonstrated.

Recently, Peddle *et al.* (1995) have attempted to apply the method of Spectral Mixture Analysis (SMA) to coral reefs in Fiji. The method was originally developed for terrestrial systems and aims to address sub-pixel variability (mixels). The seascape is divided into a few distinct components – in this case, deep water and coral reef. Spectral characteristics of pure deep water and homogeneous coral reef are then described in each of several spectral bands. The radiance of each pixel in the image is then assumed to be divisible into its constituent fractions i.e. the amounts contributed by deep water and coral. In this case, Peddle *et al.* (1995) used extensive field data to translate the radiance at a given depth into its constituent 'deep water likeness' and 'coral likeness'. Once the model is developed, coral area can be predicted for each pixel provided that the depth is known. As such, the method is limited to areas where good bathymetry data exist. Another possible problem is the use of a 'pure' coral signature that is derived for shallow (or emergent) coral. For straightforward biological considerations, such as photo-adaptation and species zonation with depth, it does not necessarily follow that shallow-water coral reefs will resemble those of deep water. However, the authors concede that the method is still under development and that it will need to incorporate a greater number of component habitats (e.g. sand, algae) to realise its potential.

The descriptive resolution of satellite imagery

This section assumes that water column correction is conducted on multispectral imagery and that contextual editing is conducted on all images. Even though the focus of this chapter is coral reefs and macroalgae, seagrass habitats will also be mentioned. This is because maps of coral reefs and algae are often intrinsically linked to seagrass (e.g. due to misclassification errors and contextual editing). Seagrass habitats are treated fully in the following chapter.

Coarse descriptive resolution

A pronounced and usually significant drop in accuracy was found consistently between coarse, intermediate and fine habitat discrimination. For mapping at coarse descriptive resolution (i.e. four habitat classes; sand, coral, macroalgae, seagrass), Landsat TM was significantly more accurate than other satellite sensors (overall accuracy 73%; Figure 11.4, Plate 12). SPOT XS also achieved a relatively high overall accuracy (67%). The accuracy of merged Landsat TM/SPOT Pan was not significantly different from SPOT XS ($Z = 1.82$, $P > 0.05$). The maps derived from SPOT Pan and Landsat MSS had an accuracy of < 60% (Figure 11.5).

Coral and sand habitats were generally more accurately distinguished than macroalgal and seagrass habitats (Table 11.6). This is because macroalgal and seagrass habitats were spectrally and spatially confused with one

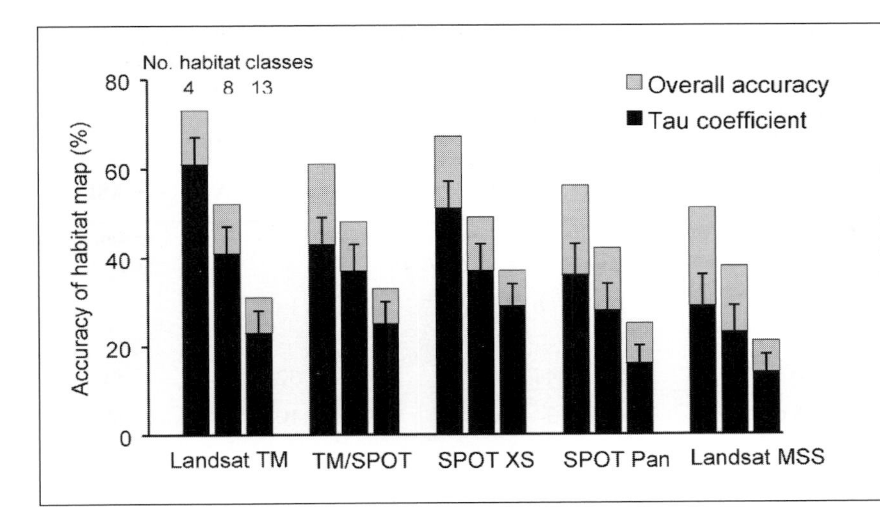

Figure 11.5 Comparison of satellite sensors for mapping marine habitats of the Caicos Bank for three levels of descriptive resolution; coarse (4 habitat classes), medium (8 habitat classes) and fine (13 habitat classes). The upper part of each bar represents the overall accuracy and the lower (solid) part, the Tau coefficient and its upper 95% confidence limit. Water column correction was conducted on multispectral imagery. Contextual editing was carried out in all cases.

another, which is not unusual (e.g. Kirkman *et al.* 1988) and has several causes. Whilst the photosynthetic pigments in seagrass and algae (e.g. chlorophyll, phycoerythrin and fucoxanthin) have different albedo characteristics, satellite spectral bands are generally unsuitable for distinguishing them (see Maritorena *et al.* 1994). This is because the range of wavelengths that penetrate water is small (< 580 nm) and may not encompass the characteristic reflectance minima and maxima for a particular pigment. For example, whilst most photosynthetic pigments show reflectance minima below 450 nm, the maxima ('red edges') lie between 670–700 nm, which is beyond the range of water-penetrating irradiance. Where distinguishing minima and maxima exist within the water-penetrating spectrum, satellite bands may be too broad to distinguish them. For example, the reflectance minima for both green and brown algae are below 500 nm and their maxima are 550 nm and 575 nm respectively (Maritorena *et al.* 1994). SPOT XS band 1 cannot differentiate these maxima because it detects radiance within the range 500–590 nm. Landsat TM can, in principle, distinguish these maxima because band 1 is sensitive to 450–520 nm and band 2 detects within the range 520–600 nm. However, given sensor noise and light attenuation problems, precise discrimination is unlikely.

Coral habitats also possess a high cover of macroalgae and the corals themselves contain photosynthetic pigment-bearing zooxanthellae. Whilst they may be spectrally confused with seagrass and some algal habitats, coral reefs may be spatially distinguished. This is because their location (context) within the reef landscape is usually confined to the seaward margin of the coastal zone (i.e. fringing reef) where wave exposure is moderate to high. In this study, contextual discrimination of seagrass and algal habitats was more difficult because gradients of exposure were generally less obvious and it was not so easy to predict the location of habitats.

Intermediate and fine descriptive resolution

Overall, the difference in accuracy between intermediate and fine descriptive resolution was considerably larger than the variation between sensors for either level of habitat discrimination. In practical terms, if the objective is to map more detail than coral reef, macroalgae, sand, seagrass, then the accuracy of a habitat map is more sensitive to the choice of descriptive resolution than the choice of sensor. It should be borne in mind, however, that overall accuracies for both higher descriptive resolutions were low; intermediate (8 habitats) 38–52%, fine (13 habitats) 21–37% and showed marked variability (Table 11.6). We conclude that satellite imagery is not well suited to detailed mapping of benthic habitats. This conclusion is in agreement with Bainbridge and Reichelt (1989) who concluded that satellite imagery is more appropriate for studying reef geomorphology than reef biology. The combined spatial and spectral resolutions of satellite sensors were not capable of reliably distinguishing many habitats that had a high inter-habitat similarity (Bray-Curtis Similarity, 60–80%). This was borne out by the high variability in accuracy associated with individual habitat classes. The poor separability of spectra rendered the supervised classification unable to assign pixels to appropriate habitat classes and resulted in large and variable allocation errors. It must be borne in mind, however, that mapping with fine descriptive resolution is an ambitious objective for any remote sensing method – some reef habitats look similar even to the field surveyor underwater!

Satellite imagery versus aerial photography

This section is separated into 1) a comparison of processed satellite imagery against aerial photography and 2) the visual interpretation of satellite imagery and aerial photography. The latter procedure is necessary if digital analysis of imagery is not possible (e.g. due to

Table 11.6 User accuracies of habitat classes for all sensors and at two levels of descriptive resolution (coarse – habitats in bold – and fine). *Note*: details of habitat types do not apply directly to API categories that were described by Sheppard *et al.* (1995); however, categories of Sheppard and co-authors are broadly analogous to those described in the table. The most accurate satellite sensors for each habitat are shaded to facilitate comparison with airborne remote sensing.

Habitat type	Accuracy of sensor (%)						
	MSS	TM	TM/Pan	XS	Pan	API	CASI
Living and dead stands of *Acropora palmata*						52	90
Microdictyon marinum (77%), *Sargassum* spp (4%), medium soft coral density (5 m^{-2}) and rubble (10%)	0	0	69	19	13		
Bare substratum (40%), low soft coral density (3 m^{-2}), *Microdictyon marinum* (30%), *Lobophora variegata* (12%)	32	44	57	54	32		81
Bare substratum (80%), medium soft coral density (5 m^{-2})	4	34	44	39	10	48	80
Bare substratum (60%), high soft coral density (8 m^{-2}), *Lobophora variegata* (14%), high live coral cover (18%) of which ~ 9% is *Montastraea* spp.	33	18	36	51	47	66	83
Lobophora variegata (76%) and branching red/brown algae (9%)	13	41	31	41	0	38	82
Amphiroa spp. (40%), sand (30%), encrusting sponge (17%), sparse *Thalassia testudinum* and calcareous green algae	11	25	24	75	7		
Sand and occasional branching red algae (< 6%)	11	45	64	46	50	73	75
Thalassia testudinum of low standing crop (5 $g.m^{-2}$) and *Batophora* spp (33%)	100	8	14	22	14		
Thalassia testudinum of low standing crop (5 $g.m^{-2}$) and sand	49	50	35	73	36		
Medium–dense colonies of calcareous algae – principally *Halimeda* spp. (25 m^{-2}); *Thalassia testudinum* of low standing crop (~10 $g.m^{-2}$)	6	3	5	8	0		
Dense colonies of calcareous algae – principally *Penicillus* spp. (55 m^{-2}) and *Halimeda* spp. (100 m^{-2}); *Thalassia testudinum* of medium standing crop (~80 $g.m^{-2}$)	0	12	8	0	0	68	77
Thalassia testudinum and *Syringodium filiforme* of 5–80 $g.m^{-2}$ standing crop	6	15	37	2	15	40	72
Thalassia testudinum and *Syringodium filiforme* of 80–280 $g.m^{-2}$ standing crop	48	40	34	56	46	100	93
Coral	67	86	80	81	53	76	93
Algae	14	47	28	41	21	58	92
Sand	46	83	74	78	80	73	75
Seagrass	32	59	44	45	65	63	87

inadequate computing facilities). In this case, prints of an image can be used as a surrogate for aerial photography. Habitat maps may be created by tracing polygons on to an acetate overlay. Where field data are not available, delineation and identification of polygons is a subjective process which is guided by the colour, contrast, texture and context of areas in the imagery. Comparisons are made between the authors' satellite data of the Caicos Bank and aerial photography of Anguilla which was interpreted by Sheppard *et al.* (1995). The habitats mapped in each area are moderately similar (Table 11.7), allowing broad comparisons to be made.

Note: The accuracies reported for satellite imagery in this section are greater than those reported earlier because fewer habitat classes were used in the comparison with aerial photography.

1. Processed satellite imagery versus aerial photography

On the basis of the results from Sheppard *et al.* (1995), satellite sensors compared favourably to aerial photography for coarse and intermediate levels of habitat discrimination. Aerial photography was found to be inferior to Landsat TM and similar to merged Landsat TM/SPOT

Table 11.7 Habitat types from the Caicos Bank and Anguilla (fine descriptive resolution)

Caicos Bank	Anguilla (Sheppard *et al.* 1995)
Living and dead stands of *Acropora palmata*	Dead elkhorn reef: merge of living and dead elkhorn coral (*A. palmata*), often dominated by fleshy algae and calcareous reds.
Microdictyon marinum (77%), *Sargassum* spp. (4%), medium soft coral density (5 m^{-2}) and rubble (10%)	
Bare substratum (40%), low soft coral density (3 m^{-2}), *Microdictyon marinum* (30%), *Lobophora variegata* (12%)	
Bare substratum (80%), medium soft coral density (5 m^{-2})	Soft coral reefs: Contain hard corals of various composition but not separable from photographs.
Bare substratum (60%), high soft coral density (8 m^{-2}), *Lobophora variegata* (14%), high live coral cover (18%) of which ~ 9% is *Montastraea* spp.	*Montastraea* reefs: *Montastraea* cover 10%– 35%, rarely more. Most support 5%–15% soft coral too.
Lobophora variegata (76%) and branching red/brown algae (9%)	Algal reefs (mostly browns): deeper (> 3 m) rocky substrates dominated by brown algae. The rocky 'algae reef' is a clear and large biological category. They usually have scattered soft and hard corals but not in appreciable abundance.
Sand and occasional branching red algae (<6%)	Bare sublittoral sand: limestone sand, commonly clearly mobile and abrasive.
Amphiroa spp. (40%), sand (30%), encrusting sponge (17%), sparse *Thalassia testudinum* and calcareous green algae	
Thalassia testudinum of low standing crop (5 g.m^{-2}) and *Batophora* spp. (33%).	
Thalassia testudinum of low standing crop (5 g.m^{-2}) and sand	
Medium dense colonies of calcareous algae – principally *Halimeda* spp. (25 m^{-2}). *Thalassia testudinum* of low standing crop (< 10 g.m^{-2})	
Dense colonies of calcareous algae – principally *Penicillus* spp. (55 m^{-2}) and *Halimeda* spp. (100 m^{-2}). *Thalassia testudinum* of medium standing crop (~80 g.m^{-2})	High fleshy/calcareous greens: Two biologically distinct groups: one dominated by fleshy algae, the second by calcareous greens. Mainly on soft substrates but examples of both occur on stony or rocky substrates. They cannot be distinguished except by close inspection in the water, so cannot be differentiated on photographs.
Thalassia testudinum and *Syringodium filiforme* of standing crop 5–80 g.m^{-2}	Traces of algae/seagrass: poor cover (< 5%) of seagrasses or algae on otherwise bare substrate.
Thalassia testudinum and *Syringodium filiforme* of standing crop 80–280 g.m^{-2}	High seagrass: typical seagrass beds, commonly mixed with algae (> 50% plant cover).

Pan, SPOT XS and SPOT Pan (Figure 11.6). This may appear surprising given the superior spatial resolution of aerial photography. However, digital satellite sensors have better spectral resolution and at coarse and intermediate descriptive resolutions, the habitats were sufficiently dissimilar to one another (Bray-Curtis Similarity, 10–15% and 30–50% respectively) that a crude discrimination of their spectra was possible. This conclusion did not apply to maps from Landsat MSS which were less accurate than those from aerial photography. For fine descriptive resolution, aerial photography was more accurate than all satellite sensors (Figure 11.6, Table 11.6).

2. Visual interpretation of satellite imagery versus aerial photography

Depth compensation was carried out for all satellite imagery except SPOT Pan which only has a single spectral band. The results of visual interpretation presented in Figure 11.7 reflect the visual identification of habitat types from satellite images rather than the delineation of habitat maps which can involve additional errors in boundary demarcation. As such, the comparison with aerial photographs is not entirely fair because the accuracies reported for aerial photographs incorporate these additional errors which result from habitat delineation. For fur-

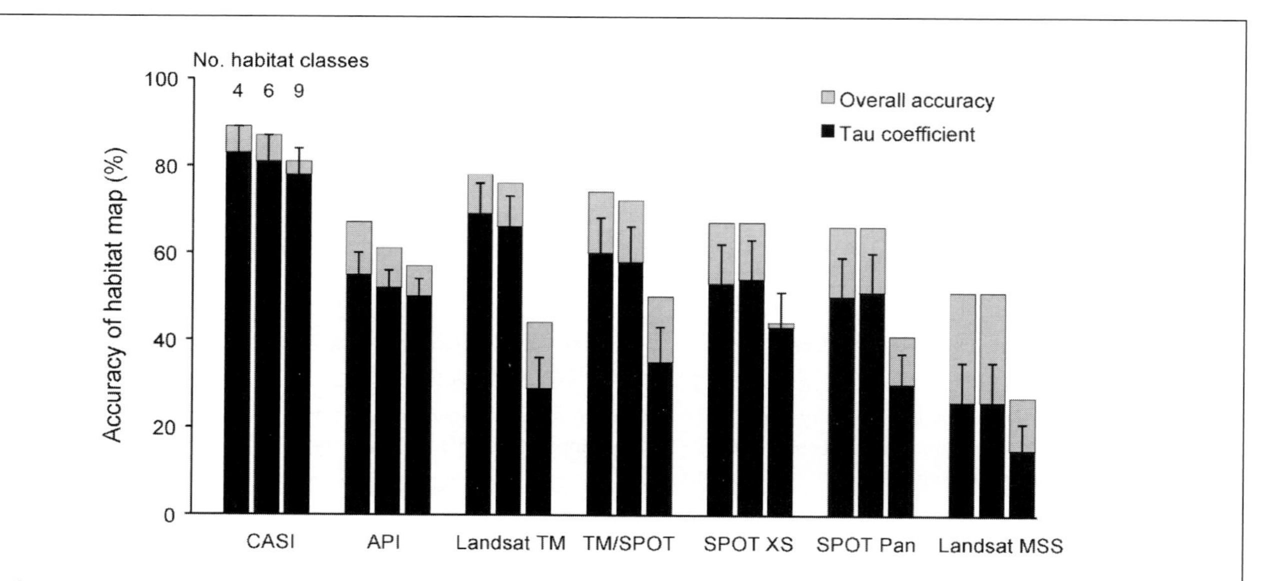

Figure 11.6 Comparison of satellite sensors, aerial photography and airborne multispectral (CASI) data for mapping marine habitats at three levels of descriptive resolution using supervised classification. Data for aerial photography were recalculated from Sheppard *et al.* (1995). The upper part of each bar represents the overall accuracy, and the lower (solid) part, the Tau coefficient and its upper 95% confidence limit. Water column correction was conducted on multispectral imagery. Contextual editing was carried in all cases.
Note that accuracies for satellite sensors are higher than described elsewhere (Figures 11.1, 11.4 and 11.5). This results from the exclusion of rarer (and less accurately mapped) categories that were not present within the area mapped using CASI and thus could not be used in this comparison. Inclusion of these categories would have biased the comparison in favour of the airborne sensors.

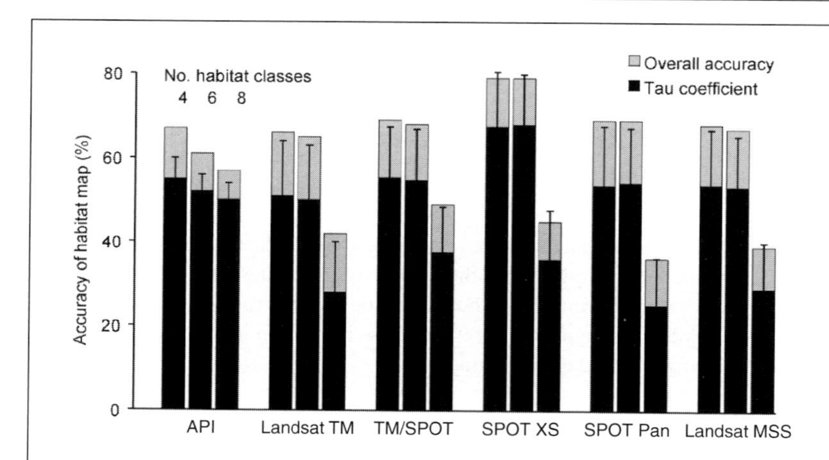

Figure 11.7 Comparison of satellite sensors and aerial photography for mapping marine habitats at three levels of descriptive resolution using visual interpretation. Data for aerial photography were recalculated from Sheppard *et al.* (1995). The upper part of each bar represents the overall accuracy, and the lower (solid) part, the Tau coefficient and its upper 95% confidence limit. Water column correction was conducted on multispectral imagery. Contextual editing was carried in all cases.

ther details on visual interpretation, refer to Chapter 10.

The existence of boundary errors notwithstanding, for coarse and intermediate levels of habitat discrimination, aerial photography (Sheppard *et al.* 1995) was found to be slightly poorer than all forms of satellite imagery and significantly less accurate than SPOT XS, which had the greatest visual contrast. By contrast, aerial photography was superior at mapping habitats with fine detail, in which case it was significantly better than all satellite image types.

In conclusion, if new imagery is to be acquired for visual interpretation and the choice includes aerial photography and satellite imagery, it is best to acquire a (processed) SPOT XS image for coarse and intermediate habitat mapping and aerial photography for fine habitat mapping.

CASI imagery versus satellite imagery and aerial photography

For all three levels of descriptive resolution, CASI imagery gave significantly more accurate results than satellite sensors and aerial photography (Figure 11.6, $P < 0.001$). The accuracy with which individual habitats were mapped was more consistent than that found for aerial photography or satellite sensors (Table 11.6) and even fine habitat discrimination was possible with an accuracy of 81% (almost double that achieved with any satellite; Figure 11.8, Plate 13). CASI has the advantage of offering tremendous flexibility to the user. In this case, four narrow spectral bands were set to penetrate water, which increased the likelihood of distinguishing habitat spectra (band settings; 402.5–421.8 nm, 453.4-469.2 nm, 531.1–543.5 nm, 571.9–584.3 nm). For other band settings refer to Table 5.10 of Chapter 5.

Whilst CASI was found to provide significantly greater accuracies than aerial photography, the comparison was not entirely balanced. Although the habitat categories were comparable, Sheppard *et al.* (1995) mapped a much larger area than that tested for CASI (14,600 ha versus 100 ha). Until carefully controlled comparisons can be conducted, it is perhaps safest to conclude that for comparable areas, CASI would be at least as good as aerial photography.

As a practical tool for mapping at high resolution, the relative merits of CASI and aerial photography require closer inspection. Whereas photographs offer greater spatial resolution than CASI, making use of this resolution is not straightforward. It is highly unlikely that a photo-interpreter would delineate features smaller than several metres because to do so would be too time consuming. This statement is borne out in the trace illustrated by Sheppard *et al.* (1995) in which the minimum polygon size was probably several metres or more. In contrast, polygon size does not constrain a digital classification of pixels. Thus, the spatial resolution of CASI may, in effect, be finer than the practical resolution of aerial photography. In addition, the delineation of habitats is likely to be both faster and more objective. It would be interesting to make an explicit evaluation of this issue – i.e. how does efficiency of each method vary with area covered? Furthermore, it would be useful to compare the effectiveness of digital remote sensing (i.e. CASI) and thematic classification of photographs which have been digitised using a scanner. Scanned aerial photographs have been used to great effect for mapping small areas (Thamrongnawasawat and Hopley 1995). A similar argument may be made for the new generation of digital cameras that are capable of taking images in a few broad spectral bands. Intuitively, however, CASI would be expected to fare better because of its greater spectral resolution (up to 21 spectral bands available to distinguish habitats).

Conclusions regarding coral and macroalgal habitat mapping

If new imagery is required for a site, the most cost-effective solution depends on the mapping objectives, required accuracy, size of the area, climate of the area (e.g. persistence of cloud cover), and the availability of technical expertise and equipment. These issues are discussed further in Part 5 but a few simple rules emerge here:

- For maximal gain in accuracy per unit processing time, multispectral imagery should be subjected to water column correction and post-classification contextual editing.
- For areas greater than 60 km in any direction (the size of a single SPOT scene), Landsat TM is likely to be the most cost-effective option. While being approximately US $1,700 per scene more expensive than SPOT XS, it covers nine times the area of SPOT and offers greater accuracy. If more than one SPOT scene is required to cover an area, SPOT will be more expensive than Landsat TM.
- SPOT XS is more cost-effective and accurate than SPOT Pan.
- Merged Landsat TM/SPOT Pan is not a cost-effective option. It is particularly expensive and does not provide a corresponding improvement in accuracy.
- If image processing is not possible and maps must be made by visual interpretation, it is likely to be cheaper to purchase a pre-processed satellite image than mount a new campaign of aerial photography. This statement makes the assumption that imagery is optimised for visual interpretation (i.e. that the spectral contrast has been stretched and (preferably) that depth-invariant processing has been carried out).
- CASI is significantly ($P < 0.01$) more accurate than satellite imagery for all levels of reef habitat mapping.
- CASI is at least as accurate, if not more so, than aerial photography for the same purpose.
- Satellite sensors cannot be used for mapping reefs with fine habitat discrimination if an accuracy exceeding 37% is required.

Field survey methods

Introduction

The subject of sampling coral reefs is vast. Methods for describing coral reefs have been developed for decades and it remains an active area of research even today. Like any sampling problem, the choice of method is intrinsically linked to the spatial and/or temporal scales of the sampling objective. For example, methods for measuring fluctuations in living coral cover are far more intense than those for describing the main characteristics of reef along a continental shelf. A full description of sampling methodologies is beyond the scope of this *Handbook* and readers are referred to existing texts for a more thorough treatment of the subject (e.g. Stoddart and Johannes 1978, English *et al.* 1997, Rogers *et al.* 1994).

This section will provide a general overview of reef sampling methods and comment on their suitability for use in a remote sensing context. The chapter ends with a brief section on statistical issues such as determining the number of samples required.

Survey methods for broad-scale characterisation of coral reefs

If the mapping objective focuses on reef geomorphology and/or major habitat types, then the survey methods should be simple, quick and able to cover large areas. Given that the descriptive resolution of satellite imagery is confined to this type of general information on coral reefs, these techniques are probably best suited to most satellite remote sensing operations.

Broad-scale characterisation of coral reefs often makes use of an ordinal abundance (cover) scale of reef categories which are assigned by a diver or snorkeller. The survey is conducted by being towed behind a vessel (the manta tow technique) or simply swimming across the reef profile (plotless belt transects).

The manta tow technique

The manta tow method has been widely used in Micronesia and the Great Barrier Reef for assessing broad-scale changes in reef cover due to cyclone damage, coral bleaching and outbreaks of the Crown-of-thorns starfish, *Acanthaster plancii*. A good synopsis of the method is given in English *et al.* (1997) which forms the basis of the following description.

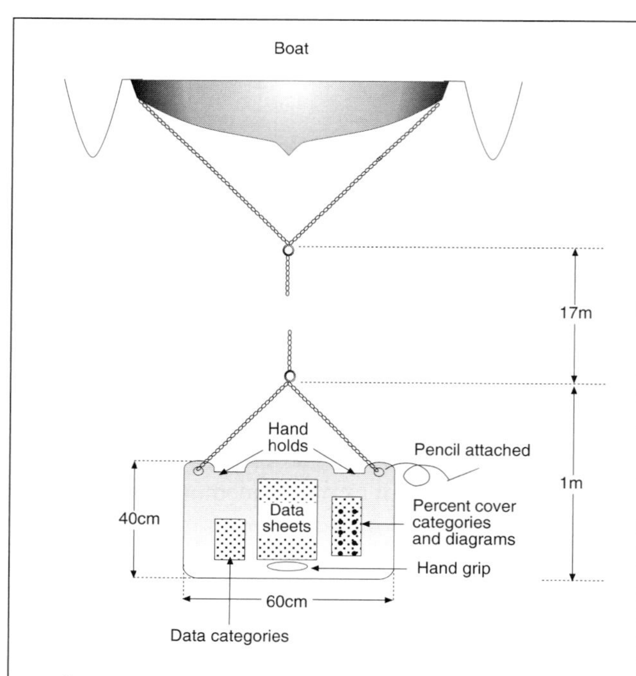

Figure 11.9 Detail of the manta board and associated equipment. It is recommended that the board be made from marine ply and painted white. Two indented handgrips are positioned towards both front corners of the board and a single handhold is located centrally on the back of the board. Redrawn from: English, S., Wilkinson, C., and Baker, V., 1997, *Survey Manual for Tropical Marine Resources*, 2nd Edition. (Townsville: Australian Institute of Marine Science).

The manta board (Figure 11.9) is attached to a motor boat with a 17 m length of rope which has buoys placed at distances of 6 m and 12 m from the board. A snorkeller grips the board and is towed for approximately 2 minutes, at the end of which the boat pauses to allow the surveyor to record data (usually on water-resistant paper). The coverage of bottom features may be recorded on a percentage scale (for an example, see Figure 11.10) or on a scale of 1–5, where 5 indicates the greatest cover and 0 is used for absence. However, a scale of 1–5 does have the shortcoming that observers may be tempted to place a disproportionately large number of values in the middle category (i.e. 3), thus creating observer bias. If possible, a scale of 1–6 or 1–4 is more desirable (see Kenchington 1978).

Features which should be amenable to this type of survey include:

Living biotic features	Substrata	Others
live hard corals	sand	geomorphology
soft corals	mud	visibility
macroalgae	bedrock	depth
sponge	rubble	
	dead coral	

Plotless belt transects

The principle of plotless belt transects is similar to the manta tow, the main difference being that observers are not towed behind a boat. This affords a useful means of independence allowing the use of scuba and permitting very shallow areas to be surveyed safely. Since the observer can get much closer to the sea bed, it is possible to record more detailed data on bottom features. In the species-rich Indo-Pacific, this may include coral and algal lifeforms (for examples, see English *et al.* 1997) and species or genera in the Caribbean.

Like the manta tow method, the logistical requirements for surveys are fairly simple and given adequate training, useful data can be collected by relatively inexperienced individuals. The UK-based organisation, Coral Cay Conservation has successfully adopted this principle to conduct surveys of reef habitat throughout Belize and the Philippines. After seven days intensive training, volunteer divers were able to conduct species-level surveys of Belizean coral reefs with a moderate degree of accuracy and precision (Mumby *et al.* 1995b) which is probably all that is required for most habitat mapping purposes.

Surveys are conducted along profiles which tend to run either parallel or perpendicular to the reef crest. The former approach is useful for describing individual reef habitats (zones) and the latter provides data on habitat

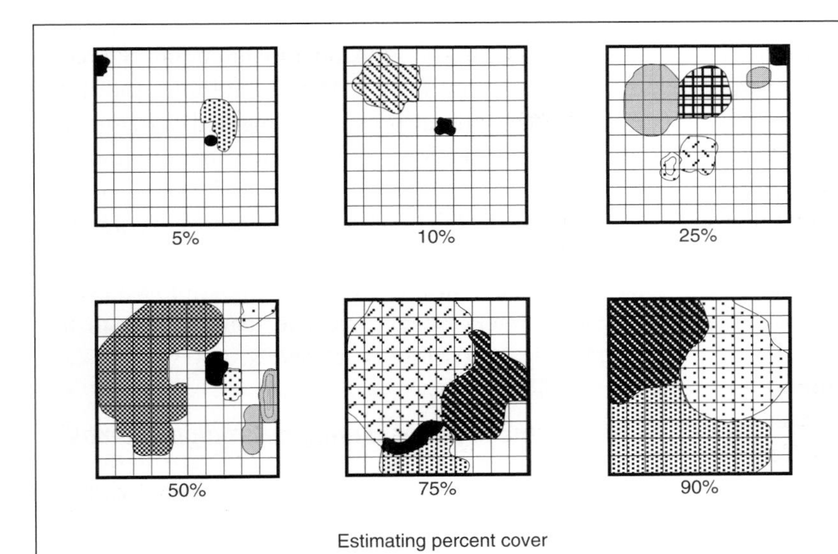

Figure 11.10 Estimating percent cover on the benthos using a metre-square quadrat with string or fishing line strung across at 10 cm intervals. In each example, shaded regions represent different types of substratum that are included collectively in the total percent cover estimate. Based on a drawing in: Rogers, C., Garrison, G., Grober, R., Hillis, Z-M., and Franke, M.A., 1994, *Coral Reef Monitoring Manual for the Caribbean and Western Atlantic.* (St. John: National Park Service, Virgin Islands National Park).

zonation. Surveys are directed using a compass bearing. The width of the surveyed area (i.e. the belt width) is estimated visually but should not exceed 4–5 m beyond which, recognition of benthic cover becomes difficult. Distances along the survey can be estimated by counting fin strokes and multiplying these by the average distance travelled per kick cycle. However, it is easier and more accurate to use a 10 m line and survey in short 'hops'.

Survey teams of 2–4 divers are used for safety reasons and to delegate survey responsibilities (e.g. into corals and macroalgae). Plotless belt transects and manta tow methods may use the same recording scales.

Additional data to record

1. *Location*: The start and end locations of surveys should be located with a global positioning system (preferably with a differential capability, see Chapter 4).
2. *Water visibility*: Visibility of the water should be noted because it may influence the interpretation of imagery and help explain areas which prove difficult to map. Visibility can be estimated in the horizontal plane by noting the visibility of buoys along the manta tow line or by holding a Secchi disc 0.5 m below the surface and seeing at what horizontal distance it remains visible.
3. *Date and time*: Recording the date and time allows data to be catalogued and tidal heights to be compensated for (e.g. in bathymetry studies, Chapter 15).
4. *Depth*: Depth can be measured using a weighted plumb-line (cheap approach) or an echosounder (more expensive option). A number of manufacturers (e.g. Scubapro) produce cheap hand-held sounders which give accuracies within about 0.10 m (see equipment costs, Chapter 19).

Survey methods for estimating coral and macroalgal cover

For more detailed habitat mapping where habitats need to be described quantitatively, a suite of methods are available. These are more likely to be appropriate for high resolution imagery such as CASI or aerial photography. With adequate replication, most of these methods can also be used to monitor changes in bottom cover. The methods are all fairly time-consuming and the most accurate and precise methods tend to take longer. As such, quantitative sampling regimes are usually limited to smaller areas. There are four principal groups of methods for measuring coral and macroalgal cover; quadrats, photo-quadrats, line intercept transects, and video. These will be considered in turn.

Quadrats

Quadrats are extensively used for sampling in all branches of ecology and many approaches are available (reviewed by Greig-Smith 1983). For coral reef assessment, quadrats usually have a minimum area of 1 m^2 and are divided into a uniform grid of 100 segments (i.e. if the quadrat has a side of 1 m, each cell in the grid has dimensions of 10 x 10 cm). Every cell represents 1% coverage in the sampling unit (see Figure 11.10 and Figure 11.11, Plate 14). Quadrat size can be reduced to (say) 0.25 m^2 if the main sampling objective is to estimate the cover of macroalgae.

Compared to most other quantitative sampling techniques, quadrats have the advantage that data are acquired relatively rapidly and cheaply in the field. The main disadvantages of quadrat sampling are (i) quadrats cannot be used to measure spatial relief (rugosity), (ii) large branching corals such as elkhorn coral (*Acropora palmata*) are difficult to sample, and (iii) quadrats only provide data on a two-dimensional surface thus underestimating coverage of features which have a predominant orientation in the vertical plane (e.g. soft corals). However, although the latter limitation is pertinent to ecological assessment, it may in fact be advantageous in the context of remote sensing where the sensor also samples a two-dimensional (flat) surface.

Photo-quadrats
Photo-quadrats are good for coral monitoring programmes but are not recommended for the purposes of field survey. Although they can provide accurate information on reef cover (particularly if taken in stereo pairs), analysis of photographs can be time consuming. Readers interested in the technique should refer to Rogers *et al.* (1994).

Line intercept transects
Like quadrats, line intercept transects are fairly rapid to deploy in the field. A fibreglass tape measure is laid close to the reef contour and the length (cover) of each reef category is recorded. A faster variant is the point intercept transect in which only the type of reef category is noted at equidistant points along the line (e.g. every 20 cm). The cover of each category is calculated by the ratio of number of points per category to the total number of points. The main limitation with line and point intercept transects is that they tend to under-sample heterogeneous areas with low cover of reef categories (e.g. areas of scattered corals).

Video transects
Underwater video is well-suited to field survey because large areas can be covered fairly quickly and the method can be used without extensive training. It also has the additional advantage of producing a permanent visual record of the data. A variety of methods exist for using video; the following description is taken from Carleton and Done (1995) who have used the method for monitoring coral reefs of the Great Barrier Reef.

A National video camcorder (Model NV-MC5A) with an in-water depth of field ranging from a few centimetres to infinity is used with VHS C-format video tapes. The camera is mounted in a 'Video Sea' underwater housing which is attached to a manta board for ease of use by scuba divers. The video is pointed directly at the seabed and held between 1 m and 1.5 m from the substratum. Diver and video are towed on a 30 m line behind a boat at ~ 1 $m.s^{-1}$.

Data are analysed using the software 'VIPS' (Video Point Sampling) which allows the operator to pause the tape at random or evenly spaced intervals and place points at random or fixed locations of the monitor. To improve image quality, the output from the video can be 'gen-locked' (paused) with an Amiga Commodore 500 computer. Carleton and Done (1995) found that, for 200 m transects, the optimal processing method was five subsamples of 110 random points or one subsample of 550 points.

The main drawbacks of video are the cost of equipment and processing facilities. It should also be borne in mind that species recognition is not usually possible with video. Most interpreters might expect to identify taxa to life-form, family or possibly genus levels.

Survey methods for measuring macroalgal biomass

Macroalgal biomass may show marked natural variation over a reef and seasonally. However, a dramatic increase in macroalgal biomass may be indicative of increased levels of nutrients in the water (e.g. from sewage effluent) or reduced grazing pressure by overfishing of herbivorous fish (e.g. parrotfish).

Average macroalgal biomass can be estimated by collecting and weighing algae in a random sample of 15 or more 0.25 m^2 quadrats (Rogers *et al.* 1994). For most purposes, wet weight is sufficient.

Determining sample size

A pilot study should be used to determine the number of replicates required to sample each of the main habitats adequately. This phase can be avoided if relevant details are available in the literature. Calculating the desired sample size depends on the objective of the sampling. If the aim is simply to represent the presence of most benthic features and species, a species area curve can be plotted (Figure 11.12) from a preliminary sample of (say) 20 quadrats. Samples are selected at random and the cumulative number of species (and substrata if necessary) plotted against quadrat number. The asymptote (levelling off) marks the required number of samples to represent most of the species of interest. Species area curves can also be used to determine the required length for line transects (in which case, cumulative species number is plotted against number of metres of transect traversed).

While being useful for general guidance, species area curves do not describe the accuracy or precision with which reef parameters are estimated from the samples. For example, sampling might be used to estimate the percentage cover of coral on a reef. The confidence of the estimate would be expected to grow as more of the reef is sampled (i.e. as sample size increases). In many cases, the required confidence is specified before undertaking sampling. Precision refers to the degree of concordance among a number of measurements for the same population (Andrew and Mapstone 1987). Precision depends on the variability of the item(s) being sampled and the degree

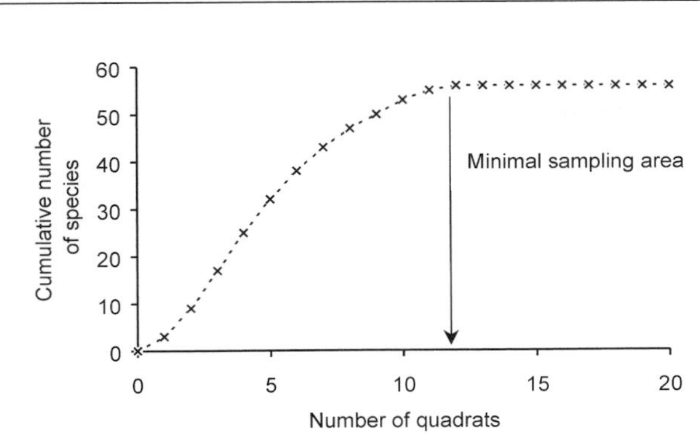

Figure 11.12 Species area curve for 20 quadrats showing the minimal sampling area to represent all species.

of confidence required (Type I error). There are several equations available to calculate sample size based on these criteria and an example is given in Box 11.1

If sampling is being planned with a specific hypothesis in mind, the pilot study is extremely important. Such a consideration often arises in a monitoring context where (say) coral cover is being compared over time or between sites. In this case, the investigator may wish to know the sample size required to test for a specified change in the parameter being measured and with a desired level of confidence in doing so correctly. For example, one might ask: 'How many quadrats must I sample to have a 95% chance of correctly detecting a 10% change in coral cover between two reefs?' This type of question is addressed using 'power analysis' and depends on the power of the statistical test being proposed, the variability in the data, the size of change being detected and the confidence of avoiding a Type I error. Since this is unlikely to be an issue in remote sensing, readers are directed to Zar (1996) for further discussions and methods.

BOX 11.1

Estimation of sample size needed to achieve a stated precision in estimating a population mean (modified from Zar, 1996)

An estimate of population variance (s^2) is required which can be obtained from a pilot study or the literature. The method requires access to basic statistical tables and uses an iterative approach:

$$n = \frac{s^2 t^2_{\alpha 2,(n-1)}}{d^2}$$

where n is the sample size, s^2 is sample variance (calculated with $v = n - 1$ degrees of freedom), $t^2_{\alpha 2,(n-1)}$ is the two-tailed critical value of Student's t with $v = n - 1$ degrees of freedom and d is the half-width of the desired confidence interval.

Worked example (adapted from Zar 1996): suppose we wish to estimate the density of soft corals with a 95% confidence interval no wider than 0.5 m^{-2}, then $d = 0.25\ m^{-2}$, $1-\alpha = 0.05$ and suppose the variance (s^2) of soft coral density was found to be 0.4008.

We begin by guessing that a sample size of 40 is necessary in which $t^2_{0.05(2),39} = 2.023$

We estimate n to be:

$$n = \frac{(0.4008)(2.023)^2}{(0.25)^2} = 26.2$$

Next we estimate n to be 27, for which $t^2_{0.05(2),26} = 2.056$ giving:

$$n = \frac{(0.4008)(2.056)^2}{(0.25)^2} = 27.1$$

Since the output agrees with our suggested sample size, we conclude that an adequate sample size is 27.

Conclusion

Satellite sensors seem to be capable of mapping the geomorphology of coral reefs but poor at mapping ecological habitats including assemblages of reef-dwelling species and macroalgal beds. Infra-red aerial photography can be used to assess coral and macroalgal cover in shallow water environments (< 1 m depth) such as reef flats. However, to achieve such detail, the aircraft must fly at low altitude and this technique is best-suited to small areas (e.g. individual reefs).

Compared to infra-red film, true-colour aerial photography has a much greater penetration of water (up to approximately 25 m in clear water) and can be used for habitat mapping. If habitats are being defined in detail, such aerial photography appears to permit more accurate mapping than satellite sensors. However, for more general habitat mapping, multispectral classification of satellite imagery and visual interpretation of true-colour aerial photography achieve similar accuracies. In which case, satellite imagery is likely to be more cost-effective because multispectral classification of digital data is a much faster method of habitat mapping than visual interpretation of photographs. In addition, satellite imagery is much cheaper than the acquisition of new aerial photography. Even if habitat maps must be created using visual interpretation of imagery (e.g. if image processing facilities are unavailable), satellite imagery would be more appropriate than aerial photography for coarse and medium-detail habitat mapping. Although map accuracies are similar for both image types, satellite images cover a much larger area than individual aerial photographs, thus increasing the availability of ground control points and facilitating geometric correction of the map.

The airborne digital sensor, CASI, provides significantly more accurate habitat maps than satellite sensors and is at least as accurate as aerial photography. With the exception of infra-red aerial photography of shallow waters, CASI is the only method discussed here that achieves high accuracies (> 80%) for coral and macroalgal habitats, and habitat spectra can be recorded to a depth of at least 18 m.

Field survey methods for coral reefs and macroalgae correspond to the scale of remotely sensed imagery. For general-purpose habitat mapping with imagery of medium to coarse spatial resolution such as Landsat TM, SPOT XS or Landsat MSS, broad-scale survey methods such as manta tow or plotless belt-transects are appropriate. The cover of benthic features is usually recorded on a categorical scale. Imagery with greater spatial resolution, such as CASI, warrants more intensive sampling methods such as quadrats, line intercept transects or underwater video.

References

References referred to by number in Table 11.1 are indicated by numbers in square brackets, [ ].

[1] Ahmad, W., and Neil, D.T., 1994, An evaluation of Landsat Thematic Mapper (TM) digital data for discriminating coral reef zonation: Heron Reef (GBR). *International Journal of Remote Sensing*, **15**, 2583–2597.

Andrew, N.L., and Mapstone, B.D., 1987, Sampling and the description of spatial pattern in marine ecology. *Oceanography & Marine Biology Annual Reviews*, **25**, 39–90.

[2] Bainbridge, S.J., and Reichelt, R.E., 1989, An assessment of ground truthing methods for coral reef remote sensing data. *Proceedings of the 6th International Coral Reef Symposium*, Townsville, **2**, 439–444.

[3] Biña, R.T., 1982, Application of Landsat data to coral reef management in the Philippines. *Proceedings of the Great Barrier Reef Remote Sensing Workshop*, James Cook University, Townsville, Australia, 1–39.

[4] Biña, R.T., Carpenter, K., Zacher, W., Jara, R., and Lim, J.B., 1978, Coral reef mapping using Landsat data: follow-up studies. *Proceedings of the 12th International Symposium on Remote Sensing of the Environment*, (Michigan: ERIM), **1**, 2051–2070.

[5] Biña, R.T., and Ombac, E.R., 1979, Effects of tidal fluctuations on the spectral patterns of Landsat coral reef imageries. *Proceedings of the 13th International Symposium on Remote Sensing of the Environment*, (Michigan: ERIM), **3**, 1293–1308.

[6] Bour, W., 1989, SPOT images for coral reef mapping in New Caledonia. A fruitful approach for classic and new topics. *Proceedings of the 6th International Coral Reef Symposium*, Townsville, **2**, 445–448.

[7] Bour, W., Dupont, S., and Joannot, P., 1996, Establishing a SPOT thematic neo-channel for the study of hard-to-access lagoon environments. Example of application on the growth areas of the New Caledonian reefs. *Geocarto International*, **11**, 29–39.

[8] Bour, W., Nosmas, P., and Joannot, P., 1990, The setting up of a live madrepora or coral index through remote sensing for reef bionomic mapping. *Proceedings of an International Workshop on Remote Sensing & Insular Environments in the Pacific: Integrated Approaches*, ORSTOM/IFREMER, pp. 247–254.

[9] Bour, W., Loubersac, L., and Rual, P., 1986, Thematic mapping of reefs by processing of simulated SPOT satellite data: application to the Trochus niloticus biotope on Tetembia Reef (New Caledonia). *Marine Ecology Progress Series*, **34**, 243–249.

Carleton, J.H., and Done, T.J., 1995, Quantitative video sampling of coral reef benthos: large-scale application. *Coral Reefs*, **14**, 35–46.

[10] Catt, P., and Hopley, D., 1988, Assessment of large scale photographic imagery for management and monitoring of the Great Barrier Reef. *Proceedings of the Symposium on Remote Sensing of the Coastal Zone, Gold Coast, Queensland*, (Brisbane: Department of Geographic Information), pp. III.1.1–1.14.

[11] de Vel, O.Y., and Bour, W., 1990 The structural and thematic mapping of coral reefs using high resolution SPOT data: application to the Tétembia reef (New Caledonia). *Geocarto International*, **5**, 27–34.

Done, T.J., 1983, Coral zonation: its nature and significance. In *Perspectives on Coral Reefs*, edited by D.J. Barnes (Canberra: Brian Clouston Publishing), pp. 107–147.

Dubinsky, Z, 1990, *Ecosystems of the World 25: Coral Reefs* (New York: Elsevier).

[12] Eckardt, F., 1992, Mapping coral reefs using Landsat TM. A feasibility study for Belize, Central America. MSc dissertation, Cranfield Institute of Technology, UK.

English, S., Wilkinson, C., and Baker, V., 1997, *Survey Manual for Tropical Marine Resources*, 2nd Edition. (Townsville: Australian Institute of Marine Science).

Fagerstrom, J.A., 1987, *The Evolution of Reef Communities* (New York: John Wiley and Sons).

Green, E.P., Mumby, P.J., Edwards, A.J., and Clark, C.D., 1996, A review of remote sensing for the assessment and management of tropical coastal resources. *Coastal Management*, **24**, 1–40.

Greig-Smith, P., 1983, *Quantitative Plant Ecology*. (Oxford: Blackwell Scientific Publications).

[13] Guerin, P.R., 1985, Reducing remoteness with remote sensing. *Proceedings of the 27th Australian Survey Congress*, Alice Springs, **1**, 119–125.

[14] Hopley, D., and Catt, P.C., 1989, Use of near infra-red aerial photography for monitoring ecological changes to coral reef flats on the Great Barrier Reef. *Proceedings of the 6th International Coral Reef Symposium*, Townsville, **3**, 503-508.

[15] Hopley, D., and van Stevenick, A.L., 1977, Infra-red aerial photography of coral reefs. *Proceedings of the 3rd International Coral Reef Symposium*, Miami, Florida, **1**, 305–311.

[16] Jupp, D.L.B., 1986, *The Application and Potential of Remote Sensing in the Great Barrier Reef Region*. (Townsville: Great Barrier Reef Marine Park Authority Research Publication).

[17] Jupp, D.L.B., Mayo, K.K., Kuchler, D.A., Claasen, D.R., Kenchington, R.A., and Guerin, P.R., 1985, Remote sensing for planning and managing the Great Barrier Reef Australia. *Photogrammetrica*, **40**, 21–42.

Jupp, D.L.B, Mayo, K.K., Kuchler, D.A., Heggen, S.J., Kendall, S.W., Radke, B.M., and Ayling, T., 1986, *Landsat based inter-*

pretation of the Cairns section of the Great Barrier Reef Marine Park. Natural Resources Series No. 4. (Canberra: CSIRO Division of Water and Land Resources).

Kenchington, R.A., 1978, Visual surveys of large areas of coral reefs. In *Coral Reefs: Research Methods*, edited by D.R. Stoddart and R.E. Johannes (Paris: UNESCO), pp. 149–161.

Kenchington, R.A., and Claasen, D.R., 1988, Australia's Great Barrier Reef – management technology. *Proceedings of the Symposium on Remote Sensing of the Coastal Zone, Gold Coast, Queensland*, (Brisbane: Department of Geographic Information), pp. KA.2.2–2.13.

Kirkman, H., Olive, L., and Digby, B., 1988, Mapping of underwater seagrass meadows. *Proceedings of the Symposium on Remote Sensing of the Coastal Zone, Gold Coast, Queensland*, (Brisbane: Department of Geographic Information), pp. VA.2.2–2.9.

[18] Kuchler, D., Bour, W., and Douillet, P., 1988, Ground verification method for bathymetric satellite image maps of unsurveyed coral reefs. *ITC Journal*, **2**, 196–199.

[19] Kuchler, D.A., Jupp, D.L.B., Claasen, D.R., and Bour, W., 1986, Coral reef remote sensing applications. *Geocarto International*, **4**, 3–15.

Littler, D.S., Littler, M.M., Bucher, K.E., and Norris, J.N., 1989, *Marine Plants of the Caribbean: a Field Guide from Florida to Brazil* (Shrewsbury: Airlife Publishing Ltd).

[20] Loo, M.G.K., Lim, T.M., and Chou, L.M., 1992, Landuse changes of a recreational island as observed by satellite imagery. In *Third ASEAN Science and Technology Week Conference Proceedings, Vol. 6*, edited by L.M. Chou and C.R. Wilkinson (Singapore: University of Singapore), pp. 401–404.

[21] Loubersac, L., Dahl, A.L., Collotte, P., LeMaire, O., D'Ozouville, L., and Grotte, A., 1988, Impact assessment of Cyclone Sally on the almost atoll of Aitutaki (Cook Islands) by remote sensing. *Proceedings of the 6th International Coral Reef Symposium*, Townsville, **2**, 455–462.

[22] Luczkovich, J.J., Wagner, T.W., Michalek, J.L., and Stoffle, R.W., 1993, Discrimination of coral reefs, seagrass meadows, and sand bottom types from space: a Dominican Republic case study. *Photogrammetric Engineering & Remote Sensing*, **59**, 385–389.

[23] Manieri, R., and Jaubert, J., 1985, Traitements d'image et cartographie de récifs coralliens en Mer Rouge (Golfe d'Aqaba). *Oceanologica Acta*, **8**, 321–329.

[24] Manieri, R., and Jaubert, J., 1984, Coral reef mapping in the Gulf of Aqaba (Red Sea) using computer image processing techniques for coral reef survey. *Proceedings of the Symposium on Coral Reef Environments of the Red Sea*, Jeddah, pp. 614–623.

Maritorena, S., Morel, A., and Gentili, B., 1994, Diffuse reflectance of oceanic shallow waters: influence of water depth and bottom albedo. *Limnology & Oceanography*, **37**, 1689–1703.

McNeill, S.E., 1994, The selection and design of marine protected areas: Australia as a case study. *Biodiversity & Conservation*, **3**, 586–605.

Meulstee, C., Nienhuis, P.H., and Van Stokkom, H.T.C., 1986, Biomass assessment of estuarine macrophytobenthos using aerial photography. *Marine Biology*, **91**, 331–335.

[25] Michalek, J.L., Wagner, T.W., Luczkovich, J.J., and Stoffle, R.W., 1993, Multispectral change vector analysis for monitoring coastal marine environments. *Photogrammetric Engineering & Remote Sensing*, **59**, 381–384.

[26] Mumby, P.J., Baker, M.A., Raines, P.S., Ridley, J.M., and Phillips, A.T., 1994, The potential of SPOT Panchromatic imagery as a tool for mapping coral reefs. *Proceedings of the 2nd Thematic Conference on Remote Sensing for Marine & Coastal Environments, New Orleans, Louisiana* (Michigan: ERIM), **I**, 259–267.

[27] Mumby, P.J., Gray, D.A., Gibson, J.P., and Raines, P.S., 1995a, Geographic Information Systems: a tool for integrated coastal zone management in Belize. *Coastal Management*, **23**, 111–121.

Mumby, P.J., Harborne, A.R., Raines, P.S., and Ridley, J.M., 1995b, A critical assessment of data derived from Coral Cay Conservation volunteers. *Bulletin of Marine Science*, **56**, 742–756.

[28] Mumby, P.J., Clark, C.D., Green, E.P., and Edwards, A.J., 1998a, The practical benefits of water column correction and contextual editing for mapping coral reefs. *International Journal of Remote Sensing*, **19**, 203–210.

[29] Mumby, P.J., Green, E.P., Clark, C.D., and Edwards, A.J., 1998b., Digital analysis of multispectral airborne imagery of coral reefs. *Coral Reefs*, **17**, 59–69.

[30] Mumby, P.J., Green, E.P., Edwards, A.J., and Clark, C.D., 1997, Coral reef habitat mapping: how much detail can remote sensing provide? *Marine Biology*, **130**, 193–202.

Peddle, D.R., LeDrew, E.F., and Holden, H.M., 1995, Spectral mixture analysis of coral reef abundance from satellite imagery and in situ ocean spectra, Savusavu Bay, Fiji. *Proceedings of the 3rd Thematic Conference Remote Sensing for Marine & Coastal Environments, Seattle* (Michigan, ERIM), **II**, 563–575.

Rogers, C., Garrison, G., Grober, R., Hillis, Z-M., and Franke, M.A., 1994, *Coral Reef Monitoring Manual for the Caribbean and Western Atlantic*. (St. John: National Park Service, Virgin Islands National Park).

[31] Sheppard, C.R.C, Matheson, K., Bythell, J.C., Murphy, P., Blair Myers, C., and Blake, B., 1995, Habitat mapping in the Caribbean for management and conservation: use and assessment of aerial photography. *Aquatic Conservation: Marine & Freshwater Ecosystems*, **5**, 277–298.

[32] Siswandono, 1992, The microBRIAN method for bathymetric mapping of the Pari reef and its accuracy assessment. *Proceedings of the Regional Symposium on Living Resources in Coastal Areas. Manila, Philippines*. (Manila: University of the Philippines), pp. 535–543.

Smith, V.E., Rogers, R.H., and Reed, L.E., 1975, Thematic mapping of coral reefs using Landsat data. *Proceedings of the 10th International Symposium Remote Sensing of Environment* (Michigan: ERIM) **1**, 585–594.

Stoddart, D.R., and Johannes, R.E., 1978, *Coral Reefs: Research Methods* (Paris: UNESCO).

[33] Sullivan, K.M., and Chiappone, M., 1992, A comparison of belt quadrat and species presence/absence sampling of stony coral (Scleractinia and Milleporina) and sponges for evaluating species patterning on patch reefs of the Central Bahamas. *Bulletin of Marine Science*, **50**, 464–488.

Tiefang, L., Jianchun, Y., Huai, L., Hongda, F., Yuguo, D. and Xuelian, C., 1990, Applications of remote sensing for *Sargassum* on Daya Bay. *Proceedings of the 11th Asian Conference on Remote Sensing,* Guangzhou, China, November 15-21 1990, **2**, J-5-1–J-5-7.

[34] Thamrongnawasawat T., and Catt P., 1994, High resolution remote sensing of reef biology: the application of digitised air

photography to coral mapping. *Proceedings of the 7th Australasian Remote Sensing Conference, Melbourne* pp. 690–697.

[35] Thamrongnawasawat, T., and Hopley, D., 1994, Digitised aerial photography: a new tool for reef research and management. *Proceedings of the 3rd ASEAN-Australia Regional Conference*. Bangkok, Thailand.

[36] Thamrongnawasawat, T., and Hopley, D., 1995, Digitised aerial photography applied to small area reef management in Thailand. In *Recent Advances in Marine Science and Technology 94*. edited by O. Bellwood, J.H. Choat, and N. Saxena (Townsville: James Cook University), pp. 385–394.

[37] Thamrongnawasawat, T., and Sudura, S., 1992, Image processing techniques for studying fringing reefs in Thailand using Landsat and SPOT data. In *Third ASEAN Science & Technology Week Conference Proceedings*, edited by L.M. Chou and C.R. Wilkinson (Singapore: University of Singapore).

[38] Thamrongnawasawat, T., Sudura, S., and Tangjaitrong, S., 1994, Summary of remote sensing of the coastal zone: the ASEAN region. *Proceedings of the 3rd ASEAN-Australia Regional Conference*, Bangkok, Thailand.

UNESCO, 1995, *Contending With Global Change*. Meeting report on coral reef assessment and status evaluation workshop (Jakarta: UNESCO).

[39] Vousden, D., 1988, The Bahrain marine survey: a study of the marine environment of Bahrain using remote sensing as a rapid assessment methodology. *Proceedings of the ROPME Workshop on Coastal Area Development*. UNEP Regional Seas Reports and Studies No. 90 ROPME Publication No. GC-5/006, pp. 3–34.

[40] Vousden, D., 1996, Remote sensing and habitat mapping. Chapter 5, unpublished PhD thesis, University College of North Wales.

[41] Wagle, B.G., and Hashimi, N.H., 1994, Coastal geomorphology of Mahe Island, Seychelles. *International Journal of Remote Sensing*, **11**, 281–287.

[42] Wouthuyzen, S., Gotoh, K., Uno, S., and Yutoh, Y., 1990, Mapping of sedimentation area in coral reef zone of Okinawa main island using satellite data. *Pacific Congress on Marine Science and Technology*, Tokyo pp. 1–5.

[43] Zainal, A.J.M., Dalby, D.H., and Robinson, I.S., 1993, Monitoring marine ecological changes on the east coast of Bahrain with Landsat TM. *Photogrammetric Engineering & Remote Sensing*, **59**, 415–421.

[44] Zainal, A.J.M., 1994, New technique for enhancing the detection and classification of shallow marine habitats. *Marine Technology Journal*, **28**, 68–74.

Zar, J.H., 1996, *Biostatistical Analysis* (New Jersey: Prentice-Hall).

12

Mapping Seagrass Beds

Summary *Seagrass beds can be described in terms of their species composition, standing crop, shoot density and associated species. The location and extent of seagrass beds can be mapped reasonably accurately with satellite imagery (~60%). Landsat TM appears to be the most appropriate satellite sensor although mis-classification (through spectral confusion) may occur between seagrass habitats, coral reefs, blue-green algae, detritus and deep water. To improve the accuracy of habitat maps, contextual information should be added (e.g. re-code mis-classifications between seagrass beds and fringing reef classes). Airborne multispectral imagery (CASI) allows seagrass beds to be mapped with high accuracies (80–90%) and permits seagrass boundaries to be monitored. Similarly, the interpretation of aerial photography can generate reasonably high accuracies (~60–70%) but map-making may be time consuming.*

A variety of field survey techniques exist for measuring seagrass density and standing crop. A non-destructive visual assessment technique for standing crop is recommended and described.

☞ ***Remote sensing can be used to predict seagrass standing crop. This is discussed later in Chapter 16 in the context of quantitative measurement of ecological parameters.***

Introduction

This chapter sets out methods of mapping the location, extent and nature of seagrass beds. It begins with a summary of the descriptive resolution of different sensors to highlight their suitability for particular objectives. The main constraints to seagrass mapping are described and advice given on appropriate image processing methods.

A variety of destructive and non-destructive field survey methods exist for sampling seagrass beds. These methods are reviewed and their relative advantages discussed. The last part of the chapter describes time, effort and cost considerations when conducting seagrass mapping.

Descriptive resolution for mapping seagrass and sensors

Many studies have not attempted to describe seagrass beds in any greater detail than whether they are present or absent (e.g. Ackleson and Klemas 1987, Monaghan and Williams 1988, Thamrongnawasawat and Sudura 1992b, Ferguson *et al.* 1993, Luczkovich *et al.* 1993). More detailed studies have attempted to distinguish exposed and submerged seagrass (Table 12.1) and others have assigned subjective qualitative density classes such as sparse, medium and dense seagrass (Savastano *et al.* 1984, Robblee *et al.* 1991).

Seagrass biomass has been evaluated using Landsat TM imagery (Armstrong 1993). Although the intial mapping of seagrass is discussed here, the use of remote sensing for predicting seagrass standing crop (above-ground biomass) is explored in Chapter 16.

Seagrass species identification appears to be beyond the capability of remote sensing. Even the high resolution airborne scanner, CASI, cannot distinguish different monospecific beds of different seagrass species (Jernakoff and Hick 1994). In fact, these authors found that the spectra of seagrass species could not be distinguished even when a radiometer was suspended just above the seagrass canopy *in situ*. This evidence suggests that seagrass species identification is simply not possible even at the finest remote sensing scales (see also Greenway and Fry 1988, Lennon and Luck 1990).

Comparisons of the performance of image types for seagrass mapping in the Turks and Caicos Islands are summarised in Table 12.2 and can be used to guide selection of remote sensing methods. Unfortunately however,

Table 12.1 The descriptive resolution achieved for seagrass beds. Numbers denote papers in the reference list at the end of the chapter.

Level of classification achieved	Landsat MSS	Landsat TM	SPOT XS	Airborne MSS	API
Seagrass presence/absence	1	1, 10, 14, 19, 22, 23	19		3, 5, 8
Exposed/submerged seagrass		4*, 6, 7	4	4*	
Seagrass density		18, 20, 21	2, 13, 15, 16	17	4, 11, 12

(*) denotes exposed seagrass only. Source: Green *et al.* (1996).

Table 12.2 User accuracies (%) for seagrass mapping using satellite sensors and CASI in the Turks and Caicos Islands and 1:10,000 colour aerial photography (API) of Anguilla (Sheppard *et al.* 1995). For digital imagery, accuracies are quoted for supervised classification of water column corrected data (where appropriate) plus the addition of contextual information. Overall accuracies for entire habitat maps (reef, sand algae and seagrass) are given to put the values for individual seagrass classes into context. Ranges of standing crop represent seagrasses of the Turks and Caicos Islands only. MSS = Landsat MSS, TM = Landsat TM, XS = SPOT XS, Pan = SPOT Panchromatic.

Habitat type	Accuracy for each sensor (%)						
	MSS	TM	TM/Pan	XS	Pan	API	CASI
Overall accuracy for seagrass, sand, coral and algal habitats	51	73	61	67	56	67	89
Seagrass (general)	32	59	44	45	65	63	87
Thalassia testudinum and *Syringodium filiforme* of low to medium standing crop (5-80 gm^{-2})	6	15	37	2	15	40	72
Thalassia testudinum and *Syringodium filiforme* of medium to high standing crop (80–280 gm^{-2})	48	40	34	56	46	100	93

Source: Mumby *et al.* (1997b), Sheppard *et al.* (1995).

interpretation of user accuracies for seagrass habitats is not as straightforward as it might at first seem. The accuracies cannot be interpreted in isolation because they are intrinsically linked to the accuracies of other habitats in the map. For example, the final map of reef, sand, algal and seagrass habitats had an overall accuracy of 56% when created from SPOT Pan imagery. Confusion between classes was therefore considerable. However, the user accuracy of seagrass was 65%. This tells us that, while the map was fairly inaccurate as a whole (56%), the seagrass component was moderately accurate. Most of the other habitats must have had lower accuracies to give an overall accuracy of 56%. In other words, SPOT Pan mapped seagrass fairly well but other habitats poorly.

If seagrass mapping is the only goal of a remote sensing project, then our results would suggest that Landsat TM is probably the best source of satellite imagery to use. Although the accuracy of seagrass maps is slightly lower than that of SPOT Pan, one Landsat image covers 9 times the area of a SPOT scene. Aerial photography and most satellite sensors were poor at mapping seagrass beds of low to medium standing crop but their performance was considerably improved for higher biomass beds. CASI was the only sensor capable of mapping seagrass of low biomass with an accuracy exceeding 50% (actually 72%).

If seagrass mapping will form part of a larger habitat mapping exercise (as is usually the case), greater emphasis must be placed on the overall accuracy for all habitats represented on the map (e.g. reefs, sand, algae and seagrass). In this case, SPOT XS and Landsat TM fared better than other sources of satellite imagery. These results are in general agreement with the literature. Several authors favour the use of SPOT because it possesses greater spatial resolution than Landsat (Ben Moussa *et al.* 1989, Kirkman 1990, Robblee *et al.* 1991).

If accuracies greater than 70% are required, it seems that CASI is the best option. Aerial photography should also give high accuracies (Kirkman *et al.* 1988) but, with the exception of a study by Sheppard *et al.* (1995), few studies have quantified map accuracy of seagrasses from aerial photographs.

Image processing methods for mapping seagrass beds

Most studies have used supervised classification to map seagrass habitats. Two mis-classification problems arise. Firstly, several authors have reported difficulties in distinguishing seagrasses from spectrally similar features such as mussel beds (Greenway and Fry 1988), algal blooms and detritus (Kirkman *et al.* 1988), deep water (Ackleson and Klemas 1987) and coral reefs (Luczkovich *et al.* 1993, Zainal *et al.* 1993, Mumby *et al.* 1994). The second problem is specific to low-density seagrass beds, which seem to be difficult to distinguish from bare substratum (Cuq 1993; and the results presented above). Spectral confusion will increase with depth because greater light attenuation reduces the range and intensity of light frequencies recorded by the sensor. This reduces the chance of distinguishing different substrata.

Methods of reducing mis-classification

It is difficult to generalise on the most appropriate image processing options for seagrass mapping. The accuracy of a seagrass map will vary according to study site, sensor, processing method and the nature of other habitats which might be spectrally confused with seagrass. For most habitat mapping activities, we found water column compensation and contextual editing to be advantageous; for a detailed discussion of these benefits, readers are directed to the appropriate sections in Chapter 11 (Mapping coral reefs and macroalgae). The best advice is to attempt more than one processing method and use the habitat map with the greatest accuracy.

Water column compensation and contextual editing are described in Chapters 8, 10 and 11. The following methods have also been used to improve the accuracy of seagrass mapping.

Zonation of the imagery into distinctive areas
Zainal *et al* (1993) attempted several methods to improve classification accuracy of seagrasses and reefs in Bahrain. They found the most successful approach to be zonation of the original image on the basis of major habitat types. Four areas were identified on the basis of visual interpretation of the imagery, bathymetric data and field data: (i) seagrass, (ii) *Sargassum*, (iii) sparse corals, and (iv) dense corals. Classification was then carried out in each area independently, which improved the overall classification accuracies of each class (although the accuracies were not shown). This is an appealing method for reducing inter-habitat misclassification during multispectral classification of spectral data. However, the method is only feasible where good ancillary data are available. If the designation of perceived 'seagrass' or 'coral' zones is incorrect, the implications for map accuracy are dire.

This method might also be used to improve habitat mapping where the water colour shows sharp discontinuities. Variations in water colour resulting from (say) plumes of suspended sediments will affect multispectral classification: the spectra of seagrass in turbid water will differ from those of seagrass in clearer water. Habitat map accuracy should be improved if the study site is segregated into water quality zones and then multispectral classification is conducted independently in each zone.

Zonation of the imagery according to depth
Ackleson and Klemas (1987) found spectral confusion between the temperate eelgrass, *Zostera marina*, and deep water. To increase the accuracy of mapping dense aquatic vegetation in shallow water, the authors used ancillary bathymetric data to mask out water whose depth exceeded 1.9 m. Similarly, Zainal *et al.* (1993) found that masking deep water (> 7 m) slightly improved classification accuracy but considerable confusion remained in shallow water habitats.

Visual interpretation of images for mapping seagrass beds

We conducted visual interpretation of satellite imagery of the Turks and Caicos Islands and concluded that visual interpretation was significantly ($P < 0.01$) more accurate at mapping seagrass beds than supervised multispectral classification of reflectance data. However, if contextual editing was carried out after classification, visual interpretation failed to convey a significant improvement in accuracy. In light of the time required to digitise and interpret every polygon on an image, the most efficient means of mapping seagrass habitats is supervised multispectral classification (preferably of depth-invariant bottom-indices) together with contextual editing.
Note: visual interpretation methods are described in Chapter 10.

Field survey methods

Remote sensing studies of seagrass beds are only able to provide information on the above-ground component of seagrass structure (i.e. shoots and leaves but not roots and rhizomes; see Figure 12.1, Plate 14). Fortuitously, it is the above-ground component that responds most quickly to perturbation (Kirkman 1996), thus making it amenable to monitoring and sensitivity mapping.

The parameters of most relevance to seagrass assessment using remote sensing are shoot density and standing crop (above-ground biomass). A wide variety of techniques exist for measuring these including destructive and non-destructive approaches. These methods are discussed below but emphasis is placed on non-destructive methods because they allow repeated monitoring, are less harmful to the environment and are generally faster to implement than destructive methods. The latter point is important because it allows a greater number of samples to be acquired in a given survey period. This means a wider area can be surveyed and/or greater replication achieved at each site (which in turn, improves the precision of population estimates and the statistical power of analyses).

Good texts on seagrass sampling methods include UNESCO (1990), English *et al.* (1997) and Kirkman (1996). For general information on sampling design such as estimating the number of quadrats, refer to Chapter 11.

Measuring seagrass shoot density

Measures of shoot density are necessary for estimating a variety of plant parameters such as productivity per unit area of seabed. Shoot density (shoots m^{-2}) has been related to shoot size, sediment nutrient concentrations and sediment type (Dennison 1990) and responds quickly to perturbations such as shading (Bulthius 1984).

Destructive sampling
Shoot density is usually calculated by:
(i) clipping a quadrat of shoots at the sediment surface,
(ii) obtaining a sample of sediments and associated plants with a corer, or
(iii) uprooting shoots within a quadrat (if the substratum is soft enough).

The advantage of destructive sampling is that measurements are accurate and laboratory processing can yield data on biomass and leaf area index. The main disadvantage is the lengthy processing time which means that fewer samples can be obtained than with visual assessment techniques (Mellors *et al.* 1993, Long *et al.* 1994).

Non-destructive sampling
Estimates of shoot density can be made by visually inspecting shoots within quadrats (Figure 12.1, Plate 14). While this may be accurate, it can take a long time. Kirkman (1996) advocates the use of transects and subjective estimates of density such as 'sparse', 'medium' and 'dense'. Harvesting a small number of reference quadrats may enable the limits of the scale to be quantified.

Estimating percentage cover of seagrass

Percentage cover of seagrass is estimated in relation to visible substratum (e.g. Long *et al.* 1995). This is quite repeatable when currents are weak and seagrass blades have a vertical orientation. However, the method becomes potentially unreliable when current strength increases and forces the seagrass canopy into a progressively horizontal (flattened) plane. Under these circumstances, percentage cover of the seagrass may appear to be greater than would be expected under calmer conditions.

Measuring seagrass standing crop

Standing crop refers to the above-ground component of seagrass biomass. Standing crop is generally measured in units of dry-weight seagrass per unit area ($g.m^{-2}$) and is correlated to shoot density and leaf area index.

Destructive sampling
Perhaps the most widely used method for measuring seagrass standing crop is the manual harvest of macrophytic tissue within quadrats (Downing and Anderson 1985). As mentioned above, the method is time consuming and collection of large sample sizes may not be feasible. If biomass estimates are to include root and rhizome material, corers and grabs are appropriate. Long *et al.* (1994) used a modified 'orange-peel' grab to collect seagrass samples in Moreton Bay, Queensland. They demonstrated that measurements of seagrass biomass from the grab were indistinguishable from those using a coring technique. Furthermore, the grab was easier to use, quicker and could be operated by one person in a survey vessel (rather than by diving). While grabs may be more efficient than both corers and quadrat harvest, perhaps their greatest set back is inconsistent response to sediment type. The firmer the substratum, the less efficiently the grab sampler will operate. Long *et al.* (1994) found the sample area to vary between 0.016 m^2 in hard-packed sands to 0.047 m^2 in softer muddy sands. Standardisation of biomass estimates to unit area thus constitutes a source of error if samples are drawn from different sediment types. While corrections can be made by categorising the substratum type in each sample, Long *et al.* (1994) suggest designing an even heavier grab which is less influenced by the hardness of substratum.

Non-destructive sampling
Regardless of which of the above techniques is adopted, the need to wash, sort, dry and weigh each sample remains (Plate 18). To improve the speed of data collection further, Mellors (1991) designed a non-destructive visual assessment method for estimating standing crop *in situ*. The technique uses a linear scale of biomass categories which are assigned to seagrass samples in 0.25 m^2 quadrats. A reference scale of 5 evenly-spaced biomass categories is used (1–5) and biomass scores are interpolated to a precision of one decimal place (i.e. there are 50 potential biomass scores, e.g. 1.3, 4.2). We found this precision to be a little optimistic for seagrass beds of the Western Atlantic because mis-assignment of standing crop categories was found to occur on a six-point integer scale (1–6), thus casting doubt on the use of the decimal scale advocated by Mellors (1991). A modified technique (see Plate 18) is summarised in Figure 12.2 and an analysis of its errors and limitations is given by Mumby *et al.* (1997a). Provided that the survey will exceed 24 samples, visual assessment conveys considerable savings in time over conventional quadrat harvest (Figure 12.3). For 500 samples, the saving in time is 13-fold.

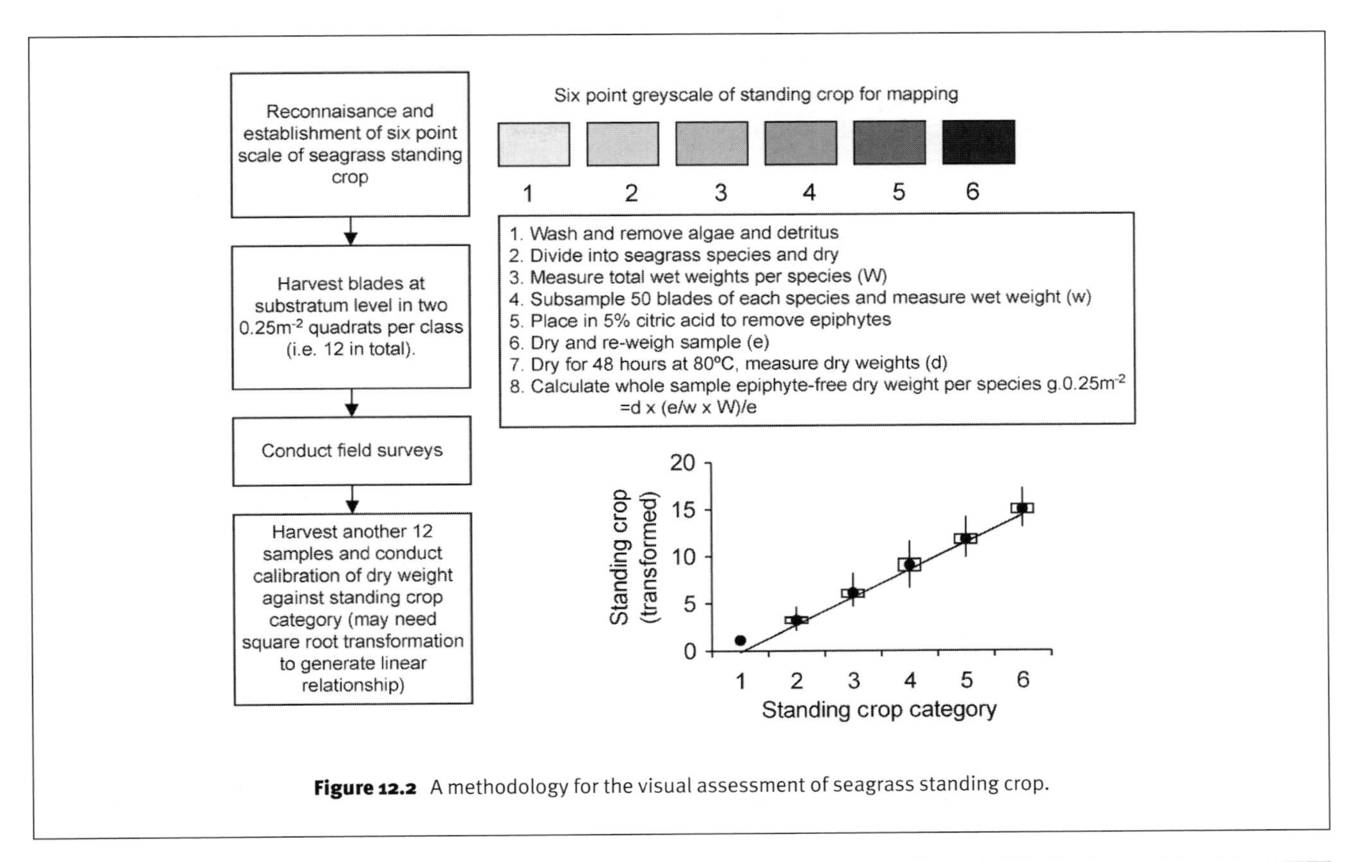

Figure 12.2 A methodology for the visual assessment of seagrass standing crop.

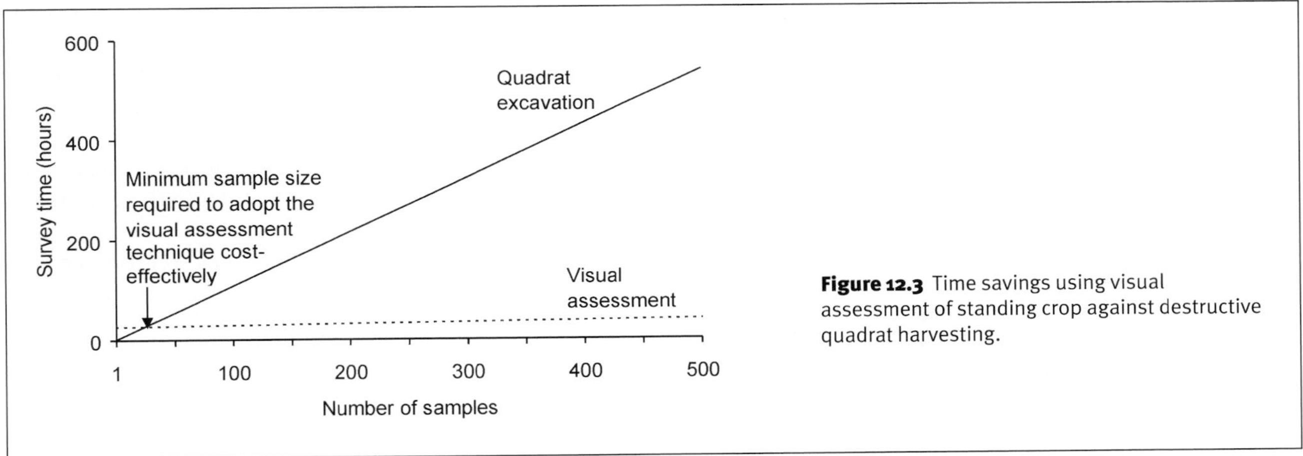

Figure 12.3 Time savings using visual assessment of standing crop against destructive quadrat harvesting.

Conclusion

Satellite sensors, airborne digital imagery (e.g. CASI) and aerial photography allow the boundaries of seagrass beds to be mapped reasonably accurately (~60% or better). However, less dense (lower biomass) areas of seagrass are more difficult to map than denser (higher biomass) beds. This is chiefly due to misclassification with sand habitats and greater patchiness in the seagrass canopy. If a high map accuracy is required, CASI and possibly aerial photography are the most appropriate methods to use.

Denser seagrass habitats are prone to spectral confusion with coral reefs, blue-green algae, detritus and deep water. This confusion can be reduced to some extent by using a range of image processing methods such as water column correction and contextual editing. The output map with greatest accuracy should then be selected for use.

Non-destructive methods of seagrass sampling are recommended because they are less harmful to the environment, allow repeated monitoring and are generally quicker to implement. Shoot density can be categorised in quadrats using terms like, 'sparse', 'medium' and 'dense'. The limits of the scale can be quantified later using several reference quadrats. Seagrass standing crop can also be estimated using a visual assessment method. A categorical scale of standing crop can be calibrated and may convey substantial savings in time compared to destructive quadrat-harvesting methods (e.g. a 13-fold saving in time for 500 samples).

References

References referred to by number in Table 12.1 are indicated by numbers in square brackets, [ ].

[1] Ackleson, S. G., and Klemas, V., 1987, Remote sensing of submerged aquatic vegetation in Lower Chesapeake Bay: A Comparison of Landsat MSS to TM imagery. *Remote Sensing of Environment*, **22**, 235–248.

Ben Moussa, H., Violler, M., and Belsher, T., 1989, Télédétection des algues macrophytes de l'Archipel de Molëne (France): radiométrie de terrain et application aux données du satellite SPOT. *International Journal of Remote Sensing*, **10**, 53–69.

Bulthius, D., 1984, Control of the seagrass Heterozostera tasmanica by benthic screens. *Journal of Aquatic Plant Management*, **22**, 41–43.

[2] Cuq, F., 1993, Remote sensing of sea and surface features in the area of Golfe d'Arguin, Mauritania. *Hydrobiologica*, **258**, 33–40.

Dennison, W.C., 1990, Shoot density. In *Seagrass Research Methods. Monographs on Oceanographic Methodology*: 9 edited by Philips, R.C. and McRoy, C.P. (Paris: UNESCO), pp. 61–64.

Downing, J.A, and Anderson, M.R., 1985, Estimating the standing biomass of aquatic macrophytes. *Canadian Journal of Fisheries and Aquatic Science*, **42**, 1860–1869.

English, S., Wilkinson, C., and Baker, V., 1997, *Survey Manual for Tropical Marine Resources,* 2nd Edition. (Townsville: Australian Institute of Marine Science).

[3] Ferguson, R. L., Wood, L.L., and Graham, D.B., 1993, Monitoring spatial change in seagrass habitat with aerial photography. *Photogrammetric Engineering and Remote Sensing*, **59**, 1033–1038.

Green, E.P., Mumby, P.J., Edwards, A.J., and Clark, C.D., 1996, A review of remote sensing for the assessment and management of tropical coastal resources. *Coastal Management*, **24**, 1–40.

[4] Greenway, M., and Fry, W., 1988, Remote sensing techniques for seagrass mapping. *Proceedings of the Symposium on Remote Sensing in the Coastal Zone, Gold Coast, Queensland*, (Brisbane: Department of Geographic Information), pp. VA.1.1–VA.1.12.

[5] Hopley, D., and van Stevenick, A.L., 1977, Infra-red aerial photography of coral reefs. *Proceedings of the 3rd International Coral Reef Symposium, Miami, Florida* pp. 305–311.

[6] Hyland, S.P., Lennon, P., and Luck, P., 1989, An assessment of the Landsat Thematic Mapper for monitoring seagrasses in Moreton Bay, Queensland, Australia. *Asian-Pacific Remote Sensing Journal*, **2**, 35–42.

[7] Hyland, S.P., Lennon, P., and Luck, P., 1988, An assessment of the Landsat thematic mapper to monitor seagrasses in Moreton Bay, Southern Queensland. *Proceedings of the Symposium on Remote Sensing in the Coastal Zone, Gold Coast, Queensland* (Brisbane: Department of Geographic Information), pp. VA.4.1–VA.4.15.

Jernakoff, P., and Hick, P., 1994, Spectral measurements of marine habitat: simultaneous field measurements and CASI data. *Proceedings of the 7th Australasian Remote Sensing Conference*, pp. 706–713.

Kirkman, H., 1990, Seagrass distribution and mapping. In *Seagrass Research Methods. Monographs on Oceanographic Methodology*: 9 edited by Philips, R.C. and McRoy, C.P. (Paris: UNESCO), pp. 19–27.

Kirkman, H., 1996, Baseline and monitoring methods for seagrass meadows. *Journal of Environmental Management*, **47**, 191–201.

[8] Kirkman, H., Olive, L., and Digby, B., 1988, Mapping of underwater seagrass meadows. *Proceedings of the Symposium Remote Sensing Coastal Zone, Gold Coast, Queensland.* (Brisbane: Department of Geographic Information), pp. VA.2.2–2.9.

[9] Lennon, P., and Luck, P., 1990, Seagrass mapping using Landsat TM data: a case study in Southern Queensland. *Asian-Pacific Remote Sensing Journal,* **2**, 1–6.

Long, B.G., Skewes, T.D., and Pointer, I.R., 1994, An efficient method for estimating seagrass biomass. *Aquatic Botany,* **47**, 277–291.

Long, B., Skewes,T., and Poiner, I., 1995, Torres Strait marine geographic information system. In *Recent Advances in Marine Science and Technology '94*, edited by O. Bellwood, J.H. Choat, N. Saxena, (Townsville: James Cook University), pp. 231–239.

[10] Luczkovich, J.J., Wagner, T.W., Michalek, J.L., and Stoffle, R.W., 1993, Discrimination of coral reefs, seagrass meadows, and sand bottom types from space: a Dominican Republic case study. *Photogrammetric Engineering & Remote Sensing*, **59**, 385–389.

[11] Manieri, R., and Jaubert, J., 1985, Traitements d'image et cartographie de récifs coralliens en Mer Rouge (Golfe d'Aqaba). *Oceanologica Acta*, **8**, 321–329.

[12] Manieri, R., and Jaubert, J., 1984, Coral reef mapping in the Gulf of Aqaba (Red Sea) using computer image processing techniques for coral reef survey. *Proceedings of the Symposium on Coral Reef Environments in the Red Sea, Jeddah, January 1984*, pp. 614–623.

[13] Meinesz, A., Belsher, T., Boudouresque, C-F., and Lefevre, J-R., 1991, Première evaluation des potentiales du satellite SPOT pour la cartographie des peuplements benthiques superficiels de Méditerranée occidentale. *Oceanologica Acta*, **14**, 299–307.

Mellors, J.E., 1991, An evaluation of a rapid visual technique for estimating seagrass biomass. *Aquatic Botany*, **42**, 67–73.

Mellors, J.E., Marsh, H., and Coles, R.G., 1993, Intra-annual changes in seagrass standing crop, Green Island, Northern Queensland. *Australian Journal of Marine and Freshwater Research*, **44**, 33–41.

[14] Monaghan, J.M., and Williams, R.J., 1988, The monitoring of estuarine fish habitats in New South Wales: the remote sensing of seagrass beds within Lake Macquarie with Landsat Thematic Mapper imagery. *Proceedings of the Symposium on Remote Sensing in the Coastal Zone, Gold Coast, Queensland* (Brisbane: Department of Geographic Information), pp. VIA.1.1–VIA. 1.16.

Mumby, P.J., Baker, M.A., Raines, P.S., Ridley, J.M., and Phillips, A.T., 1994, The potential of SPOT Panchromatic imagery as a tool for mapping coral reefs. *Proceedings of the 2nd Thematic Conference on Remote Sensing for Marine and Coastal Environments, New Orleans, Louisiana* (Michigan: ERIM) **I**, 259–267.

Mumby, P.J., Edwards, A.J., Green, E.P., Anderson, C.W., Ellis, A.C., and Clark, C.D., 1997a, A visual assessment technique for estimating seagrass standing crop on a calibrated ordinal scale. *Aquatic Conservation: Marine and Freshwater Ecosystems*,**7**, 239–251.

Mumby, P.J., Green, E.P., Edwards, A.J., and Clark, C.D., 1997b, Coral reef habitat mapping: how much detail can remote sensing provide? *Marine Biology*, **130**, 193–202.

[15] Populus, J., and Lantieri, D., 1991, Use of high resolution satellite data for coastal fisheries. *FAO RSC Series No. 58, Rome*.

[16] Robblee, M.B., Barber, T.R., Carlson, P.R., Jr, Durako, M.J., Fourqurean, J.W., Muehlstein, L.K., Porter, D., Yarbro, L.A., Zieman, R.T., and Zieman, J.C., 1991, Mass mortality of the tropical seagrass Thalassia testudinum in Florida Bay (USA). *Marine Ecology Progress Series*, **71**, 297–299.

[17] Savastano, K.J., Faller, K.H., and Iverson, R.L., 1984, Estimating vegetation coverage in St. Joseph Bay, Florida with an airborne multispectral scanner. *Photogrammetric Engineering and Remote Sensing*, **50**, 1159–1170.

Sheppard, C.R.C., Matheson, K., Bythell, J.C. Murphy, P., Blair Myers, C., and Blake, B., 1995, Habitat mapping in the Caribbean for management and conservation: use and assessment of aerial photography. *Aquatic Conservation: Marine and Freshwater Ecosystems*, **5**, 277–298.

[18] Thamrongnawasawat, T., and Sudura, S., 1992a, The morphology of seagrass habitats at Samui reefs. In *Third ASEAN Science and Technology Week Conference Proceedings*, edited by L.M. Chou and C.R. Wilkinson (Singapore: University of Singapore), pp. 415–422.

[19] Thamrongnawasawat, T., and Sudura, S., 1992b, Image processing techniques for studying fringing reefs in Thailand using Landsat and SPOT data. In *Third ASEAN Science & Technology Week Conference Proceedings*, edited by L.M. Chou and C.R. Wilkinson (Singapore: University of Singapore).

UNESCO, 1990, *Seagrass Research Methods. Monographs on Oceanographic Methodology, 9*. (Paris: UNESCO).

[20] Vousden, D., 1988, The Bahrain marine survey: a study of the marine environment of Bahrain using remote sensing as a rapid assessment methodology. *Proceedings of the ROPME Workshop on coastal area development*. UNEP Regional Seas Reports and Studies No. 90 ROPME Publication No. GC-5/006 pp. 3–34.

[21] Vousden, D., 1996, Remote sensing and habitat mapping. Chapter 5, unpublished PhD thesis, University College of North Wales.

[22] Zainal, A.J.M., Dalby, D.H., and Robinson, I.S., 1993, Monitoring marine ecological changes on the east coast of Bahrain with Landsat TM. *Photogrammetric Engineering & Remote Sensing*, **59**, 415–421.

[23] Zainal, A.J.M., 1994, New technique for enhancing the detection and classification of shallow marine habitats. *Marine Technology Journal*, **28**, 68–74.

13

Mapping Mangroves

Summary *Mangrove habitat maps have been used for three general management applications: resource inventory, change detection and the selection and inventory of aquaculture sites. The types of image data and processing techniques that have been used are reviewed in light of the descriptive resolution and accuracy achieved. The image processing techniques can be categorised into five main types: 1) visual interpretation, 2) vegetation indices, 3) unsupervised classification, 4) supervised classification, and 5) band ratios.*

These five techniques, and a new technique whereby two types of satellite data are merged together, are compared in a case study using mangrove data from the Turks and Caicos Islands. Mangroves there cannot be distinguished from adjacent thorn scrub using SPOT XS but can be distinguished from terrestrial vegetation using Landsat TM or CASI data. Landsat TM data are found to be more cost-effective than CASI if all that is required is to map 'mangrove' (but not classify different types of mangrove). An image processing method based on ratios of different red and infra-red bands gave the most accurate results.

Airborne digital imagery, such as CASI data, are superior when it comes to mapping at fine levels of habitat discrimination. Using CASI, nine different mangrove habitats could be mapped to an overall accuracy of 85%. By contrast only two mangrove habitats could be distinguished using Landsat TM data.

A protocol for selecting the most appropriate image processing technique for mangrove mapping is presented.

Introduction

This chapter provides an overview of the use of remote sensing in mangrove management, reviews the sensors and image processing techniques that have been used for the mapping of mangroves and provides a detailed case-study from the Turks and Caicos Islands. The main difference between mapping mangroves and other coastal habitats such as reefs and seagrass is that mangrove foliage is terrestrial. No compensation has to be made for variations in water depth and colour. In addition the information provided by wavebands in the infra-red portion of the electromagnetic spectrum can be used. Mangrove areas, especially the interior of mangrove stands, are often difficult to access. Remote sensing allows information to be gathered from areas that would otherwise be logistically and practically very difficult to survey.

☞ **Canopy characteristics such as leaf area index and canopy closure may also be estimated from remotely sensed data. This is the subject of Chapter 17.**

Review of the use of remote sensing in mangrove management

Useful types of image data

A variety of sensors and image processing methods have been used in the remote sensing of mangroves (Table 13.1). Landsat TM and SPOT XS are the most common sensors but aerial photography (both black and white, and colour) is heavily under-represented in Table 13.1. Despite more than 70 years of application – the first aerial photographic surveys of mangroves were made in the 1920s – it is difficult to obtain an overview of aerial photography for the assessment of mangroves because published accounts are scarce. This in itself is no reason to doubt the success of aerial photography, instead it probably reflects the low emphasis that governmental departments, aid agencies or consultancy firms place on the publication of results in scientific literature. Some use has also been made of video (Everitt and Judd 1989, Everitt *et al.* 1991, 1996) and colour infra-red photography (Reark 1975, Ross 1975, Saenger and Hopkins 1975, Sherrod and

Table 13.1 A summary of the sensors and image processing techniques which have been used for habitat mapping of mangroves. Further discussion of the techniques is given in the text. Studies have been arranged in five groups according to the image processing technique employed – only 12% of them include an assessment of accuracy for the classification.

Processing method	Sensor(s)	Validation	Acc?	Level of discrimination achieved
I. Visual Interpretation				
Gang and Agatsiva (1992)	SPOT XS	Field data	No	Five classes (labelled after dominant species or associations of species)
Roy (1989)	MK6 KATE-140	Field data	No	Seven classes (labelled after associations of dominant mangrove species)
Paterson and Rehder (1985)	Aerial photos	Field data	No	Four classes (fringing, black, mixed and riverine mangrove)
Untawale *et al.* 1982	Aerial photos	Field data	No	Ten classes (labelled by species or genera)
II. Vegetation Index				
Blasco et *al.* (1986)	SPOT*	Aerial photographs	No	Two classes (fringing and cleared mangrove)
Jensen *et al.* (1991)	SPOT XS	Field data	No	Percentage canopy closure
Chaudhury (1990)	Landsat TM	Aerial photographs	No	Two classes (labelled according to dominant species)
III. Unsupervised Classification				
Vits and Tack (1995)	SPOT XS Landsat TM	Field data	95% 97%	Four classes (2 fringing, mixed, shrub and logged mangrove)
Loo *et al.* (1992)	Landsat TM SPOT XS	Field data	No	Three classes (dense, less dense and cleared coastal vegetation)
Woodfine (1991)	Landsat TM	Field data	N/A	Failed to distinguish mangrove and forest satisfactorily
Chaudhury (1990)	SPOT XS	Aerial photographs	No	Four classes (labelled according to dominant species)
IV. Supervised Classification				
Dutrieux *et al.* (1990)	SPOT XS	Field data	No	Four classes (labelled according to dominant species and species associations)
Vits and Tack (1995)	SPOT XS	Field data	91%	Four classes (2 fringing, mixed, shrub and logged mangrove)
Aschbacher *et al.* (1995)	SPOT XS Landsat TM MESSR JERS-1 ERS-1 SAR	Field data, maps	No	Four qualitative density classes (dense and medium, low and very low density)
Mohamed *et al.* (1992)	Landsat MSS	Field data	No	Two classes of wetland vegetation
Eong *et al.* (1992)	Landsat TM	Field data	No	Three classes (2 labelled according to dominant species, cleared mangrove)
Palaganas (1992)	SPOT XS	Field data	81%	Two classes (primary and secondary mangrove)

Table 13.1 (cont'd)

IV. Supervised Classification (cont'd)				
Vibulsresth *et al.* (1990)	Landsat TM SPOT Pan	Aerial photographs	No	Six classes (4 labelled according to dominant species, 2 mixed mangrove)
Biña *et al.* (1980)	Landsat MSS MESSR	Field data, aerial photographs, maps	85%	Mangrove (as separate from non-mangrove vegetation)
Lorenzo *et al.* (1979)	Landsat MSS	Field data	No	Mangrove (as separate from non-mangrove vegetation)
Green *et al.* (1998)	CASI	Field data	78%	Six classes (defined from hierarchical cluster analysis of field data)
Woodfine (1991)	Landsat TM	Field data	No	Five classes (mixed community, complex community, transitional to freshwater, transitional to upland vegetation, cleared mangrove with secondary invasion)
V. Band Ratioing				
Gray *et al.* (1990)	Landsat TM	Field data	No	Three height classes (tall (>10 m), medium (4–10 m), and dwarf (<4 m) mangrove)
Kay *et al.* (1991)	Landsat TM	Field data	N/A	Mangrove (as separate from non-mangrove vegetation)
Long and Skewes (1994)	Landsat TM	Aerial photographs	No	Mangrove (as separate from non-mangrove vegetation)
Populus and Lantieri (1991)	Landsat TM SPOT XS	Field data	No	Two classes (high density, mature mangrove and low density, young mangrove)
Ranganath *et al.* (1989)	Landsat TM	Field data	No	Mangrove (as separate from non-mangrove vegetation)

SPOT XS = multispectral mode imagery from Satellite Pour l'Observation de la Terre. MK6 is a Russian multispectral camera carried on the Salyut-7 satellite. KATE-140 is a Soviet panchromatic large format camera. Landsat TM = Landsat Thematic Mapper. MOS-1 MESSR = Multispectral Electronic Self-Scanning Radiometer carried on the Marine Observation Satellite. JERS 1 = Japanese Earth Resources Satellite. ERS-1 SAR = Synthetic Aperture Radar carried on the European Remote Sensing Satellite, Landsat MSS = Landsat Multispectral Scanner, SPOT XP = panchromatic mode imagery from SPOT. CASI = Compact Airborne Spectrographic Imager. N/A = not applicable and Acc? = was an accuracy assessment performed? If the answer is yes, then the overall % accuracy achieved is given.

McMillan 1981, Everitt and Judd 1989) for monitoring mangroves. A comparison between aerial photography, video and airborne multispectral sensors would be very instructive. However, it was unfortunately not possible to obtain aerial photography or video for the Turks and Caicos Islands and so this chapter concentrates on satellite and aerial multispectral (CASI) imagery.

The Landsat and SPOT sensors operate in the visible and infra-red portion of the electromagnetic spectrum and suffer from the problem of cloud cover hampering image acquisition. For this reason some effort has recently been made to utilise radar images, specifically SAR (Synthetic Aperture Radar) from which it is possible to acquire data in any season, irrespective of cloud cover. Initial work would seem to indicate that it is harder to extract mangrove information from SAR than from optical data. Pons and Le Toan (1994) using SAR in Guinea, West Africa, produced maps of mangrove broadly split into two height classes (low and high mangrove). Similarly, analysis of SAR to yield information on homogeneity of tree height was used by Aschbacher *et al.* (1996) to complement classification of SPOT XS data. This approach enabled more mangrove classes to be identified than with SPOT XS alone, but unfortunately the degree of success of neither of these studies was quantified. Mangrove mapping using SAR is still in its infancy and further studies are required to verify and refine the identification of mangrove characteristics obtainable from radar. If these are found to be useful for resource management purposes then SAR would offer great potential, particularly because of its independence from climatic conditions. Unfortunately SAR imagery was unavailable for our study area and could not therefore be evaluated.

Main image processing techniques for mangroves

Analysis of the literature reveals that the approaches taken to classify remotely sensed data on mangroves fall into five methodological groups (Table 13.1):

- visual interpretation,
- vegetation indices,
- unsupervised classification,
- supervised classification, and
- band ratios.

The imagery was processed using each of these five approaches, which are described below, and a new technique whereby Landsat TM and SPOT XP data are merged together (see image processing methods, below).

Table 13.2 Summary of satellite remote sensing applications for the management of mangrove areas, with selected example papers

Management use	Landsat MSS	Landsat TM	SPOT XS
RESOURCE INVENTORY AND MAPPING			
Mangrove forests	Lorenzo *et al.* (1979) Shines (1979) Siddiqui and Shakeel (1989) Vibulsresth (1990)* Mohammed *et al.* (1992) Vits and Tack (1995)	Ranganath *et al.* (1989) Vibulsresth (1990)* Gray *et al.* (1990) Eong *et al.* (1992) McPadden (1993) Vits and Tack (1995)	Blasco *et al.* (1986) Siddiqui and Shakeel (1989) Chaudhury (1990) Thollot (1990) Populus and Lantieri (1991) Palaganas (1992)
CHANGE DETECTION			
Deforestation of mangrove	Biña *et al.* (1978) Lorenzo *et al.* (1979) Biña *et al.* (1980) Chaudhury (1990) Chitsamphandvej (1991)	Ibrahim and Bujang (1992) Ibrahim and Yosuh (1992) Loo *et al.* (1992)	Populus and Lantieri (1990) Loo *et al.* (1992) Palaganas (1992)
AQUACULTURE MANAGEMENT			
Inventory of aquaculture sites in mangroves			Biña (1988) Loubersac *et al.* (1986) Loubersac and Populus (1990) Chitsamphandvej (1991) Siswandono *et al.* (1991)
Site assessment for shrimp aquaculture		Kapetsky *et al.* (1987; 1990) Woodfine (1991) Woodfine (1993)	Loubersac *et al.* (1990) Loubersac and Populus (1986) Chitsamphandvej (1991)

* Also used SPOT Pan

Use of mangrove habitat maps in management

The use of remotely sensed data for mangrove mapping is dependent on an fundamental step, the ability of the sensor and image processing techniques to distinguish accurately between mangrove vegetation and other habitats such as non-mangrove vegetation, bare soil and water. Clearly, a resource map that fails in this, is of limited use in the management of mangroves. Table 13.2 lists several case studies in which habitat maps have been used in mangrove management.

The application of remote sensing falls into three broad categories.

Mangrove inventory and mapping

This has been the main use of remote sensing and has sought to answer general questions such as 'what area of mangrove is present in this area and how are different mangrove habitats distributed?' The end-product has usually been a mangrove habitat map (e.g. Vibulsresth 1990, Palaganas 1992, Vits and Tack 1995), occasionally with supplementary quantitative information such as tree height (Gray *et al.* 1990).

Change detection (deforestation)

Remote sensing has proved particularly useful in monitoring the clearance of mangroves. Biña *et al.* (1980) compared mangrove coverage calculated from Landsat MSS with historical data to estimate the deforestation of mangroves in the Philippines. Loubersac *et al.* (1990) were able to survey rapidly areas within mangroves which had been cleared for aquaculture. It is unfortunate that in most cases these case studies concentrate on the details of the remote sensing campaign rather than the application of habitat maps to management issues. For example, Ibrahim and Yosuh (1992) found that a developer had cleared an area of mangroves in the Pulau Redang marine park (Malaysia) twice the size that was permitted. It is not clear, however, whether this finding was sufficient evidence to support mitigation or compensation procedures.

Management of aquaculture activities

It has proved possible both to monitor current aquaculture activity in mangrove forests and to select areas suitable for new sites. Monitoring of aquaculture activity has to be performed indirectly – remote sensing will not determine whether shrimp ponds are stocked, but the number of ponds and the number of full versus empty ponds, can be assessed quite successfully (Loubersac *et al.* 1990). Several criteria derived from remote sensing, such as degree of shelter from prevailing winds, water depth and distance from market, were analysed by Kapetsky *et al.* (1987, 1990) in the process of selecting new sites suitable for shrimp aquaculture in the Gulf of Nicoya, Costa Rica.

Descriptive resolution of different sensors for mangrove mapping

By and large the discrimination of mangrove and non-mangrove vegetation can be achieved using all the sensors in Table 13.1 and 13.2. In fact failure to do so has only been reported in two instances. Kay *et al.* (1991) found that it was difficult to distinguish rainforest vegetation from fringing coastal mangroves using Landsat MSS and aerial photography. Similar confusion between mangroves and inland vegetation was reported by Woodfine (1991) when an unsupervised classification technique was used on Landsat TM data in the Philippines. The failure of these two examples are presumably case-specific because others have succeeded elsewhere using the same sensors and techniques. In fact Woodfine was successful when she changed her image processing technique to use a supervised classification routine.

In many cases it is also desirable to obtain information on the spatial distribution of different mangrove habitats. Table 13.1 summarises the level of mangrove habitat discrimination that has been achieved using various processing methods. In interpreting Table 13.1 it is important to bear in mind two points.

1. Mangrove habitats have been commonly described by dividing mangrove areas into 2–7 categories which are named according to dominant or characteristic species (Table 13.1). Category names such as '*Bruguiera* zone' must be clearly defined (i.e. the percent composition of *Bruguiera* and associated species should be quantified) if users need to interpret maps in the field. Although a habitat may be labelled using a species or genus name, this does not mean that remote sensing enabled the analyst to identify individual mangrove trees belonging to that genus. The term is simply used as a label. Similarly when mangrove classes are described by assigning a qualitative measure of a particular mangrove characteristic (e.g. 'high', 'medium' and 'low' density or 'tall' and 'low') a user does not know how dense or how tall. Such maps have great descriptive value but cannot be used quantitatively. A few studies have obtained quantitative measurements of mangrove height (Gray *et al.* 1990) and height, density and species composition (Green *et al.* 1998).

☞ **Statistical methods for defining habitats are discussed in Chapter 9. A classification scheme for the mangrove habitats of the Turks and Caicos Islands is presented in Chapter 10. Mangrove field survey techniques are described later in this chapter.**

2. It is unfortunate that published reports rarely include an assessment of accuracy (Table 13.1) and as a result the literature is not a good guide to selecting the appropriate image processing technique for a specific application. The few reports that do include accuracy assessments suggest that high accuracies are possible (78–97%).

Case study: mangrove mapping in the Turks and Caicos Islands

In the eastern Caribbean the steep shorelines of the high islands, the limited freshwater run off of the low dry islands and the exposure of a large portion of the shoreline to intense wave action imposes severe limits on the development of mangroves. These typically occur in small stands at protected river mouths or in narrow fringes along the most sheltered coasts (Bossi and Cintron 1990). As a result, most of the mangrove forests in this region are small. Nevertheless they occur in up to 50 different areas where they are particularly important for water quality control, shoreline stabilisation and as aquatic nurseries. Mangroves in the Turks and Caicos Islands are typical of this area but completely different to the large mangrove forests that occur along continental coasts and at large river deltas. The results and conclusions of this case study must therefore be interpreted in that light.

Field survey techniques and mangrove habitat classification scheme

Three species of mangrove, *Rhizophora mangle*, *Laguncularia racemosa* and *Avicennia germinans* grow with *Conocarpus erectus* in mixed stands along the inland margin of the islands fringing the Caicos Bank (Plates 15 and 21). The field survey was divided into two phases. Calibration data were collected in July 1995, accuracy data in March 1996. Species composition, maximum canopy height and tree density were recorded at all sites which were 5 m^2 plots marked by a tape measure. Species composition was visually estimated. Tree height was measured using a 5.3 m telescopic pole. Tree density was measured by counting the number of tree trunks at breast height. When a tree forked beneath breast height (~1.3 m) each branch was recorded as a separate stem (after English *et al.* 1997). The location of each field site was determined using DGPS with a probable circle error of 2–5 m (Trimble Navigation Ltd 1993).

A habitat classification was developed for the mangrove areas of the Turks and Caicos using hierarchical agglomerative clustering with group-average sorting applied to the calibration data. The calibration data were 4th root transformed in order to weight the contribution of tree height and density more evenly with species composition (the range of data was an order of magnitude higher for density and height and would cluster accordingly). This identified seven classes which separated at a Bray-Curtis similarity of 85% (Figure 10.10). Categories were described in terms of mean species composition (percentage of each species), mean tree height and mean tree density (Table 10.3). One category, *Laguncularia*-dominated mangrove, was discarded because white mangrove was rare in this area – in both the calibration and accuracy field phases it was observed at only two locations. Three other ground cover types were recorded, though no quantitative information was obtained beyond the location of 'typical' sites: (i) sand, (ii) saline mud crust, and (iii) mats of the halophytic succulents *Salicornia perennis* and *S. portulacastrum*. These nine habitat categories (six mangrove, three other) were used to direct the image classification of the CASI data and the collection of accuracy data in 1996.

Image processing methods

In the Turks and Caicos Islands mangroves grow at spatial scales which are inappropriate for remote sensing using data with a spatial resolution less than Landsat TM. Therefore, Landsat TM, SPOT XS, SPOT Pan (XP) and CASI image data were used but the pixel size of Landsat MSS was considered too coarse to be useful.

Each of these images, with the exception of SPOT Pan, was analysed using the five image processing methods listed in Table 13.1. Visual interpretation, unsupervised and supervised classification techniques are covered in detail in Chapter 10: vegetation indices and band-ratios are specific to this chapter, further details are provided in boxes.

Visual interpretation (Method I, Table 13.1)

The imagery was enhanced with a linear contrast stretch (Chapter 9). This was then interpreted visually with reference

Box 13.1.

The use of vegetation indices in the remote sensing of mangroves

Vegetation indices are complex ratios involving mathematical transformations of spectral bands. As such vegetation indices transform the information from two bands or more into a single index. The normalised difference vegetation index or NDVI is a common vegetation index. It is calculated from red and infra-red bands:

$$\text{NDVI} = \frac{infrared - red}{infrared + red}$$

Vegetation indices may be broadly categorised into three types: ratio indices (e.g. NDVI, Jensen 1986), orthogonal indices (e.g. tassled cap transformation, Crist and Cicone 1984) and others (Logan and Strahler 1983, Perry and Lautenschlager 1984). Of these, ratio indices have been most frequently applied to mangrove data. Jensen (1991) correlated field data with four different ratio indices derived from SPOT XS data. He found that correlation between all indices and species composition was poor, but that all correlated well with percentage canopy cover (NDVI had the highest correlation). In other vegetation studies NDVI has been empirically correlated to variables such biomass, cover, leaf area index and carbon in the standing biomass – however this index is quite sensitive to atmospheric conditions and the geometric relationship of the sun and sensor. Therefore, a comparison of different NDVI images in time would require large amounts of information about the composition of the atmosphere at the time of collection (to perform the necessary corrections), information which is not always easily obtainable.

More atmospherically robust indices are available. GEMI (Global Environment Monitoring Index) has been designed specifically to reduce the relative effects of undesirable atmospheric perturbations while maintaining information about vegetation cover. The Angular Vegetation Index (AVI) does the same and is also unaffected by variations in soil brightness. Further details can be found in Pinty and Verstraete (1991) and Plummer *et al.* (1994) respectively. GEMI is calculated from the visible red (λ_1) and near infra-red (λ_2) bands:

$$\text{GEMI} = \eta(1-0.25\eta) - \left(\frac{\lambda_1 - 0.125}{1-\lambda_2}\right) \quad \text{where } \eta = \frac{2(\lambda_2^2 - \lambda_1^2) + 1.5\lambda_2 + 0.5\lambda_1}{\lambda_2 + \lambda_1 + 0.5} \quad \text{and } \lambda_i = \text{reflectance in band } i.$$

AVI is suitable for any sensor with green (λ_1), red (λ_2) and infra-red (λ_3) bands:

$$\text{AVI} = tan^{-1}\left\{\frac{\lambda_3 - \lambda_2}{\lambda_2}\left[\rho(\lambda_3) - \rho(\lambda_2)\right]^{-1}\right\} + tan^{-1}\left\{\frac{\lambda_2 - \lambda_1}{\lambda_2}\left[\rho(\lambda_1) - \rho(\lambda_2)\right]^{-1}\right\}$$

where $\rho(\lambda_i)$ is the radiance in band *i*, λ_i is the centre wavelength of band *i*. AVI values are scaled to 0–1 range by subtracting from 180 and dividing by 90.

to field data and UK Ordnance Survey maps (Series E8112 DOS 309P).

Vegetation indices (Method II, Table 13.1)
Normalised difference vegetation index (NDVI, Box 13.1) was calculated using red and near infra-red bands. This was performed for Landsat TM using bands 3 and 4, SPOT XS bands 2 and 3, and CASI bands 6 and 7 (Table 5.10). An unsupervised classification of the NDVI image produced 50 classes at a convergence level of 99%. Mangrove classes were identified, separated from non-mangrove classes, edited with reference to field data, then merged into a single mangrove category.

Unsupervised classification (Method III, Table 13.1)
An unsupervised classification of the image produced 50 classes at a convergence level of 99%. Mangrove classes were identified, separated from non-mangrove classes, edited with reference to field data, then merged into a single mangrove category.

Supervised classification (Method IV, Table 13.1)
Woodfine (1991) applied a 3x3 edge enhancement filter to her Landsat TM data to accentuate the boundaries between different mangrove areas and experimentation showed the application of a 3x3 filter helped in most cases. A 3x3 edge enhancement filter was thus applied to the imagery. The mangrove habitat classification scheme (see previous section) was used to direct a supervised classification using the maximum likelihood decision rule.

☞ **Other workers have found that the application of a smoothing filter prior to classification is helpful. Experimentation with a particular image is the best way to decide whether a filter may be usefully applied prior to classification.**

Band ratios (Method V, Table 13.1)
A method was used based on the approach of Gray *et al.* (1990), further details are provided in Box 13.2. TM bands 3, 4 and 5 produced the best visual discrimination of the mangrove/non-mangrove boundary in the Turks and Caicos Islands, so ratios of 3/5 and 5/4 were calculated. For SPOT XS ratios of bands 2/3 and 3/1 were used, for CASI 5/8 and 8/7 (see Table 5.10 for wavelengths of these bands). For all three sensors, principal components PC1, PC2 and PC4 were combined in a false colour composite image because (i) PC1 and PC2 accounted for more than 95% of the variability in the data, and (ii) a composite of these principal components provided the best visual discrimination of the mangrove areas. A supervised classification scheme was performed on the principal component composite image using the maximum likelihood decision rule and field data for training.

Resolution merge between Landsat TM and SPOT Pan
To the best of our knowledge this is a new image processing method for mangroves. To recap: the Landsat TM sensor records electromagnetic radiation in six bands (ignoring the thermal band, band 6) at a spatial resolution of 30 m, whilst the HRV sensor on SPOT satellites collects data with a spatial

Box 13.2

The use of band ratios in the remote sensing of mangroves

The process of dividing the pixels in one image by the corresponding pixels in a second image is known as ratioing. It is one of the most common transformations applied to remotely sensed data, for two reasons. Firstly, because differences in the spectra of different surfaces can frequently be emphasised by ratioing. Secondly, because the effects of variations in topography can sometimes be reduced.

There is considerable variation in the ratios that have been used on mangrove data. Populus and Lantieri (1991) used a ratio of SPOT XS bands 3/2, and Ranganath *et al.* (1989) a ratio of Landsat TM bands 1/5. Long and Skewes (1994) ratioed TM bands 4/3 and classified a false colour composite image of this ratio and TM bands 4 and 3. Similarly Kay *et al.* (1991) combined ratios of TM bands 5/2 and 7/4 with TM band 3. This variety in approach with band ratios makes it difficult to generalise about their use and readers who may consider a similar technique are advised to experiment with several different options.

In a study of mangroves in Belize, Gray *et al.* (1990) found that TM bands 3, 4 and 5 allowed clear visual discrimination of the mangrove/non-mangrove boundary. In order to maximise the contrast across this boundary and within the mangrove area itself, they calculated two band-ratios, namely bands 3/5 and 5/4. Data from these five bands (3, 4, 5, 3/5, 5/4) were used as input to a principal components analysis.

Principal components analysis is a useful data compression method, the mathematics of which is dicussed in Mather (1987). It allows redundant data to be compressed into fewer bands – in other words the dimensionality of the data is reduced. The bands produced by principal components analysis (the principal components) are non-correlated and independent, and are often more easily interpreted than the original data (Jensen 1986, Faust 1989). Gray *et al.* (1990) performed an unsupervised classification of mangrove habitats on an image consisting of the first three principal components.

resolution of 10 m in panchromatic (Pan) mode, but only in one broad band (510–730 nm). Resolution merges are attractive because they resample Landsat TM data to the spatial resolution of SPOT Pan while retaining the spectral resolution of the TM sensor. Merging the two images creates a six band image with a spatial resolution of 10 m and attempts to combine the most desirable characteristics of both sensors. Two merges were used, the intensity-hue-saturation (IHS) merge and principal components (PC) merge (Box 13.3). Each merged image was then classified using image processing methods I–IV (Table 13.1).

The ability to discriminate mangrove from non-mangrove vegetation

The accuracy of each image processing method in discriminating mangrove from non-mangrove vegetation (mainly thorn scrub and salt marsh vegetation), for each image type, is given in Figure 13.1.

Accurate discrimination between mangrove and non-mangrove vegetation was not possible using SPOT XS data (Figure 13.1). Overall accuracy was low for each of the five image processing methods (35–57%) and there were no significant differences between the tau coefficients. Values of tau for methods I–V ranged between 0.03 and 0.29, which suggests that very few more pixels were being classified correctly than would be expected from chance alone. SPOT XS data therefore does not appear suitable for mapping the mangroves of the Turks and Caicos Islands, though as Table 13.1 shows clearly they have been used successfully elsewhere.

Accurate discrimination was possible using Landsat TM data (Figure 13.1) and classification accuracy improved as more intensive image processing methods were applied. Visual interpretation (I) and classification of a NDVI image (II) were the least accurate processing methods (overall accuracy 42 and 57% respectively). Image processing methods III–V produced an overall accuracy of more than 70% and were significantly more accurate than methods I and II ($P < 0.001$). There was no significant difference in the tau coefficient between unsupervised (III) and supervised (IV) classification ($t = 0.12$). The most accurate classification of Landsat TM data was obtained using band ratios (V), 92%, and this was significantly more accurate than either method III or IV ($P < 0.001$).

Similar results were obtained for CASI (Figure 13.1). Accurate discrimination between mangrove and non-mangrove vegetation was not possible using either visual interpretation (I) or classification of a NDVI image (II): although the overall accuracy of the visually interpreted image was 71% there was no statistical difference between the tau coefficients for methods I and II ($t = -2.82$). Image processing

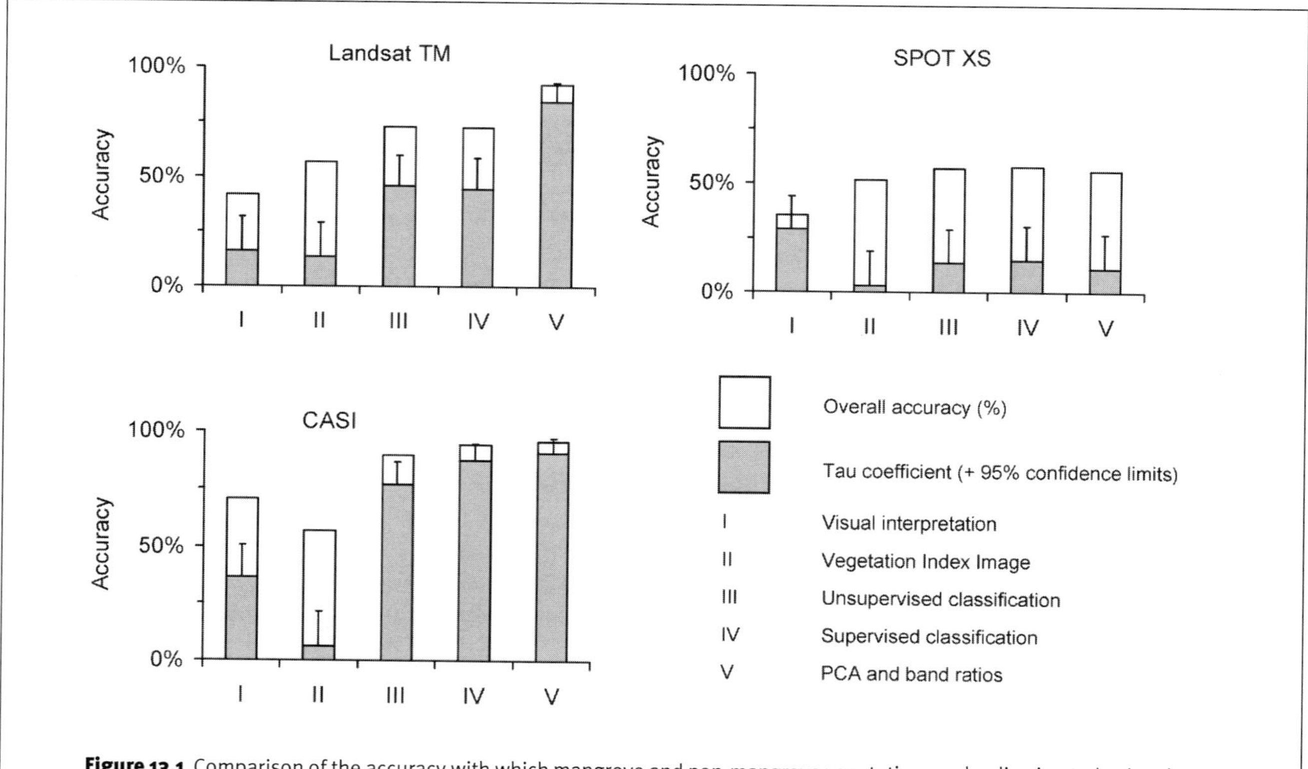

Figure 13.1 Comparison of the accuracy with which mangrove and non-mangrove vegetation can be discrimated using three different data types – Landsat Thematic Mapper (TM), SPOT HRV multispectral (XS), and Compact Airborne Spectrographic Imager(CASI). The effect of five different image processing methods on accuracy is shown for each sensor. Overall accuracy is presented as a percentage. Tau coefficients have been multiplied by 100 to fit on the same scale. Vertical error bars are the positive 95% confidence interval for the tau coefficient.

Box 13.3

Resolution merges

Intensity-hue-saturation (IHS) merge. Three TM bands are selected, forward transformed to IHS and the intensity component replaced with the SPOT Pan (XP) data. All data are then reverse IHS transformed (for details, see Welch and Ehlers 1987). This method assumes that the intensity component is spectrally equivalent to the SPOT Pan image, and that all the spectral information is contained in the hue and saturation components. TM bands 4, 5 and 7 were selected for transformation.

Principal components (PC) merge. The main aim of this method is to retain the spectral information of all TM bands. These are forward transformed into principal components. PC1 is removed: the SPOT Pan is stretched to the same numerical range (minimum-maximum) and substituted for PC1. The data are then reverse transformed (for further details, see Chavez 1991). This method assumes that PC1 contains only the scene luminance, is spectrally equivalent to the SPOT Pan image and that all the spectral information is contained in the other principal components.

☞ **Resolution merges are exceedingly time-consuming and require a significant amount of free disk space. It was necessary to have 0.5 Gbytes free disk space in order to merge a SPOT Pan scene with a Landsat TM scene of the same area.**

methods III–V produced an overall accuracy of more than 89% and were significantly more accurate than methods I and II ($P < 0.001$) and there was no significant difference in the tau coefficient between unsupervised (III) and supervised (IV) classification ($t = 1.69$). The most accurate classification of CASI data was obtained using band ratios (V), 96%, and this was significantly more accurate than method III ($P < 0.05$) but not method IV ($t = 0.71$).

Interestingly it seems that by using band ratios the mangroves of the Turks and Caicos Islands can be distinguished from non-mangrove vegetation just as accurately with Landsat TM data as with CASI. Slightly higher overall accuracy was obtained using CASI data (96% compared to 92% for Landsat TM) but there was no significant difference between tau coefficients ($t = 0.88$).

Merging Landsat TM and SPOT Pan greatly facilitated the visual interpretation of mangrove/non-mangrove vegetation in the TM data (Figure 13.2), almost doubling the overall accuracy from 42% to 71% for the IHS merge and 79% for the PC merge. There were significant increases in accuracy ($t = 2.39$, $P < 0.05$ for the IHS merge; $t = 4.00$, $P < 0.01$ for the PC merge). However a supervised classification of data merged by either method was not more accurate than a supervised classification of the original Landsat TM data (Figure 13.2). In fact visual interpretation appears to be the best way of distinguishing mangrove from non-mangrove vegetation in merged data. Visual interpretation was more accurate than supervised classification for both IHS merged data ($t = 1.99$, $P < 0.05$) and PC merged data ($t = 2.56$, $P < 0.01$). This is probably because the assumption that the Pan image is spectrally similar to the intensity component of the TM data is not always valid (Chavez 1991). The differences between the intensity of Landsat TM bands 4, 5 and 7 and the Pan image is sufficient to distort the spectral characteristics of the TM data. A PC merge distorts the TM data less than a IHS merge because all TM bands are used and as a result the first principal component is more similar to the Pan image than the intensity component in a IHS merge (Chavez 1991). However PC merged data were not significantly more accurate than IHS merged data, whether processed using visual interpretation ($t = -1.51$) or supervised classification ($t = -0.24$).

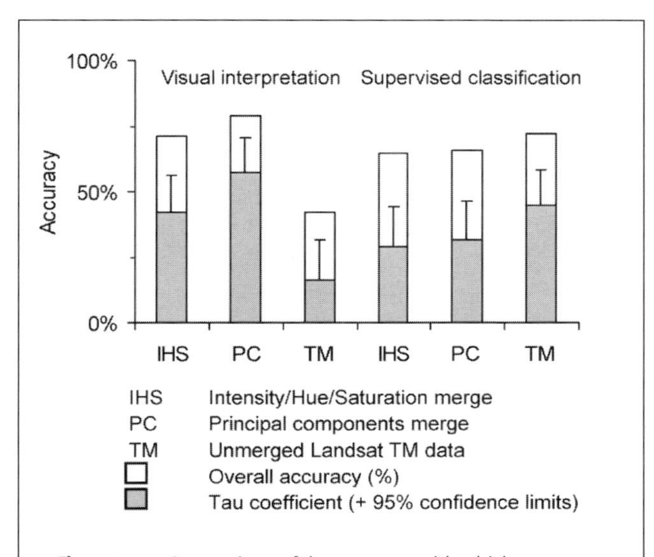

Figure 13.2 Comparison of the accuracy with which mangrove and non-mangrove vegetation can be discriminated using Landsat TM data alone and TM data which has been merged with SPOT panchromatic data (10 m spatial resolution). Accuracies are compared between two types of merges (intensity/hue/saturation and principal components) and unmerged Landsat TM data. Two types of classification were used: visual interpretation and supervised classification of the imagery. Overall accuracy is presented as a percentage. Tau coefficients have been multiplied by 100 to fit on the same scale. Vertical error bars are the positive 95% confidence interval for the tau coefficient.

Summary of case study results for discriminating mangrove and non-mangrove vegetation

- In the Turks and Caicos Islands it does not appear possible to discriminate accurately between mangrove and non-

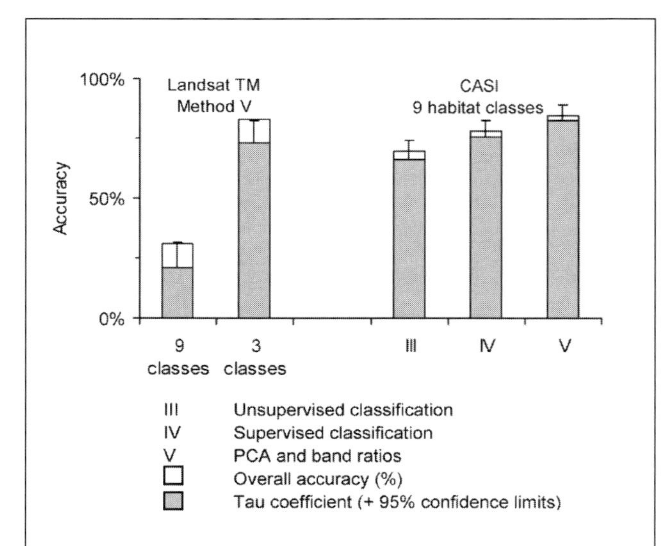

Figure 13.3 *Left:* Comparison for Landsat TM data of the accuracy of mapping either 3 or 9 mangrove habitat classes using the most accurate method tested (see Figure 13.1). *Right:* Comparison for CASI data of the accuracy of mapping nine different mangrove habitat classes using the three most accurate methods identified in Figure 13.1. Overall accuracy is presented as a percentage. Tau coefficients have been multiplied by 100 to fit on the same scale. Vertical error bars are the positive 95% confidence interval for the tau coefficient.

mangrove vegetation using SPOT XS (Figure 13.1). This sensor is therefore inappropriate for mapping mangrove in the Eastern Caribbean. However, in larger mangrove areas SPOT XS can undoubtedly be used (Table 13.1).

- The band ratio/principal components analysis (Box 13.2) is the most accurate method of discriminating between mangrove and non-mangrove vegetation in the Eastern Caribbean (Figure 13.1).
- Discrimination between mangrove and non-mangrove vegetation in the Eastern Caribbean is achieved equally well using CASI and Landsat TM data with 92–96% accuracy.
- Merging Landsat TM and SPOT Pan data does not improve accuracy over processing the original TM data alone. If a merged dataset is available then visual interpretation is the most accurate way of interpreting it.

The ability to discriminate different mangrove habitats

The habitat classification scheme used here to direct the classification of the CASI imagery was based on six mangrove categories, and three mangrove-associated categories, which separated at the 85% level (see Chapter 10). At higher levels of similarity the field data do not define ecologically meaningful categories.

Figure 13.3 compares the classification accuracy for mapping the nine different mangrove habitats of the Turks and Caicos Islands using Landsat TM and CASI data. Accurate discrimination between classes was not possible using the satellite data with an accuracy of only 31% being achieved. Tall *Rhizophora* was mapped accurately (77%) but the user's accuracies of all other mangrove habitats were very low (0–29%). The classification was performed at a coarser (less similar) level of habitat discrimination using just three classes: tall *Rhizophora*, 'other' mangrove (i.e. mangrove vegetation which is not 100% *Rhizophora* and more than 2 m in height) and non-mangrove vegetation. Reduction to just three classes significantly increases accuracy ($t = 7.18$, $P < 0.001$). Figure 13.4 (Plate 15) is a classified map of the south-eastern coast of North and Middle Caicos derived from Landsat TM data. This area was designated a wetland of international importance under the Ramsar Convention (UNESCO, 1971) and contains the most extensive mangrove growth of the Caicos Bank.

The mangrove habitats in Table 10.3 can only be accurately mapped using CASI data (Figure 13.3). Band ratios (V) had significantly improved accuracy over unsupervised classification (III) ($t = 3.26$, $P < 0.01$) but not over supervised (IV) ($t = 1.77$). There was no significant increase in accuracy using supervised classification as opposed to unsupervised (no significant difference in tau coefficient between methods III and IV, $t = 1.50$). In other words if supervised classification is to be used then an improvement in accuracy will only be obtained if it is performed on a PCA/band ratio image (method V). Figure 10.9 (Plate 11) is a map of the mangroves growing on South Caicos that was derived from CASI data (Figure 10.8).

The user's accuracies of the nine mangrove habitats produced from CASI data processed using methods III–V are reasonably high, with the exception of short mixed mangrove (Table 13.3). Only six accuracy sites were surveyed in this category, so this low accuracy may be a function of small sample size. User's accuracy was improved by using a supervised instead of an unsupervised classification routine: the classification of a PCA/band ratio image produced the highest user's accuracy for the majority of habitats (Table 13.3).

These results also suggest a possible correlation between image processing effort and accuracy. Unsupervised classification (III) is relatively rapid, requiring the operator to do little more than edit the final classes but was the least accurate (overall accuracy 70% for CASI, Figure 13.5). The supervised classification procedure in image processing method IV requires greater effort from the operator during the process of signature editing (overall accuracy 78% for CASI, Figure 13.5). Principal component analysis is computationally intensive, and different combinations of band ratios must be experimented with. As a result, image processing method V requires the most effort, but is the most accurate classification procedure (overall accuracy 85% for CASI, Figure 13.5). Figure 13.5 shows that although band ratios and principal components analysis are computationally intensive they are also cost-effective: the extra investment of time produces a significant increase in accuracy.

Table 13.3 The user accuracies of the nine mangrove habitats of the Turks and Caicos Islands (described in Table 10.3) produced from CASI data processed using three different methods (described in Table 13.1 and the text).

User accuracy for each method in Table 13.1 (%)	*Conocarpus erectus*	*Avicennia germinans*	Short *Rhizophora*	Tall *Rhizophora*	Short mixed mangrove	Tall mixed mangrove	Sand	Mud crust	*Salicornia* spp.	Overall accuracy
Band ratios (V)	71.4	83.3	94.7	93.1	16.7	82.6	81.0	96.6	82.4	85
Supervised (IV)	71.4	80.0	78.9	93.1	16.7	69.6	61.9	96.6	76.5	78
Unsupervised (III)	64.3	66.7	63.2	75.9	16.7	73.9	47.6	86.2	88.2	70

Summary of case study results for discriminating different mangrove habitats

- In the Turks and Caicos Islands it does not appear possible to discriminate accurately between mangrove and non-mangrove vegetation using SPOT XS (Figure 13.1). This sensor is therefore inappropriate for mapping mangrove habitats in the Eastern Caribbean. However, in larger mangrove areas SPOT XS can undoubtedly be used (Table 13.1).
- Only two classes of mangrove habitat can be reliably discriminated from Landsat TM data in this region, tall *Rhizophora* and 'other' mangrove.
- Five of the six mangrove and all of the three mangrove-associated habitats identified in the Turks and Caicos Islands can be accurately discriminated using CASI data (user accuracies 71–95%).
- Supervised classification (either based directly on the reflectance values or on principal components and band ratios) is the most accurate method of mapping mangroves at the finest level of habitat discrimination (Table 13.3).

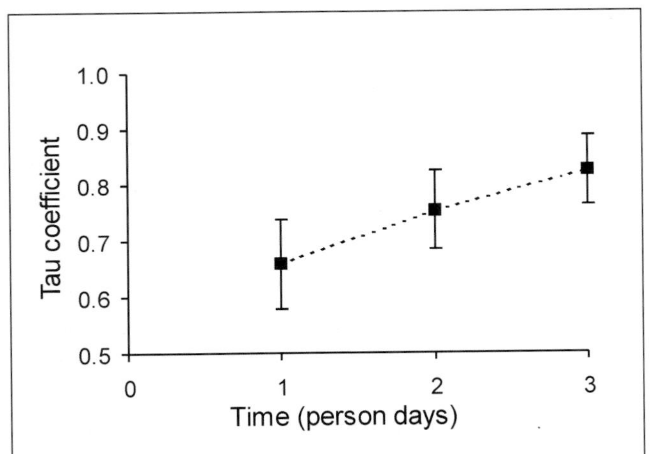

Figure 13.5 The accuracy of classifications (expressed in terms of the tau coefficient) of CASI data using image processing methods III (*ca.* 1 day to carry out), IV (*ca.* 2 days operator time) and V (*ca.* 3 days to complete) expressed as a function of processing effort (time taken in person days). Vertical error bars are the 95% confidence intervals for the tau coefficient and show that method V gives a significant gain in accuracy over method III, but not method IV. See text for details of overall accuracies for each method.

Time, effort and cost considerations

No specialist field equipment was needed for habitat mapping, therefore equipment costs are restricted to general field equipment only (see Chapter 19). Fieldwork was divided between two phases, survey and accuracy assessment. A total of 81 sites were visited during survey, 121 sites during accuracy. Species composition, tree height and tree density were recorded at each site. The time costs of fieldwork in the Turks and Caicos Islands are given below:

Survey		Accuracy assessment	
Field time (person days)	17	Field time (person days)	17
Computing (person days)	1	Computing (person days)	1
Total km travelled	15.8	Total km travelled	17

Conclusions

East Caribbean mangrove areas differ from the majority of mangroves elsewhere in the world which may form true forests covering hundreds of square kilometres with trees up to 30 m in height. Nonetheless they are important coastal features over a large geopolitical area (Bossi and Cintron 1990) and remote sensing is capable of being a valuable tool in their management (Green *et al.* 1996). However the relatively small scale of East Caribbean mangroves (Plate 15) make them less amenable to remote sensing than, say, the mangroves of the Sunderbans or Niger delta. Consequently the Turks and Caicos Islands can be considered a difficult case study and results achievable here should be bettered elsewhere.

Mapping the spatial distribution of mangroves is only possible if mangrove areas can be accurately distinguished from surrounding non-mangrove vegetation. The failure of all classifications of SPOT XS data to distinguish mangrove and non-mangrove in the Turks and Caicos is probably simply an example of site-specific failure (given the success of XS in other areas, e.g. Vits and Tack 1995). It may be a consequence of the limited spectral resolution of the SPOT HRV sensor.

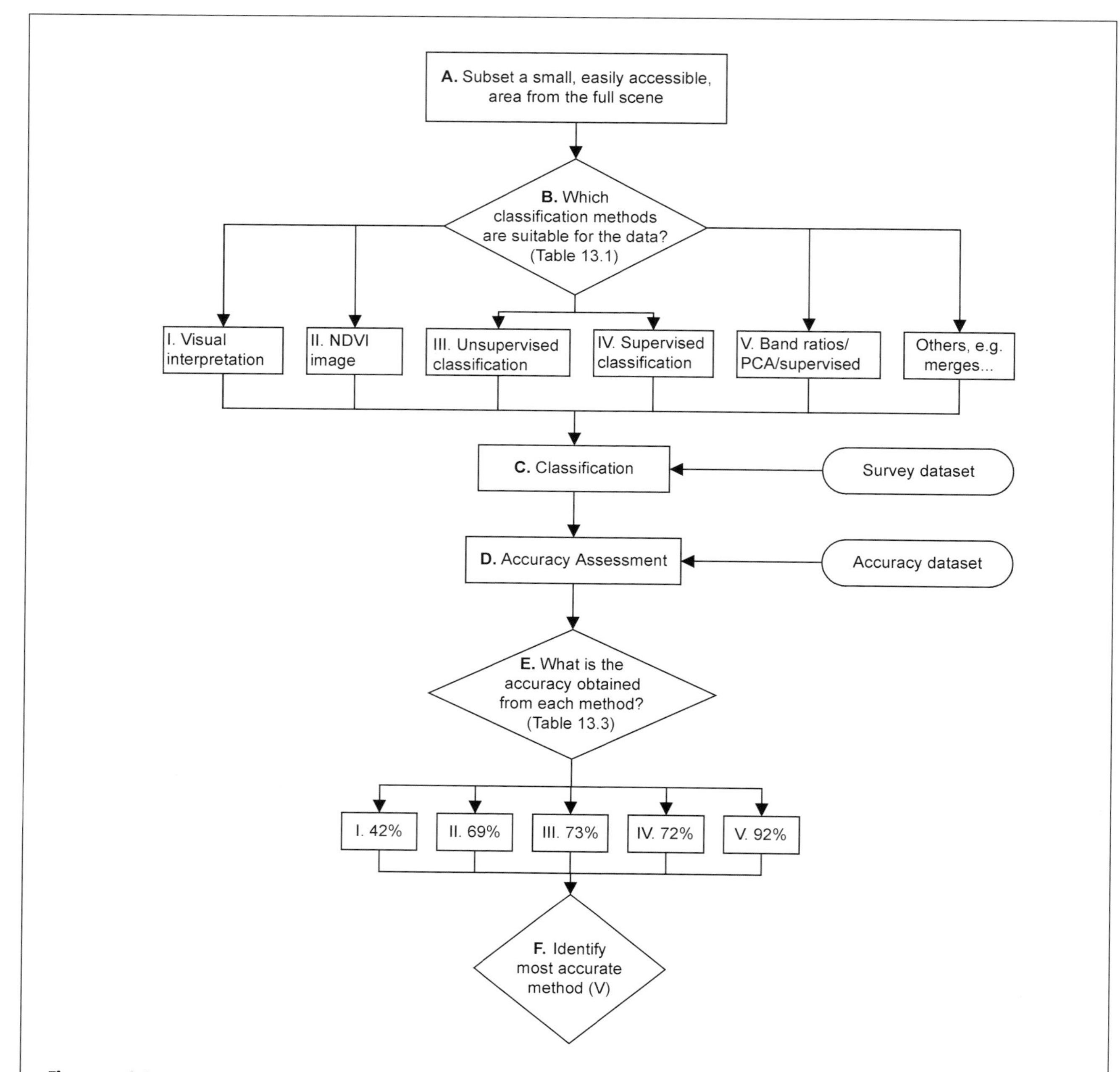

Figure 13.6 An experimental protocol, illustrated with reference to Landsat TM, for selecting the best image processing method for mangroves. A small subset is chosen from the full image for which field data can be obtained relatively quickly. This is tested against a suite of image processing techniques using two independent field datasets, one for classification, the other for accuracy assessment. The classification accuracy of each image processing method is compared, and the most accurate method identified. The whole satellite image may then be classified using this technique, with further field data if necessary.

The most accurate classification of Landsat TM necessitates the calculation of principal components and band ratios. Band ratios both reduce the effect of topography on recorded radiance and emphasise differences in spectral reflectance of different cover types (Mather 1987). Landsat TM with six bands enables the analyst to achieve this by providing a greater variety of spectrally dissimilar ratios than does SPOT XS with three. Much of the error in methods II–IV was due to dense terrestrial vegetation growing at the foot of steep slopes being classified as mangrove. Under a Landsat TM band ratio procedure this was correctly classified as thorn scrub. The consistently poor accuracies obtained from classification of NDVI images derived from both Landsat TM and SPOT XS are probably a result of the low correlation between species composition and many ratio indices (Jensen 1991).

Although mangroves can be distinguished from non-mangrove vegetation fairly accurately with Landsat TM

(92%), only the two most dissimilar mangrove categories can be discriminated. This is probably due to the spatial resolution of Landsat TM (30 m) and the growth forms of mangrove stands in the Turks and Caicos Islands: as already been noted, mangroves grow there in relatively small patches and linear stands along inlets typical of the eastern Caribbean islands. The relatively low spatial resolution of Landsat TM can only be partially compensated for by merging with SPOT Pan data. The increased spatial resolution derived from the SPOT Pan data facilitates visual interpretation of small mangrove stands, but the spectral information of the data is compromised. The superior spatial and spectral resolution of airborne multispectral data such as CASI allows mangrove areas to be assessed to a greater level of detail and accuracy than with satellite sensors. Similarly, Panapitukkul *et al.* (1998) have recently shown the use of black and white aerial photography for mapping mangrove progression over mudflats in Thailand. Quite simply, most of the mangrove categories described in Table 10.3 occur in this region at scales that cannot be resolved by satellite data.

It must be emphasised that both SPOT XS and Landsat TM have regularly been used to map mangroves: the failure of one sensor and several image processing techniques to map mangroves accurately in the Turks and Caicos Islands does not necessarily preclude their applicability elsewhere. Having said that, it is clear that the type of data and classification procedure used can substantially affect the accuracy of the final mangrove map. In the eastern Caribbean islands, the most accurate combinations of sensor and image processing method are:

1. Landsat TM using PCA/band ratio method (V), if discrimination between mangrove and non-mangrove vegetation is required over a large area, at relatively low cost, or
2. Airborne digital imagery (e.g. CASI) using PCA/band ratio method (V), if discrimination between different mangrove classes is required.

Figure 13.6 outlines a protocol for readers wishing to select the best image processing method experimentally for themselves.

References

Aschbacher, J., Ofren, R.S., Delsol, J.P., Suselo, T.B., Vibulsresth, S. and Charrupat, T., 1995, An integrated comparative approach to mangrove vegetation mapping using remote sensing and GIS technologies, preliminary results, *Hydrobiologia*, **295**, 285–294.

Biña, R.T., 1988, An update on the application of satellite remote sensing data to coastal zone management activities in the Philippines. *Proceedings of the Symposium on Remote Sensing of the Coastal Zone*, Gold Coast, Queensland, September 1988, **I**,2.1–I.2.7.

Biña, R.T., Jara, R.B., and B.R. de Jesus, Jr., 1978, Mangrove inventory of the Philippines using Landsat multispectral data and the Image 100 system. *NRMC Research Monograph*, Series 2, 1978.

Biña, R.T., Jara, R.B., and Roque, C.R., 1980, Application of multilevel remote sensing survey to mangrove forest resource management in the Philippines. *Proceedings of the Asian Symposium on Mangrove Development, Research and Management*. University of Malaya, Kuala Lumpar, Malaysia, August 28–29th, 1980.

Blasco, F., Lavenu, F., and Baraza, J., 1986, Remote sensing data applied to mangroves of Kenya coast. *Proceedings of the 20th International Symposium on Remote Sensing of the Environment*, **3**, 1465–1480.

Bossi, R., and Cintron, G., 1990, *Mangroves of the Wider Caribbean: toward Sustainable Management* (Caribbean Conservation Association, the Panos Institute, United Nations Environment Programme).

Chaudhury, M.U., 1990, Digital analysis of remote sensing data for monitoring the ecological status of the mangrove forests of Sunderbans in Bangladesh. *Proceedings of the 23rd International Symposium on Remote Sensing of the Environment*, **1**, 493–497.

Chavez, P.S., 1991, Comparison of three different methods to merge multiresolution and multi-spectral data: Landsat TM and SPOT Panchromatic. *Photogrammetric Engineering and Remote Sensing*, **57**, 3, 295–303.

Chitsamphandvej, S., 1991, Digital coastal resources inventory map of Pak Panang region, Thailand, using digital mapping system and Landsat MSS data. *Proceedings of the Regional Symposium on Living Resources in Coastal Areas, Manila, Philippines.* Marine Science Institute, University of the Philippines, 561.

Crist, E., and Cicone, R., 1984, A physically based transformation of thematic mapper data – the TM Tassled Cap. *IEEE Transactions on Geoscience and Remote Sensing*, **GE–22 3**, 256–263.

Dutrieux, E., Denis, J., and Populus, J., 1990, Application of SPOT data to a base-line ecological study the Mahakam Delta mangroves East Kalimantan, Indonesia). Oceanologica Acta, 13, 317–326.

English, S., Wilkinson, C., and Baker, V., 1997, *Survey Manual for Tropical Marine Resources*. (Townsville: Australian Institute of Marine Science).

Eong, O.J., Khoon, G.W., Ping, W.Y., and Kheng, W.H., 1992, Indentification of mangrove vegetation zones using MicroBRIAN and Landsat imagery. In *Third ASEAN Science and Technology Week Conference Proceedings*, edited by L.M. Chou and C.R. Wilkinson, Marine Science, Living Coastal Resources, **6**, University of Singapore, September 1992, 383–389.

Everitt, J.H., and Judd, F.W., 1989, Using remote sensing techniques to distinguish black mangrove (Avicennia germinans). *Journal of Coastal Research,* **5**, 737–745.

Everitt, J.H., Escobar, D.E., and Judd, F.W., 1991, Evaluation of airborne video imagery for distinguishing black mangrove (Avicennia germinans) on the Texas coast. *Journal of Coastal Research*, **7**, 1169–1173.

Everitt, J.H., Judd, F.W., Escobar, D.E. and Davis, M.R., 1996, Integration of remote sensing and spatial information technologies for mapping black mangrove on the Texas coast. *Journal of Coastal Research*, **12**, 64–69.

Faust, N.L., 1989, Image enhancement. In *Encyclopedia of Computer Science and Technology*, **20**, 5, edited by Kent, A., and Willimas, J.G. (New York: Marcel Dekker, Inc.).

Gang, P.O., and Agatsiva, J.L., 1992, The current status of mangroves along the Kenyan coast, a case study of Mida Creek mangroves based on remote sensing. *Hydrobiologia*, **247**, 29–36.

Gray, D., Zisman, S., and Corves, C., 1990, Mapping of the mangroves of Belize. *University of Edinburgh Technical Report* (Edinburgh: University of Edinburgh).

Green, E.P., Mumby, P.J., Edwards, A.J., and Clark, C.D., 1996, A review of remote sensing for the assessment and management of tropical coastal resources. *Coastal Management*, **24**, 1, 1–40.

Green, E.P., Mumby, P.J., Edwards, A.J., Clark, C.D, and Ellis, A.C., 1998, The assessment of mangrove areas using high resolution multispectral airborne imagery. *Journal of Coastal Research*, **14**, 433–443.

Ibrahim, M. and Bujang, J.S., 1992, Application of remote sensing in the study of development impact of a marine park in Malaysia. *Proceedings of the 18th Annual Conference of the Remote Sensing Society*, Dundee, 235–243.

Ibrahim, M., and Yosuh, M., 1992, Monitoring the development impacts on the coastal resources of Pulau Redang marine park by remote sensing, In *Third ASEAN Science and Technology Week Conference Proceedings,* edited by L.M. Chou and C.R. Wilkinson, Marine Science: Living Coastal Resources, **6**, University of Singapore, September 1992, 407–413.

Jensen, J.R., Ramset, E., Davis, B.A., and Thoemke, C.W., 1991, The measurement of mangrove characteristcs in south-west Florida using SPOT multispectral data. *Geocarto International*, **2**, 13–21.

Jensen, J.R., 1986, *Introductory Digital Image Processing*. (New Jersey: Prentice Hall).

Kapetsky, J.M., Hill, J.M., Worthy, L.D., and Evans, D.L., 1990, Assessing Potential for Aquaculture Development with a Geographic Information System. *Journal of the World Aquaculture Society*, **21**, 4, 241–249.

Kapetsky, J.M., McGregor, L., and Nanne, H. E., 1987, A geographical information system to plan for aquaculture: a FAO-UNEP/GRID study in Costa Rica. *FAO Fisheries Technical Paper* **287**. (Rome: Food and Agriculture Organisation of the United Nations).

Kay, R.J., Hick, P.T., and Houghton, H.J., 1991, Remote sensing of Kimberley rainforests. In *Kimberley Rainforests*, edited by N.I. McKenzie, R.B. Johnston and P.G. Kendrick. (Surrey Beatty & Sons: Chipping Norton), pp. 41–51.

Logan, T.L., and Strahler, A.H., 1983, Optimal Landsat transforms for forest applications. *Proceeedings on the Symposium on Machine Processing of Remotely Sensed Data 1983*, (West Lafayette, Indiana: Purdue University), pp. 146–153.

Long, B.G., and Skewes, T.D., 1994, GIS and remote sensing improves mangrove mapping. *Proceedings of the 7th Australasian Remote Sensing Conference*, Melbourne, March 1994, 545–550.

Loo, M.G.K., Lim, T.M., and Chou, L.M., 1992, Land use changes of a recreational island as observed by satellite imagery. In *Third ASEAN Science and Technology Week Conference Proceedings*, edited by L.M. Chou and C.R. Wilkinson, Marine Science, Living Coastal Resources, **6**, University of Singapore, September 1992, 401–405.

Loubersac, L., and Populus, J., 1986, The applications of high resolution satellite data for coastal management and planning in a Pacific coral island. *Geocarto International* **2**, 17–31.

Loubersac, L., Populus, J., Grotte, A., and Burban, P.Y., 1990, The Alias Project, inventorying aquaculture sites in New Caledonia. *Proceedings of the International Workshop on Remote Sensing and Insular Environments in the Pacific: Integrated Approaches*, ORSTOM/IFREMER: 205–216.

Lorenzo, R., de Jesus, B.R., and Jara, R.B., 1979, Assessment of mangrove forest deterioration in Zamboanga Peninsula, Philippines, using Landsat MSS data. *Proceedings of the 13th International Symposium on Remote Sensing of the Environment.*

Mather, P.M., 1987, *Computer Processing of Remotely-sensed Images: An Introduction*. (Chichester: John Wiley & Sons).

McPadden, C.A., 1993, *The Malacca Straits Coastal Environment and Shrimp Aquaculture in North Sumatra Province. North East Sumatra Prawn Project*. (London: Overseas Development Administration).

Mohamed, M.I.H., Hussein, N.A. Ibrahim, M.I., and Kuchler, D.A., 1992, Coastal resources mapping of Besut (Terrengganu coast), Malayasia, using microBRIAN digital mapping technology and Landsat MSS satellite data. *Proceedings of the Regional Symposium on Living Resources in Coastal Areas, Manila, Philippines*, Marine Science Institute, University of the Philippines, 545–555.

Palaganas, V.P., 1992, *Assessing changes in mangrove forest of Infanta-Real, Quezon Province (Philippines) using remote sensing*. MSc dissertation, University of Newcastle upon Tyne, 106pp.

Panapitukkul, N., Duarte, C.M., Thampanya, U., Kheowvongsri, P., Srichai, N., GeertzHansen, O., Terrados, J., and Boromthanarath, S., 1998, Mangrove colonization: Mangrove progression over the growing Pak Phanang (SE Thailand) mud flat. *Estuarine Coastal and Shelf Science*, **47**, 51–61.

Patterson, S.G., and Rehder, J.B., 1985, An assessment of conversion and loss of mangroves using remote sensing imagery on Marco Island, Florida. *Proceedings of the American Society for Photogrammetry and Remote Sensing*, (Reston, Virginia: ASPRS), pp. 728–735.

Perry, C.R., and Lautenschlager, L.F., 1984. Functional equivalence of spectral vegetation indices. *Remote Sensing of Environment*, **14**, 169–182.

Pinty, B., and Verstraete, M.M., 1991, GEMI: a non-linear index to monitor global vegetation from satellites. *Vegetatio*, **101**, 15–20.

Plummer, S.E., North, P.R., and Briggs, S.A., 1994, The angular vegetation index: an atmospherically resistant index for the second along track scanning radiometer (ATSR-2), *Proceedings of the 6th International Symposium on Spectral Signatures and Physical Measurements in Remote Sensing*, Val d'Isere, France, 717–722.

Pons, I., and Le Toan, T., 1994, Assessment of the potential of ERS-1 data for mapping and monitoring a mangrove ecosystem. *Pro-*

ceeding of the First ERS-1 Pilot Project Workshop, Toledo, Spain, June 1994 (European Space Agency SP-365), 273–282.

Populus, J., and Lantieri, D., 1991, Use of high resolution satellite data for coastal fisheries. *FAO RSC Series No. 58*, Rome.

Populus, J., and Lantieri, D., 1990, High resolution satellite data for assessment of tropical coastal fisheries. Case study in the Philippines. *Proceedings of the International Workshop on Remote Sensing and Insular Environments in the Pacific: Integrated Approaches*, ORSTOM/IFREMER, 523–536.

Ranganath, B.K., Dutt, C.B.S., and Manikan, B., 1989, Digital mapping of mangrove in middle Andamans of India. *Proceedings of the 6th Symposium on Coastal and Ocean Management*, **1**, 741–750.

Reark, J.B., 1975, A history of the colonisation of mangroves on a tract of land on Biscayne Bay, Florida. In *Proceedings of the International Symposium on the Biology and Management of Mangroves*, edited by G.E. Walsh, S.C. Snedaker and H.J. Teas, Gainesville, Florida, Institute of Food and Agricultural Sciences, University of Florida, 776–804.

Ross, P., 1975, The mangroves of South Vietnam, the impact of military use of herbicides. In *Proceedings of the International Symposium on the Biology and Management of Mangroves*, edited by G.E. Walsh, S.C. Snedaker and H.J.Teas, Gainesville, Florida, Institute of Food and Agricultural Sciences, University of Florida, 695–709.

Roy, P.S., 1989, Mangrove vegetation stratification using Salyut 7 photographs. *Geocarto International*, **3**, 31– 47.

Saenger, P., and Hopkins, M.S. 1984, Obervations on the mangroves of the southwestern Gulf of Carpentaria, Australia. In *Proceedings of the International Symposium on the Biology and Management of Mangroves,* edited by G.E. Walsh, S.C. Snedaker and H.J.Teas, Gainesville, Florida, Institute of Food and Agricultural Sciences, University of Florida, 126–136.

Sherrod, C.L., and McMillan, C., 1981, Black mangrove, Avicennia germinans, in Texas, past and present distribution. *Contributions in Marine Science*, **24**, 115–131.

Shines, J.E., 1979, *Distribution of mangrove communities in the State of Florida by Landsat MSS* (Las Vegas: US Environmental Protection Agency).

Siddiqui, M., and Shakeel., M., 1989, Digital techniques applied to high resolution satellite data in the identification and mapping of mangrove forests along the coast near Karachi. *Report of the Symposium on Remote Sensing Applications for Resource Development and Environmental Management.* ESCAP/UNDP Regional Remote Sensing Programme, Karachi, Pakistan, 159–167.

Siswandono, Utaminingsih, S., Purwadhi, F.S.H., Imami, S., Rosyid, Ongkosongo, O.S.R., and Kuchler, D., 1991, Digital inventory of fishpond resources of eastern Jakarta Bay (Indonesia) using MicroBRIAN digital mapping system and Landsat MSS image data. *Proceedings of the Regional Symposium on Living Resources in Coastal Areas, Manila, Philippines.* Marine Science Institute, University of the Philippines, 562.

Thollot, P., 1990, Remote sensing data contribution to the knowledge of inshore fisheries resources: mangrove fishes of the South West lagoon of New Caledonia. *Proceedings of the International Workshop on Remote Sensing and Insular Environments in the Pacific: Integrated Approaches*, ORSTOM/IFREMER, 233–244.

Untawale, A.G., Sayeeda, W. and Jagtap, T.G., 1982, Application of remote sensing techniques to study the distribution of mangroves along the estuaries of Goa. In *Wetlands Ecology and Management* (India: Lucknow Publishing House), pp. 52–67.

Vibulsresth, S., Downreang, D., Ratanasermpong, S., and Silapathong, C., 1990, Mangrove forest zonation by using high resolution satellite data. *Proceedings of the 11th Asian Conference on Remote Sensing*, D-1-6.

Vits, C., and Tack, J., 1995, The use of remote sensing as information source for environmental accounting of coastal areas in Kenja. *Feasibility Study Reference No. T3/02/603*, University of Ghent, 1–45.

Welch, R., and Ehlers, W., 1987, Merging multiresolution SPOT HRV and Landsat TM data. *Photogrammetric Engineering and Remote Sensing*, **53**, 301–303.

Woodfine, A.C., 1991, North east Sumatra prawn project, remote sensing component. *Final Report to NRI/ODA*, December 1991.

Woodfine, A.C., 1993, Mangrove mapping – a review of potential sources in North Sumatra. In *The Malacca Straits Coastal Environment and Shrimp Aquaculture in North Sumatra Province*, edited by C.A. McPadden (London: Overseas Development Administration), pp. 80–93.

PLATE 9

Figure 10.1 The process of visual interpretation.

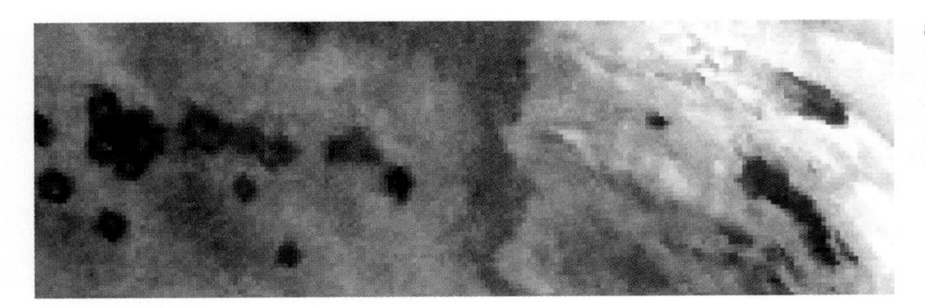

a. Part of a Landsat Thematic Mapper image (with TM bands 3, 2 and 1 displayed on the monitor's red, green and blue guns respectively) from a 1.6 x 0.6 km area of the Caicos Bank. The image contains reef, sand, seagrass and algal dominated areas.

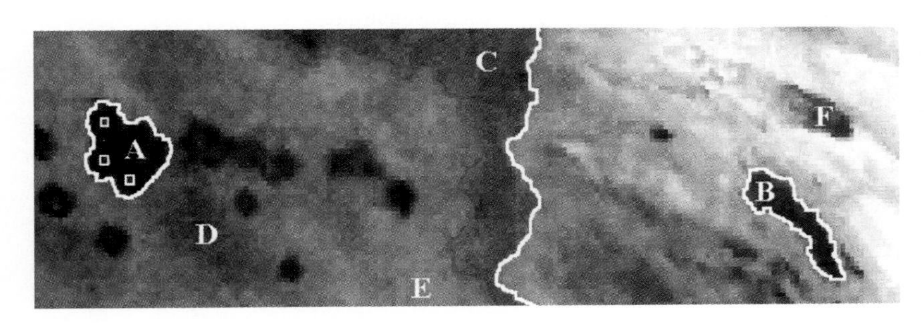

b. Visual interpretation is used to delineate some habitat boundaries on the Landsat image above. Polygon A = a reef patch; polygon B = an algal dominated patch; C and D overlie seagrass areas; E overlies sand; F overlies an algal patch.

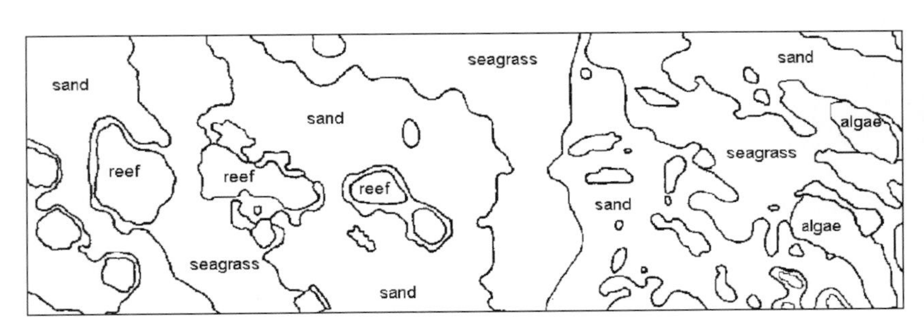

c. Habitat map derived from the Landsat image by visual interpretation. Boundaries between the habitat polygons have been digitised by hand and therefore the map is in vector format, although conversion to a raster format would be straightforward with most image processing software. The type of habitat within each polygon has to be interpreted from the colour and texture of the original image (with reference to field data where available).

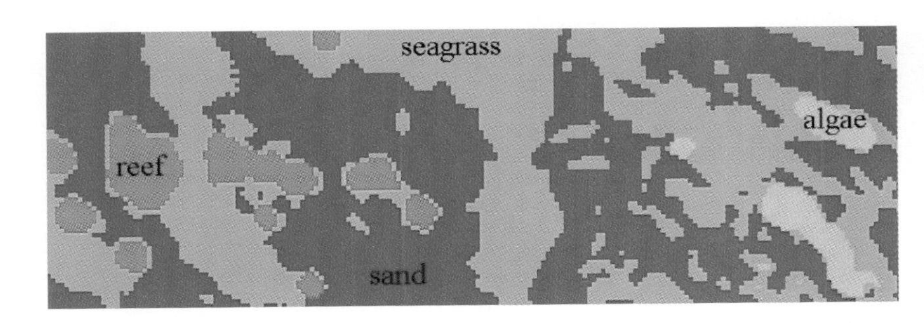

d. A rasterised version of (c) with different habitats displayed in different colours.

Figure 10.4 Unsupervised classification of the area shown in Figure 10.1 (Plate 9).

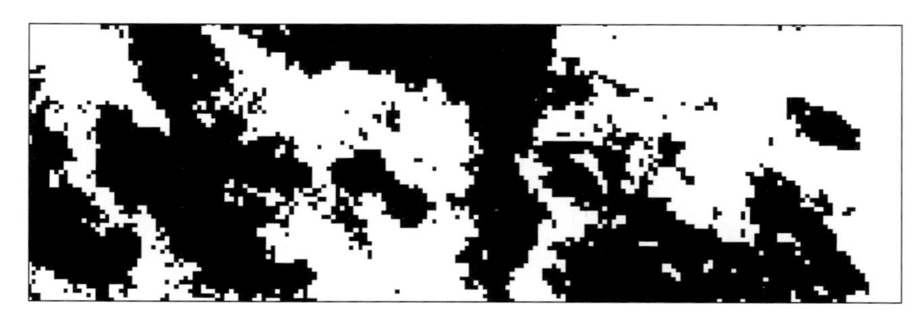

a. Unsupervised classification into two classes.

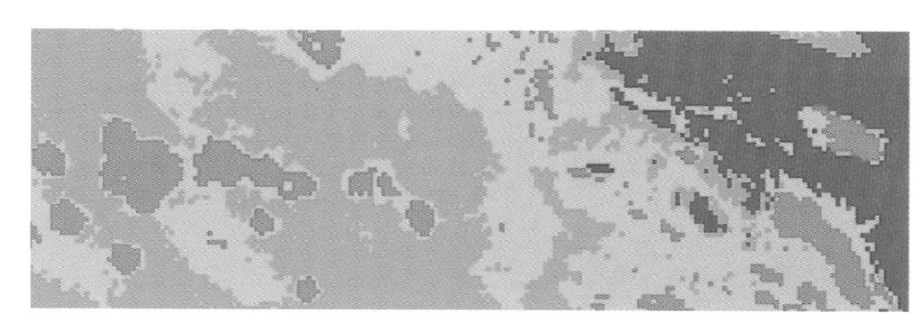

b. Unsupervised classification into four classes.

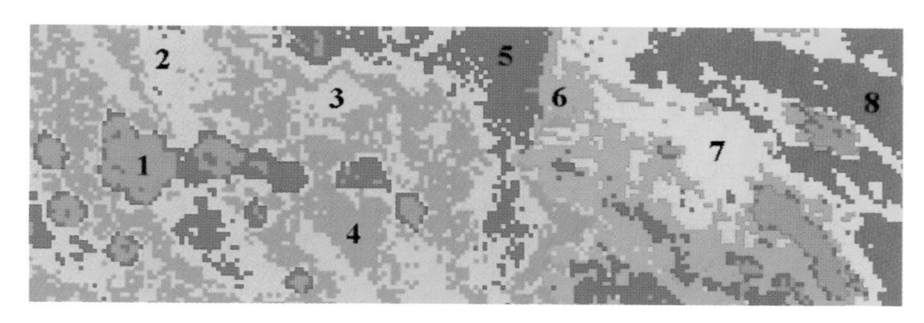

c. Unsupervised classification into eight classes (labelled 1 to 8). See Table 10.1 for further information on hypothetical field data collected from 20 sites in each class.

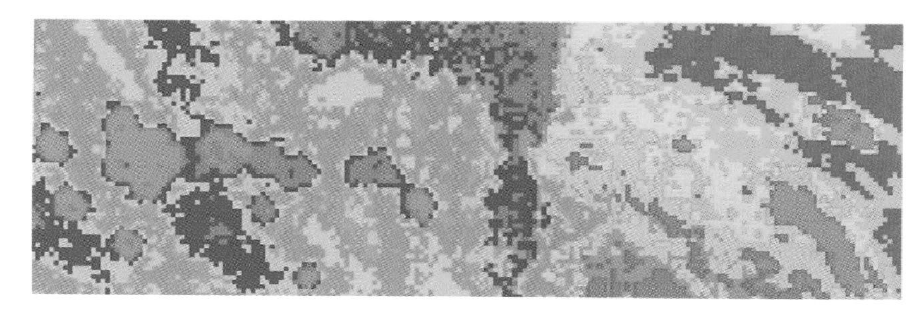

d. Unsupervised classification into twelve classes.

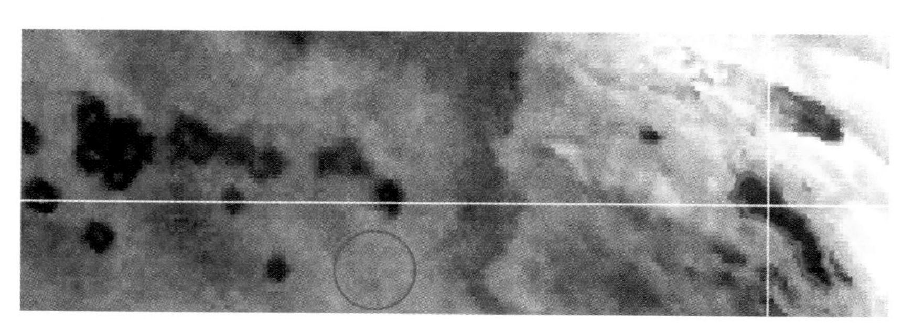

Figure 10.5 The selection of training samples for supervised classification of the area in Figure 10.1 (Plate 9). These can be selected by seeding a pixel at a known location and growing a training area around it (red polygon). An alternative method is to select pixels from within a vector layer: in this case the pixels within a (red) circle 0.5 km from a known location are known to be sand.

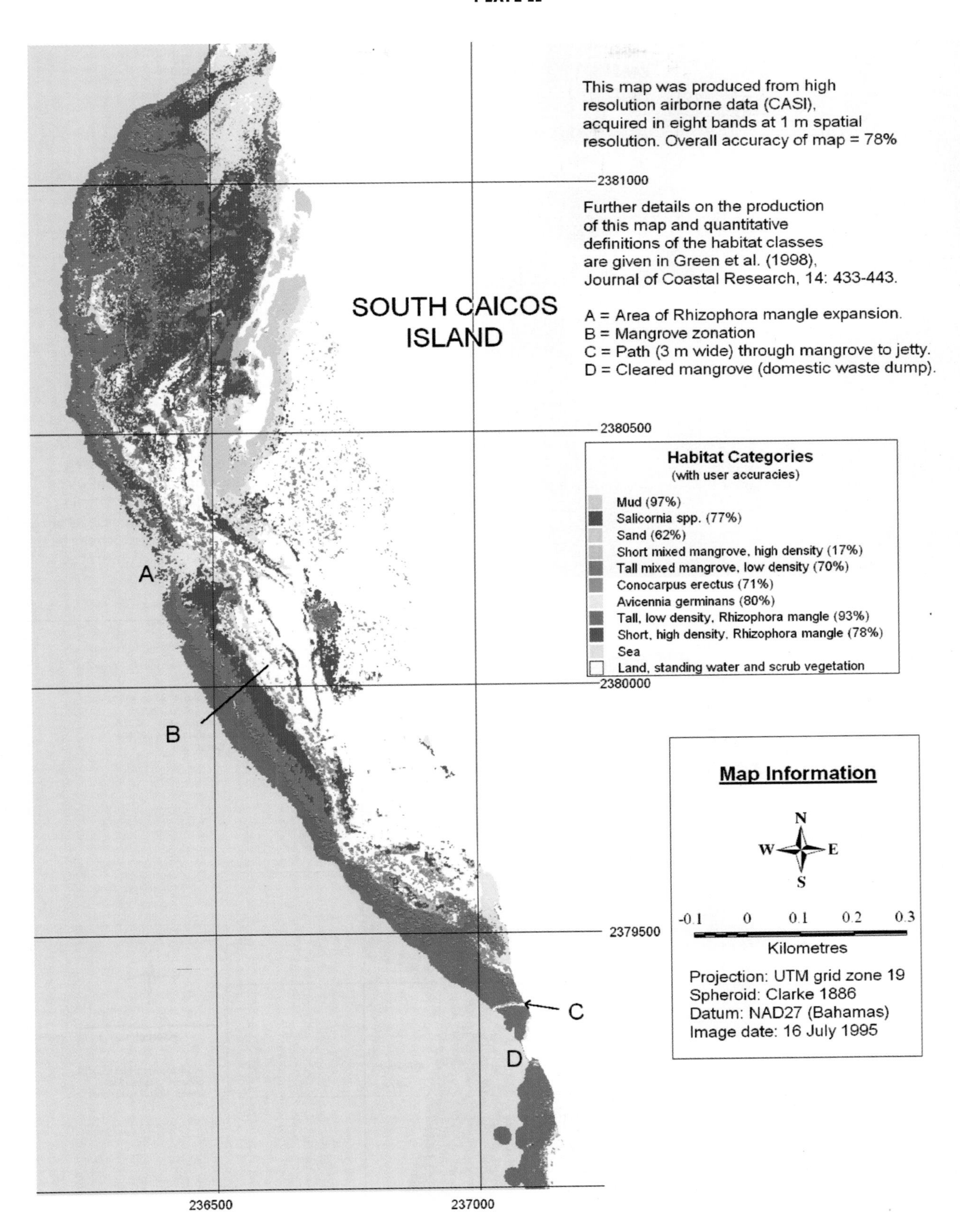

Figure 10.9 Mangrove habitat map derived from Compact Airborne Spectrographic Imager (CASI) image shown in Figure 10.8 using the signature file in Table 10.7.

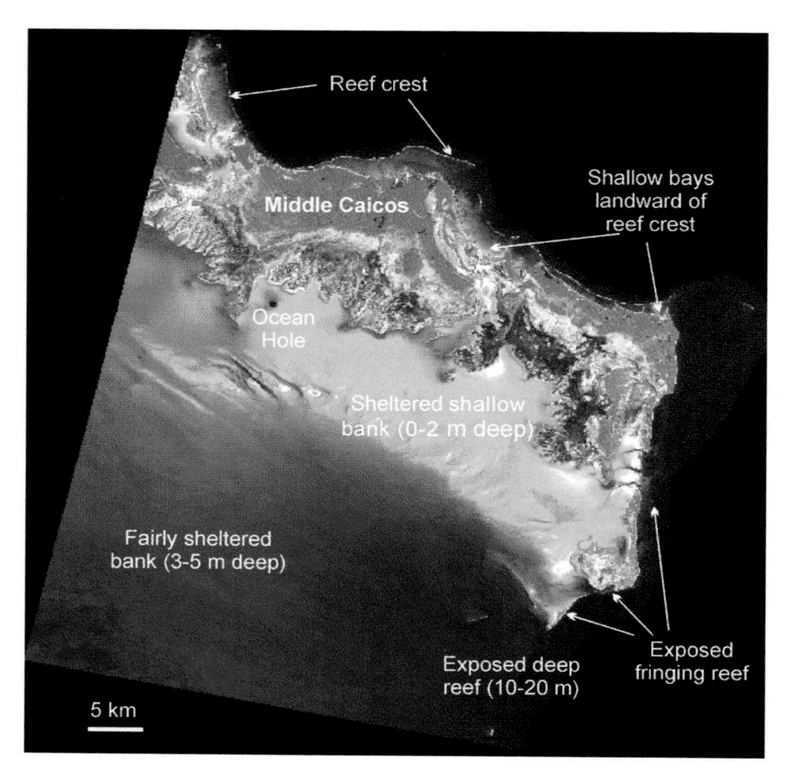

Figure 11.3 The principal physical environments of the Caicos Bank. Colour composite image (XS1 in blue, XS2 in green, XS3 in red) derived from full SPOT XS scene.

Figure 11.4 Broadscale habitat map (4 classes only) based on a supervised classification of the whole Caicos Bank using Landsat TM imagery that has been subject to water column correction and contextual editing. Even at this coarse level of habitat discrimination the best overall accuracy which can be achieved is only 73%. For SPOT XS an overall accuracy of 67% was achieved for the same level of processing.

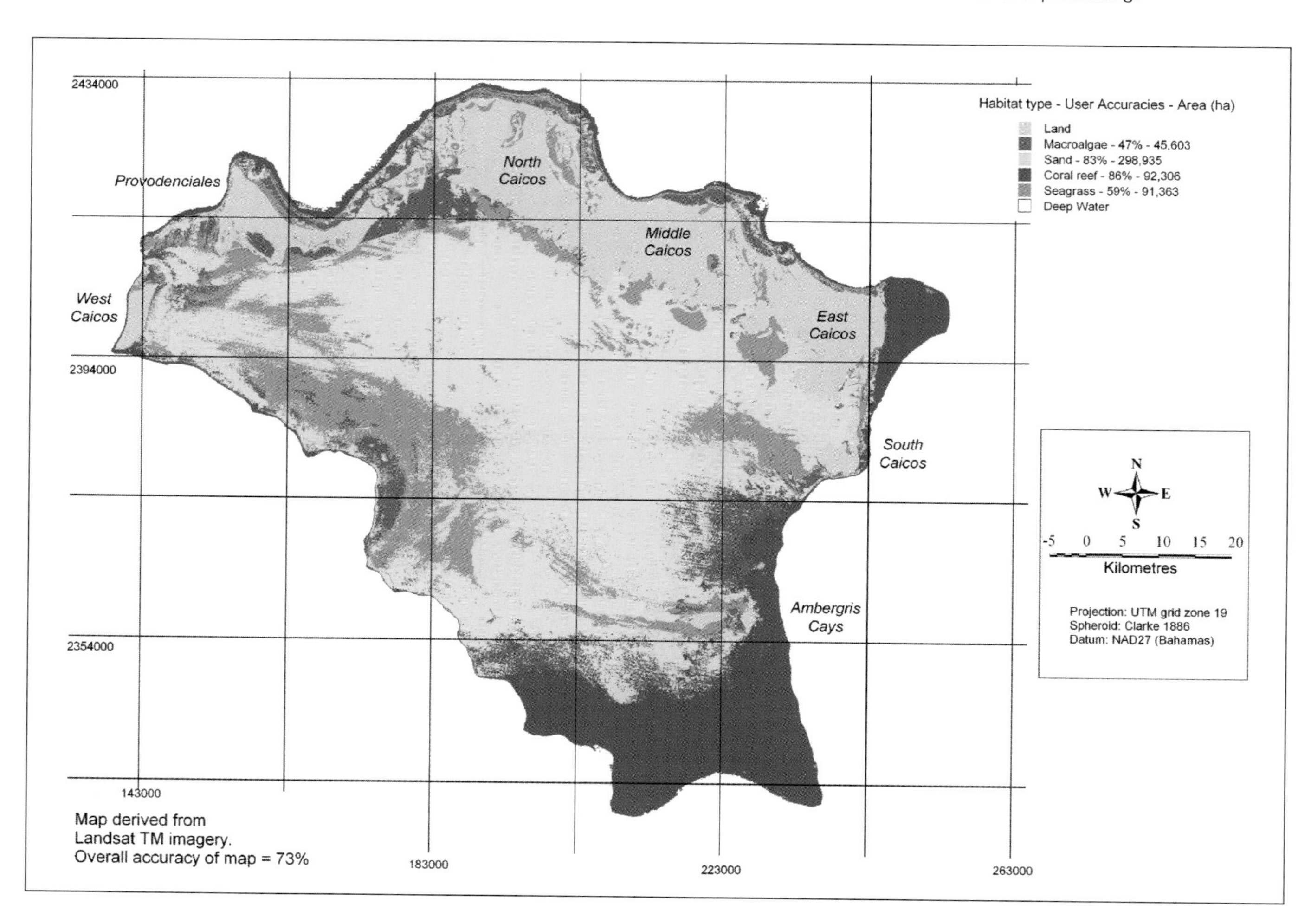

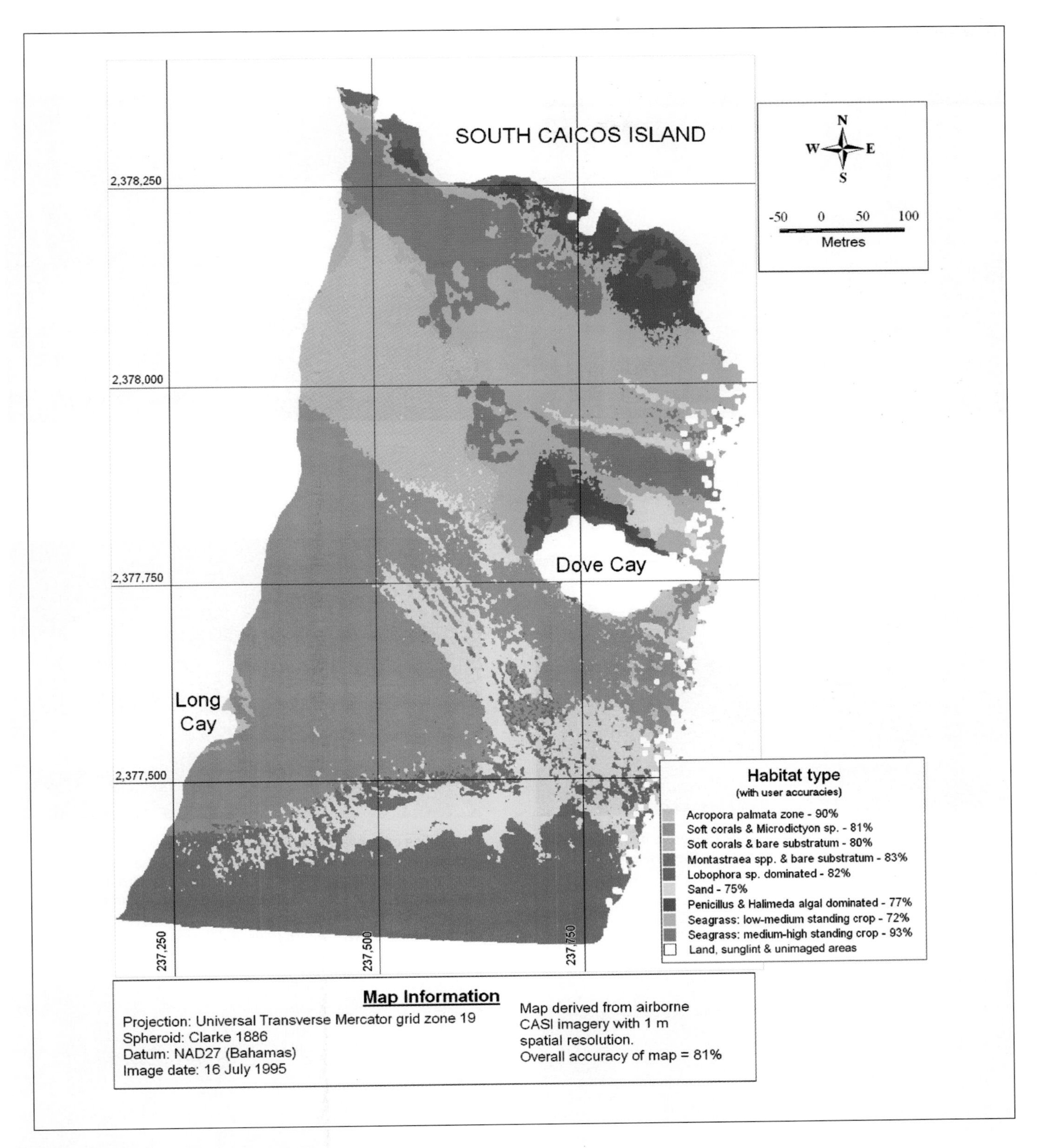

Figure 11.8 Habitat map of Cockburn Harbour, South Caicos based on a supervised classification of CASI data which has been subjected to water column correction and contextual editing. Land and sunglint (black rectangles) have been removed. Overall accuracy of map = 81%.

Figure 11.11 Estimating percentage cover using one-metre quadrats. Each 10 cm x 10 cm grid represents 1% of cover. [*Photographs*: E.Green, A. Edwards]

Figure 12.1 Seagrass habitats

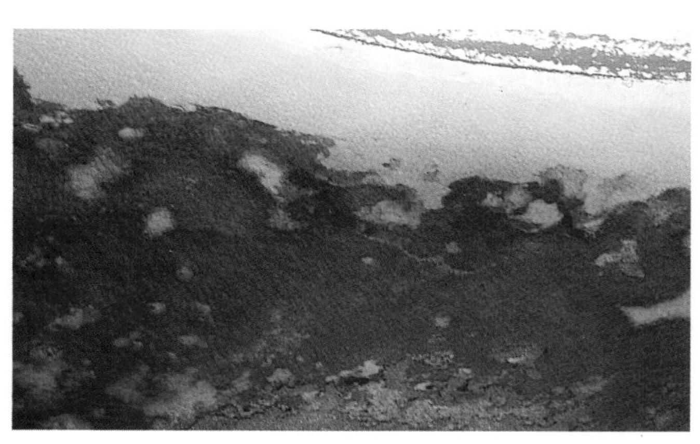

An aerial view of dense *Thalassia*-dominated seagrass beds off the east coast of South Caicos Island at depths of 1–3 m. Seagrass leaf-litter can be seen on the beach and a patch reef offshore. [*Photograph:* P. Mumby]

The eroding edge of a dense *Thalassia* (Turtle grass) meadow in Cockburn Harbour off South Caicos to emphasise that the standing crop (amenable to remote sensing) represents only about 25% of the total biomass of the seagrass. The remainder of the biomass is comprised of the extensive rhizome and root network, the edge of which is here exposed. [*Photograph:* A. Edwards]

A *Syringodium* (Manatee grass) dominated seagrass meadow at about 2 m depth on the central Caicos Bank. [*Photograph:* A. Edwards]

A dense mixed meadow of Turtle grass (*Thalassia*) and Manatee grass (*Syringodium*) with *Thalassia* dominating. [*Photograph:* A. Edwards]

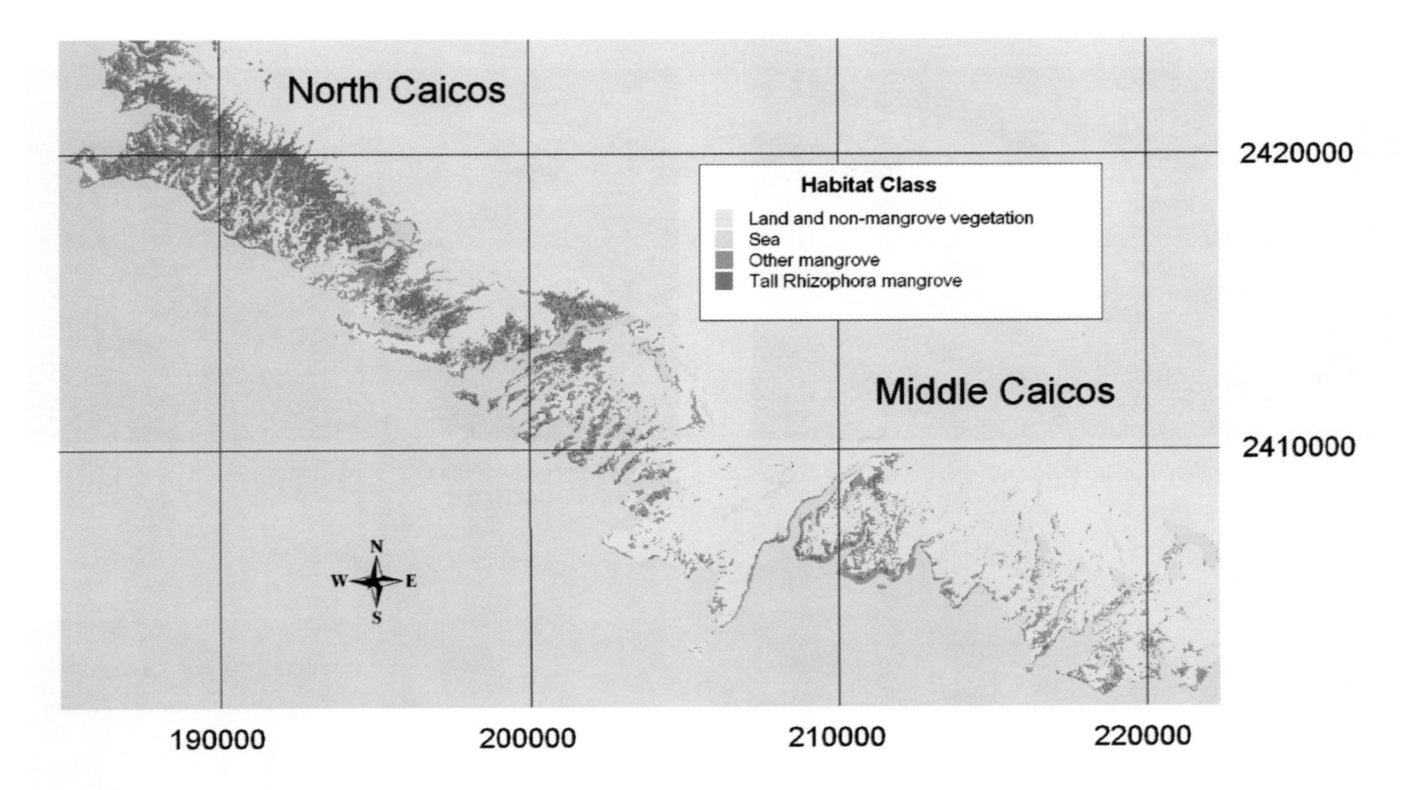

Figure 13.4 A classified map of the mangroves of North and Middle Caicos derived from Landsat TM data using method V. This area is a Ramsar (Unesco 1971) wetland reserve and contains the most extensive mangrove stands to be found on the Caicos Bank. The coordinates are Universal Transverse Mercator and so each square represents 10 x 10 km. With only 3 classes (tall *Rhizophora*, other mangrove (not 100% *Rhizophora*, <2 m high), and non-mangrove covered land), overall accuracy was about 83%. However, trying to map 9 mangrove habitat classes with this satellite sensor resulted in only 31% accuracy.

An oblique aerial photograph of mangrove habitats typical of the sheltered south coast of Middle and East Caicos. [*Photograph*: P. Mumby]

Well-developed tall *Rhizophora* on a cay in Belize serving as a roosting area for Frigate birds and Brown Boobies. [*Photograph*: P. Mumby]

PLATE 16

Freshly spilled crude oil spreading close to the coast. [*Photograph:* ITOPF]

Coastal seagrass beds can be mistaken for oil. [*Photograph:* ITOPF]

Part of the Gulf oil spill of 1991 being blown south past a headland on the Saudi Arabian coast. [*Photograph:* Jon Moore, Cordah Ltd]

An oiled stand of *Avicennia* mangrove which is dying as a result. The oil-coated breathing-roots (pneumatophores) are clearly visible. [*Photograph:* ITOPF]

Thick crude oil coating a beach following the Gulf oil spill in 1991. The thin oil sheen on the waters offshore is clearly visible. [*Photograph:* Jon Moore, Cordah Ltd]

Freshly oiled *Rhizophora* mangrove with prop-roots thickly coated in oil. [*Photograph*: ITOPF]

Part 4

Mapping Depth and Water Quality

14

Monitoring Coastal Water Quality

Summary *This chapter reviews the extent to which remote sensing can contribute to the monitoring of pollutants (substances introduced directly or indirectly into the marine environment by Man which result in deleterious effects) in coastal waters. The pollutants reviewed include sediments, oil, industrial wastes, sewage and thermal discharges. Toxic algal blooms which may be promoted by Man's activities are also discussed. On the one hand, there is a mismatch between the temporal and spatial scale of the majority of pollution events and the information provided by satellite sensors which makes these of limited use for monitoring. On the other hand, although airborne sensors can deliver data of an appropriate temporal and spatial scale, such data tend to be expensive to collect and may often not be cost-effective. Thus, whether to use remote sensing for monitoring coastal water quality needs to be assessed very carefully.*

Relative concentrations of sediment in coastal waters can be fairly easily mapped using remote sensing but retrieving absolute concentrations is much harder. However, remote sensing clearly provides useful synoptic information on sediment plumes and transport along the coast. Considerable research has been carried out into the remote sensing of marine oil pollution using satellite and airborne optical, infra-red, ultra-violet and radar sensors. Optical sensors on satellites are not good at discriminating oil reliably and apart from AVHRR tend to have too poor temporal resolution to be of much operational use. Thermal infra-red sensors can be effective under certain conditions with AVHRR bands being used to discriminate large slicks in the Gulf War of 1991. Infra-red (IR) sensors can detect thicker oil spills but cannot detect thin oil (< 10–70 μm) or sheen. Airborne ultra-violet (UV) sensors can detect sheens and a combination of IR and UV sensors may allow better discrimination of oil from substances which behave similarly. In terms of routine monitoring of oil pollution, Synthetic Aperture Radar (SAR) appears to offer the most promising way forward. It can operate through cloud and, although there are still some problems in terms of temporal resolution and the reliability of discriminating oil from other phenomena, these problems seem likely to be resolved in the near future.

For point source discharges, such as those of industrial wastes, sewage or thermal effluents, the main problem from an operational point of view is the lack of an appropriate combination of temporal and spatial resolution in satellite imagery and the cost of routine airborne remote sensing. Although remote sensing can contribute usefully to monitoring such discharges into coastal waters and provide a synoptic view to aid interpretation of field data and interpolate to unsampled areas, it is essentially an add-on. For monitoring, enforcement and legislative purposes, environmental protection agencies or pollution control boards will need in situ *measurements of any pollution. This is costly in itself and the economic case to augment these data with remote sensing needs careful scrutiny.*

Toxic algal blooms of dinoflagellates and cyanobacteria appear to be increasing in incidence and can lead to the death of finfish and shellfish resources and have major economic impacts on coastal aquaculture. The blooms are readily detectable by a wide range of satellite and airborne sensors but, although existing technology is more or less capable of operational monitoring to allow early warning of blooms, it is unclear who would pay the substantial costs involved.

The use of remote sensing to monitor sea-surface temperature anomalies, such as those correlated with widespread 'bleaching' of corals during years of strong El Niño Southern Oscillation (ENSO) events, is also briefly reviewed.

Introduction

This chapter presents a review of how remote sensing may aid coastal managers in assessing and monitoring coastal water quality. Pollution of coastal waters (leading to poor water quality) by, for example:

- sediment run-off as a result of deforestation or agricultural development,
- fertilisers and pesticides from agriculture,
- oil spills from land-based or marine installations,
- thermal discharges from power-stations or other industry requiring cooling water,
- discharges of insufficiently treated sewage,
- toxic and other industrial wastes, and
- nutrient-rich effluents from coastal aquaculture ponds,

are just some of the issues which coastal managers may be called upon to assess or monitor. Such pollutants (loosely defined here as substances in the water column which are likely to have detrimental effects on the environment at the levels observed – see Box 14.1 for a more formal definition) may have adverse impacts on coastal ecosystems and the economically important resources that these support. Thus, monitoring of 'water quality' is in practice about detecting, measuring and monitoring of pollutants.

BOX 14.1

Definition of Marine Pollution

Pollution means the introduction by Man, directly or indirectly, of substances or energy into the marine environment (including estuaries) resulting in such deleterious effects as harm to living resources, hindrance to marine activities including fishing, impairment of quality for use of seawater and reduction of amenities (GESAMP, 1991).

Satellite remote sensing has been shown in Chapters 9–12 to be a useful tool in mapping broadscale coastal habitats which, apart from during catastrophic events such as tropical cyclones, tend to change over a timescale of months, to years or decades. Water quality on the other hand is far more ephemeral and dynamic, often changing dramatically over a tidal cycle, or after a monsoon downpour, or following the flushing of an aquaculture pond or the discharge of a tanker's ballast water. Thus the usefulness of water quality measurements at a single point in time (as available from a single satellite image) is likely to be far less than, say, knowledge of the habitat type at a particular location.

Since most incidents of pollution are likely to be short-lived (hours to days) and relatively small in size (tens to hundreds of metres across) one needs high temporal resolution and moderately high spatial resolution in order to monitor them (see Chapter 3). Further, since the pollutants can only be detected and measured if they can be differentiated from other substances in the seawater, one usually also needs a sensor of high spectral resolution or LIDAR (Light Intensity Detection and Ranging) that uses powerful lasers to stimulate fluorescence and detects and analyses the re-emitted light, which will be characteristic of the substances illuminated.

Such a combination of characteristics is at present only available from airborne remote sensors. Satellite sensors are thus only likely to be useful in coastal water quality assessment where mapping large scale sediment plumes from estuaries or huge oil spills such as that released during the Gulf War of 1990/1991. This means that application of effective remote sensing techniques in pollution monitoring is likely to be expensive (costing one thousand to several thousand dollars a day) because of the costs of hiring aircraft to fly the required sensors and the probable need for repeat surveys to establish the dynamics of the pollution in space and time. These high costs, the relatively small areas which need to be investigated (particularly where point-sources are involved), and the need to collect field survey data for calibration (establishing a quantitative relationship between the amount of pollutant and some characteristic measured by the remote sensor) anyway, may make the cost-effectiveness of using remote sensing to aid pollution monitoring questionable.

Further it must be made clear that remote sensing can, at present, only directly detect some of the pollutants listed above, for example, oil, sediment and thermal discharges. The others can in some cases be detected by proxy but only via ephemeral site-specific correlations between remotely sensed features and the pollutants in question. Thus fertilisers, pesticides, nutrients and colourless toxic and industrial wastes (containing, for example, heavy metals, aromatic hydrocarbons, acid, etc.) cannot in general be detected but in some instances their concentrations (measured by field survey at the time of image acquisition) may be correlated with levels of substances such as suspended particulate matter, *Gelbstoff* (yellow substance), or phytoplankton pigments which can be detected. Such experimentally demonstrated ephemeral relationships are still far from allowing operational detection and monitoring of these pollutants in a cost-effective way. The particulate matter and yellow substance in untreated sewage and aquaculture pond effluents allow these to be mapped by proxy using remote sensing but considerable field survey will be required to produce useful quantitative data and this is likely to be site specific. Finally, it must be emphasised that remote sensing only sees the surface layers of a complex three-dimensional environment and major assumptions have to be made in extrapolating from what is measured at the surface to what is in the water column as a whole. If there is stratification or other heterogeneity, such extrapolations may be misleading.

Whilst remembering these shortcomings, the primary benefits from adding a remote sensing perspective are:

- rapid survey for pollutants over large or remote areas which could not feasibly be surveyed from the surface,
- a synoptic overview of the pollution and clear demarcation of its boundaries, and
- once reliable calibration has been achieved, the potential for measurement of the pollutant without the need for more field survey.

We suggest that coastal managers seeking to employ the techniques described below to assist in water quality monitoring should carefully weigh up these benefits and the added costs, against the costs of a conventional field-based techniques before embarking on such a survey.

Bearing this *caveat* in mind, for each of the major types of pollution listed above, the ways in which remote sensing technologies have been applied to aid monitoring are discussed (Figure 14.1). Since the water quality around our field study area in the Turks and Caicos Islands was more or less uniformally excellent, we did not have the opportunity to test these techniques and so cannot offer practical guidance in this field. This chapter thus reviews previous work in a coastal management context and builds on the reviews of Muller-Karger (1992) and Clark (1993).

Given the recent widespread coral 'bleaching' during 1998 which has been related to warm-water anomalies (correlated with a strong El Niño Southern Oscillation (ENSO) event), a section on satellite monitoring of sea-surface temperature anomalies is also included at the end of the chapter.

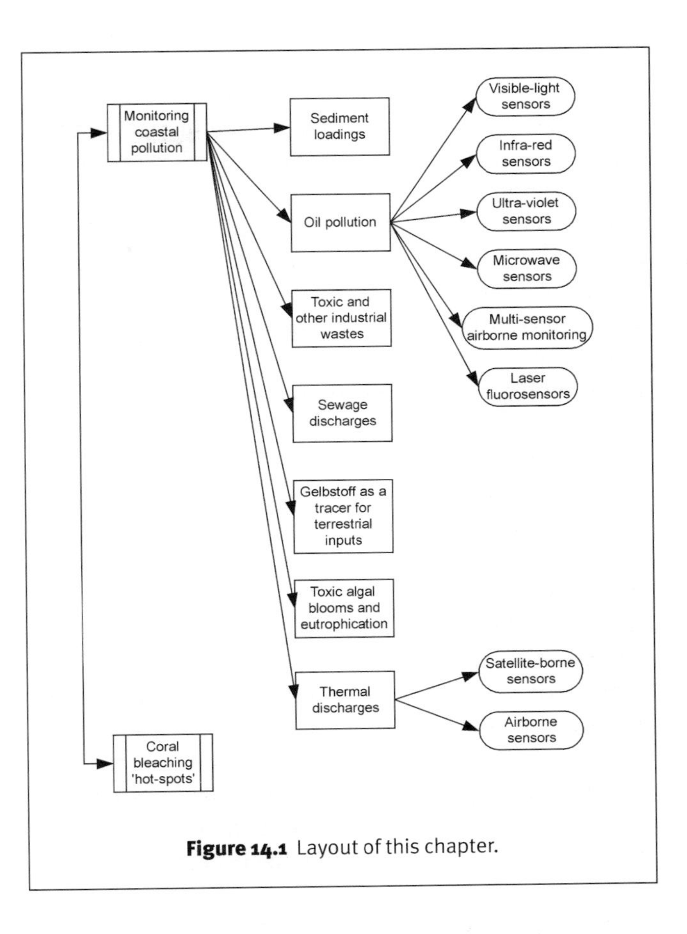

Figure 14.1 Layout of this chapter.

Monitoring coastal pollution

There are a number of ways in which remote sensing may assist in the monitoring of marine pollution:

- detection (presence or absence of a pollutant, e.g. detect whether a tanker has discharged oil-contaminated ballast water),
- quantitative mapping (where a significant relationship between a signal at the sensor and the concentration of the pollutant measured in the water can be established),
- tracking, or pattern of dispersal (where repeated imaging allows the movement of the pollutant to be mapped and perhaps inform response decisions, e.g. in an oil spill),
- damage assessment (for instance, mapping of the extent of oiling of beaches or the extent of toxic algal blooms).

Each potential use should be weighed against alternative methods of achieving the same goals. For example, good software is available for predicting the tracks of oil spills given a knowledge of wind and current patterns. It may be more cost-effective to model than to measure.

Sediment loadings

High sediment loads in normally clear coastal waters can be a problem to sea-bottom communities not adapted to such conditions. For example, anthropogenic increases in sedimentation on coral reefs may lead to death of corals and reef degradation. Such rises in sediment loadings may be the result of land-based activities such as urban development, deforestation or increased agricultural activity in river catchments (Lavery *et al.* 1993), aquaculture development (Populus *et al.* 1995), clear-cutting of coastal mangroves or poorly planned coastal development close to reef areas (Ibrahim and Bujang 1992). They may also result from marine activities such as nearshore dredging, loss of seagrass beds, or increased boat traffic in very shallow water stirring up sediment. The eventual deposition of such sediments may also have impacts such as siltation of harbours, accumulation of sand bars to create navigational hazards, or seasonal blockage of estuaries with associated impacts on estuarine communities (e.g. Minji and Yawei 1990, Caixing *et al.* 1990).

Marked reductions in sediment loadings in normally turbid coastal waters may also have adverse environmental impacts, leading to erosion. Damming of rivers reduces sediment supply to estuaries and can have major erosional impacts on adjacent shorelines. Thus detection by remote sensing of sediment transport (Peck *et al.* 1994) and changes in the amount of sediment suspended in the water would clearly be of benefit to coastal managers, whether these changes are increases or decreases. Can remote sensing help monitor such changes in suspended sediments?

Satellite remote sensing can readily detect suspended sediment and the size of areas one might wish to monitor is often large enough for it to be useful. Sediment plumes should generally be easy to detect on AVHRR, SeaWiFS, Landsat TM and MSS and SPOT imagery (see Chapter 3 for details of sensors). Landsat MSS synoptic views have been used to infer water circulation patterns and sediment transport and then identify areas of the coast susceptible to erosion (Welby 1978). In comparison to ship-borne sampling which may yield accurate concentration measurements but extremely poor spatial coverage, remote sensing is excellent at mapping the pattern and extent of sediment plumes. However, deriving accurate concentrations is difficult. Ideally water sampling to measure sediment concentrations should be undertaken during the satellite overpass in order to calibrate the image. However, this is logistically problematic and seldom an operational option although successfully done for several experimental studies (e.g. Braga *et al.* 1993, Pattiaratchi *et al.* 1994).

If accurate concentration measurements are not required then it is relatively easy to process a visible waveband of an image by simple contrast stretching and to density slice the resulting data so that a gradation from clear water through slightly turbid to maximum concentration is displayed. This provides useful information on location and extent of sediment plumes at the time the image was acquired but this will of course vary with tides and currents. It also provides a relative assessment of concentration.

To retrieve absolute concentration data from radiances is much harder. The most common approach has been to measure suspended sediment concentrations in the field, at the time of the satellite or airborne overpass, and to seek statistical relationships between these values and the image radiance values. Firstly, atmospheric correction (see Chapter 7) is required to convert the radiance received at the sensor to the actual upwelling radiance from the surface. This introduces a variable degree of uncertainty as precise atmospheric conditions are rarely known. Secondly, regression equations are sought that permit the image radiance values to be converted to a suspended sediment concentration measurement. Whitlock *et al.* (1982) provide some useful guidelines for the use of regression analysis for remote sensing of suspended sediments and pollutants which we recommend should be consulted before planning any such study. Also, Chen *et al.* (1991) provide useful advice on the form of the relationship between concentration and radiance that permits a better description of the regression if only limited samples of concentration data can be collected.

Relationships between suspended sediment concentrations (usually obtained by filtering water samples) or a related parameter Secchi disk depth (a measure of water clarity) and upwelling radiances have been investigated using a variety of sensors and algorithms, for example, Landsat MSS (Johnson 1978, Delu and Shouren 1989), Landsat TM (Yu-ming and Hou 1990, Braga *et al.* 1993, Lavery *et al.* 1993, Pattiaratchi *et al.* 1994, Populus *et al.* 1995), Landsat MSS and TM (Minji and Yawei 1990), airborne MSS (Johnson and Bahn 1977), airborne TM (Rimmer *et al.* 1987), airborne CASI (Vallis *et al.* 1996), CZCS (Tassan and Sturm 1986), SPOT XS (Froidefond *et al.* 1991, Populus *et al.* 1995), and NOAA AVHRR (Althuis and Shimwell 1994, Woodruff and Stumpf 1997). However, such relationships tend to be sensor, site and season specific and no universal algorithm has been forthcoming. Topliss *et al.* (1990) reported some success in developing an algorithm using airborne data that may be reasonably applied to Landsat MSS and CZCS data, thus overcoming the sensor dependency, but not the site and season dependency. Similarly, Tassan (1993) reported an improved algorithm for Landsat TM data in coastal waters which seems likely to cope with seasonal fluctuations but remains sensor and site dependent. Further, he recently presented a numerical model which uses Landsat TM data to determine sediment concentration in stratified river plumes (Tassan 1997).

It is perhaps not surprising that a general relationship between sediment concentration and the sensor signal cannot be found. There is likely to be a great deal of variation in the lithology (rock type) and grain size of suspended sediment, which will be further confounded by other factors which affect ocean colour (what the sensor sees), such as phytoplankton and dissolved organic matter. Hinton (1991) accepted that different algorithms would be required for different locations and proposed the use of eigenvector analysis of radiance spectra to determine the type and composition of suspended sediment which may be used to select an appropriate algorithm, without actually sampling the sediment itself. In a slightly different approach, Mertes *et al.* (1993) developed a method based on spectral mixture analysis to estimate suspended sediment concentrations in wetland surface waters. Their results focused on the Amazon using Landsat TM and MSS images but the methodology can be applied universally if the optical properties of the water and sediment are known.

Nanu and Robertson (1990) sought to abandon the empirical approach of regression and considered the physical processes of light penetration at different depths. However, at present for practical purposes the relationship between suspended sediment concentrations and upwelling radiance should be regarded as site and season specific, so that if absolute concentration values are required then field sampling must be incorporated in the study.

Case study: resort development on Pulau Redang, Malaysia

In a coastal management related case study an attempt was made to use remote sensing to monitor the impact of a commercial tourist resort development within a marine park around Redang Island in eastern peninsular Malaysia (Ibrahim and Bujang 1992). The substantial resort development comprised an 18-hole golf course, several 100–250 room hotels, a horse ranch, 45 holiday villas, condominium complex, presidential chalets, staff quarters, tourist centre, sports complex and supporting infrastructure on 265.7 ha of the approximately 9 x 12 km island. There was considerable concern that the scale of earthworks associated with the development would cause mangrove loss and widespread erosion and sedimentation on the sensitive fringing coral reef ecosystems. The latter are a prime attraction for tourists as well as being within a marine park.

Two Landsat TM images, one from prior and one during development, were geometrically corrected so that they could be co-registered and used for change detection analysis. Unfortunately the key factor (sedimentation) is most important during the rains of the North-east Monsoon (November to March) at which time cloud-cover is high and images were not available. The only reasonable post-development image that could be obtained was in August when there is negligible terrigenous run-off and little sedimentation, hence it was of little use in terms of assessing sedimentation changes. However, it was useful in indicating that 6 ha (or more) of mangrove had been cut and landfilled in the golf course construction as against 2.9 ha allowed in the conditions for approval of the development.

Ibrahim and Bujang (1992) concluded that remote sensing provided an effective, inexpensive and timely means of monitoring the impact on the ecosystems. However, their paper indicates that it was their field survey which provided the effective monitoring with respect to sedimentation. This showed around ten fold increases in sedimentation rates at one site close to the development area during the course of construction activities and a general increase of 2–10 fold in sedimentation rates near the development from what they were prior to construction. Sites distant from the development area did not suffer these changes.

This case study illustrates two points:

1. the ephemeral nature of water quality data means that the probability of obtaining a useful image (in terms of appropriate timing and acceptably low cloud cover) using visible wavelengths is much lower than that for habitat mapping, and
2. where land-based pollutants are primarily a problem during heavy rains (true for sediment and other agricultural run-off), the time when images are needed is the time when you are least likely to obtain them because of cloud cover. This is a particular problem in the humid tropics.

We consider that this case study is most useful in illustrating the limitations of remote sensing in measuring impacts and the need for extensive field survey.

Oil pollution

About 4 million tonnes of oil contaminate the world's oceans each year (Lean and Hinrichsen 1992). About half is derived from shipping and other marine activities and half from land-based sources. Some 75% of this pollution falls into five categories: operational discharges from tankers (22%), municipal waste (22%), tanker accidents (12.5%), atmospheric inputs from industry (9.5%), and bilge and fuel oils (9.0%). Overall it is estimated that almost 1.1 million tonnes are deliberately discharged every year from tankers pumping out their tanks before taking on new cargoes. Oil can have significant impacts on the quality and productivity of coastal marine environments (notably mangroves, Plate 16), species dependent on those environments (e.g. seabirds), and endangered species such as turtles and dugong. A spill of 200 tonnes in the Wadden Sea in 1969 killed 40,000 seabirds. It can also have major economic impacts on tourism, desalination plants, aquaculture and fisheries and be very expensive to clear up. In a recent cost-analysis of major oil spills reported by the International Tanker Owners Pollution Federation the costs of clean-up operations after tanker accidents was found to range from $US 5000–30,000 per tonne of oil spilled. Claims for environmental damage to the marine environment often extended to tens of millions of dollars.

Remote sensing has been widely used for the detection and mapping of oil slicks (e.g. Massin 1984). It has been estimated that by using a multi-sensor remote sensing approach, clean-up costs could be reduced by at least 15–20% and damage to the marine environment significantly reduced (Barrett and Curtis 1992). Fingas and Brown (1997) provide a concise review of the use of remote sensing for detecting and monitoring oil pollution in the marine environment from an operational viewpoint. Remote sensing also has a role (via habitat mapping) in oil spill contingency planning and environmental sensitivity mapping prior to spills (Jensen *et al.* 1990).

It is useful to separate applications into two categories:

1. mapping of oil slicks after the event to aid oil-spill response teams and clean-up operations and hopefully reduce environmental impacts, and
2. routine monitoring of coastal waters to detect spills with the aim of improving policing and prosecution, and thus reducing inputs to the sea.

Publications relating to different satellite and airborne sensors and image processing techniques for oil pollution studies are far too numerous to review fully here. Most of the publications report on experimental results and the development of new techniques. An outline of these with some key references will be provided, but the operational use of remote sensing related to recent and major oil spills in the Arabian Gulf and Prince William Sound, Alaska, will be dealt with in more detail.

Oil is reasonably amenable to remote sensing investigation at a wide range of wavelengths from the ultra-violet to the microwave. Aircraft borne sensors have been used most frequently for operational studies of oil spills because the revisit time for most satellite borne sensors is too infrequent. However, a range of satellite sensors have been used to detect oil slicks using visible, infra-red and microwave wavelengths, with Synthetic Aperture Radar (SAR) being considered promising in an operational monitoring role by some. The goal is to be able to discriminate between oil slicks and other features with a high degree of certainty. The various techniques will be discussed with reference to this goal for the different classes of sensor (visible sensors, ultra-violet sensors, infra-red sensors, microwave sensors, and laser fluorosensors).

The emissivity of oil slicks differs from that of a calm sea surface, and there appears to be a relationship between emissivity and the specific gravity of oil (Barrett and Curtis 1992).

Visible-light sensors

Visible wavelengths are not thought particularly useful for the detection of oil owing to the often poor contrast between oil-covered and oil-free water surfaces. The reflectance of oil in the visible part of the electromagnetic spectrum is generally greater than that of clear water so that it can often be distinguished from the background. However, its reflectance is fairly flat across the visible wavelengths without distinguishing features that would allow positive identification. This makes it difficult to distinguish oil from sun-glint, wind sheens, and biogenic surface films. Also, in shallow water, it can be difficult from the air to distinguish floating oil from subsurface seagrass or macroalgal patches (Plate 16). Whilst recognising the use of airborne visible remote sensing to document oil spills, the consensus appears to be that its use will decrease, and that ultra-violet, infra-red and microwave sensors are more promising (Estes and Senger 1972, Lo 1986, Fingas and Brown 1997).

Optical remote sensing of oil using satellite sensors suffers both from the poor discrimination possible using visible wavelengths, the low frequency of overpasses of most satellites except NOAA/AVHRR (see Chapter 3), and the the problem of cloud cover. Despite the drawbacks, visible wavebands from Landsat MSS and TM (e.g. Stumpf and Strong 1974, Otterman *et al.* 1974, Deutsch and Estes 1980, Deustch *et al.* 1980, Stringer *et al.* 1992), and from meteorological satellites (Hayes 1980, Cracknell *et al.* 1983) have been used to map oil slicks.

Landsat MSS images exist from 1972 until 1993. As such they provide the longest time-series of image data that can be used to determine trends in oil pollution. Wald *et al.* (1984) demonstrated the capability of Landsat MSS data for producing a large-scale inventory of oil spills by examining 800 images of the Mediterranean region. The optical properties of an oil spill at sea are complex and dependent upon many factors including thickness, type, age and thermodynamic state of the oil, and sun height. As reflectance is also affected by the influence the oil has on sea state (localised calming) this also aids in identification. Ideally multispectral analysis and digital image processing should be used to identify slicks, but because the study by Wald *et al.* (1984) needed to cover the whole Mediterranean between 1972 and 1975, a compromise was made and visual interpretation of single band (MSS band 7) photographic prints was performed. They were able to identify oil slicks on the imagery although possible confusion between anthropogenic slicks and natural organic films was acknowledged. Their results indicated that most pollution arose from tanker rinsing rather than from large-scale spill events, and they produced a synoptic map of oil spill frequency that highlighted areas of maximum pollution hazard. Such studies provide base-lines against which longer-term changes can be compared.

Infra-red sensors

Dependent upon the time of day, thermal infra-red sensing can be effective at delineating an oil slick. This may be for a number of reasons: different heat capacity of oil and water; reduction of heat transfer across the air-water interface as a result of surface tension changes; or the infra-red energy may be modified by transmission through the surface film of oil (Lo 1986). Thick oil tends to absorb solar radiation and re-emit it as thermal infra-red energy in the 8–14 μm region of the electromagnetic spectrum. It thus appears hot relative to seawater during the day. Intermediate thickness slicks appear cooler than seawater during the day. This relative thickness information may be useful. Thin oil (< 10–70 μm thick) or sheens are not detectable in the infra-red. Infra-red sensors on both aircraft and satellites have been utilised for slick detection and mapping. As with visible sensors, there may be confusion with non-oil features. An infra-red camera and ancillary equipment for aerial surveys can be purchased off the shelf for under US$ 100,000 and would weigh about 50 kg. The AVHRR sensor carried on NOAA polar orbiting satellites (Chapter 3) provides global coverage on a daily basis in the infra-red which can detect large slicks (in the order of tens of square kilometres or larger) and, to those with receivers, is free.

An instructive case study of its use was provided during the Gulf War in 1991 (Legg 1991a, Cross 1992). Large quantities of oil were discharged into the north-western Arabian Gulf from the coast of Kuwait, reports at the time estimating the slick to be the largest in history (Cross 1992). The resultant slicks posed a major threat to desalination plants of adjacent Saudi Arabia and Bahrain as well as to a range of unique habitats along the Gulf coastline (Plate 16). It was assessed that a million birds and their habitat, 600 dugongs, coral reef islands, green turtles, shrimping grounds, fisheries and humpback dolphins were at risk (WCMC 1991). Near-real-time monitoring of slick movement was required and satellite remote sensing proved particularly effective in this case, as other means of monitoring its movements were problematic as it was in the heart of a war zone.

The United Nations Environment Programme (UNEP) coordinated an international effort to assess the potential consequence of the disaster and support the protection and recovery of the habitats. As part of this, the Global Resource Information Database (GRID) acquired the necessary remote sensing and map data, and Cross (1992) reports on the methods and results of the remote sensing investigation. For such an investigation the revisit period, or temporal resolution, is critical and the only appropriate sensor was the AVHRR on board the NOAA polar orbiting satellites. These provide five wavebands in the visible and infra-red (including thermal IR), a spatial resolution of 1.1 km and a temporal resolution of 12 hours. For both the Legg (1991a, 1991b) and Cross (1992) investigations it was apparent that the slick was only visible in the thermal wavebands of daytime images. It was suspected that the thick parts of the oil slick behaved as a black-body heating up during the day and showing maximum heat differential with surrounding water by mid-afternoon, and cooling in the evening and early morning to such an extent that the slick was no longer distinguishable (Legg 1991b).

Only the thicker portions of the slick were detectable by AVHRR, the much more extensive thin film of more volatile oils could not be observed (although they were possibly visible on one image). In both investigations about 20 images were acquired, over half of which were useless due to cloud cover. Legg (1991a, 1991b) produced false colour composites using AVHRR bands 1, 3 and 4 and visually interpreted slick location, whereas Cross (1992) interactively density sliced band 4 into thermal thresholds to enhance the detectability of the oil slick. These studies were based on *a priori* knowledge of the expected location of at least part of the slick and so these techniques are not universally appropriate for detecting unknown slicks. By comparing successive images, the position, extent and rates of movement were assessed and maps of slick trajectory produced.

The consensus appears to be that infra-red sensors are more useful than visible ones and that there is some role for both airborne and satellite based sensors in oil spill mapping. For satellite sensors, the lesson is that oil is only likely to be detectable during the daytime on relatively cloud-free days.

Ultra-violet sensors

Oil slicks are highly reflective in the ultra-violet (UV) part of the electromagnetic spectrum. Even a sheen (Plate 16) as thin as 0.01 μm will reflect UV. There can be a problem of confusing oil with sun glint but on the whole any returns in the 350–380 nm range indicate petroleum hydrocarbons. At present, airborne UV sensors are not commonly used but, if used in combination with IR sensors to provide overlay maps, are useful in providing relative thickness maps for oil spills. Also, the two sensors together may allow better discrimination of oil from confusing substances than either sensor alone.

Microwave sensors

Oil on water, smooths the surface and under certain sea states and weather conditions (not too calm and not too rough) can dramatically affect the surface roughness generated by waves and ripples. This parameter is measured by microwave remote sensing, of which Side-Looking Airborne Radar (SLAR) and Synthetic Aperture Radar (SAR), such as that on board the ERS satellites, is particularly well suited. Practitioners in the field consider SLAR easier to interpret and a more cost-effective option for oil spill monitoring, whilst recognising the superior resolution of SAR. The former is older but cheaper technology costing around US$ 0.7–1.0 million. The latter, which is considered vastly superior to SLAR (but see *caveat* above), would cost about US$ 2–4 million to mount on a dedicated aircraft (Fingas and Brown 1997). Use of ERS satellite SAR imagery would seem a more cost-effective option than dedicated aircraft survey with each of the two

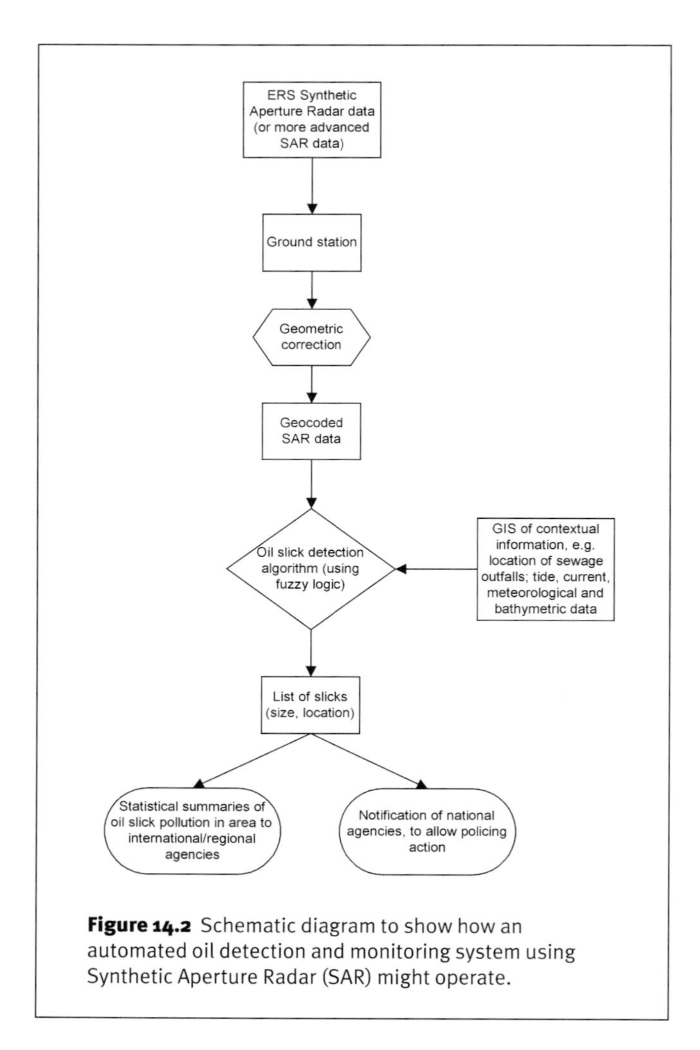

Figure 14.2 Schematic diagram to show how an automated oil detection and monitoring system using Synthetic Aperture Radar (SAR) might operate.

satellites giving repeat coverage once every three days. The great advantage over the previous sensors is the all weather, day and night, capability of microwave sensors which can penetrate cloud.

Guinard (1971) examined oil slicks using a dual polarised airborne SAR system at three different frequencies and found that in general, oil detectability increased with increasing sea state. Fingas and Brown (1997) suggest that slicks will only be detectable between wind speeds of 1.5–6.0 $m.s^{-1}$, which is a very narrow range (Beaufort Force 1 to lower end of Force 4), but Bos *et al.* (1994) indicate success up to Force 6 (11–13.5 $m.s^{-1}$) and possibly higher wind speeds. As with other sensors, there is a problem of confusing oil slicks with other phenomena that reduce backscatter such as plankton blooms, coral mass spawning, fish and whale sperm, sewage outfalls, shelter from wind by the coastline, flattening of the sea surface caused by interactions between tidal currents and shallows, oceanic fronts, and calming effects of subsurface seaweed or seagrass beds. Distinguishing between actual oil slicks and these many different potential false alarms is the main obstacle to routine use of SAR for oil spill detection and monitoring.

Sloggett and Jory (1995) discuss the development of a sophisticated automated detection algorithm that seeks to overcome this problem. It uses information on the reduction in backscatter, the smooth edges characteristic of oil slicks, and the nature of the dispersal pattern in relation to sea state and integrates this with contextual information e.g. location of sewage outfalls, tide and current data, meteorological data, and bathymetry in a GIS using fuzzy logic. In this way they suggest that satellite-based SAR imagery may be able to monitor oil pollution routinely (Figure 14.2). In another approach, Yan and Colón (1997) describe the use of Maximum Similarity Shape Matching (MSSM) techniques to track oil spills using ERS and RADARSAT SAR imagery.

At present there is considerable optimism that satellite-based SAR imagery may present a feasible and cost-effective method of monitoring and policing oil spills larger than about 0.3 km^2. Apart from the SAR carried by ERS, there are similar instruments on the Canadian RADARSAT (used in the *Sea Empress* spill off Wales) and ENVISAT, which will between them provide approximately daily coverage. Solberg and Theophilopoulos (1997) describe a prototype system (ENVISYS) for the automatic detection of oil spills in the Mediterranean using SAR imagery. The aim is to detect 98% of oil slicks with a minimal number of false alarms.

Further details of the use of microwave remote sensing to detect and monitor oil slicks can be found in Kraus *et al.* (1977), Crosswell and Alvarado (1984), Fontanel (1984), Hollinger and Mennella (1984), Lacoste (1984), Thompson *et al.* (1984), Fujita *et al.* (1986), and Huhnerfuss *et al.* (1986).

Multisensor airborne monitoring

The promise of using a combination of infra-red and UV sensors was noted above. The UK and other European countries have adopted the use of several types of sensor mounted in aircraft to patrol their EEZs and monitor for oil discharges at sea. In the UK a combination of Side-Looking Airborne Radar (SLAR), and infra-red and UV sensors has been found to be very effective and deliver useable data in most weather conditions. SLAR maps the broad area of oil slicks whilst the infra-red sensor delimits major concentrations of oil where clean up is likely to be required. About 90% of the volume of oil in a slick will lie in about 10% of the area and infra-red sensors can detect this thicker oil well. UV sensors detect even very thin petroleum hydrocarbon oil sheens, as noted above, and can distinguish between other oils such as fish and palm oils which may be discharged at sea. SLAR and infra-red sensors can operate at night as well as by day but UV sensors cannot operate effectively in bright sunlight or at all at night.

Routine airborne monitoring, which needs to be carried out day and night, tends to target major shipping lanes and is a costly exercise. Equipping an aircraft with the three sensors may cost about £1 million. The aircraft itself (e.g. a Cessna 404 or 406) may cost about £250,000 or £500,000 respectively and flying costs may be in the order of £800–1000 per hour including crew costs. It is difficult to recoup such costs unless fines are substantial and prosecutions successful. In Europe an average of roughly 10 prosecutions per year per patrolling country seem to come to court at present and the costs of patrolling greatly exceed fines accruing. Changes in both the legal process and the level of fines would probably be needed to make this important environmental protection use of remote sensing pay for itself. However, use of the equipment for applications other than just oil pollution monitoring might make the operation more cost-effective. For example, the sensors and aircraft could also play a useful role in fisheries monitoring, search and rescue, and drug interdiction operations (e.g. detecting wakes of high speed boats moving under cover of darkness). Such integrated EEZ surveillance may be the way forward.

For contact details for further information on multi-sensor airborne monitoring of oil pollution see Appendix 2.

Exxon Valdez case study

Perhaps the best example to date of operational remote sensing applied to oil pollution is that reported by Stringer *et al.* (1992) on the detection of petroleum spilled from the *MV Exxon Valdez*. When this vessel ran aground in Prince William Sound, Alaska in 1989 over eleven million gallons of crude oil were spilled. Multi-sensor remote sensing investigation of the area shortly after the spill was able to provide valuable information to those organising and operating the massive clean-up campaign, and much was also learnt about the spectral properties and the patterns of spilled oil. Landsat TM, AVHRR and SPOT panchromatic and multispectral images were acquired for before and after the spill. Side-Looking Airborne Radar (SLAR) imagery was acquired when weather permitted the aircraft to be flown and large volumes of airborne sightings and shipborne sightings and samples were also collected.

Of the imagery available only the AVHRR and SLAR data were useful as a tool in near-real-time monitoring. Processed AVHRR data were telefaxed to a research ship and onshore station within eight hours of the satellite overpass. The data were processed by contrast stretching the thermal infra-red waveband (channel 4) and density slicing it into four grey levels so that it was suitable for telefaxing. Both the SLAR and AVHRR images could be interpreted for oil-related patterns and currents and used in conjunction with other data for planning the clean-up campaign and keeping investigators informed of the current disposition of the slick.

Using the Landsat TM and SPOT multispectral data the spectral signature of oil was examined. The digital numbers (DN) representing oil were typically only a few values higher than for the surrounding water, making for a very subtle spectral signature, which was not unique. Sensor noise at these DN levels (especially in the SPOT image) proved problematic and required the application of algorithms to reduce this effect. As the DN range that distinguished oil from water was small compared to the full dynamic range, then quite severe contrast stretches were required to enhance the images. False colour composites and principal components analysis proved useful for displaying the results. Despite the oil being widespread for about a month only one reasonable Landsat image was obtainable; considerable image processing was then needed to reveal the presence of oil in spite of knowing where it was from ground survey.

The large volumes of data collected during the clean-up campaign by airborne and shipborne sighting and sampling were utilised to make detailed interpretations of the satellite imagery. In summary, there was considerable agreement in terms of general distribution of oil. However, the high resolution imagery (Landsat and SPOT) did not exhibit the large areas of 'oil sheen' reported by the field observers, but showed the oil to consist of filament structures interpreted as oil windrows. The lower spatial resolution (1.1 km) of the AVHRR imagery represented oil as structureless masses. Mapping of sea surface temperatures and suspended sediment plumes permitted the position and direction of major currents to be inferred which assisted in assessing the trajectory of the slick. Suspended sediment or turbidity patterns expressed on the images were also useful for considering where tar balls might develop and sink, as high concentrations of suspended sediments can act as a catalyst for tar ball formation.

Conclusions from this investigation are several. Firstly, the need for near-real-time data on slick position could only realistically be met by AVHRR and airborne sensors especially flown. Secondly, the Landsat TM and SPOT imagery were not very useful due to the time between repeat coverage, the interference from cloud cover, and the general inadequacies of visible sensors in detecting oil discussed above. Thirdly, the development of satellite SAR images (instead of the airborne SLAR used then) could prove useful for detecting the detailed structure and characteristics of a slick.

Laser fluorosensors
One of the drawbacks of the sensors so far discussed has been the difficulty of positively identifying oil from substances or phenomena which have similar properties with respect to a given sensor. A technology which has significant potential and overcomes this drawback is the use of airborne UV lasers (300–340 μm) to excite fluorescence in the oil which can then be detected by a sensor (fluorosensor). Oil compounds fluoresce between 400–650 nm with peak fluorescence at 480 nm. Few other compounds likely to be on the sea surface show this tendency to fluoresce. Possibly confusing substances such as chlorophyll with a sharp peak at 685 nm and yellow substance (*Gelbstoff*) with broad return centred at 420 nm can be satisfactorily separated from the oil, and in good conditions it may even be possible to distinguish heavy and light oils on the basis of their fluorescence. The instrument is large (weighing around 200 kg) and it is estimated that a production unit would cost around US$ 500,000 (Fingas and Brown 1997). This type of sensor, which is still being evaluated for operational use, is mentioned because it does appear to offer significant potential for the future.

Toxic and other industrial wastes

As noted in the introduction to this chapter, most toxic and industrial wastes are at present not directly detectable using remote sensing, but their concentrations may be closely correlated with features such as suspended sediments, yellow substance (if associated with freshwater outflows), or other substances which can be detected by remote sensing (e.g. Coskun and Örmeci 1994). In some cases, such as the discharges that emanated from chemical plants making titanium dioxide (a white pigment) using the sulphate process, the acid-iron wastes had a pronounced rust colour readily detectable by red sensitive sensors allowing plumes to be mapped by satellites with <100 m spatial resolution multispectral sensors (e.g. Landsat MSS in the Humber estuary, England).

The use of remote sensing in monitoring will in general be to supply a synoptic view of plume dispersion to augment *in situ* monitoring of discharges by environmental authorities or pollution control boards. As for other discrete discharges discussed below (sewage, thermal effluent) there is in general a mismatch between the temporal and spatial resolution of satellite imagery and operational needs. This leaves airborne sensors, which can be deployed as needed and can provide sufficient spatial resolution. The main problem with these is likely to be the cost since the remote sensing studies are only augmenting the field survey measurements, which are essential in terms of licences to discharge ('consents'), monitoring and enforcement of breaches of statutory limits, and legal proceedings.

Remote sensing has been used to monitor the dumping and dispersal of toxic and acidic wastes offshore, and to record the nature of dispersal of plumes (e.g. Johnson 1980, Klemas and Philpot 1981, Bowker and LeCroy 1985, Elrod 1988). In a study of ocean dumping of industrial waste about 60 km from the Delaware coast, sixteen Landsat MSS images were used to monitor the behaviour of the plumes that developed (Klemas and Philpot 1981). Principal components analysis on the four wavebands proved successful in discriminating the iron-acid industrial waste from surrounding waters. Knowledge of the time of dumping and the subsequent satellite overpass permitted drift speeds and spreading rates to be calculated. Bowker and LeCroy (1985), also using Landsat MSS, advised that spectral analysis be restricted to pixels displaying maximum concentration of wastes as the mixing effects with water in the dispersed zone will be misleading. They concluded that the time-decay of the spectral response is predictable and can be used to estimate the age of the plume, and that for acid-iron wastes it is impossible to detect their presence after about 100 hours.

Sewage discharges

Remote sensing has the potential to assist in the monitoring of sewage outfall plumes, dispersion of sewage from sludge dump sites and in studying water circulation patterns in environmental impact studies for the siting of sewage outfalls. Sewage plumes are detectable on optical imagery with elevated levels of both particulate and dissolved organic matter contributing to discrimination. The calming effect on the sea surface (visible as a surface 'slick') seen at outfalls will in certain sea conditions (see earlier discussion of microwave remote sensing of oil) be detectable as reduced backscatter (dark areas) on SAR imagery. Furthermore, outfalls may be detectable in the thermal infra-red when the effluent is at a significantly higher temperature than the receiving waters.

The area of monitoring of a small outfall might be a few square kilometres whilst for a large outfall it might be up to approximately 20 x 10 km. Thus for a small outfall only higher resolution satellite imagery or airborne sensors can be of use. The former is likely to be operationally all but useless because of poor temporal resolution (exacerbated by cloud cover) and cost, and the latter is unlikely to be cost-effective. For a large outfall, a sensor such as AVHRR or SeaWiFS would give improved temporal resolution (cloud cover permitting) but would not resolve features of the plume smaller than *ca.* 1 km^2 or provide much subsurface information. Svejkovsky and Haydock (1997), in a study of a large outfall off California, concluded that AVHRR could provide useful synoptic information on larger scale events such as upwelling, coastal runoff, red tides and current patterns, against which to interpret monthly monitoring data from the field.

Johnson *et al.* (1979) studied the dispersion of a plume from a sewage sludge dump in the New York Bight. Sea - truth data collection and imaging by an airborne multi-spectral scanner permitted a regression equation to be established and the pattern and rate of dispersion of sewage to be determined. This kind of study is useful for determining the effectiveness of the dispersal system and for providing information that may be useful for planning future disposal schemes. A sewage outfall system in the Firth of Forth, Scotland, was studied by an airborne multi-spectral scanner (Davies and Charlton 1987). Classification techniques and principal components analysis were used to enhance the structure and position of the plume and it proved possible to assess the performance of individual outlets within the overall system.

In summary, remote sensing may be of limited operational use to augment existing monitoring programmes of sewage outfall plumes, but the key monitoring data will remain water samples collected on field survey. These deliver the necessary information which relate to legislative needs and national or international water quality standards (e.g. coliform bacterial counts).

Gelbstoff *as a tracer for terrestrial inputs*

Dissolved organic matter (DOM) mainly enters coastal waters from river discharges and also from urban and industrial effluents (Bricaud *et al.* 1981). It is also produced *in situ* from the natural decay of phytoplankton or other marine organisms (Rashid 1985). A fraction of the heavily polymerised DOM compounds is commonly referred to as humic substances, or yellow substance/ *Gelbstoff* (Yentsch 1983) because of the yellow colour they have when dissolved in water. They are one of the key components that determine the colour of coastal water. The spectral signature of such waters is a combination of reflectance from suspended sediment, chlorophyll and accessory pigments of phytoplankton, and yellow substance (Fischer *et al.* 1991). Nichol (1993) used Landsat TM imagery to map plumes from peat swamp catchments dominated by *Gelbstoff* and to classify coastal waters (e.g. as to suitability for fish culture or primary contact recreation) in the Singapore-Johor-Riau area of SE Asia.

DOM is not a pollutant *per se* but *Gelbstoff* acts as a useful marker for freshwater inputs to coastal waters, being inversely correlated with salinity. It can thus act as a surrogate to monitor dispersal of sewage, agricultural or polluted estuarine inputs to the coastal marine environment. Algorithms that retrieve quantitative measures of suspended sediment, chlorophyll and yellow substance from upwelling radiances are much sought after, but are complicated by many problems (e.g. Tassan 1988, Carder *et al.* 1989,1991, Ferrari and Tassan 1992). Such work is still in its experimental phase and the relatively low spectral resolution of most current satellite sensors prevents accurate discrimination between the three elements (Ferrari and Tassan 1992). However, hopefully SeaWiFS, launched in 1997, will allow much better discrimination of the three.

Toxic algal blooms and eutrophication

In temperate waters plankton blooms occur naturally every spring and in tropical coastal waters may be triggered by the onset of monsoon rains or wind-driven upwelling. In all cases a build up of nutrient levels in surface waters allows the phytoplankton to multiply rapidly producing bloom conditions. Problems can arise when unseasonal inputs of nutrients from anthropogenic sources (e.g. agriculture, sewage) cause such large blooms that eutrophication ensues with dissolved oxygen levels dropping and animals dying. Eutrophication tends only to occur in semi-enclosed or enclosed water bodies or where water circulation is poor. For example, in the early 1990s in the shallow (average depth 1 m) eutrophic lagoon of Venice the green seaweed *Ulva rigida* bloomed extraordinarily from March to April/May each year, hindering tidal water exchange and leading to dystrophic conditions followed by mass die-off. This led to the need for frequent removal of the seaweed by mechanical means. Tassan (1992) reported an algorithm which allowed mapping of the benthic algal coverage in the 550 km^2 lagoon using Landsat TM data.

Another problem results from the type of algal bloom. Over the last thirty years blooms of toxic dinoflagellates such as *Gyrodinium aureolum*, *Prorocentrum minimum* and *Chrysochromulina polylepis* and cyanobacteria such as *Trichodesmium* (=*Oscillatoria*) and *Nodularia spumigena* appear to have increased in coastal waters (e.g. Rud and Kahru 1994). The huge concentrations of dinoflagellates or cyanobacteria discolour surface waters and, depending on the pigments in the species involved, the blooms may be reported as 'red' (e.g. *Trichodesmium erythraeum*), 'yellow' or 'gray tides' (Dupouy 1992). Where bloom species produce toxins these 'tides' can kill fish or accumulate in filter-feeding shellfish such as mussels rendering them toxic to consumers. Blooms may also cause mortality of marine animals as a result of physical blockage of the gills (a particular problem with cyanobacteria) or as a result of deoxygenation of the water during bloom decay. Major blooms have recently been been reported along the coasts of USA, Canada, Venezuela, Korea, Taiwan, Tasmania, Guatemala, England and Norway (Johannessen *et al.* 1989a).

With increased aquaculture of fish and shellfish in coastal waters such blooms can have devastating economic consequences. For example, in Norway in May 1988 a *Chrysochromulina* bloom wiped out 480 tons of caged fish (mainly salmon) with a market value of 30 million Norwegian Kroner before precautions could be taken. Subsequently, 130 seafarms with an insurance value of 1.0–1.5 billion Kroner were towed to safety in low salinity water in fjords (Johannessen *et al.* 1989b). Similar kills have occurred in SE Asia, for example, a *Trichodesmium* bloom caused US$ 1.16 million worth of damage to fish farms in the Gulf of Thailand in 1983 (Suvapepun 1992). Cyanobacterial blooms in lakes and rivers can lead to fish kills through eutrophication (Jupp *et al.* 1994) and toxins (Hawser and Codd 1992), and consequent closure of lakes for recreational use. They are also causing increasing concern in brackish and coastal waters. Toxic blooms of the cyanobacterium *Nodularia spumigena* have been reported in coastal and estuarine habitats in Australia (Lavery *et al.* 1993), New Zealand, the Baltic Sea (Rud and Kahru 1994) and the United Kingdom and have led to animal poisoning and human illness. In the Caribbean, Hawser and Codd (1992) found that *Trichodesmium thiebautii* possessed a neurotoxin which was highly toxic to mammals but *T. erythraeum* did not appear to be toxic. Furthermore, Jones (1992) reported that *Trichodesmium* blooms in the Great Barrier Reef lagoon led to increased bio-availability of trace metals. This was correlated with significant and sustained increases of certain trace elements in the black lip oyster *Saccostrea amassa* such that zinc, cadmium and copper levels in the oyster exceeded health guidelines by 8000%, 4000% and 3000% respectively. Thus both toxic and non-toxic blooms may have adverse impacts on cultured or wild-harvested shellfish.

Remote sensing can be used to monitor toxic or other algal blooms (e.g. Johannessen *et al.* 1989a, 1989b, Sørensen *et al.* 1989, Borstad *et al.* 1992, Dupouy 1992, Furnas 1992, Lavery *et al.* 1993) and could help direct precautionary measures. Johannessen's and Sørensen's teams relied primarily on AVHRR imagery and Sørensen *et al.* (1989) found Landsat MSS and TM and SPOT XS satellite imagery also to be useful for mapping the bloom. Borstad *et al.* (1992) used Coastal Zone Color Scanner (CZCS) imagery and Dupouy (1992) worked with both CZCS and SPOT imagery. However, to be of operational use imagery would need to be available during a bloom on a more or less real-time basis (preferably at least daily). Among these sensors only AVHRR would thus seem to be useful operationally.

Phytoplankton in surface water is readily detected by remote sensing due to the strong spectral signature of chlorophyll-*a* (and accessory pigments such as phycocyanin, phycoerythrin, fucoxanthin and peridinin) although suspended sediment and yellow substance in coastal waters complicate matters making it difficult to obtain accurate chlorophyll-*a* values. However, it is fairly easy to map the location and pattern of chlorophyll-*a* anomalies and recent work indicates that accessory pigments should allow the types of bloom to be identified (e.g. Malthus *et al.* 1997), although Garver *et al.* (1994) considered it unlikely that robust global algorithms to distinguish phytoplankton groups could be developed using optical remote sensing.

López García and Caselles (1990) used Landsat TM imagery in a multispectral study of chlorophyll-*a* concentration of a Spanish Lagoon, and Bagheri and Dios (1990) also used TM data in a study of the eutrophication of coastal waters of New Jersey. A range of protocols and algorithms to identify and characterise algal blooms have been developed in recent years for a range of sensors (Borstad *et al.* 1989, 1992, Jupp *et al.* 1994, Subramaniam and Carpenter 1994, Tassan 1995, Malthus *et al.* 1997). New satellite sensors such as OCTS and SeaWiFs and airborne scanners such as CASI seem to offer considerable scope for detecting developing blooms. Whether operational early warning of toxic blooms can be achieved is unclear. The technology is almost there but it is unclear who would pay the substantial costs involved.

Thermal discharges

Coastal and estuarine waters are useful sources of cooling water for industry and in particular for power-stations. Once this cooling water has been used it is returned at an elevated temperature into the environment. Modern power stations have thermal efficiencies of about 30–35%, so that for every 1 MW of electricity generated, about 2 MW of waste heat must be dispersed in the cooling water. These thermal discharges may be 10°C or more above ambient water temperatures and may thus have environmental impacts in the vicinity of the outfalls. Temperatures typically decrease to less than 1°C above ambient within 1–2 km of the outfall so the area of impact is fairly restricted. Since the discharges are from known point sources and restricted in area they can be mapped fairly easily using boat-based surveys. To put the problem in perspective, in an example cited by Vaughan *et al.* (1995) for a power-plant on the southern coast of Spain, cooling water was discharged at a rate of about 7000 cubic metres per hour at a temperature of about 30.6°C into receiving water with an ambient temperature of approximately 17°C. The resultant plume warmed the sea by 0.5°C or more over an area of about 1.6–2.5 km^2, but only about 0.1–0.5 km^2 were warmed by more than 2°C. In coastal waters surrounding a plume, the discharged heat is gradually dispersed alongshore and offshore and to the atmosphere. These waters which may be a few tenths of a degree Celsius above ambient may extend alongshore the distance of a tidal excursion (typically 10–20 km either side of a power station depending on the speed of tidal streams) and 1–3 km offshore. Impacts in these peripheral areas are likely to be negligible.

Satellite-borne sensors
Clearly the satellite sensors which measure sea-surface temperature best, the NOAA AVHRR sensor which has a spatial resolution of 1.1 km and the new ATSR sensor on board ERS-1/2 which resolves at 1 km, are not suitable for measuring such small scale thermal plumes along coasts.

However, the thermal waveband of Landsat TM (band 6) may be suitable as it provides temperature measurements at a spatial resolution of 120 m. A few attempts have been made to use it to map thermal plumes. Pilling *et al.* (1989) describe an attempt to determine thermal plume dynamics under different wind conditions for a Welsh lake that provides cooling for a nuclear power station. This study gave a qualitative feel for the extent of the thermal plume but found a poor and non-significant correlation between DN recorded in Landsat TM Band 6 and *in situ* water temperature measurements. Despite selectively excluding half of the data points in an effort to improve the correlation the authors were unable to demonstrate a significant relationship between DN and water temperature. Errors of predicted temperatures from those measured *in situ* by a thermistor ranged from 0.5–1.9°C (for the non-excluded points). These results do not inspire much confidence but may have been in part due to the very limited ground survey data and the lack of atmospheric correction.

A more careful study by Gibbons *et al.* (1989) of a thermal plume from Diablo Canyon power plant on the Californian coast, where radiometric and atmospheric correction using local radiosonde data were carried out, gave better results. The differences between ground-truthed water temperature measurements and those predicted from Landsat-5 TM band 6 data were only 0.4–0.6°C (absolute accuracy ±10%) and the satellite imagery provided a useful map of the plume to within ±0.5°C. Poli *et al.* (1997) in a Landsat TM study of the thermal plume resulting from industrial and power-station discharges into the harbour of Portovesme, Sardinia found a maximum temperature elevation of 7.5°C and suggested that 0.5°C temperature differences could be detected.

Whether such studies are of much operational use is another matter. The areal extent of the thermal plume will tend to vary with the prevailing currents and wind and can be expected to change markedly over a tidal cycle. Satellite remote sensing, with the possibility of perhaps one over-pass every 16 days, does not offer the temporal resolution to contribute synoptic information on the plume dynamics. In addition, the area of interest on an image is likely to be a minute part (< 0.02%) of a Landsat scene (or quarter scene), the relative resolution of temperature seems unlikely to be better than 0.5°C, and the spatial resolution of 1.44 ha too coarse. Considerable field survey will of course be required to calibrate any imagery obtained. Although the Enhanced Thematic Mapper on Landsat 7 has improved spatial resolution (60 m) and should deliver better temperature estimates, mapping of thermal plumes is unlikely to be a sensible application for it.

In conclusion, satellite-borne sensors at present appear to offer few realistic benefits to thermal plume mapping.

Airborne sensors
Airborne multispectral scanning has been utilised for thermal plume mapping in a number of studies and does not suffer from the temporal and spatial limitations of satellite imagery.

Kuo and Talay (1979) used an airborne multispectral scanner and boat surveys to study heated water discharges from the Surrey nuclear power station on the tidal section of the James river, Virginia. Schott (1979) drew attention to the effects of thermal discharges from power stations on the aquatic ecosystem and noted that plumes can extend for over 1.5 km out to sea, with temperature increases in excess of 9°C close to the source. He proposed airborne multispectral scanning as a timely and cost-effective means of monitoring thermal plumes and showed, by means of a case study, that the accuracy of an airborne-only study could match that of sampling from a boat, and of course provided much better spatial coverage.

Wilson and Anderson (1984) used an airborne thermal IR scanner to detect heat plumes emanating from a power station and sewage outlet in the Tay Estuary, Scotland. They concluded that this is a suitable technique if the plume temperatures are of the order of 10°C higher than ambient water temperatures for power stations and 5°C higher for sewage plumes.

In a study of the thermal plume from a power-plant in southern Spain using a Daedalus Airborne Thematic Mapper (ATM), Vaughan *et al.* (1995) were able to map seawater temperature increases in 0.5°C increments from 13.5°C above ambient (17°C) at the outfall to 0.5°C above ambient at distance of a hundred metres or more. Radiometric and atmospheric correction and ground sampling along a transect away from the outfall were used to achieve these results. The spatial resolution of the imagery varied from 2 m (flown at an altitude of 1000 m) to 4 m (flown at 2000 m). The main conclusion of this survey was that attempts to model the plume were rather inadequate, demonstrating the need for a remote sensing approach.

In conclusion, as an adjunct to field sampling, airborne multispectral scanners sensitive to thermal IR seem to offer considerable potential for mapping thermal plume behaviour and potential impact area.

Coral bleaching 'hot-spots'

In this section we discuss the use of remote sensing to monitor large-scale natural changes in coastal water temperature which can have dramatic impacts on coral reefs.

Since 1982 when multi-waveband infra-red data became available on the US National Oceanographic and Atmospheric Administration (NOAA) polar-orbiting satellites carrying the Advanced Very High Resolution Radiometer (AVHRR) sensor (see Chapter 3), reasonably accurate sea-surface temperature (SST) data (with absolute accuracy of around 1°C and relative accuracy of about 0.2°C), corrected for atmospheric conditions, have been made available by NOAA (McClain *et al.* 1985, Reynolds 1988, Reynolds *et al.* 1989, Strong 1989, Reynolds and Marisco 1993, Reynolds and Smith 1994). Since mid-1996, with a decade of satellite SST measurements providing a baseline, it has been possible to generate 'daily' charts of SST anomalies for the world every few days using satellite data alone. You can view these over the Internet at URL http://psbsgi1.nesdis.noaa.gov:8080/PSB/EPS/SST/climo.html. SST monthly anomaly charts for 1984–1995 (except 1991–1992, which are omitted due to interference from volcanic aerosols released into the atmosphere by the Mount Pinatubo eruption in Philippines: see Reynolds 1993) are also available for downloading at http://psbsgi1.nesdis.noaa.gov:8080/PSB/EPS/SST/al_climo.html. The 'daily' data are interpolated to a spatial resolution of 50 km whilst the monthly historical dataset is provided at 36 km resolution. Temperatures are displayed with 0.25°C isotherms.

During the 1980s, the early 1990s and most recently in 1998 there were strong El Niño Southern Oscillation (ENSO) events and sea-surface temperature anomalies were recorded widely in the tropical Pacific and Indian Oceans and Caribbean. These involved warming of surface waters to 1–3°C above normal monthly mean temperatures and were associated with widespread 'bleaching' of corals – loss of symbiotic algae called zooxanthellae which live inside the corals' tissues (Goreau and Hayes 1994, Montgomery and Strong 1994, Gleeson and Strong 1995). In some areas in 1998 the bleaching events were of unprecedented severity with over 90% death of shallow water branching corals resulting (e.g. Maldives). In 1998 the build-up of warm water anomalies at different locations around the world as the seasons progressed could be followed on the 'Coral Reef Hotspots' (http://manati.wwb.noaa.gov/orad/sub/noaarsrc.html) and 'Coral Bleaching Hotspots' (http://psbsgi1.nesdis.noaa.gov:8080/PBS/EPS/SST/climohot.html) web pages of NOAA. These provide Western and Eastern Hemisphere colour coded 'hotspot' charts at 50 km spatial resolution (Plate 17). These are updated every few days. In addition reports from field-based coral researchers on bleaching events are also posted on the Coral Bleaching Hotspots page. Anomalies of greater than 1°C are displayed in shades of yellow to red as 0.25°C isotherms and animated displays of up to 6 months of anomaly charts are available on the Coral Reef Hotspots page.

Unfortunately, forewarning of warm water anomalies and potential bleaching is of limited use to coastal managers as there is nothing they can do about it. However, awareness of the potential for warm water anomaly induced bleaching does mean that managers are less likely to waste time looking for other culprits to explain bleaching and death of corals during such events. Also it may stimulate field-based monitoring surveys of coral reefs before, during and after such events so that we can build up better information on the magnitude of impacts and subsequent recovery of reefs. Such information can then feed into management.

References

Althuis, A., and Shimwell, S., 1994, Interpretation of remote sensing imagery for suspended matter monitoring in coastal waters. In *EARSeL Workshop on Remote Sensing and GIS for Coastal Zone Management*, edited by L.L.F. Janssen and R. Allewijn (Delft, The Netherlands: Rijkswaterstaat, Survey Department), pp. 7–15.

Bagheri, S., and Dios, R.A., 1990, Chlorophyll–*a* estimation in New Jersey's coastal waters using Thematic Mapper data. *International Journal of Remote Sensing*, **11**, 289–299.

Barrett, E.C., and Curtis, L.F., 1992, *Introduction to Environmental Remote Sensing*. 3rd edition (London: Chapman and Hall).

Borstad, G.A., Gower, J.F.R., Carpenter, E.J., and Reuter, J.G., 1989, Development of algorithms for remote sensing of marine Trichodesmium. *IGARSS '89 Remote Sensing: an Economic Tool for the Nineties*, Vancouver, Canada July 10–14, 1989 (Vancouver, Canada: IGARSS '89 12th Canadian Symposium on Remote Sensing), **4**, 2451–2455.

Borstad, G.A., Gower, J.F.R., Carpenter, E.J., 1992, Development of algorithms for remote sensing of *Trichodesmium* blooms. In *Marine Pelagic Cyanobacteria: Trichodesmium and Other Diazotrophs*, edited by E.J. Carpenter, D.G. Capone and J.G. Rueter (Dordrecht: Kluwer Academic Publishers), pp. 193–210.

Bos, W.G., Konings, H., Pellemans, A.H.J.M., Janssen, L.L.F., and van Swol, R.W., 1994, The use of spaceborne SAR imagery for oil slick detection at the North Sea. In *EARSeL Workshop on Remote Sensing and GIS for Coastal Zone Management*, edited by L.L.F. Janssen and R. Allewijn (Delft, The Netherlands: Rijkswaterstaat, Survey Department), pp. 57–66.

Bowker, D.E., and LeCroy, S.R., 1985, Bright spot analysis of ocean dump plumes using Landsat MSS. *International Journal of Remote Sensing*, **6**, 759–772.

Braga, C.Z.F., Setzer, A.W., and de Lacerda, L.D., 1993, Water quality assessment with simultaneous Landsat-5 TM data at

Guanabara Bay, Rio de Janeiro, Brazil. *Remote Sensing of Environment*, **45**, 95–106.

Bricaud, A., Morel, A., and Prieur, L., 1981, Absorption by dissolved organic matter yellow substance, in the UV and visible domains. *Limnology and Oceanography*, **26**, 43–53.

Caixing, Y., Jiamin, H., Hui, F., and Yongliang, X., 1990, Remote sensing for harbour planning. *Remote Sensing for Marine Studies. Report of the Seminar on Remote Sensing Applications for Oceanography and Fishing Environment Analysis*, Beijing, China, 7 to 11 May 1990 (Bangkok: ESCAP/UNDP Regional Remote Sensing Programme (RAS/86/141) ST/ESCAP/897), pp. 11–22.

Carder, K.L, Hawes, S.K., Baker, K.A., Smith, R.C., Steward, R.G., and Mitchell, B.G., 1991, Reflectance model for quantifying chlorophyll-*a* in the presence of productivity degradation products. *Journal of Geophysical Research*, **96**, 20599–20611.

Carder, K.L., Steward, R.G., Harvey, G.R., and Ortner, P.B., 1989, Marine humic and fulvic acids: their effects on remote sensing of ocean chlorophyll. *Limnology and Oceanography*, **34**, 68–81.

Chen, Z., Hanson, J.D., and Curran, P.J., 1991, The form of the relationship between suspended sediment concentration and spectral reflectance: its implications for the use of Daedalus 1268 data. *International Journal of Remote Sensing*, **12**, 215–222.

Clark, C.D., 1993, Satellite remote sensing for marine pollution investigations. *Marine Pollution Bulletin*, **26**, 357–367.

Coskun, H.G., and Örmeci, C., 1994, Water quality monitoring in Haliç (Golden Horn) using satellite images. In *EARSeL Workshop on Remote Sensing and GIS for Coastal Zone Management*, edited by L.L.F. Janssen and R. Allewijn (Delft, The Netherlands: Rijkswaterstaat, Survey Department), pp. 84–93.

Cracknell, A.P., Muirhead, K., Callison, R.D., and Campbell, N.A., 1983, Satellite remote sensing, environmental monitoring and the offshore oil industries. In *Proceedings of an EARSeL/ESA Symposium on Remote Sensing Applications for Environmental Studies*, Brussels-Paris, ESA, 163–171.

Cross, A.M., 1992, Monitoring marine pollution using AVHRR data: observations off the coast of Kuwait and Saudi Arabia during January 1991. *International Journal of Remote Sensing*, **13**, 781–788.

Crosswell, W.F., and Alvarado, U.A. 1984, An assessment of the use of space technology in monitoring oil spills. In *Remote Sensing for the Control of Marine Pollution*, edited by J.M. Massin (New York: Plenum Press), pp. 331–340.

Davies, P.A. and Mofor, L.A., 1990, Observations of flow separation by an isolated island. *International Journal of Remote Sensing*, **11**, 767–782.

Davies, P.A., and Charlton, J.A., 1987, The determination of the internal structure of an effluent plume using MSS data. *International Journal of Remote Sensing*, **8**, 75–83.

Delu, P. and Shouren, L., 1989, Remote sensing the suspended sediment distribution on the Mingjiang river mouth in different tide phases as shown by MSS and TM imagery. *IGARSS '89 Remote Sensing: an Economic Tool for the Nineties*, Vancouver, Canada July 10–14, 1989 (Vancouver, Canada: IGARSS '89 12th Canadian Symposium on Remote Sensing), **4**, 2623–2626.

Deustch, M., and Estes, J.E., 1980, Landsat detection of oil from natural seeps. *Journal of American Society of Photogrammetry*, **66**, 1313–1320.

Deustch, M., Vollers, R., and Deustch, J.P., 1980, Landsat tracking of oil slicks from the 1979 Gulf of Mexico oil well blowout. *Proceedings of the 14th International Symposium on Remote Sensing of the Environment*, 23–30 April, San José, Costa Rica (Ann Arbor: Environment Research Institute of Michigan), **2**, 1197–1211.

Dupouy, C., 1992, Discoloured waters in the Melanesian Archipelago (New Caledonia and Vanuatu). The value of the Nimbus-7 Coastal Zone Color Scanner observations. In *Marine Pelagic Cyanobacteria: Trichodesmium and Other Diazotrophs*, edited by E.J. Carpenter, D.G. Capone and J.G. Rueter (Dordrecht: Kluwer Academic Publishers), pp. 177–191.

Elrod, J.A., 1988, CZCS view of oceanic acid waste dumps. *Remote Sensing of Environment*, **25**, 245–254.

Estes, J.E., and Senger, L.W., 1972, The multispectral concept as applied to marine oil spills. *Remote Sensing of Environment*, **2**, 141–163.

Ferrari, G.M., and Tassan, S., 1992, Evaluation of the influence of yellow substance absorption on the remote sensing of water quality in the Gulf of Naples: a case study. *International Journal of Remote Sensing*, **13**, 2177–2189.

Fingas, M.F., and Brown, C.E., 1997, Oil spill remote sensors: review, trends and new developments. *Proceedings of the Fourth International Conference on Remote Sensing for Marine and Coastal Environments*, Orlando, Florida, 17–19 March 1997, **1**, 371–380.

Fischer, J., Doerffer, R., and Grassl, H., 1991, 2. Remote sensing of water substances in rivers, estuarine and coastal waters. In *SCOPE 42: Biogeochemistry of Major World Rivers*, edited by E.T. Degens, S. Kempe and J.E. Richey (John Wiley, Chichester), pp. 25–55.

Fontanel, A., 1984, Detection of oil slicks using real aperture and synthetic aperture imaging radars: experimental results. In *Remote Sensing for the Control of Marine Pollution*, edited by J.M. Massin (New York: Plenum Press), pp. 217–226.

Froidefond, J.M., Castaing, P., Mirmand, M., and Ruch, P., 1991, Analysis of the turbid plume of the Gironde (France) based on SPOT radiometric data. *Remote Sensing of Environment*, **36**, 149–163.

Fujita, M., Masuko, H., Yoshikado, S., Okamoto, K., Inomata, H., and Fugono, N., 1986, SIR-B experiments in Japan: sensor calibration and oil pollution detection over oceans. *IEEE Transactions Geoscience and Remote Sensing*, **24**, 567–574.

Furnas, M.J., 1992, Pelagic *Trichodesmium* (= *Oscillatoria*) in the Great Barrier Reef region. In *Marine Pelagic Cyanobacteria: Trichodesmium and Other Diazotrophs*, edited by E.J. Carpenter, D.G. Capone and J.G. Rueter (Dordrecht: Kluwer Academic Publishers), pp. 265–272.

Garver, S.A., Siegel, D.A., and Mitchell, B.G., 1994, Variability in near-surface particulate absorption spectra: What can a satellite ocean color imager see? *Limnology and Oceanography*, **39** (6), 1349–1367.

GESAMP, 1991, *The State of the Marine Environment*. (Oxford: Blackwell).

Gibbons, D.E., Wukelic, G.E., Leighton, J.P., and Doyle, M.J., 1989, Application of Landsat Thematic Mapper data for coastal thermal plume analysis at Diablo Canyon. *Photogrammetric Engineering and Remote Sensing*, **55** (6), 903–909.

Gleeson, M.W., and Strong, A.E., 1995, Applying MCSST to coral reef bleaching, *Advances in Space Research*, **16**, 151–154.

Goreau, T.J., and Hayes, R.L., 1994, Coral bleaching and ocean 'hotspots'. *Ambio*, **23**, (3), 176–180.

Guinard, N.W., 1971, The remote sensing of oil slicks. *Proceedings of the 7th International Symposium on Remote Sensing of Environment* (Ann Arbor: Environment Research Institute of Michigan), pp. 1005–1026.

Hawser, S.P., and Codd, G.A., 1992, The toxicity of Trichodesmium blooms from Caribbean waters. In *Marine Pelagic Cyanobacteria: Trichodesmium and Other Diazotrophs*, edited by E.J. Carpenter, D.G. Capone and J.G. Rueter (Dordrecht: Kluwer Academic Publishers), pp. 319–329.

Hayes, R.M., 1980, Operational use of remote sensing during the Campeche Bay oil well blowout. In *Proceedings of the 14th International Symposium on Remote Sensing of Environment*, 23–30 April, San José, Costa Rica (Ann Arbor: Environment Research Institute of Michigan) **2**, 1187–1196

Hinton, J.C., 1991, Application of eigenvector analysis to remote sensing of coastal water quality. *International Journal of Remote Sensing*, **12**, 1441–1460.

Hollinger, J.P., and Mennella, R.A., 1984, Measurements of the distribution and volume of sea surface oil spills using microwave radiometry. In *Remote Sensing for the Control of Marine Pollution*, edited by J.M. Massin (New York: Plenum Press), pp. 267–276.

Huhnerfuss, M., Alpers, W., and Richter, K., 1986, Discrimination between crude oil spill and monomolecular sea slicks by airborne radar and infrared radiometer: possibilities and limitations. *International Journal of Remote Sensing*, **10**, 1893–1906.

Ibrahim, M., and Bujang, J.S., 1992, Application of remote sensing in the study of development impact of a marine park in Malaysia. *Proceedings of the 18th Annual Conference of the Remote Sensing Society* (Dundee: Remote Sensing Society), pp. 235–243.

Jensen, J.R., Ramsey III, E.W., Holmes, J.M., Mitchel, J.E., Savistsky, B., and Davis, B.A., 1990, Environmental sensitivity index ESI, mapping for oil spills using remote sensing and geographic information system technology. *International Journal of Geographical Information Systems*, **4**, 181–201

Johannessen, J.A., Johannessen, O.M., and Haugan, P.M., 1989a, Remote sensing and model simulation studies of the Norwegian coastal current during the algal bloom in May 1988. *International Journal of Remote Sensing*, **10**, 1893–1906.

Johannessen, O.M., Pettersson, L.H., Johannessen, J.A., Haugan, P.M., Olaussen, T.I., and Kloster, K., 1989b, The toxic *Chrysochromulina polylepis* bloom in Scandinavian waters – May/June 1988. *IGARSS '89 Remote Sensing: an Economic Tool for the Nineties,* Vancouver, Canada July 10–14, 1989 (Vancouver, Canada: IGARSS '89 12th Canadian Symposium on Remote Sensing), **4**, 2048–2051.

Johnson, R.W., 1978, Mapping of chlorophyll–*a* distribution in coastal zones. *Photogrammetric Engineering and Remote Sensing*, **44**, 617–624.

Johnson, R.W., 1980, Remote sensing of plumes from ocean dumping. *Remote Sensing of Environment*, **9**, 197–209.

Johnson, R.W., and Bahn, G.S., 1977, Quantitative analysis of aircraft data and mapping of water quality in the James River in Virginia. *NASA Technical Report*, 1021.

Johnson, R.W., Glasgow, R.M., Duedall, I.W., and Proni, J.R., 1979, Monitoring the temporal dispersion of a sewage sludge plume. *Photogrammetric Engineering and Remote Sensing*, **45**, 763–768.

Johnson, R.W., and Munday Jr, J.C., Carter, V., Kemmerer, A.J., Kendall, B.M., Legeckis, R., Polcyn, F.C., Proni, J.R., and Walter, D.J., 1983, The marine environment. In *Manual of Remote Sensing*, edited by R.N. Colwell, (Virginia, USA: Sheridan Press), **2**, 1371–1496.

Jones, G.B., 1992, Effect of *Trichodesmium* blooms on water quality in the Great Barrier Reef lagoon. In *Marine Pelagic Cyanobacteria: Trichodesmium and Other Diazotrophs*, edited by E.J. Carpenter, D.G. Capone and J.G. Rueter (Dordrecht: Kluwer Academic Publishers), pp. 273–287.

Jupp, D.L.B., Kirk, J.T.O., and Harris, G.P., 1994, Detection, identification and mapping of Cyanobacteria – using remote sensing to measure the optical quality of turbid inland waters. *Australian Journal of Marine and Freshwater Research*, **45**, 801–828.

Klemas, V. and Philpot, W.D., 1981, Drift and dispersion studies of ocean dumped waste using Landsat imagery and current drogues. *Photogrammetric Engineering and Remote Sensing*, **47**, 533–542.

Kraus, S.P., Estes, J.E., Atwater, S.G., Jensen, J.R., and Vollmers, R.R., 1977, Radar detection of surface oil slicks. *Photogrammetric Engineering and Remote Sensing*, **43**, 1523–1531.

Kuo, C.Y., and Talay, T.A., 1979, Remote monitoring of a thermal plume. *NASA TM 80125*, Washington D.C.

Lacoste, F., 1984, Oil spill detection using a multipurpose synthetic aperture radar. In *Remote Sensing for the Control of Marine Pollution*, edited by J.M. Massin (New York: Plenum Press), pp. 227–230.

Lavery, P., Pattiaratchi, C., Wyllie, A., and Hick, P., 1993, Water quality monitoring in estuarine waters using the Landsat Thematic Mapper. *Remote Sensing of Environment*, **46**, 268–280.

Lean, G., and Hinrichsen, D., 1992, *WWF Atlas of the Environment* (Oxford: Helicon Publishing Ltd).

Legg, C.A., 1991a, The Arabian Gulf oil slick, January and February 1991. *International Journal of Remote Sensing*, **12**, 1795–1796.

Legg, C.A., 1991b, An investigation of the Arabian Gulf oil slick using NOAA AVHRR imagery. SP(91) WP19. (Farnborough, UK: National Remote Sensing Centre), 8 pp.

Lo, C.P., 1986, *Applied Remote Sensing*. (Harlow, UK: Longman).

López García, M.J., and Caselles, V., 1990, A multi-temporal study of chlorophyll–*a* concentration in the Albufera lagoon of Valencia, Spain, using Thematic Mapper data. *International Journal of Remote Sensing*, **11**, 301–311.

Malthus, T.J., Grieve, L., and Harwar, M.D., 1997, Spectral modeling for the identification and quantification of algal blooms: a test of approach. *Proceedings of the Fourth International Conference on Remote Sensing for Marine and Coastal Environments*, Orlando, Florida, 17–19 March 1997, **1**, 223–232.

Massin, J.M., (editor) 1984, *Remote Sensing for the Control of Marine Pollution* (New York: Plenum Press).

McClain, E.P., Pichel, W.G., and Walton, C.C., 1985, Comparative performance of AVHRR-based multichannel sea-surface temperatures. *Journal of Geophysical Research – Oceans*, **90**, 1587–1601.

Mertes, L.A.K., Smith, M.O., and Adams, J.B., 1993, Estimating suspended sediment concentrations in surface waters of the Amazon River wetlands from Landsat images. *Remote Sensing of Environment*, **43**, 281–301.

Minji, L., and Yawei, D., 1990, The use of satellite remote sensing for studying suspended sand for a seagoing channel renovation project. *Remote Sensing for Marine Studies. Report of the Seminar on Remote Sensing Applications for Oceanography and Fishing Environment Analysis*, Beijing, China, 7 to 11 May 1990 (Bangkok: ESCAP/UNDP Regional Remote Sensing Programme (RAS/86/141) ST/ESCAP/897), pp. 129–140.

Montgomery, R.S., and Strong, A.E., 1994, Coral bleaching threatens ocean life. *EOS*, **75**, 145–147.

Muller-Karger, F.E., 1992, Remote sensing of marine pollution: a challenge for the 1990s. *Marine Pollution Bulletin*, **25**, 54–60.

Muller-Karger, F.E., McClain, C.R., and Richardson, P.L., 1988, The dispersal of the Amazon's water. *Nature* **333**, 56–59.

Nanu, L. and Robertson, C., 1990, Estimating suspended sediment concentrations from spectral reflectance data. *International Journal of Remote Sensing*, **11**, 913–920.

Nichol, J.E., 1993, Remote sensing of water quality in the Singapore-Johor-Riau growth triangle. *Remote Sensing of Environment*, **43**, 139–148.

Otterman, J., Ohring, G., and Ginsberg, A., 1974, Results of the Israeli multidisciplinary data analysis of ERTS-1 imagery. *Remote Sensing of Environment*, **3**, 133–148.

Pattiaratchi, C., Lavery, P., Wyllie, A., and Hick, P., 1994, Estimates of water quality in coastal waters using multi-date Landsat Thematic Mapper data. *International Journal of Remote Sensing*, **15** (8), 1571–1584.

Peck, T.M., Sweet, R.J.M., Southgate, H.N., Boxall, S., Matthews, A., Nash, R., Aiken, J., and Bottrell, H., 1994, COAST (Coastal Earth Observation Application for Sediment Transport). In *EARSeL Workshop on Remote Sensing and GIS for Coastal Zone Management*, edited by L.L.F. Janssen and R. Allewijn (Delft, The Netherlands: Rijkswaterstaat, Survey Department), pp. 260–266.

Pilling, I., Van Smirren, J., Newlands, A.G., and Rodgers, I., 1989, Collaborative study on the application of remote sensing techniques to the spatial and thermal mapping of plumes resulting from the outfall of power station cooling waters. *SP1 Division Working Paper*, SP(89) WP24 (Farnborough, UK: National Remote Sensing Centre), 29 pp.

Poli, U., Ippoliti, M., Venturini, C., Falcone, P., and Marino, A., 1997, Coastal water quality from remote sensing and GIS. A case study on south west Sardinia (Italy). *Proceedings of the Fourth International Conference on Remote Sensing for Marine and Coastal Environments*, Orlando, Florida, 17–19 March 1997, **2**, 406–414.

Populus, J., Hastuti, W., Martin, J.-L. M., Guelorget, O., Sumartono, B., and Wibowo, A., 1995, Remote sensing as a tool for diagnosis of water quality in Indonesian seas. *Ocean & Coastal Management*, **27** (3), 197–215.

Rashid, M.A., 1985, *Geochemistry of Marine Humic Compounds*. (New York: Springer Verlag).

Reynolds, R.W., 1988, A real-time global sea surface temperature analysis. *Journal of Climate*, **1**, 75–86.

Reynolds, R.W., 1993, Impact of Mount Pinatubo aerosols on satellite-derived sea-surface temperatures. *Journal of Climate*, **6**, 768–774.

Reynolds, R.W., Folland C.K., and Parker, D.E., 1989, Biases in satellite-derived sea-surface temperature data. *Nature*, **341**, 728–731.

Reynolds, R.W., and Marisco, D.C., 1993, An improved real-time global sea-surface temperature analysis. *Journal of Climate*, **6**, 114–119.

Reynolds, R.W., and Smith, T.M., 1994, Improved global sea-surface temperature analyses using optimum interpoltion. *Journal of Climate*, **7**, 929–948.

Rimmer, J.C., Collins, M.B., and Pattiaratchi, C.B., 1987, Mapping of water quality in coastal waters using airborne Thematic Mapper data. *International Journal of Remote Sensing*, **8**, 85–102.

Rud, O., and Kahru, M., 1994, Long-term time series of NOAA AVHRR imagery reveals large interannual variations of surface cyanobacterial accumulations in the Baltic Sea. In *EARSeL Workshop on Remote Sensing and GIS for Coastal Zone Management*, edited by L.L.F. Janssen and R. Allewijn (Delft, The Netherlands: Rijkswaterstaat, Survey Department), pp. 287–293.

Schott, J.T., 1979, Temperature measurement of cooling water discharged from power plants. *Photogrammetric Engineering and Remote Sensing*, **45**, 753–761.

Sloggett, D.R., and Jory, I.S., 1995, An operational, satellite-based, European oil slicks monitoring system. In *Sensors and Environmental Applications of Remote Sensing*, edited by J. Askne (Rotterdam: Balkema), pp. 183–187.

Sørensen, K., Lindell, T., and Nisell, J., 1989, The information content of AVHRR, MSS, TM and SPOT data in the Skagerrak Sea. *IGARSS '89 Remote Sensing: an Economic Tool for the Nineties*, Vancouver, Canada July 10–14, 1989 (Vancouver, Canada: IGARSS '89 12th Canadian Symposium on Remote Sensing), **4**, 2439–2442.

Solberg, R., and Theophilopoulos, N., 1997, ENVISYS – a solution for automatic oil spill detection in the Mediterranean. *Proceedings of the Fourth International Conference on Remote Sensing for Marine and Coastal Environments*, Orlando, Florida, 17–19 March 1997, **1**, 3–12.

Stringer, W.J., Dean, K.G., Guritz, R.M., Garbeil, H.M., Groves, J.E., and Ahlnaes, K., 1992, Detection of petroleum spilled from the *MV Exxon Valdez*. *International Journal of Remote Sensing*, **13**, 799–824.

Strong, A.E., 1989, Greater global warming revealed by satellite-derived sea-surface temperature trends. *Nature*, **338**, 642–645.

Stumpf, H.G. and Strong, R.E., 1974, ERTS-1 views on oil slicks. *Remote Sensing of Environment*, **3**, 87.

Subramaniam, A., and Carpenter, E.J., 1994, An empirically derived protocol for the detection of blooms of the marine cyanobacterium *Trichodesmium* using CZCS imagery. *International Journal of Remote Sensing*, **15** (8), 1559–1569.

Suvapepun, S., 1992, *Trichodesmium* blooms in the Gulf of Thailand. In *Marine Pelagic Cyanobacteria: Trichodesmium and Other Diazotrophs*, edited by E.J. Carpenter, D.G. Capone and J.G. Rueter (Dordrecht: Kluwer Academic Publishers), pp. 343–348.

Svejkovsky, J., and Haydock, I., 1997, Satellite remote sensing as part of an ocean outfall environmental monitoring program. *Proceedings of the Fourth International Conference on Remote Sensing for Marine and Coastal Environments*, Orlando, Florida, 17–19 March 1997, **1**, 1–2.

Tassan, S., 1988, The effect of dissolved 'yellow substance' on the quantitative retrieval of chlorophyll and total suspended sediment concentrations from remote measurements of water colour. *International Journal of Remote Sensing*, **9**, 787–797.

Tassan, S., 1992, An algorithm for the identification of benthic algae in the Venice lagoon from Thematic Mapper data. *International Journal of Remote Sensing,* **13** (15), 2887–2909.

Tassan, S., 1993, An improved in-water algorithm for the determination of chlorophyll and suspended sediment concentration from Thematic Mapper data in coastal waters. *International Journal of Remote Sensing*, **14** (6), 1221–1229.

Tassan, S., 1995, SeaWiFS potential for remote sensing of marine *Trichodesmium* at sub-bloom concentration. *International Journal of Remote Sensing*, **16**, 3619–3627.

Tassan, S., 1997, A numerical model for the detection of sediment concentration in stratified river plumes using Thematic Mapper data. *International Journal of Remote Sensing*, **18** (12), 2699–2705.

Tassan, S., and Sturm, B., 1986, An algorithm for the retrieval of sediment content in turbid coastal waters from CZCS data. *International Journal of Remote Sensing*, **7**, 643–655.

Thompson, V., Neville, R.A., Gray, L. and Hawkins, R.K., 1984, Observation of two test oil spills with a microwave scatterometer and a synthetic aperture radar. In *Remote Sensing for the Control of Marine Pollution*, edited by J.M. Massin (New York: Plenum Press), pp. 257–266

Topliss, B.J., Almos, C.L. and Hill, P.R., 1990, Algorithms for remote sensing of high concentration, inorganic suspended sediment. *International Journal of Remote Sensing*, **11**, 947–966.

Vallis, R., Currie, D., and Dechka, J., 1996, Assessment of *casi* data for monitoring chlorophyll–*a* and suspended solids in the Northumberland Strait, P.E.I. *Proceedings of the Sixth (1996) International Offshore and Polar Engineering Conference*, **2**, 380–383.

Vaughan, R.A., Wen, C.-C., and Mansell, W., 1995, Thermal plume monitoring from ATM data. In *Sensors and Environmental Applications of Remote Sensing*, edited by J. Askne (Rotterdam: Balkema), pp. 319–325.

Wald, L., Monget, J.M., Albuisson, A., and Byrne, H.M., 1984, A large scale monitoring of the hydrocarbon pollution from the Landsat satellite. In *Remote Sensing for the Control of Marine Pollution*, edited by J.M. Massin (New York: Plenum Press), pp. 347–358.

WCMC, 1991, *Environmental Effects of the Gulf War*. (Cambridge, UK: World Conservation and Monitoring Centre).

Welby, C.W., 1978, Application of Landsat imagery to shoreline erosion. *Photogrammetric Engineering and Remote Sensing*, **44** (9), 1173–1177.

Whitlock, C.H., Kuo, C.Y., and Le Croy, S.R., 1982, Criteria for the use of regression analysis for remote sensing of suspended sediment and pollutants. *Remote Sensing of Environment*, **12**, 151–168.

Wilson, S.B., and Anderson, J.M., 1984, A thermal plume in the Tay Estuary detected by aerial thermography. *International Journal of Remote Sensing*, **5**, 247–249.

Woodruff, D.L., and Stumpf, R.P., 1997, Remote estimation of Secchi depth, suspended sediments, and light attenuation: impact of storm events on algorithm effectiveness. *Proceedings of the Fourth International Conference on Remote Sensing for Marine and Coastal Environments*, Orlando, Florida, 17–19 March 1997, **2**, 426–435.

Yan, X.-H., and Clemente-Colón, P., 1997, The Maximum Similarity Shape Matching (MSSM) method applied to oil spill feature tracking observed in SAR imagery. *Proceedings of the Fourth International Conference on Remote Sensing for Marine and Coastal Environments*, Orlando, Florida, 17–19 March 1997, **1**, 43–55.

Yentsch, C.S., 1983, Remote sensing of biological substances. In *Remote Sensing Applications in Marine Science and Technology*, edited by A.P. Cracknell (Dordrecht: D. Riedel), pp. 263–297.

Yu-ming, Y., and Hou, M., 1990, A remote sensing analysis of suspended sediment transport. *Remote Sensing for Marine Studies. Report of the Seminar on Remote Sensing Applications for Oceanography and Fishing Environment Analysis*, Beijing, China, 7 to 11 May 1990 (Bangkok: ESCAP/UNDP Regional Remote Sensing Programme (RAS/86/141) ST/ESCAP/897), pp. 45–55.

15

Mapping Bathymetry

Summary *Optical remote sensing offers an alternative to traditional hydrographic surveys for measuring water depth, with the advantage that data are collected synoptically over large areas. However bathymetry can only be derived from remote sensing to a maximum depth of 25 m in the clearest water, and considerably less in turbid water. Bathymetric mapping is also often confounded by variation in substrate albedo.*

This chapter illustrates three methods for mapping bathymetry of varying complexity, with worked examples from the Turks and Caicos Islands. The accuracy of depth prediction is tested against field data and recommendations made as to the best method. Predicted depth correlated very poorly (Pearson product moment correlation coefficient of around 0.50) with actual depth for all but one method (calibrated depth of penetration zones method of Jupp 1988). This method produced a correlation coefficient of 0.91 between predicted and actual depth and required extensive field bathymetric data. Other methods of mapping bathymetry using airborne LIDAR or echo-sounders are briefly mentioned.

Introduction

Superficial bathymetric features in an image reveal themselves at a glance. Colour is the most obvious feature of three of the habitats labelled in Figure 11.3, Plate 12 (sheltered, shallow bank; fairly sheltered, medium depth bank; exposed, deep fringing reef). The shallow areas (< 1 m deep) west of East Caicos are bright, light blue whereas the deep area (15–20 m) between Long Cay and Ambergris (the Lake) is dark blue. Medium depth areas (5–15 m) to the west of Six Hills are intermediate between the two extremes. This illustrates the theory behind all bathymetric mapping using optical remote sensing: more light is reflected from the seabed in shallow water because less has been absorbed in the water column. These areas therefore appear bright, and conversely deep areas are dark. All the methods described here attempt to relate individual pixel values to depth to produce a bathymetric chart for the image area. In this way it is possible to acquire a synoptic picture of water depth over large areas quite rapidly. The alternatives to using passive optical remote sensing are airborne LIDAR or hydrographic survey to collect point or track depth data, with echo-sounders. This chapter will focus on passive optical techniques which can utilise satellite and airborne imagery acquired for habitat mapping. Specialist techniques such as airborne LIDAR and echo-sounders are only discussed briefly at the end of the chapter.

The theory behind measuring bathymetry using remote sensing

The fundamental principle behind using remote sensing to map bathymetry is that different wavelengths of light will penetrate water to varying degrees. When light passes through water it becomes attenuated by interaction with the water column (see Chapter 8). The intensity of light, I_d, remaining after passage length p through water, is given by:

$$I_d = I_0 . e^{-pk}$$

Equation 15.1

where I_0 = intensity of the incident light and k = attenuation coefficient, which varies with wavelength. If a vertical pathway of light from surface to bottom and back is assumed then p may be substituted by the term $2d$, where d = water depth. Equation 15.1 can be made linear by taking natural logarithms (logarithms to the base e):

$$\log_e(I_d) = \log_e(I_0) - 2dk$$

Equation 15.2

Longer wavelength light (red in the visible part of the spectrum) has a higher attenuation coefficient than short wavelengths (blue). Red light therefore attenuates rapidly in water and does not penetrate further than a few metres. By contrast, blue light penetrates much further and in clear water the seabed can reflect enough light to be detected by a satellite sensor even when the depth of water approaches 30 m. The depth of penetration is dependent on water turbidity. Suspended sediment particles, phytoplankton and dissolved organic compounds will all affect the depth of penetration because they scatter and absorb light, thus increasing attenuation (and so limit the range over which optical data may be used to estimate depth).

☞ **Among sensors tested, the measurement of bathymetry from satellites can be expected to be best with Landsat TM data because that sensor detects visible light from a wider portion of the visible spectrum and in more bands. Landsat bands 1, 2 and 3 are all useful in measuring bathymetry: so too is band 4 in very shallow (< 1 m deep), clear water over bright sediment. Bands 5 and 7 are completely absorbed by even a few centimetres of water and should therefore be removed from the image before bathymetric calculations are performed. Digital airborne imagery (e.g. CASI) where a series of wavebands can be selected across the visible spectrum should also be good for bathymetric mapping.**

Radar images may also be used to map depth changes. Water is an excellent reflector of microwaves and so the characteristics of a radar image are controlled by the nature of the sea surface. Variations in centimetre-scale roughness are strongly correlated with changes in water depth of up to approximately 100 m, but only if a significant current is flowing across a smooth seabed. SAR imagery has provided information on bathymetry, although the highly specific conditions of sea state and currents (Lodge 1983, Hesselmans *et al.* 1994) necessary for such analysis have probably prevented wider application.

The use of bathymetric maps derived from remote sensing

Table 15.1 summarises the uses to which bathymetric data derived from remote sensing have been put. Remote sensing has been used to augment existing charts (Bullard 1983, Pirazolli 1985), assist in interpreting reef features (Jupp *et al.* 1985) and map shipping corridors (Kuchler *et al.* 1988). However, it has not been used as a primary source of bathymetric data for navigational purposes (e.g. mapping shipping hazards). A major limitation with respect to satellite data is inadequate spatial resolution. Hazards to shipping such as emergent coral outcrops or rocks are frequently much smaller than the sensor pixel and so will fail to be detected. Another limitation is that the accuracy of depths predicted from images may be in the order of ± 1–2 m which is generally considered inadequate for navigation purposes in shallow waters (< 25 m deep).

Table 15.1 Some uses of bathymetric data derived from remote sensing.

Application of bathymetric data	Landsat MSS	Landsat TM	SPOT XS	Airborne MSS
Mapping shipping hazards	Cracknell *et al.* 1987, Jupp *et al.* 1988	Cracknell *et al.* 1987, Lantieri 1988	Lantieri 1988	Cracknell *et al.* 1987
Mapping transportation corridors	Gray *et al.* 1988, Kuchler *et al.* 1988			
Updating/augmenting existing charts	Benny and Dawson 1983, Biña 1982, 1988, Bullard 1983, Guerin 1985, Hammack 1977, Ibrahim and Cracknell 1990, Lyons 1977, Nordmann 1990, Pirazzoli 1985, Siswandono 1992, Warne 1978	Lantieri 1988, Jiamin and Faiz 1990	Lantieri 1988, Albert and Nosmas 1990, Biña 1988, Fourgrassie 1990, Masson *et al.* 1990	
Planning hydrographic surveys	Benny and Dawson 1983, Claasen and Pirazzoli 1984, Warne 1978		Albert and Nosmas 1990, Fourgrassie 1990	
Coastal sediment accumulation/loss				Danaher and Cottrell 1987, Danaher and Smith 1988
Assisting interpretation of reef features	Jupp *et al.* 1985, Siswandono 1992	Zainal 1993		

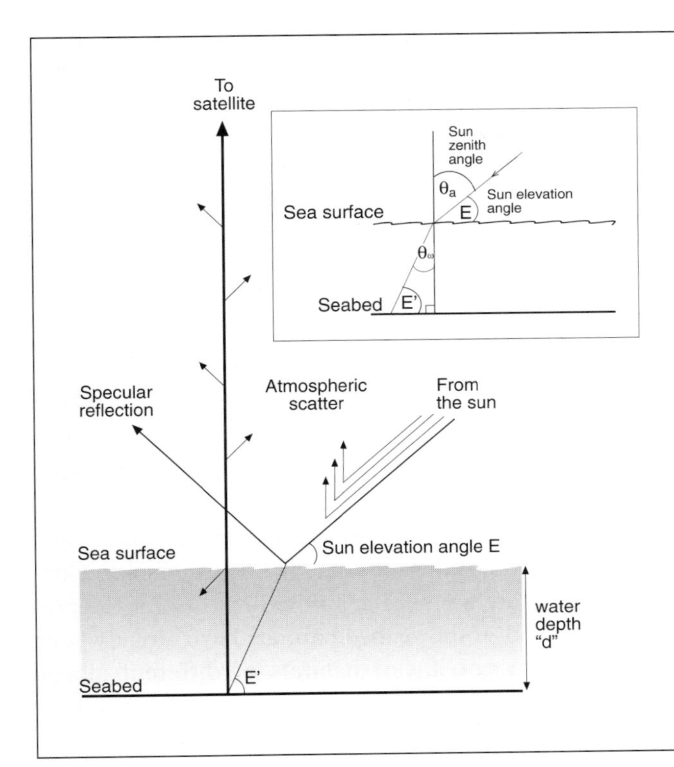

Figure 15.1 The light paths from sun to satellite assumed in the method of Benny and Dawson (1983). The sun elevation angle *E* can be obtained from the imagery header file. *E'* can be calculated from a knowledge of the refractive index of seawater (which is tabulated for different temperatures and salinity). The path of the light from the sea surface to the seabed and back up to the sea surface will be the depth *d + d.cosec(E')*.
The inset shows how to calculate *E'* knowing the refractive index of the water. From the inset it can be seen that $E' = 90^\circ - \theta_w$. $\sin\theta_w$ can be found from the following equation:

$$\frac{\sin\theta_a}{\sin\theta_w} = \frac{n_w}{n_a}$$

where n_w is the refractive index of water and n_a, the refractive index of air (1.000). For example, at 25°C and a salinity of 36 parts per thousand, $n_w = 1.339$ (from tables). If the sun elevation angle (known from the imagery) is 39°, then θ_a (the sun zenith angle) is

51° and $\frac{\sin 51^\circ}{\sin \theta_w} = \frac{1.339}{1.000}$. Thus $\sin \theta_w = \frac{\sin 51^\circ}{1.339} = 0.58039$

so $\theta_w = 35.5^\circ$ and $E' = 90^\circ - 35.5^\circ = 54.5^\circ$.

Main diagram based on: Benny, A.H., and Dawson, G.J., 1983, Satellite imagery as an aid to bathymetric charting in the Red Sea. *The Cartographic Journal*, **20** (1), 5–16.

Three methods of predicting bathymetry

In this chapter we have selected three methods for predicting bathymetry spanning a range of theoretical and processing complexity. All have been used operationally for various applications summarised in Table 15.1. The theory behind each method is explained and the results for Landsat TM are tested against an accuracy dataset from the Caicos Bank.

☞ **The depth of water computed by any method will be depth at the time of image acquisition. These depths must be adjusted for the effects of tide to a datum before any comparison can be made to depths acquired from other images, nautical charts or field data (see section below on *Compensating for the effect of tide on water depth*).**

Benny and Dawson (1983) method

There are three assumptions implicit in this method: i) light attenuation is an exponential function of depth (Equation 15.1), ii) water quality (and hence the attenuation coefficient, *k*) does not vary within an image, and iii) the colour (and therefore reflective properties or albedo) of the substrate is constant. Assumption (ii) may or may not be valid for any image but has to be made unless supplementary field data are collected at the time of image acquisition. Assumption (iii) is certainly not true for many areas and will cause dark, shallow, benthos (such as dense seagrass beds) to be interpreted as deep water. Corrections to the final bathymetry chart could be made if a habitat map exists and the depth range of, say, dense seagrass beds is known. The final assumption can in theory be circumvented if areas of seabed have previously been classified into different habitats which are of differing albedo; these can then be processed separately. However, this would be extremely time-consuming.

This method assumes that light which returns to the sensor follows a path through the water column as described by Figure 15.1. For water of depth *d*, the length of that path is equal to *d + d.cosec(E')*, which can be rewritten as *d(1 + cosec(E'))*. The *cosec* function corrects for the fact that the sun is not vertically overhead at the time of image acquisition. This pathway corresponds to the term *2d* in Equation 15.2. Therefore the amount of light remaining after passage through the water column to depth *d* and back upwards (I_d) will be:

$$I_d = R.I_0.e^{-k.d(1+cosec(E'))}$$

Equation 15.3

where *R* is the proportion scattered upwards from the seabed and I_0 is the amount of incident light.

The sensor will record both the light that has passed through the water column and light returned by specular reflection and atmospheric scattering, Figure 15.1 (see Chapter 7 for further information on specular reflection and atmospheric scatter). Assuming that the radiance of light returned to the sensor by specular reflection and atmospheric scatter is the same for deep water (where this should be the only source of radiance) and shallow water, the total light (L) recorded by the sensor is:

$$L = L_d + R.I_0.e^{-k.d(1+cosec(E'))}$$

Equation 15.4

where L_d = deep-water radiance. The ratio of the radiance from two different depths, x and y, is therefore:

$$\frac{L_x - L_d}{L_y - L_d} = \frac{R.I_0.e^{-k.x(1+cosec(E'))}}{R.I_0.e^{-k.y(1+cosec(E'))}}$$

Equation 15.5

Assuming similar bottom types and thus similar reflectances R at the two depths, this equation simplifies to:

$$\frac{L_x - L_d}{L_y - L_d} = \frac{e^{-k.x(1+cosec(E'))}}{e^{-k.y(1+cosec(E'))}}$$

Equation 15.6

If we let y = 0 (i.e. very shallow water), then L_y becomes L_0, the signal from very shallow water, and the equation reduces to:

$$\frac{L_x - L_d}{L_0 - L_d} = e^{-k.x(1+cosec(E'))}$$

Equation 15.7

because the denominator on the right hand side of Equation 15.6 becomes $e^{-k.0(1+cosec(E'))}$ which simplifies to $e^{-0} = 1$.

Equation 15.6 can be solved for x (depth) by taking natural logarithms and re-arranging:

$$depth\ x = \frac{log_e(L_x - L_d) - log_e(L_0 - L_d)}{-k(1 + cosec(E'))}$$

Equation 15.8

where L_x = the signal recorded by the sensor from water of depth x, L_d = the signal recorded by the sensor from deep (> 30 m) water, L_0 = the signal recorded by the sensor from shallow (ideally a few cm) water, k = attenuation coefficient, E' = the sun elevation angle adjusted for refraction through seawater. The sun elevation angle is either provided with the image acquisition data or can be calculated easily using several different software packages. One such is available free of charge from Dr Barry Clough at the Australian Institute for Marine Science (AIMS, Townsville, Australia 4810). The water attenuation coefficients (k) for each band can be determined using a combination of field survey and image data (see worked example below). L_d and L_0 can all be deduced from the imagery (Table 15.2) and hence the depth of water under each pixel can be estimated.

Table 15.2 Average deep-water (L_d) and shallow water (L_0) pixel values (DN) calculated from Landsat TM data of the Caicos Bank.

Band	L_d	L_0
1	52	158
2	14	87
3	9	86
4	4	21

Jupp (1988) method

Jupp's method makes the same assumptions as Benny and Dawson's, i.e. that (i) light attenuation is an exponential function of depth (Equation 15.1), (ii) water quality (and hence k) does not vary within an image, and (iii) the albedo of the substrate is fairly constant. Consequently there are similar limitations. As for the previous method, if the seabed habitats have already been classified, processing of habitats of different albedo could be carried out separately to avoid the need for assumption (iii).

There are three parts to Jupp's method: (i) the calculation of depth of penetration zones or DOP zones, (ii) the interpolation of depths within DOP zones, and (iii) the calibration of depth within DOP zones. The application of his method will be discussed in relation to Landsat MSS and TM imagery to make it less abstract.

Depth of penetration zones (DOP zones)

As red light attenuates more rapidly than green or blue, there will be a depth, the maximum depth of penetration, at which all the light detected by band 3 of the Landsat TM sensor (630–690 nm) has been fully attenuated (and which will therefore appear dark in a band 3 image). However at this depth there will still be some light which is detectable by bands 2 and 1 of the Landsat TM sensor (520–600 nm, and 450–520 nm respectively). DOP zones are delineated by the maximum depths of penetration (z_i) of successively shorter wavelengths (Figure 15.2).

This is most easily visualised if the same area is viewed in images consisting of different bands (Figure 15.3). Shallower areas are bright but the deepest areas are illuminated only in band 1 because only the shorter wavelengths penetrate to the bottom sufficiently strongly to be reflected back to the satellite.

The attenuation of light through water has been modelled by Jerlov (1976) and the maximum depth of penetration for different wavelengths calculated. Taking Landsat MSS as an example, it is known that green light (band 1, 500–600 nm) will penetrate to a maximum depth of approximately 15 m in the clear waters of the Great Barrier Reef, red light (band 2, 600–700 nm) to 5 m, near infra-red (band 3, 700–800 nm) to 0.5 m and infra-red

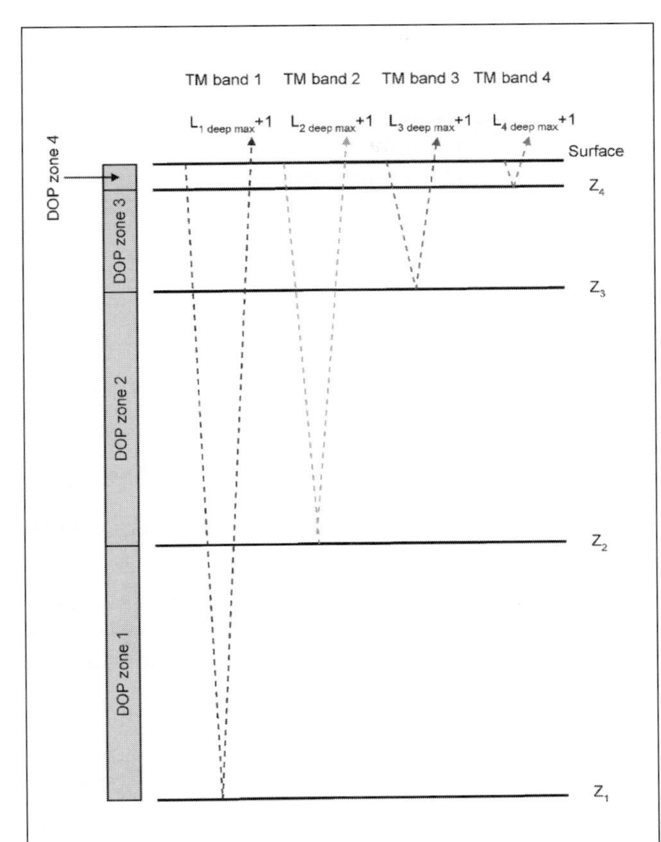

Figure 15.2 Depth of penetration zones (DOPs) for Landsat TM bands 1–4. DOP zones are named after the band of longest wavelength present at that depth. For example, the maximum depth of penetration of red light, z_3, is about 5 m on the Great Barrier Reef (but about 4 m on the Caicos Bank) and in DOP zone 3 the longest waveband penetrating is band 3 (red). Only blue and green light penetrate beyond 5 m, green only as far as about 15 m (z_2) in clear waters around coral reefs. The zone between 5 m and *ca* 15 m deep where the green (Landsat TM band 2) waveband is the longest waveband is therefore called DOP zone 2. At depths greater than z_2, until the maximum depth of penetration of the blue band (z_1), is DOP Zone 1. If using DN data, then the lowest DN (L_{min}) in each zone will be one more than the $L_{deep\ max}$ for the longest wavelength band penetrating that zone.

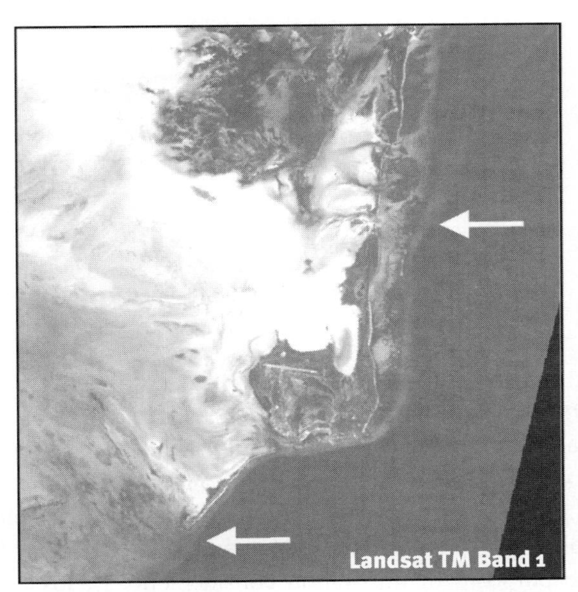

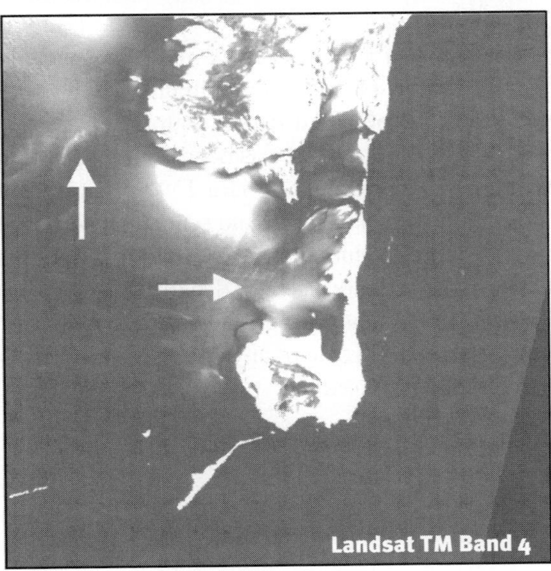

Figure 15.3 Bands 1 and 4 Landsat TM images (DN) of the area around South Caicos. Deeper areas are illuminated by shorter wavebands. There is deep (30–2500 m) water in the bottom right hand corner of the images – this is not completely dark because some light is returned to the sensor over these areas by specular reflection and atmospheric scattering. Band 1 illuminates the deep forereef areas (as indicated by arrows). The edge of the reef or 'drop-off' can be seen running towards the top right hand corner of the band 1 image. In areas of shallow (< 1 m deep) clear water over white sand, band 4 reveals some bathymetric detail (as indicated by the arrows).

(band 4, 800–1100 nm) is fully absorbed (Jupp 1988). Appropriate image processing will produce a 'depth of penetration' image which has been split into depth zones – this is the basis of determining the DOP zones. Each zone represents a region in which light is reflected in one band but not the next. For example, if light is detected in Landsat MSS band 1 but not band 2, the depth should lie between 5 m and 15 m. Using this approach, three zones can be derived from a Landsat MSS image of the Great Barrier Reef, 5–15 m, 0.5–5 m and 0 (surface)–0.5 m. The depths of penetration for Landsat TM bands 1–4 are given in Table 15.3.

Table 15.3 Mean reflectance values – $L_{deep\ mean}$ (with range in parentheses) for an area of typical deep water off the Caicos Bank and maximum depth of penetration, z_i, for Landsat TM bands over the waters of the Great Barrier Reef (from Jupp 1988).

Band (i)	Caicos Bank deep water Mean reflectance ($L_{deep\ min}$–$L_{deep\ max}$)	Great Barrier Reef Depth of penetration, z_i
1	19.2% (18.7–20.3%)	25 m
2	10.2% (8.9–11.3%)	15 m
3	5.5% (5.1–6.4%)	5 m
4	0.9% (0.7–1.6%)	1 m

Interpolation of DOP zones

Calculating DOP zones does not assign a depth to each pixel, instead it assigns a pixel to a depth range (e.g. 12–20 m). Step two of Jupp's method involves interpolating depths for each pixel within each DOP zone. For the purposes of illustration consider only Landsat TM DOP zone 2 (Figure 15.2). For a particular bottom type the DN value can vary between a minimum which is the mean deep-water DN, and a maximum which is the DN which would be obtained if that bottom type was at the surface and no attenuation was occurring. In between the maximum depth of penetration for TM band 2 (z_2) and the surface, where the substrate (and seabed albedo) remains constant the DN value is purely a function of depth (z), with the rate of decrease in DN with depth being controlled by the attenuation coefficient for band 2 (k_2). Thus, in DOP zone 2, the value of any submerged pixel in band 2 (L_2) can be expressed as:

$$L_2 = L_{2deepmean} + \left(L_{2surface} - L_{2deepmean}\right)e^{-2k_2z}$$

Equation 15.9

where $L_{2\ deep\ mean}$ is the average deep-water pixel value for TM band 2, $L_{2\ surface}$ is the average DN value at the sea-surface (i.e. with no attenuation in the water column), k_2 is the attenuation coefficient for TM band 2 wavelengths through the water column, and z is the depth. Caicos Bank deep-water reflectance values for each TM band are shown in Table 15.3. Equation 15.9 can be rewritten as:

$$L_2 - L_{2deepmean} = \left(L_{2surface} - L_{2deepmean}\right)e^{-2k_2z}$$

Equation 15.10

If we define X_2 as follows:

$$X_2 = \log_e\left(L_2 - L_{2deepmean}\right)$$

we can get rid of the exponential term in Equation 15.10, by taking natural logarithms of the right hand side of the equation thus:

$$X_2 = \log_e\left(L_{2surface} - L_{2deepmean}\right) - 2k_2z$$

Equation 15.11

For a given bottom type in TM band 2, $\log_e (L_{2surface} - L_{2deepmean})$ will be a constant. We will call this constant A_2 in order to simplify Equation 15.11, and the following linear regression relationship is produced:

$$X_2 = A_2 - 2k_2z$$

Equation 15.12

This is the basis for interpolating depths in each DOP zone. Continuing to take DOP zone 2 as an example, if a pixel is only just inside DOP zone 2 as opposed to DOP zone 1 (the zone with the next lowest coefficient of attenuation, Figure 15.2) then it has a value of $X_{2\ min}$ given by Equation 15.13 as:

$$X_{2\ min} = A_2 - 2k_2z_2$$

Equation 15.13

A_2 and the attenuation coefficient k_2 are specific to TM band 2, and z_2 is the maximum depth of penetration of TM band 2 (the depth at which minimum reflectances are obtained in DOP zone 2). Now consider another pixel which is only just inside DOP zone 2 as opposed to DOP zone 3 (the zone with the next highest coefficient of attenuation, Figure 15.2). This has a value defined by the equation:

$$X_{2\ max} = A_2 - 2k_2z_3$$

Equation 15.14

where values are as for Equation 15.13 except that z_3 marks the upper end of DOP zone 2 (the depth at which maximum reflectances are obtained in DOP zone 2). The values of $X_{2\ min}$ and $X_{2\ max}$ can be determined by simple image processing and values of z_2 and z_3 are known for different types of water (Jerlov 1976, Jupp 1988). Equa-

Table 15.4 Values of z_i derived from Figure 15.9 and values of k_i and A_i (Equation 15.17) that were used in the calibration of the Caicos Bank DOP zones. n = number of calibration sites in each DOP zone.

	Depth range (m)	n	z_i	k_i	A_i
DOP 1	9.1 – 19.8	20	20	0.113	4.696
DOP 2	1.6 – 11.7	237	12	0.193	4.807
DOP 3	0.1 – 4.1	115	4	0.618	4.946
DOP 4	0 – 0.9	12	0.9	2.345	3.998

tions 15.13 and 15.14 therefore form a pair of simultaneous equations which can be solved for k_2 (Equation 15.15). Once this is known, Equation 15.13 can be rearranged to find A_2 (Equation 15.16).

$$k_2 = \frac{(X_{2\,max} - X_{2\,min})}{2(z_2 - z_3)}$$

Equation 15.15

$$A_2 = X_{2\,min} + 2k_2z_2$$

Equation 15.16

This process can be done for each waveband's DOP zone (Table 15.4). Once A_i and k_i are known for each DOP zone i, Equation 15.12 can be inverted and the depth of water, z, for any pixel with value $X_i = log_e(L_i - L_{i\,deep\,mean})$ in band i calculated:

$$z = \frac{(A_i - X_i)}{2k_i}$$

Equation 15.17

Calibration of DOP zones

Interpolation of DOP zones uses values of z_i (Table 15.3) which are derived from Jerlov (1976) and assume specific water qualities which may not be appropriate in every case. If the specific conditions assumed by Jerlov are not valid for other areas, or did not prevail at the time of image acquisition, then the use of the values of z_i in Table 15.3 will produce erroneous estimates of depth. For example, if the water is more turbid than the Great Barrier Reef then values of z_i will be lower because the light will be attenuated more rapidly and use of the values of z_i in Table 15.3 will over-estimate depth. However, if field data are available, then each DOP may be calibrated to the water conditions more appropriate to the area of interest or time of image acquisition.

Calibration requires depth information from sites within each DOP zone. This is used to determine the maximum and minimum depth actually occurring within each DOP zone. For example, interpolation assumes that z_2 is

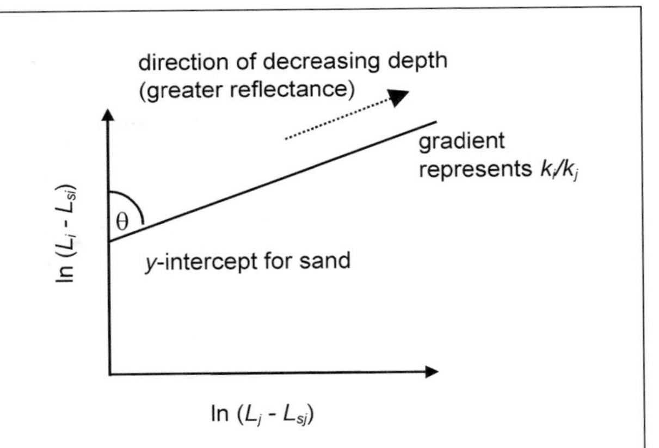

Figure 15.4 Regression line through a bi-plot of the natural logarithms of pixel radiances (atmospherically corrected using the deep water subtraction method) in bands *i* and *j* for sand at a range of depths. The slope of the line represents the ratio of the attenuation coefficients k_i/k_j, whilst the *y*-intercept is the depth-invariant bottom index for sand (see Chapter 8). L_i and L_j represent the pixel radiances recorded, and L_{si} and L_{sj} the average deep water radiances in bands *i* and *j* respectively.

15 m (Table 15.3). However if a combination of atmospheric conditions, turbidity and bottom albedo mean that the maximum depth recorded in DOP zone 2 is, say, 12 m then this value would be used in Equations 15.15 and 15.16 as z_2 to calculate more realistic values of A_2 and k_2. Depth measurements over a uniform substrate such as sand are usually used for calibration. If an image was dominated by say sand and seagrass, then separate calibrations could be performed for both habitat types.

Lyzenga (1978) method

The process of water column correction, described in Chapter 8, attempts to compensate for the effect of variable depth (bathymetry). This process can be modified to measure the effect of variable depth rather than compensate for it (Lyzenga 1978, 1981, 1985).

Lyzenga's method, like the others, assumes that light attenuation is an exponential function of depth and that water quality does not vary within an image so that the ratio of attenuation coefficients for a pair of bands i and j, k_i/k_j, is constant over an image. However by calculating the ratio of attenuation coefficients for band pairs (Chapter 8) it does not assume that the reflective properties of the substrate are constant.

Referring back to Chapter 8 on water column correction and Equation 15.2, we have seen that a transformation using natural logarithms will linearise the effect of depth on bottom reflectance (because light attenuates exponentially with depth). Theoretically, each bottom type should be represented by a parallel line, the gradient of which is the ratio of the attenuation coefficients for each band

(k_i/k_j). In the example of Figure 15.4, for one bottom type, sand, all pixels lie along the same line. Those pixels further up the line have greater reflectance and are found in shallower water. Those pixels nearest the y-intercept are found in deeper water.

The y-intercept can be used as an index of bottom type. Relative depth can be inferred by the position of a pixel along a line of gradient k_i/k_j. Depth varies along the line so that the shallowest (brightest) pixels have the highest values, the deepest (darkest) pixels the lowest. Once the gradient k_i/k_j is known, the axes of the bi-plot can be rotated through the angle θ (Figure 15.4) so that the new y-axis lies parallel to k_i/k_j. Positions of pixels along this new y-axis are indicative of depth.

Note: For simplicity, only two dimensions are shown in the example above (Figure 15.4). In reality for an n-band image, pixel values would be plotted and the axes rotated mathematically in n-1 dimensions.

The transformation described above only provides the relative depth of pixels. To calculate the actual depth, the relationship must be calibrated using bathymetric data (i.e. regression of new y-value against actual depth). In practice, you do not necessarily need to understand the mathematics underlying the method but just need to know how to implement it as described in the examples below.

Compensating for the effect of tide on water depth

Water depth calculated from an image or measured by a hand-held echo-sounder is depth beneath the surface. This depth will depend on the height of the tide at the time of satellite overpass. Field depth data will also have been collected at different times of the day, at different points in the tidal cycle. Image and field depths may be standardised by reducing to depth below datum. The datum, which is published in pilots and navigational manuals for the various nautical regions of the world, is normally calculated from the Lowest Astronomical Tide (LAT) level (the lowest point that the tide ever recedes to). Depths are recorded on nautical charts as depth below datum: this standardisation must be carried out before image, chart and field depths can be compared.

Simple software programmes are available to calculate height of tide above datum for any location, at any time of day, on any date. The *Tidal Prediction by the Admiralty Simplified Harmonic Method* NP 159A Version 2.0 software of the UK Hydrographic Office was used to calculate tidal heights during each day of field survey. Admiralty or other tide tables could be used if this or similar software is not available. The height of tide at North Caicos (the reference port for the Turks and Caicos Islands) at 09:30 hours (time of satellite overpass) on Thursday 22/11/1990 (date of Landsat TM image acquisition) was 0.73 m above datum. The depth of water, z, calculated (by any method) from the TM data is equal to the depth below datum plus tidal height (Figure 15.5). Therefore the depth of water below datum is z - 0.73 m. In other words, 0.73 must be subtracted from all depths derived from these data in order to standardise them to datum.

Similarly, the depth of water, z, measured in the field with an echo-sounder is equal to the depth below datum plus tidal height (Figure 15.5). The tidal height for each depth recording, h, must therefore be calculated and subtracted from the measured depth to give the depth of water below datum (z - h) in metres. Thus it is essential to record the time when each calibration depth is measured so that it can be corrected to datum later.

Accuracy of bathymetric mapping using Landsat TM

The accuracy of these three methods of predicting bathymetry were tested using data from 750 sites on the Caicos Bank, with the exception of calibrated DOP zones. In the latter case the data were split: 384 sites were used to calibrate the DOP zones, and 366 used to test accuracy.

The strength of correlation between predicted depth and actual depth was taken as a measure of accuracy. Correlation coefficients were low (Pearson product moment

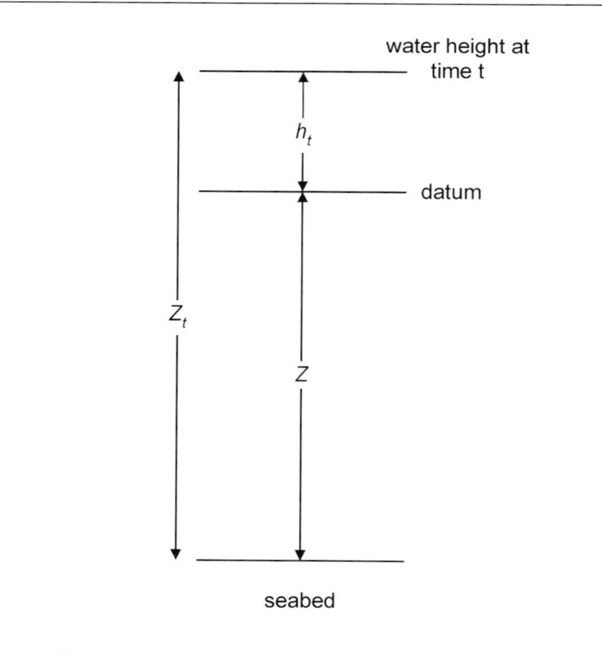

Figure 15.5 Converting depths to depths below datum. z_t = the water depth at time t (in the field or estimated from remotely sensed data), h_t = the height of the tide above datum at time t. The depth of the seabed below datum (z) is therefore z_t-h_t.

Table 15.5 The correlation between predicted and measured depths (standardised to depth below datum) for a series of sites on the Caicos Bank. n = number of sites.

Method	Pearson Product Moment Correlation	n
Benny and Dawson (1983)	0.52	750
Interpolated DOP (Jupp 1988)	0.71	750
Calibrated DOP (Jupp 1988)	0.91	384
Lyzenga (1978)	0.53	750

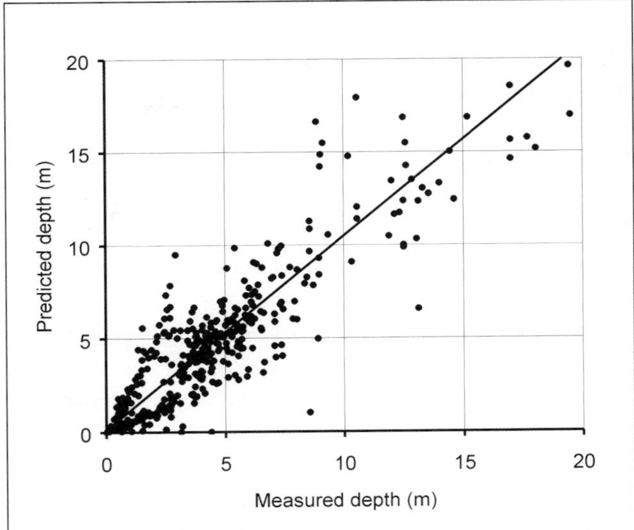

Figure 15.6 Plot of depth predicted from satellite imagery using the method of Jupp (1988) and depth measured in the field. Both have been corrected to chart datum.

correlation coefficients of around 0.5) for the Benny and Dawson method and for Lyzenga's method (Table 15.5). These are not recommended for predicting bathymetry where bottom type is at all variable, unless all that is required is a very crude bathymetric map. A good correlation (0.91) was obtained between predicted depth in calibrated DOP zones and actual depth, and therefore this method is recommended as the most appropriate for mapping bathymetry. However this is the only method which requires substantial amounts of field data. The need for field data is emphasised further by comparing the theoretical values for maximum depth of penetration for the Landsat TM bands in Table 15.3 with the maximum depth observed in each DOP zone (Table 15.4). Band 1, for example, penetrates to a depth of 25 m in the waters of the Great Barrier Reef but a maximum depth of 20 m was found for DOP zone 1 calibration pixels on the Caicos Bank. Values of z_i should be estimated for each site using a combination of image and field data.

Much of the error in predicting depth is caused by variation in the albedo and reflective properties of the substrate. The methods of Benny and Dawson and of Jupp assume a fairly constant substrate albedo, which is generally not the case in habitats as different as dense seagrass and white sand. Dark areas, such as seagrass, will be interpreted as deep water with these methods. Lyzenga's method results in a single band, which should account for depth and not bottom type. However, while depth appears to be accentuated in places, visual inspection of imagery shows that distinct bottom features are still visible (e.g. seagrass beds) which confound the estimates of depth.

We found that the average difference between depths predicted from imagery and ground-truthed depths ranged from about 1.0 m in shallow water (< 2.5 m deep) to about 2.7 m in deeper water (10–20 m deep) using Jupp's method (Figure 15.6). However, this was with a wide range of substrate types and thus seabed albedos being present.

Lyzenga's depth indices were plotted against the actual depth (tide-corrected) (Figure 15.7). There was considerable spread in the values and the Pearson Product-Moment correlation between depth indices and depth was only 0.53. This correlation was considered too weak to estimate depths and therefore, the method is not recommended for mapping bathymetry.

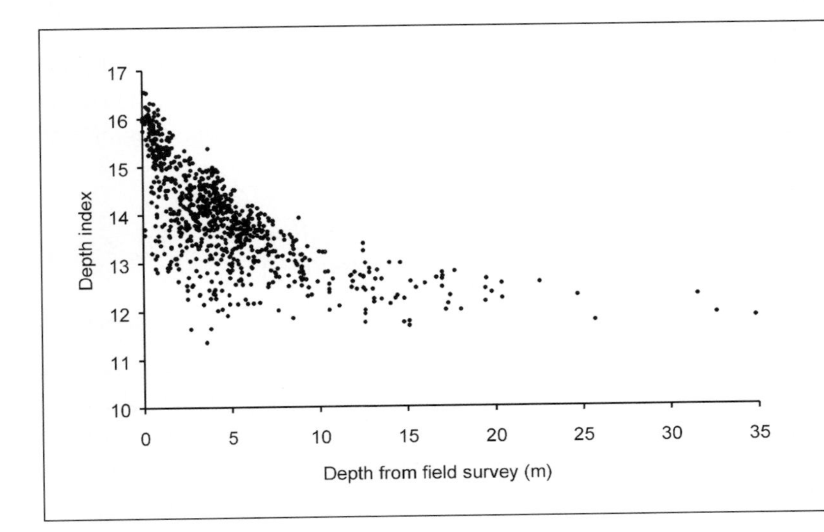

Figure 15.7 Calibration of depth indices on a Landsat TM image derived using the method of Lyzenga (1978) against field measurements of depth from 750 sites.

Examples of how to carry out bathymetric mapping using Landsat TM data

Benny and Dawson's method

Input image data can be DN, radiance or reflectance data. Remove all infra-red bands and mask out land areas.

1. Estimate the attenuation coefficients k_i for each band i. This requires field data, specifically a series of sites of known and varying depth in the same habitat and therefore bottom cover. When present, light sand is best because it occurs over a wider range of depths and is more uniform in colour than seagrass, for example. Digital data in each band are logarithmically transformed and regressed against depth – the regression line will have a negative gradient whose value is equal to k. Figure 15.8 is an example of Landsat TM band 3 data regressed against depth for 50 sand sites in the Turks and Caicos Islands: $k_3 = 0.547$. Similar plots determined $k_1 = 0.136$ and $k_2 = 0.339$. Blue light does not attenuate in water as fast as green, which does not attenuate in water as fast as red: therefore one would expect $k_1 < k_2 < k_3$.
2. Estimate L_d for each waveband. An area of deep water is selected and the mean pixel value calculated for each band. L_d values for Landsat TM data of the Caicos Bank are given in Table 15.2.
3. Estimate L_o for each waveband. An area of shallow water is selected and the mean pixel value for each band calculated. L_o values for Landsat TM data of the Caicos Bank are given in Table 15.2.
4. A good image processing software package will allow you to construct mathematical models. This may involve writing programmes in the processing software or the construction of 'flow-diagrams' using a graphical interface such as the Model Maker module in ERDAS Imagine. These flow-diagrams are automatically implemented to carry out the mathematical manipulations required to all pixels in the image. Equation 15.18 is thus carried out in the image processing software so that each pixel is assigned a depth x. A depth image is thereby created from each band: pixels with value equal to or less than L_d will be assigned a value of infinity and represent deep water, for that band.

The selection of the most appropriate band to use requires some knowledge of the likely range of depths. For example, imagine a situation in which large areas of the seabed were between 15 and 25 m deep. Landsat TM band 2 has a maximum penetration of 15 m (Jupp 1988) and so, in this case, there is no point in using band 2 to map bathymetry (because L_x will be less than or equal to L_d for nearly all pixels). Band 1 would have to be used instead. On the other hand in an area of shallow water, say 5 m maximum, bands 2 or 3 are probably best because pixels are more likely to be saturated in band 1 (resulting in very little variation in L_x), especially if the bottom is covered in highly reflective material such as white carbonate sand. In practice most areas to be mapped will range from shallow inshore water to deep water off the reef edge. Therefore it is best to compute depth images from all bands. Take a pair of adjacent wavebands, i and j, where i has a shorter wavelength than j. Some light in band j is being reflected from the seabed to the sensor if the signal in band j from water of depth x is more than the signal in band j from deep water (i.e. if L_x is > L_d). In this case Equation 15.18 is used:

$$depth\ x = \frac{\log_e(L_{x_j} - L_{d_j}) - \log_e(L_{0_j} - L_{d_j})}{-k_j(1 + cosec(E'))}$$

Equation 15.18

If the depth is such that no light in band j is being reflected from the seabed to the sensor (i.e. if L_{xj} is $\leq L_{dj}$) then depth must be calculated using the next (shorter wavelength) band available, band i:

$$depth\ x = \frac{\log_e(L_{x_i} - L_{d_i}) - \log_e(L_{0_i} - L_{d_i})}{-k_i(1 + cosec(E'))}, \text{ if } L_{xj} \text{ is} <= L_{dj}$$

Equation 15.19

where L_{xj} is the data value of the same pixel in band j (e.g. TM band 2), and L_{xi} the data value of a pixel in band i, a band of shorter wavelength and therefore less attenuation (e.g. TM band 1). In essence, this ensures that depths are calculated using data from the appropriate band.

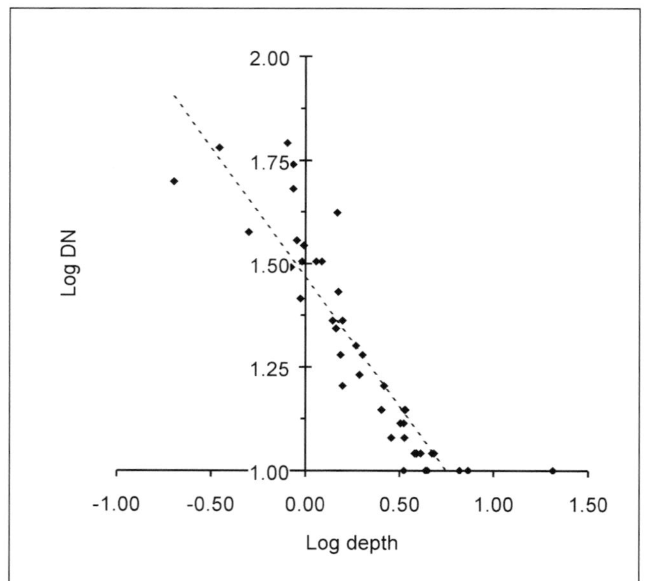

Figure 15.8 DN values from Landsat TM band 3 regressed against logged depth from 50 sand sites on the Caicos Bank. Data have been transformed using natural logarithms. The regression equation is $y = 1.425 - 0.547x$ ($r^2 = 0.69$, $df = 1,49$, $P < 0.001$). The gradient of the regression line is equal to the coefficient of attenuation for band 3.

Jupp's method

Creating depth of penetration (DOP) zones

Data can be used as DN, radiance or reflectance. DN or radiance data require either dark pixel subtraction (see Chapter 7) or some other form of atmospheric correction. For the example here, Landsat TM data were radiometrically corrected to percentage reflectance values. Remove all infrared bands and mask out land areas.

1. Choose an area of deep water with properties you believe to be typical for the area.
2. Calculate the maximum, minimum and mean deep-water pixel values for each band (Table 15.3). Let the minimum in band *i* be $L_{i\ deep\ min}$, the maximum $L_{i\ deep\ max}$ and the mean $L_{i\ deep\ mean}$.
3. Jupp infers the maximum depth of penetration, z_i, for each Landsat TM band over the Great Barrier Reef from the work of Jerlov (1976) (Table 15.3). If a pixel value in band *i*, L_i, is > $L_{i\ deep\ max}$ then some light in band *i* is being reflected from the seabed to the sensor. The depth is therefore less than the maximum depth of penetration, denoted by z_i for band *i*. If, for the same pixel, L_j (the data value in the next band) is > $L_{j\ deep\ min}$ (rules out zero values) and ≤ $L_{j\ deep\ max}$, then the depth of that pixel is between z_i and z_j. In the Turks and Caicos Islands example, if a pixel had a reflectance of more than 20.3% in TM band 1 then it was assumed to be shallower than 25 m: if its relectance in band 2 was also less than or equal to 11.3% then it was assumed to be deeper than 15 m (Table 15.3). Such a pixel would thus be placed in DOP zone 1, 15–25 m deep. DOP zones for all pixels were calculated using the following conditional statements (based on the deep-water pixel values for each band in Table 15.3).

TM Band	1	2	3	4	
Deep-water maximum	20.3%	11.3%	6.4%	1.6%	DOP
If L_i value is	≤ 20.3%	≤ 11.3%	≤ 6.4%	≤ 1.6%	then depth > 25 m
If L_i value is	> 20.3%	≤ 11.3%	≤ 6.4%	≤ 1.6%	then depth = 15–25 m (DOP zone 1)
If L_i value is	> 20.3%	> 11.3%	≤ 6.4%	≤ 1.6%	then depth = 5–15 m (DOP zone 2)
If L_i value is	> 20.3%	> 11.3%	> 6.4%	≤ 1.6%	then depth = 1–5 m (DOP zone 3)
If L_i value is	> 20.3%	> 11.3%	> 6.4%	> 1.6%	then depth = 0–1 m (DOP zone 4)

If none of these conditions apply, then the pixel was coded to 0.

4. A few erroneous pixels will have higher values in band 2 than band 1, which should not happen in theory. They were coded to zero and filtered out of the final depth image.

Interpolating DOP zones

1. Use the DOP images to make 'DOP zone masks' for all bands. This is done by coding all pixels within DOP zone 1 to a value of 1 and all other pixels in the image to 0, and repeating with DOP zones 2-4, generating four masks. Note that the land mask delimits the landward margin of DOP zone 4.
2. Multiply the original file by each DOP zone mask. This will preserve the original data in only those pixels which lie within each DOP zone. Data in pixels outside the DOP zone will be recoded to 0. For example, multiplication of a TM image by the DOP zone 1 mask will result in a four band image (because bands 5, 6 and 7 have been discarded): pixels within DOP zone 1 will have the original data in four bands, whilst pixels outside the DOP zone will have data consisting of zeros in all four bands.

☞ **This stage requires a great deal of free disk space. A four band Landsat TM image is approximately 370 MByte. Each DOP zone image will be the same size as the original: to create DOP zone images from bands 1–4 would therefore require 1.3 GByte of free disk space.**

3. Use the image processing software to produce histograms of pixel values in each DOP zone and thus estimate $L_{i\ max}$ and $L_{i\ min}$ for each DOP zone *i*. $X_{i\ min}$ and $X_{i\ max}$ can then be calculated since $L_{i\ deep\ mean}$ is known and $X_{i\ min} = log_e(L_{i\ min} - L_{i\ deep\ mean})$ and $X_{i\ max} = log_e(L_{i\ max} - L_{i\ deep\ mean})$. The results for the Turks and Caicos Islands are presented in Table 15.6. k_i and subsequently A_i can then be calculated from Equations 15.15 and 15.16.
4. A good image processing software package will allow you to construct mathematical models. This may involve writing programmes in the operating language or the construction of 'flow-diagrams' using a graphical interface such as the Model Maker module in ERDAS Imagine. Equation 15.17 is written using the image processing software and a depth assigned to each pixel in each DOP band to produce four separate interpolated DOP depth images. These are added together to produce a depth image for the area of interest.

☞ **The computed depth of water will be depth** at the time of image acquisition **and must be tidally compensated before any comparison can be made to depths acquired from other images, nautical charts or field data. See *Compensating for the effect of tide on water depth* for details.**

Calibration of DOP zones

1. Depth data typically consist of echo-sounder readings at a series of positions. Calculate which sites lie within each DOP and plot frequency distribution histograms of known depths in each DOP (Figure 15.9). Note that too few depth measurements have been taken at depths greater than

Table 15.6 $L_{i\,max}$ and $L_{i\,min}$ for each DOP zone, *i*, derived from Landsat TM of the Caicos Bank. k_i and A_i were calculated using Equations 15.15 and 15.16. Note that $L_{i\,min}$ values for each DOP zone are $L_{i\,deep\,max}$ values in Table 15.3 + 0.1% reflectance (see Figure 15.2).

Landsat TM	Band 1	Band 2	Band 3	Band 4
	$L_{1\,min}$–$L_{1\,max}$	$L_{2\,min}$–$L_{2\,max}$	$L_{3\,min}$–$L_{3\,max}$	$L_{4\,min}$–$L_{4\,max}$
DOP 1	20.4–26.5%			
DOP 2		11.4–36.4%		
DOP 3			6.5–51.7%	
DOP 4				1.7–55.4%
k_i	0.090	0.154	0.479	2.111
A_i	4.696	4.807	4.791	3.998

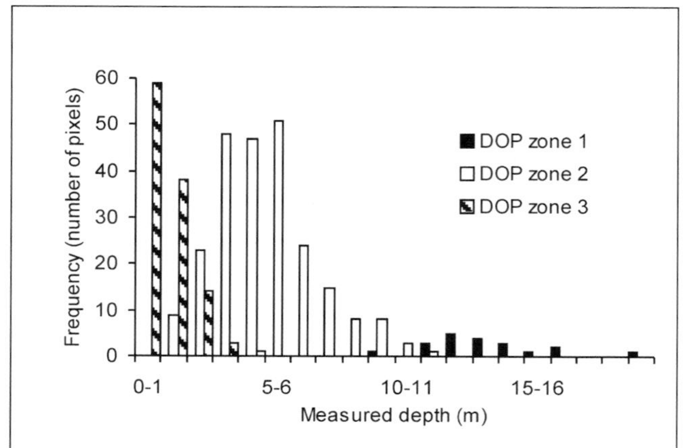

Figure 15.9 The frequency (number of pixels in each one-metre depth interval) for calibration of pixels in DOP zones 1, 2 and 3 (Landsat TM) plotted against depth. Inspection of the histogram suggests z_1 = 20 m, z_2 = 12 m and z_3 = 5 m. However, inspection of measured depth values (Table 15.5) indicates that z_3 is closer to 4 m (the deepest DOP zone 3 pixel actually being at 4.05 m below LAT). Allowing for noise and using the histogram intersections (Jupp 1988) would suggest z_1 = 17 m, z_2 = 11 m and z_3 = 3 m. Bathymetry maps could be generated using both sets of values then checked for accuracy (Figure 15.6), with the most accurate being utilised.

12 m, making it unclear whether the DOP zone 1 sample of calibration pixels is adequately representative.

☞ **The depth of water will be have been measured in the field at varying points on the tidal cycle and at a different time of day to image acquisition. Variation in water depth must be compensated for by calculating the depth of water below datum. See *Compensating for the effect of tide on water depth* for details of how to do this. Also make sure that the full range of depths are adequately sampled.**

2. Jupp indicates that the point of intersection between histograms is the best decision value for the depth separating each DOP zone. The considerable overlap between zones in Figure 15.9 makes this quite difficult and values of z_3 and z_2 of 3 m and 11 m respectively rather than 4 m and 12 m, as suggested, would be equally valid. Results for the Caicos Bank DOPs are given in Table 15.5.
3. New values of z_i, k_i and A_i are calculated and Equation 15.17 written to assign depths to each pixel in each DOP.

Lyzenga's method

This is a modification by Van Hengel and Spitzer (1991) of Lyzenga's method. Data can be used as DN, radiance or reflectance. Remove all infra-red bands and mask out land areas.

1. *Overview.* If a three-band image, such as Landsat TM band 1–3, is used then all bands are used to calculate two rotation parameters *r* and *s* (Equations 15.22 and 15.23). These rotation parameters are based on the geometry of a three dimensional plot of all bands for a constant substratum (such as sand). The rotation is achieved by applying the rotation parameters to all bands to create a single index (band) of depth (Equation 15.24).
2. *Determination of rotation parameters.* Pixels are selected which represent a uniform bottom type at various depths (usually sand). Bands are transformed with natural logarithms to remove pixels that show saturation or complete absorption in any band (the same principle as in water column correction, Chapter 8). We have bands 1, 2 and 3 available using Landsat TM: the pixel values in each band are denoted X_1, X_2 and X_3 after transformation. X_i is therefore equal to $\log_e$(DN or radiance or reflectance).

☞ **It is difficult to follow the derivation of the rotation parameters intuitively. Indeed the mathematics of rotating a three dimensional matrix (a three band image) are complex and beyond the scope of this *Handbook*. The four key equations which are necessary to calculate a depth index are presented here. Readers who require more information on the mathematics of this method are referred directly to Van Hengel and Spitzer (1991).**

3. Two intermediate terms, u_r and u_s, are calculated from the variance and covariance of the transformed data:

$$u_r = \frac{(Var\,X_2 + Var\,X_1)}{2\,Var\,X_2X_1}$$

Equation 15.20

$$u_s = \frac{(Var\,X_3 + Var\,X_1)}{2\,Var\,X_3X_1}$$

Equation 15.21

where $Var\,X_i$ is the variance of X_i and $Var\,X_iX_j$ is the covariance of bands X_i and X_j.

4. Two rotation parameters, *r* and *s*, are created from u_r and u_s:

$$r = arctan\left(u_r + \sqrt{(u_r^{\,2} + 1)}\right)$$

Equation 15.22

$$s = arctan\left(u_s + \sqrt{(u_s^{\,2} + 1)}\right)$$

Equation 15.23

5. The image is multiplied by the rotation parameters, *r* and *s*, to create the depth index as follows:

$$Depth\ index = \left[(cos(r) \times sin(s)) \times X_1\right] + \left[(sin(r) \times cos(s)) \times X_2\right] + \left[sin(s) \times X_3\right]$$

Equation 15.24

6. This assigns a relative depth value (depth index) to each pixel. Field bathymetry data are necessary to convert this to actual depth. In theory there should be a strong correlation (> 0.8) between depth index and actual depth. If so, then depth index can be regressed against real depth and the regression equation used to calibrate the whole image.

☞ **We did not perform this regression because there was a poor correlation between depth index and actual depth, see Table 15.4 and Figure 15.7.**

Cost considerations

Cost of field equipment

Various battery-powered hand-held sonar devices are available, especially from suppliers of dive and fishing equipment. We used a ScubaPro PDS-2, which is accurate to 0.1 m and cost £150. Five depth recordings were taken at every survey and accuracy site and averaged.

Processing time

The time to process a TM scene was:

Benny and Dawson (1983)	0.5 days reading up on method and 0.5 days image processing
Jupp (Interpolated DOP zones)	1 day reading up on method and 1 day image processing
Jupp (Calibrated DOP zones)	1 day reading up on method and 1.5 days image processing
Lyzenga (1978)	1 day reading up on method and 1 day image processing

Other methods of mapping bathymetry

By contrast to passive optical remote sensing, airborne LIDAR can produce very accurate (±20–30 cm; Hare 1994) and detailed bathymetric charts to 2.5–3 times Secchi disc depth (50–60 m in clear waters) and about 2.5 times the depth of penetration of passive optical remote sensing technologies. The accuracy meets International Hydrographic Organisation Standards for hydrographic surveys. LIDAR is usually operated from medium to small twin-engined aircraft which has been fitted with a standard aerial photographic camera hatch because a hole (about 60 cm) in the undercarriage of the aircraft is necessary to allow the laser beams to pass out to the water surface and back. If such an aircraft is not available locally then it can be costly to bring in an appropriate survey aircraft. A LIDAR system can cover about 100 km^2 per day if the survey site is reasonably close to an airport. Quotes obtained for the Turks and Caicos Islands suggest that a small area of 100 km^2 could be surveyed for approximately US\$100,000–150,000, not including aircraft and equipment mobilisation/demobilisation costs from the US mainland, living costs and accommodation (estimated at roughly US\$75,000–85,000). This figure would include data processing. For an area of 1000 km^2 the cost would be US\$250,000–300,000, plus mobilisation/ demobilisation costs etc.

Boat-based acoustic surveys using single or multi-beam depth sounders can produce bathymetric maps of similar accuracies and can operate to depths in excess of 500 m. The drawback of such boat-based techniques is that it may not be feasible to survey shallow water less than 2–3 m deep (about a third of the Caicos Bank) because of sounder saturation and/or the draught of the survey boat. If bathymetric charts are an objective then airborne LIDAR or boat-based acoustic techniques (depending on depth range required) appear to be the remote sensing technologies to use. However, there may be instances where crude bathymetric maps which indicate major depth contours are useful (Figure 15.10, Plate 17), in such cases the method of Jupp (1988) can be implemented in a couple of days.

Conclusions

Methods which attempt to estimate bathymetry from remotely sensed imagery without field data appear, at best, to be crude. This is presumably because it is rarely valid to assume that water quality and substrate colour are constant across large areas. Field data are necessary to calibrate the imagery and test the accuracy of depth prediction. Jupp's method of interpolated DOP zones is recommended as the most accurate of the three methods tested here.

References

Albert, F., and Nosmas, P., 1990, Assessment of SPOT data for the bathymetric model of a reef area in New Caledonia. *Proceedings of the International Workshop on Remote Sensing and Insular Environments in the Pacific: Integrated Approaches*, ORSTOM/IFREMER, 245–246.

Benny, A.H., and Dawson, G.J., 1983, Satellite imagery as an aid to bathymetric charting in the Red Sea. *The Cartographic Journal*, **20** (1), 5–16.

Biña, R.T., 1982, Application of Landsat data to coral reef management in the Philippines. *Proceedings of the Great Barrier Reef Remote Sensing Workshop*, James Cook University, Townsville, Australia, May 5th 1982, 1–39.

Biña, R.T., 1988, An update on the application of satellite remote sensing data to coastal zone management activities in the Philippines. *Proceedings of the Symposium on Remote Sensing of the Coastal Zone*, Gold Coast, Queensland, September 1988, I.2.1–I.2.7.

Bullard, R.K., 1983, Detection of marine contours from Landsat film and tape. In *Remote Sensing Applications in Marine Science and Technology*, edited by A.P. Cracknell (Dordrecht: D.Reidel), pp. 373–381.

Claasen, D. B. van R., and Pirazzoli, P. A., 1984, Remote sensing: A tool for management. In *Coral Reef Handbook*, edited by R.A. Kenchington and B.E.T. Hudson (Paris: UNESCO), pp. 63–79.

Cracknell, A.P., Ibrahim, M., and McManus, J., 1987, Use of satellite and aircraft data for bathymetry studies. *Proceedings of the 13th Annual Conference of the Remote Sensing Society*, University of Nottingham, 391–402.

Danaher, T.J., and Smith, P., 1988, Applications of shallow water mapping using passive remote sensing. *Proceedings of the Symposium on Remote Sensing of the Coastal Zone*, Gold Coast, Queensland, September 1988, VII.3.1–VII.3.8.

Danaher, T.J., and Cottrell, E.C., 1987, Remote sensing: a cost-effective tool for environmental management and monitoring. *Pacific Rim Congress*, Gold Coast, Australia, 1987: 557–560.

Fourgrassie, A. J., 1990, The marine image map, a way to map Polynesian atolls. *Proceedings of the International Workshop on Remote Sensing and Insular Environments in the Pacific: Integrated Approaches*, ORSTOM/IFREMER, 329–342.

Gray, D., Holland, P.R., and Manning, J., 1988, The marine remote sensing experience of the Australian surveying and land information group. In *Marine and Coastal Remote Sensing in the Australian Tropics*, edited by K.G. McCracken and J. Kingwell, (Canberra: CSIRO Office of Space Science and Applications), 52 pp.

Guerin, P.R., 1985, Reducing remoteness with remote sensing. *Proceedings of the 27th Australian Survey Congress*, Alice Springs, 119–125.

Hammack, J.C., 1977, Landsat goes to sea. *Photogrammetric Engineering and Remote Sensing*, **43** (6), 683–691.

Hare, R., 1994, Calibrating Larsen-500 Lidar bathymetry in Dolphin and Union Strait using dense acoustic ground-truthing. *International Hydrographic Review*, Monaco, **LXXXI** (1), 91–108.

Hesselmans, G.H.F.M., Wensink, G.J., and Calkoen, C.J., 1994, The use of optical and SAR observations to assess bathymetric information in coastal areas. *Proceedings of the 2nd Thematic Conference Remote Sensing for Marine Coastal Environments*, New Orleans, **1**, 215–224.

Ibrahim, M., and Cracknell, A.P., 1990, Bathymetry using Landsat MSS data of Penang Island in Malaysia. *International Journal of Remote Sensing*, **11**, 557–559.

Jiamin, H., and Faiz, M., 1990. Mapping the central atoll of the Maldives using Landsat TM data. *Asian-Pacific Remote Sensing Journal*, **3** (1), 97–110.

Jerlov, N.G., 1976, *Marine Optics*. (Amsterdam: Elsevier)

Jupp, D.L.B., 1988, Background and extensions to depth of penetration (DOP) mapping in shallow coastal waters. *Proceedings of the Symposium on Remote Sensing of the Coastal Zone*, Gold Coast, Queensland, September 1988, IV.2.1–IV.2.19.

Jupp, D.L.B., Mayo, K.K., Kuchler, D.A., Classen, D van R., Kenchington, R.A., and Guerin., P.R., 1985, Remote sensing for planning and managing the Great Barrier Reef Australia. *Photogrammetria,* **40**, 21–42.

Kuchler, D., Biña, R.T., and Claasen, D. van R., 1988, Status of high-technology remote sensing for mapping and monitoring coral reef environments. *Proceedings of the Sixth International Coral Reef Symposium*, Townsville, Australia, **1**, 97–101.

Lodge, D.W.S., 1983, Surface expressions of bathymetry on SEASAT synthetic aperture radar images. *International Journal of Remote Sensing*, **4**, 639–653.

Lyzenga, D.R., 1978, Passive remote sensing techniques for mapping water depth and bottom features. *Applied Optics*, **17**, 379–383.

Lyzenga, D.R., 1981, Remote sensing of bottom reflectance and water attenuation parameters in shallow water using aircraft and Landsat data. *International Journal of Remote Sensing*, **2**, 71–82.

Lyzenga, D.R., 1985, Shallow water bathymetry using a combination of LIDAR and passive multispectral scanner data. *International Journal of Remote Sensing,* **6**, 115–125.

Lantieri, D., 1988, Use of high resolution satellite data for agricultural and marine applications in the Maldives. Pilot study on Laamu Atoll. *FAO Technical Report* TCP/MDV/4905. No. 45.

Lyons, K.J., 1977, Evaluation of an experimental bathymetric map produced from Landsat data. *Cartography*, **10** (2), 75–82.

Masson, P., Dubois, G., Le Mann, G., Roux, C., and Hillion, P., 1990, Bathymetric cartography of coral coasts from classification of remote sensing images. *Proceedings of the International Workshop on Remote Sensing and Insular Environments in the Pacific: Integrated Approaches*, ORSTOM/IFREMER, 355–368.

Nordmann, M.E., Wood, L. Michalek, J.L., and Christy, J.L., 1990, Water depth extraction from Landsat-5 imagery. *Proceedings of the 23rd International Symposium on Remote Sensing of the Environment*, 1129–1139.

Pirazzoli, P.A., 1985, Bathymetry mapping of coral reefs and atolls from satellite. *Proceedings of the Fifth International Coral Reef Congress*, Tahiti, **6**, 539–545.

Siswandono, 1992, The microBRIAN method for bathymetric mapping of the Pari reef and its accuracy assessment. *Proceedings of the Regional Symposium on Living Resources in Coastal Areas*, Manila, Philippines. Marine Science Institute, University of the Philippines, 535–543.

Warne, D.K., 1978, Landsat as an aid in the preparation of hydrographic charts. *Photogrammetric Engineering and Remote Sensing*, **44** (8), 1011–1016.

Van Hengel, W., and Spitzer, D., 1991, Multi-temporal water depth mapping by means of Landsat TM. *International Journal of Remote Sensing*, **12**, 703–712.

Zainal, A.J.M., Dalby, D.H., and Robinson, I.S., 1993, Monitoring marine ecological changes on the east coast of Bahrain with Landsat TM. *Photogrammetric Engineering and Remote Sensing*, **59** (3), 415–421.

Part 5

Quantitative Measurement of Ecological Parameters and Marine Resource Assessment

16

Assessing Seagrass Standing Crop

Summary *There are three ways in which ecological parameters (e.g. seagrass standing crop, mangrove leaf area index) can be predicted from remotely sensed data: empirical modelling, semi-empirical modelling and analytical modelling. The mathematics of some empirical methods are relatively straightforward, requiring only simple statistical relationships to be derived between field and image data. Analytical methods are far more complex and attempt to model mathematically the interaction between electromagnetic radiation and the ecological parameter.*

An empirical model for the measurement of seagrass standing crop is described, with an explanation of the mathematics employed. The accuracy and sensitivity of this model is assessed using data from airborne and satellite sensors.

Remote sensing is well suited to mapping seagrass of medium to high standing crop and SPOT XS, Landsat TM and CASI performed similarly overall. Accurate predictions of low standing crop were confounded by confusion with the substrate and increased patchiness at low biomass. Measurement of seagrass standing crop with the remote sensing methods described here compared favourably to direct quadrat harvest.

Practical guidelines for the monitoring of seagrass standing crop using remote sensing are presented.

Introduction

So far in this *Handbook* we have largely focused on deriving descriptive information from remote sensing, such as mapping the spatial distribution of habitat types. Here our attention is directed towards the extraction of quantitative information measured in real units. In this chapter seagrass standing crop (g.m^{-2}) is used as an example of the quantitative information that can be measured using remotely sensed data. Chapter 17 explains a method for measuring mangrove canopy characteristics using remote sensing.

Types of modelling approaches

There are three methods by which we can make quantitative measurement of ecological parameters: empirical modelling, semi-empirical modelling and analytical modelling. These are dealt with in turn below.

Empirical modelling

This is the simplest and most common approach. It involves establishing a statistical relationship (e.g. regression) between measurements made during fieldwork (Plate 18) with measurements made at the satellite. If a strong relationship exists between satellite-measured radiance and a field parameter of interest, say, seagrass biomass, then measurements of seagrass biomass can be made from the remotely sensed data. This is achieved by using a calibration equation (the regression equation) to convert remotely sensed data from radiance into units of seagrass biomass (g.m^{-2}).

Fundamental drawbacks of this technique are:

- A correlation between radiance, as measured from an image, and seagrass biomass, as measured in the field, is just a statistical relationship between two sets of data. The relationship has not been derived from knowledge of how the light measured by the sensor has interacted with areas of different seagrass biomass and so an explanation of why the relationship exists cannot be obtained.

- Extensive field measurements are required to establish a statistically valid relationship. However, these will be far less than doing all assessment by fieldwork alone!
- The relationship may be time and place specific. If so, it may be inappropriate to use a relationship (e.g. regression equation) derived from one area in a different area. For example, a relationship between seagrass biomass and image data for the Caicos Bank may not apply in Belize. Similarly, a relationship derived in mid-summer at one site may not apply at the same site in mid-winter.

In spite of these disadvantages empirical models, if applied carefully, can be of great use in estimating quantitative parameters over wide areas with a limited amount of field data collection more rapidly and easily than field measurement of the same parameters.

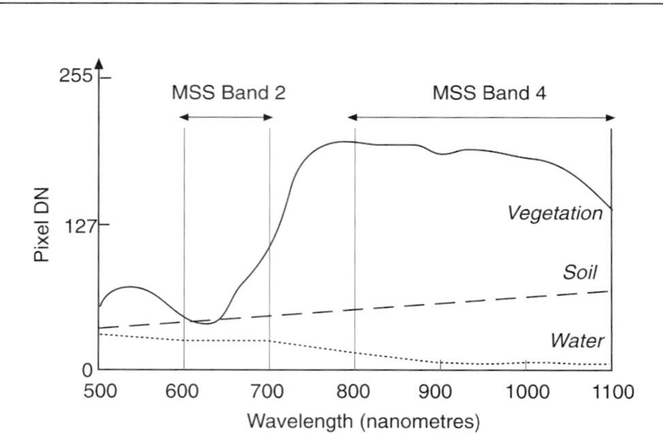

Figure 16.1 To illustrate the use of ratios this graph shows the spectra of vegetation, soil and water between 500 and 1100 nm. A ratio of Landsat MSS bands 4 and 2 (formerly 7 and 5) emphasizes the slope of the spectral reflectance curve between 800 and 1100 nm and 600 and 700 nm. It thus permits discrimination between the three surface types: vegetation (band 4/2 ratio >> 1), soil (ratio > 1) and water (ratio < 1). Based on a diagram in Mather, P.J., 1987, *Computer Processing of Remotely-Sensed Images: An Introduction*. (Chichester: John Wiley & Sons).

Semi-empirical modelling

This is a variant of the empirical approach which seeks to include some knowledge of the physical interaction of electromagnetic radiation and the Earth's surface. If we know the reflectance characteristics of our surface of interest (e.g. seagrass meadow or mangrove canopy) across the wavelengths being measured by the sensor, then spectral indices can be developed which display a stronger relationship with that surface than the bands alone. Vegetation indices (see Box 12.1) are examples of this. Figure 16.1 illustrates the spectral reflectance curves of Landsat MSS for soil, water and vegetation. If a ratio of Band 4/Band 2 is created the value for vegetation would be >>1, >1 for soil and <1 for water. In other words dividing band 4 by band 2 emphasises the spectral differences between the three surface types and a correlation of this ratio to, say, vegetation biomass would be logically robust.

Analytical modelling

From a theoretical point of view this is the purest and most reliable approach. Rather than relying on statistical correlation, analytical modelling aims to account physically for (i.e. model) the interaction between light and the surface of interest. Such interactions include scattering and absorption in the atmosphere, water column, within leaves, between components of the plant canopy, and with the background substrate. These should be modelled for both down and upwelling radiation. Analytical modelling has two stages. Firstly, given appropriate information on mangrove canopy characteristics, for example, a model should be able to simulate accurately the spectral response of mangroves to different radiation wavelengths. If this is successful the second stage is to run the model in reverse (invert it) so that remotely sensed measurements are converted into, say, leaf area index. A working model of this type has the advantage of being a rigorous solution (i.e. is not working by chance) and is applicable at different places and dates. Potentially, it is also much less demanding on the amount of field data that is required when compared to the empirical approaches. This is because field data has to be collected only once to establish the model. Unfortunately however the complexity of optical physics and the interaction of electromagnetic radiation with surfaces such as seagrass and mangroves is such that few working models exist. We know of none that could be used operationally in the context of this *Handbook*, although developments are being made rapidly and users should expect considerable advances in this area. The measurement of seagrass standing crop has been subjected to analytical modelling which has shown that it may be possible to predict this parameter from CASI data if the water depth is known (see Plummer *et al.* 1997, Malthus *et al.* 1997, Clark *et al.* 1997).

☞ **In reality empirical and analytical modelling are complementary. An empirical relationship must exist for an analytical model to be of use, whilst in the absence of a model the limitations of the empirical approach are difficult to ascertain.**

This chapter concentrates on operational techniques to derive empirical (seagrass standing crop) models with semi-empirical (mangrove leaf area index) models being discussed in the next chapter.

Seagrass standing crop

Standing crop is a useful parameter to measure because it responds promptly to environmental disturbance and changes are usually large enough to monitor (Kirkman 1996). This section describes an empirical approach to mapping seagrass standing crop using SPOT XS, Landsat TM and CASI imagery. Like most case studies in this *Handbook*, it was carried out in the Turks and Caicos Islands and, therefore, seagrass was located in shallow (< 10 m deep) clear water (horizontal Secchi distance 20–50 m) and was dominated by the species *Syringodium filiforme* and *Thalassia testudinum* (Plate 14).

Overview of methods

Field sampling

Field data on standing crop were gathered using the rapid visual assessment method (Chapter 12; Mumby *et al.* 1997a). Six visual estimates of standing crop (Plate 18) were made per site (83 sites for the Landsat TM image, 110 sites for SPOT XS and 35 for CASI).

Imagery and processing

CASI imagery was obtained in July 1995 for an area exceeding 100 ha. The CASI was configured to record data at high spatial resolution (1 m) and in 8 spectral bands. Five of these bands were designated in the region of the electromagnetic spectrum which best penetrates water (approx. 400–650 nm). In Figure 16.2 the choice of CASI bands is compared to a reflectance profile of the Caribbean seagrass *Thalassia testudinum* (Armstrong 1993). It shows that several of the bands have different reflectance characteristics for sand and seagrass which is necessary for discriminating seagrass bottom types using the water column correction method of Lyzenga (1981; see Chapter 8). Water column correction created a single depth-invariant bottom-index from each pair of spectral bands. A single bottom-index was obtained for SPOT XS (from bands 1 and 2, denoted b1_b2), three indices were created from Landsat TM bands 1, 2 and 3 (i.e. b1_b2, b1_b3, b2_b3) and ten indices were created from CASI bands 1–5 (Table 16.1). Four of these CASI indices were discarded because the data contained too much sunglint and noise. Without this processing step the coefficient of determination for seagrass standing crop regressed on reflectance (not corrected for depth) ranged from 0.15–0.51 (Landsat TM), 0.05–0.32 (SPOT XS) and 0.01–0.21 (CASI), values which are considered too low for adequate prediction of seagrass standing crop.

Where more than one depth-invariant bottom-index was available for a sensor (i.e. for Landsat TM and CASI), principal component analyses (see Box 12.2) were carried out to combine multiple indices into a single regressor index (the first principal component).

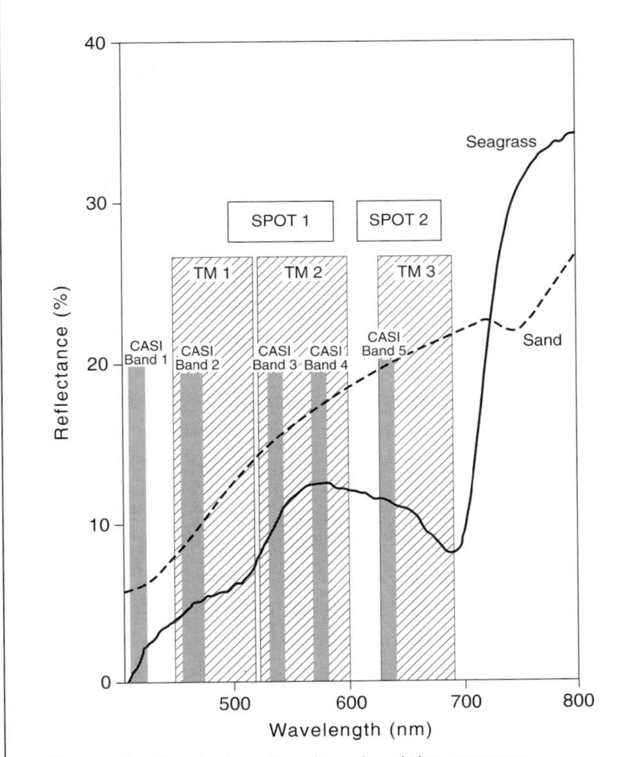

Figure 16.2 Spectral profile of sand and the seagrass *Thalassia testudinum* (epiphytes present) in relation to the spectral bands of remote sensing instruments. Spectral profile redrawn from Armstrong (1993).

Table 16.1 Band settings used on the Compact Airborne Spectrographic Imager (CASI) for applications in this chapter.

Band	Part of electromagnetic spectrum	Wavelength (nm)
1	Blue	402.5–421.8
2	Blue	453.4–469.2
3	Green	531.1–543.5
4	Green	571.9–584.3
5	Red	630.7–643.2

Statistical methods

The strength of relationship between remotely sensed image parameters (bottom-indices and first principal components) and seagrass standing crop were determined using parametric correlation. Tests for significant differences between correlations (e.g. between different bottom-indices) were carried out using the methods of Zar (1996: p. 380).

The degree to which remotely sensed imagery could predict seagrass standing crop was determined using model I linear regression and independent field data provided an estimate of the confidence of such predictions (for more detail on these methods see Mumby *et al.* 1997b).

Results

Predicting standing crop using satellite imagery

SPOT XS: the bottom-index from SPOT XS was strongly correlated to seagrass standing crop (Pearson product-moment correlation coefficient, $r = 0.83$, $n = 110$). Regression of seagrass standing crop on image data yielded a maximum coefficient of determination (r^2) of 0.79 and a minimum of 0.59 (mean 0.69). The regression equation with an r^2 of 0.79 ($P < 0.0001$) was:

Seagrass standing crop $(g.m^{-2}) = [38.6 - (5.6 \times \text{image data})]^2$

The 95% confidence levels (CLs) of bottom-index values about this regression line are shown in Figure 16.3. When interpreting this figure, it should be borne in mind that the *y*-axis has been square root transformed and that in absolute terms, the 95% CLs are symmetrical about the regression line. However, this pattern does not hold at very low standing crops ($< 2\ g.m^{-2}$) where the 95% confidence interval is asymmetric because seagrass standing crop cannot be $< 1\ g.m^{-2}$. The data set was skewed toward lower standing crop, which reduced the size of CLs in this region (because the standard error and critical-*t* values used to calculate each CL contain *n* as a denominator). This also led to greater variability among CLs for higher standing crops where the concentration of data was lower (see methods). The kind of map of seagrass standing crop which was produced using the SPOT XS imagery is shown in Figure 16.4 (Plate 19).

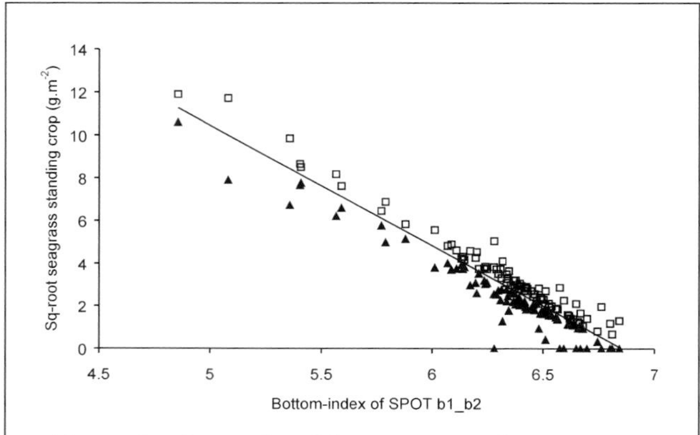

Figure 16.3 Regression of the square root of seagrass standing crop on processed SPOT XS data (depth-invariant bottom-index generated from XS bands 1 and 2) showing upper (□) and lower (▲) 95% confidence limits.

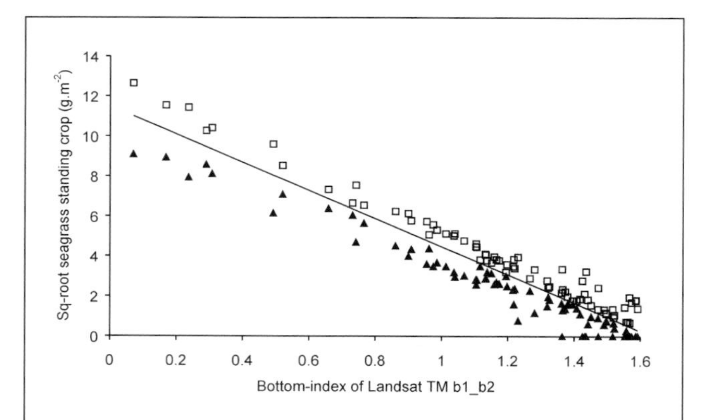

Figure 16.5 Regression of the square root of seagrass standing crop ($g.m^{-2}$) on processed Landsat TM data (depth-invariant bottom-index generated from TM bands 1 and 2) showing upper (□) and lower (▲) 95% confidence limits.

Landsat TM: the correlation between seagrass standing crop and Landsat TM data was similar to that from SPOT XS. Two of the three bottom-indices (b1_b2 and b1_b3) had a correlation of 0.80 and the third (b2_b3) achieved 0.71. The first principal component of all three bands explained 92% of their variance and had a correlation of 0.78 with seagrass standing crop. However, neither of the correlations differed significantly ($\alpha = 0.05$, $n = 83$).

Since principal component analysis constituted further image processing effort without conveying a significant correlatory advantage, it was not used for predicting seagrass standing crop. Regression of standing crop onto b1_b2 (image data) produced a wide range of r^2 (0.18 to 0.74; mean 0.56). The optimum equation is plotted in Figure 16.5 with 95% confidence limits and is described below:

Seagrass standing crop $(g.m^{-2}) = [11.5 - (7.41 \times \text{image data})]^2$

Predicting standing crop using high resolution airborne CASI imagery

Heuristic correlation analysis of all bottom-indices with seagrass standing crop found that the highest correlations (r^2) were achieved from indices derived from spectral band numbers b3_b4, b1_b5, b2_b5, b3_b5 and b4_b5 (r^2 = 0.76, 0.77, 0.81, 0.83 and 0.79 respectively). The correlations of bottom-index with seagrass standing crop (logarithmically transformed) did not differ significantly between these depth-invariant bands. However, some depth-invariant bands had significantly worse correlations (e.g. r^2 (b1_b4) = 0.37; r^2 (b1_b3) = 0.39) and were discarded from further analysis. The first principal component of the indices explained 94% of their variance and achieved a correlation of 0.83 with standing crop.

For the reasons given earlier, the first principal component was not used to predict seagrass standing crop. A logarithmic transformation of standing crop data gave the highest coefficient of determination against CASI depth-invariant bottom-indices. Figure 16.6 shows the

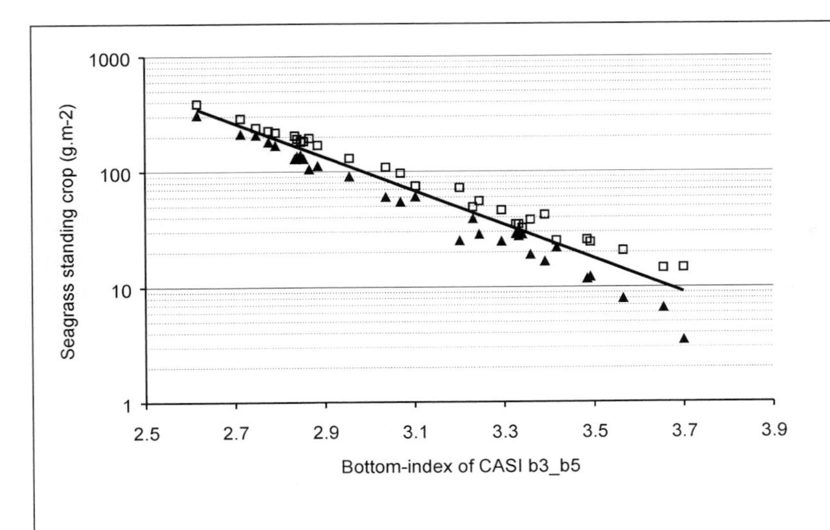

Figure 16.6 Regression of the logarithm of seagrass standing crop (g.m^{-2}) on processed CASI data (depth-invariant bottom-index generated from CASI bands 3 and 5) showing upper (□) and lower (▲) 95% confidence limits. Band 3 was set to 531.1–543.5 nm (green) and band 5 was set to 630.7–643.2 nm (red).

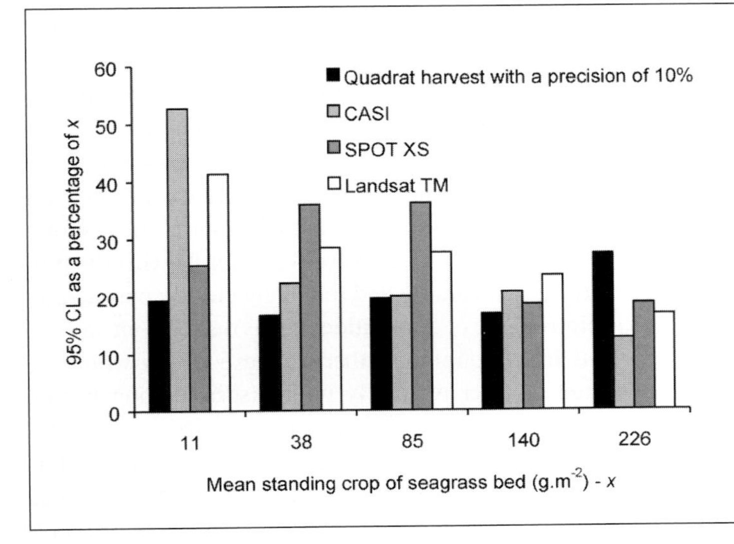

Figure 16.7 Comparison of 95% confidence limits expressed as a proportion of the mean standing crop of seagrass. 95% confidence limits for remote sensing were determined using a bootstrap procedure (Mumby *et al.* 1997b). 95% confidence limits for quadrat sampling were determined from field data (Chapter 12, Figure 12.2). The mean and range (g.m^{-2}) of standing crop categories 2 to 6 were 11 (4–21), 38 (22–67), 85 (43–134), 140 (96–200) and 226 (170–299) respectively. The number of quadrats required for a sampling precision of 10% was 20, 11, 12, 3 and 3 for each category respectively.

regression relationship for the bottom-index derived from CASI bands 3 and 5 (denoted b3_b5). The regression had a r^2 of 0.81 (minimum 0.51; mean 0.68) and the following equation:

$$\textit{Seagrass standing crop } (g.m^{-2}) = 10^{[6.36 - (1.46 \times \text{image data}]}$$

Comparison of remote sensing methods

The strength of correlations between image data (depth-invariant bottom-indices) and seagrass standing crop did not differ significantly from sensor to sensor.

To examine how 95% CLs differed between seagrass beds, five levels of standing crop were selected based on categories 2–6 of the sampling scale (Chapter 11). These categories were selected because they approximate a linear gradation of standing crop. The range of standing crop within each category (Figure 16.7) was used to select corresponding CLs from the regression analyses. The mean of the CLs was then expressed as a percentage of the mean standing crop of each category (Figure 16.7). Since this entailed unequal sample sizes, Figure 16.7 is intended to illustrate trends rather than permit hypothesis testing.

The 95% CL of standing crop prediction from CASI became proportionately smaller as the standing crop of seagrass increased. This trend was not evident for SPOT XS because the high concentration of field data of low standing crop confounded interpretation: prediction of low standing crop was made relatively accurately. The spread of field data for Landsat TM was less skewed and showed a similar trend to CASI imagery.

Comparison of remote sensing methods and direct field (quadrat) sampling

Measurement of seagrass standing crop with the remote sensing methods described here compared favourably to quadrat harvest (Figure 16.7). At low standing crop, quadrat harvest provided greater confidence although this was of the same order of magnitude to that achieved from remote sensing. Remote sensing and field sampling had comparable confidence levels for seagrass beds of medium standing crop and remote sensing generally outperformed quadrat harvest in areas of high standing crop. Clearly, this observation is dependent on the precision to which quadrat sampling is conducted. The data presented in Figure 16.7 assumed that a sampling precision (*P*) of 0.1 was specified for all seagrass beds (see Downing and Anderson 1985). The estimates for quadrat sample size and confidence were based on the calibration data from Chapter 12. Sampling precision was calculated using the equation (Andrew and Mapstone 1987):

$$\text{Sampling Precision, } P = \left(\frac{\text{Standard error}}{\text{Mean, } \bar{X}}\right)$$

Equation 16.1

which gives,

$$\text{Sampling Precision, } P = \frac{\left(\dfrac{\text{standard deviation, } s}{\sqrt{n}}\right)}{\bar{X}}$$

Equation 16.2

and

$$\text{Number of samples, } n = \left(\frac{s}{P \times \bar{X}}\right)^2$$

Equation 16.3

Production of three-dimensional maps of seagrass standing crop

To demonstrate the ability of remote sensing to map seagrass standing crop, a sequence of interpretation is illustrated in Figure 16.8 based on CASI data from Cockburn Harbour. The seagrass bed illustrated in Figure 16.8 (Plate 20) was frequently visited during fieldwork and contained a large blow-out. This feature was identified on CASI imagery using a classified habitat map from Chapter 7 and isolated. The depth-invariant bottom-index b3_b5 was extracted as a single layer and the appropriate regression equation used to predict standing crop in each pixel. The raster (pixel-based) image of standing crop was then moved to the geographic information system, IDRISI. The resulting representation of seagrass standing crop shows the structure of the blow-out in three dimensions including the presence of unattached organic matter in the blow-out (Figure 16.8, bottom).

Discussion

Performance of bottom-indices and spectral considerations

The first principal components of Landsat TM and CASI bottom-indices did not convey any significant correlatory advantage. Although this result was surprising in principle, we attribute it to the high correlation of component indices which when combined, failed to offer improved information. This explanation was borne out by the high degree of variance explained by the first principal components (92% and 94% for Landsat TM and CASI respectively).

With appropriate processing for water column effects, all three image types predicted seagrass standing crop with a 95% confidence interval that ranged from approximately 5–80 g.m^{-2}. For each pair of spectral bands, the bottom-index developed by Lyzenga (1978, 1981) distinguishes bottom types according to the difference between their reflectance characteristics and those of sand. Comparison of Armstrong's (1993) reflectance profiles for sand and seagrass in the Bahamas (Figure 16.2) shows that our CASI bands 1, 2, 3 and 4 are similar (i.e. the profiles are almost parallel and relatively closely correlated). This high degree of correlation means that bottom-indices derived from coupling either of these bands would be expected to offer relatively poor discriminating power. With the exception of b3_b4, the results verified this expectation. The reason for the surprisingly good performance of b3_b4 is not entirely clear but may be attributed to the relatively high reflectance of seagrass in these bands which allowed CASI to detect the differences in their sand-seagrass reflectance characteristics.

All depth-invariant bottom-indices that utilised CASI band 5 correlated well with standing crop. This was probably due to the marked difference in reflectance characteristics between sand and seagrass in this band.

Of the Landsat TM bands, band 3 has the greatest difference in spectral response between sand and seagrass. Bottom-indices derived from bands b1_b3 and b2_b3 were therefore expected to exhibit the greatest discriminating power. However, this was not found to be the case and may be due to the poorer light transmission through water in band 3 – i.e. the satellite sensor may lack the radiometric sensitivity required to detect signals from band 3 in water deeper than about 5 m.

Given the limited spectral envelope in which optical remote sensing of the benthos is possible (approx. 400–650 nm), it is not surprising that SPOT, Landsat and CASI performed similarly overall. The satellite sensors and the CASI (with our configuration, Table 16.1) possessed at least two bands which allowed discrimination of sand and seagrass (Figure 16.2). With hindsight, we suggest that future airborne multispectral surveys of seagrass standing crop place greater emphasis on the 580–650 nm zone of the electromagnetic spectrum. In the present study it would have been more beneficial, in terms of seagrass monitoring, if we had had two bands at 580–610 nm and 620–650 nm.

Performance of sensors and spatial considerations

Remote sensing was well suited to mapping seagrass of medium to high standing crop. Accurate predictions of low standing crop were more difficult to make for two reasons. Firstly, the spectrum of seagrass of low standing crop is easily confused with the sand spectrum because of the low cover of plants over the sediment (see Cuq 1993). Secondly, visual inspection of CASI imagery showed that seagrass of the Caicos Bank tended to be patchier at lower standing crop. Greater patchiness led to increased numbers of mixed pixels which, in turn, resulted in relatively large 95% CLs. This effect was most pertinent to CASI because its spatial resolution was an order of magnitude finer than satellite sensors. On the other hand, satellite imagery might be expected to smooth out much of the small-scale patchiness in standing crop.

Limitations of remote sensing

The satellite-based elements of this study are potentially limited by the disparity in dates of field survey and image acquisition: an empirical calibration can only be expected to give reliable absolute values if these dates are similar. When a temporal disparity exists, the calibration is affected by seagrass dynamics. If the relative standing crop of calibration sites has not changed (e.g. due to a uniform decline in biomass with season), a good empirical relationship between image data and standing crop would be expected. However, the absolute values of the calibration may be offset too high or too low. This scenario may have applied to the SPOT XS image which was acquired four months before field data were obtained. This is unlikely to be problematical where the objective is to map the pattern of standing crop.

A greater problem would be encountered if the spatial distribution of biomass changed between field survey and imagery acquisition. Under these circumstances, a proportion of the calibration sites would be misleading resulting in poor calibration.

The calibration of Landsat TM imagery may have been confounded by this effect as the intervening period between fieldwork and imagery acquisition was almost 5 years. If confounding had occurred, the effect would have been partly mitigated by the coarse spatial resolution of the sensor (30 m) which is the least sensitive to changes in the spatial distribution of standing crop. The results presented for Landsat TM are best thought of as conservative: had more recent imagery been available, the calibration might have improved.

The main limiting circumstances for optical remote sensing of standing crop are cloud cover (particularly for satellite imagery), turbid water and deep water (Green *et al.* 1996). Persistent cloud cover in a region will reduce the chance of obtaining suitable imagery and severe light attenuation in turbid water can make optical remote sensing inappropriate. Under such circumstances, acoustic (Sabol *et al.* 1997) or videographic (Norris *et al.* 1997) remote sensing methods are preferable although both are still under development.

Selection of methods for mapping and monitoring seagrass beds

SPOT XS and Landsat TM imagery date back to the mid-1980s and can be used to assess seagrass dynamics over long periods. Whilst the pixel dimensions of these sensors are 20 m and 30 m respectively, neither sensor is likely to be sensitive to changes in seagrass distribution unless the change constitutes several pixels. This is partly due to the geometric location of imagery, which (in this case) had a root mean square error of 1 pixel width. Added to this is the uncertainty of correctly classifying a pixel as seagrass – particularly at the edge of a seagrass bed where spectral confusion with neighbouring habitats is greatest. Satellite remote sensing is therefore useful for mapping large scale changes, such as the seagrass die-off in Florida Bay (4000 ha lost, 23,000 ha affected, Roblee *et al.* 1991), but will fail to distinguish small scale changes less than several tens of metres.

CASI is the only remote sensing method assessed in this study which has the capability to monitor accurately small scale (< 10 m) dynamics of seagrass. Although its ability to predict standing crop is similar to that of sensors currently mounted on satellites, it takes measurements at much finer spatial scales (1 m versus 20–30 m). It follows that seagrass dynamics can be examined in much greater detail (see Figure 16.8). The underlying caveat for these conclusions is a requirement for good geometric correction. Ideally, imagery should be registered to charts and flown with an aircraft-mounted differential global positioning system.

The order of magnitude of 95% CLs was found to be similar for remote sensing and destructive quadrat sampling of seagrass standing crop. This result was surprising given that quadrat sampling is a more accurate technique. However, in terms of estimating mean standing crop, the superior accuracy of quadrats is offset against their size. The statistical population mean of standing crop is estimated from samples of quadrats whereas remote sensing measures the entire population, albeit less accurately.

In a monitoring context, remote sensing will augment a site-specific field monitoring programme by providing a much-needed spatial dimension (Figure 16.8). The spatial dimension provides a broader understanding of seagrass dynamics and permits measurements to be made at hierarchical spatial scales (O'Neill 1989, Levin 1992). We suggest that the following procedure should be adopted for monitoring seagrass standing crop:

First assessment at time T:

1. Calibration curve is used to predict seagrass standing crop in each seagrass pixel (data layer 1; e.g. Figure 16.8).
2. Curves are fitted to the upper and lower 95% CLs of the calibration curve.
3. 95% CL curves are used to assign confidence limits to each pixel (data layers 2 and 3).
4. All three raster layers are entered into a Geographic Information System (GIS).

Second assessment at time T+1:

5. Remote sensing and re-calibration are repeated providing three new data layers.

Monitoring spatio-temporal change in seagrass standing crop:

6. CLs between times T and T+1 are compared and non-overlapping portions reveal the location and extent of significant changes in two dimensions.
7. A final GIS layer is created showing the magnitude and direction of significant changes.

References

Andrews, N.L., and Mapstone, B.D., 1987, Sampling and the description of spatial patterns in marine ecology. *Oceanography and Marine Biology Annual Review*, **25**, 39–90.

Armstrong, R.A., 1993, Remote sensing of submerged vegetation canopies for biomass estimation. *International Journal of Remote Sensing*, **14**, 10-16

Clark, C.D., Malthus, T., and Plummer, S.E., 1997, Quantitative determination of seagrass by remote sensing, Final Report to the Natural Environment Research Council of grant number GR9/02233, 10 pp.

Cuq, F., 1993, Remote sensing of sea and surface features in the area of Golfe d'Arguin, Mauritania. *Hydrobiologica*, **258**, 33–40.

Downing, J.A., and Anderson, M.R., 1985, Estimating the standing biomass of aquatic macrophytes. *Canadian Journal of Fisheries Aquatic Science*, **42**, 1860–1869.

Green, E.P., Mumby, P.J., Edwards, A.J., and Clark, C.D., 1996, A review of remote sensing for the assessment and management of tropical coastal resources. *Coastal Management*, **24**, 1–40

Kirkman, H., 1996, Baseline and monitoring methods for seagrass meadows. *Journal of Environmental Management*, **47**, 191–201.

Levin, S.A., 1992, The problem of pattern and scale in ecology. *Ecology*, **73**, 1943–1967.

Lyzenga, D.R., 1978, Passive remote sensing techniques for mapping water depth and bottom features. *Applied Optics*, **17**, 379–383.

Lyzenga, D.R., 1981, Remote sensing of bottom reflectance and water attenuation parameters in shallow water using aircraft and Landsat data. *International Journal of Remote Sensing*, **2**, 71–82.

Malthus, T.J., Ciraolo, G., La Loggia, G., Clark, C.D., Plummer, S.E., Calvo, S., and Tomasello, A., 1997, Can biophysical properties of submerged macrophytes be determined by remote sensing? *Proceedings of the 4th International Conference on Remote Sensing for Marine and Coastal Environments*, Orlando, March 1997, **1**, 562–571.

Mumby, P.J., Edwards, A.J., Green, E.P., Anderson, C.W., Ellis, A.C., and Clark, C.D., 1997a, A visual assessment technique for estimating seagrass standing crop. *Aquatic Conservation: Marine and Freshwater Ecosystems*, **7**, 239–251.

Mumby, P.J., Green, E.P., Edwards, A.J., and Clark, C.D., 1997b, Measurement of seagrass standing crop using satellite and digital airborne remote sensing. *Marine Ecology Progress Series*, **159**, 51–60.

Norris, J.G., Wyllie-Echeverria, S., Mumford, T., Bailey, A., and Turner, T., 1997, Estimating basal area coverage of subtidal seagrass beds using underwater videography. *Aquatic Botany*, **58**, 269–287.

O'Neill, R.V. 1989. Perspectives in hierarchy and scale. In *Perspectives in Ecological Theory* edited by J. Roughgarden, R.M. May, S.A. Levin (New Jersey: Princeton University Press), pp. 140–156.

Plummer, S.E., Malthus, T.J., and Clark, C.D., 1997, Adaptation of a canopy reflectance model for sub-aqueous vegetation: definition and sensitivity analysis. *Proceedings of the 4th International Conference on Remote Sensing for Marine and Coastal Environments*, Orlando, March 1997, **1**, 149–157.

Robblee, M.B., Barber, T.R., Carlson, P.R., Jr, Durako, M.J., Fourqurean, J.W., Muehlstein, L.K., Porter, D., Yarbro, L.A., Zieman, R.T., and Zieman, J.C., 1991, Mass mortality of the tropical seagrass *Thalassia testudinum* in Florida Bay (USA). *Marine Ecology Progress Series*, **71** (3), 297–299.

Sabol, B., McCarthy, E., and Rocha, K., 1997, Hydroacoustic basis for detection and characterisation of eelgrass (*Zostera marina*). *Proceedings of the 4th International Conference on Remote Sensing for Marine and Coastal Environments*, Orlando, March 1997, **1**, 679–693

Zar, J.H., 1996, *Biostatistical Analysis*. Third Edition. (New Jersey, USA: Prentice Hall)

17

Assessing Mangrove Leaf Area Index and Canopy Closure

Summary *The previous chapter introduced the different ways in which ecological parameters can be predicted from remotely sensed data, and illustrated the use of an empirical model to measure seagrass standing crop. This chapter describes the way in which a semi-empirical model can be used to measure two ecological parameters of a mangrove canopy: leaf area index (LAI) and percentage canopy closure. The accuracy and sensitivity of the mangrove models were assessed using data from airborne and satellite sensors.*

The superior spatial resolution and spectral versatility of CASI allows mangrove LAI to be measured more accurately and with greater precision than with satellite sensors. However, if large areas need to be surveyed, satellite sensors are likely to be more cost-effective. The difference between LAI measured in situ *and from remotely sensed data was typically less than 10%. Compared to direct methods of measuring LAI, remote sensing allows information to be obtained rapidly and minimises the logistical and practical difficulties of fieldwork in inaccessible mangrove areas.*

Practical guidelines for the monitoring of mangrove canopies using remote sensing are presented.

Introduction

In Chapter 13 methods for mapping mangroves using remote sensing were discussed. Airborne multispectral sensors such as Compact Airborne Spectrographic Imager (CASI) were shown to be capable of distinguishing different mangrove community types and mapping these with considerable precision (Figure 10.9, Plate 11). By contrast, satellite sensors such as Landsat TM could only separate tall *Rhizophora* from 'other mangrove', whilst SPOT XS could not distinguish mangrove from thorn scrub in the study area. However, in areas with well-developed mangroves, better results would be expected from both satellite sensors. In this chapter we describe the way in which semi-empirical models can be used to map two parameters which describe mangrove canopy structure: leaf area index (LAI, Box 17.1) and percentage canopy closure. As in the previous chapter, the method relies on gathering considerable field data both to calibrate image data – establish a significant relationship between values on an image and field measurements – and to check the accuracy of predicted values of LAI and canopy closure derived from imagery. However, the status of large areas of impenetrable mangrove can be assessed for a relatively modest investment in field survey. This chapter describes both the field survey and image processing methods required with examples based on SPOT XS and CASI imagery.

Mangrove canopy structure

Here we describe how semi-empirical models were derived to estimate LAI and percentage canopy closure from remotely sensed data of the Turks and Caicos Islands (refer to Chapter 16 for an explanation of empirical, semi-empirical and analytical models). This approach is particularly advantageous in mangrove areas where access to the interior is usually difficult or when alternative methods are laborious and difficult to replicate properly over whole forests.

Leaf area index

Many methods are available to measure LAI directly and are variations of either leaf sampling or litterfall collection techniques (Clough *et al.* 1997 and references in Chason *et al.* 1991). There are problems associated with both how-

BOX 17.1

Leaf area index (LAI)

LAI is defined as the single-side leaf area per unit ground area and as such is a dimensionless number. The importance of LAI stems from the relationships which have been established between it and a range of ecological processes (rates of photosynthesis, transpiration and evapotranspiration: McNaughton and Jarvis 1983, Pierce and Running 1988; net primary production: Monteith 1972, Norman 1980, Gholz 1982, Meyers and Paw 1986, 1987; rates of energy exchange between plants and the atmosphere: Botkin 1986, Gholz *et al.* 1991). Measurements of LAI have been used to predict future growth and yield (Kaufmann *et al.* 1982) and to monitor changes in canopy structure due to pollution and climate change (Waring 1985, Gholz *et al.* 1991). The ability to estimate leaf area index is therefore a valuable tool in modelling the ecological processes occurring within a forest and in predicting ecosystem responses.

ever: leaf sampling involves the destructive harvesting and measurement of leaf area for all leaves within a vertical quadrat down through the entire canopy. And litterfall collection is better suited to deciduous forests that have a single leaf fall as opposed to evergreen canopies. All direct methods are similar in that they are difficult, extremely labour intensive, require many replicates to account for spatial variability in the canopy and are therefore costly in terms of time and money. Consequently, many indirect methods of measuring LAI have been developed (see references in Nel and Wessman 1993). Techniques based on gap-fraction analysis assume that leaf area can be calculated from the canopy transmittance (the fraction of direct solar radiation which penetrates the canopy). This approach to measuring LAI uses data collected from along transects beneath the forest canopy (e.g. at 10 m intervals along 300–400 m transects in coniferous forests, Nel and Wessman 1993).

Mangroves are intertidal, often grow in dense stands and have complex aerial root systems which make the sampling regimes described so far difficult to carry out. However, Clough *et al.* (1997) estimated LAI from a ground area of

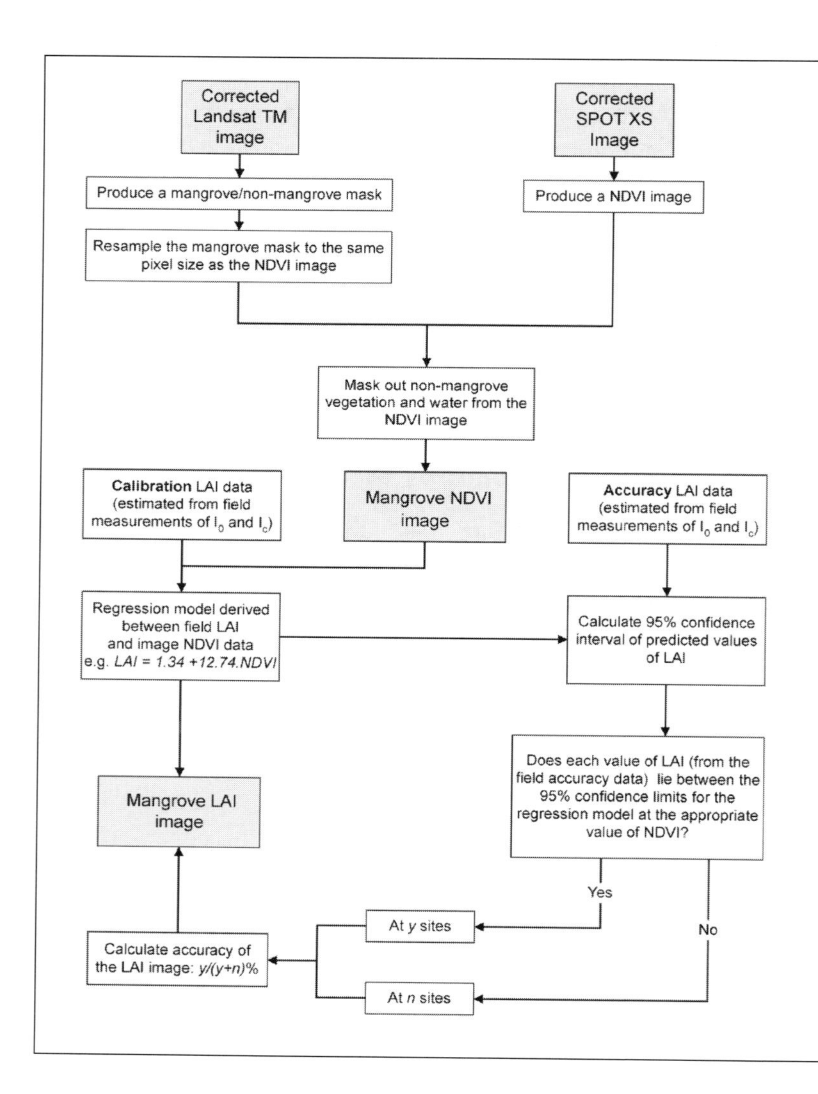

Figure 17.1 A diagrammatic representation of the semi-empirical model used to derive LAI from remotely sensed data (illustrated for SPOT XS) and assess its accuracy. Mangroves were separated from other terrestrial vegetation on the SPOT XS image using a mask derived from Landsat TM. This happened to be necessary in the Turks and Caicos but will not be necessary everywhere. Once such a mask exists (and is known to be sufficiently accurate) then, by resampling to the appropriate pixel size, it can be applied to any other imagery of the same area and map projection.

900 m^2, their objective being a comparison between an estimation of mangrove LAI based on gap-fraction analysis with a direct collection method. To measure the LAI of mangroves over large areas would require measurements at many different locations, an extremely time-consuming process. The difficulty of moving through dense mangrove stands and the general inaccessibility of many mangrove areas would clearly pose a further problem.

The interception, scattering and emission of radiation is closely related to the canopy structure of vegetation. Spatial aspects of LAI can be indirectly measured from the spectra of mangrove forests, and so many of the problems associated with obtaining LAI values for entire mangrove forests are avoided. A semi-empirical model described in this section provides an indirect estimation of mangrove LAI with gap-fraction analysis being used as ground-truthing information to calibrate remotely sensed data. Thematic maps of LAI for the entire area covered by mangroves are derived to a high level of accuracy, without the need for large numbers of ground measurements.

BOX 17.2

Image processing for the estimation of LAI in the Turks and Caicos Islands

Mangrove areas were identified in the Landsat TM scene using the band ratio/principal components analysis method (Method V, Chapter 13). This image was resampled to a pixel size of 20 m and recoded to produce a mangrove/non-mangrove mask. Using the near infra-red band 3 of SPOT, submerged areas were masked from the XS scene. Non-mangrove areas were then masked out using the mangrove mask derived from processing the Landsat TM data. It was necessary to use Landsat TM imagery as a mangrove mask for SPOT XS because in the Turks and Caicos Islands mangrove and non-mangrove vegetation cannot be separated accurately using SPOT XS imagery (Chapter 13). This process is illustrated in Figure 17.1.

Image processing methods

Landsat TM and SPOT XS imagery

Mangrove and non-mangrove vegetation in the Turks and Caicos Islands could not be separated accurately using SPOT XS imagery (Chapter 13) but, if this can achieved by other methods (Box 17.2), there is no reason why SPOT XS data cannot be used to measure LAI.

Once non-mangrove vegetation had been masked out of the Landsat TM and SPOT XS images the remaining mangrove image data was converted into LAI using the following method. Field measurements suggest a linear relationship between normalised difference vegetation index, NDVI (see Box 13.1) and mangrove LAI (Ramsey and Jensen 1995, 1996). NDVI was therefore calculated using the red (SPOT band 2, TM band 3) and infra-red (SPOT band 3, TM band 4) bands of the XS and TM data. Values of LAI estimated from *in situ* measurements of canopy transmittance (see measurement of canopy transmittance and calculation of LAI) were regressed against values of NDVI derived from the image data. The equations of this linear regression model were then used to calibrate the NDVI image into a thematic map of LAI.

CASI imagery

Water areas were removed from the image using a mask derived from band 7 (near infra-red). Since NDVI is calculated using near infra-red and red bands there were four options for calculating NDVI from the CASI data (see Table 17.1 for wavebands) with combinations of Bands 5 to 8. NDVI was calculated for all band combinations and regressed against values of LAI estimated from *in situ* measured canopy transmittance.

Measurement of canopy transmittance and calculation of LAI

LAI is a function of canopy transmittance, the fraction of direct solar radiation (Box 17.3) which penetrates the canopy. Canopy transmittance is given by the ratio I_c/I_o where I_c = light flux density beneath the canopy and I_o = light flux density outside the canopy. LAI can then be calculated, and corrected for the angle of the sun from the vertical, using the formula

$$LAI = \frac{\log_e\left(\frac{I_c}{I_0}\right)}{k} \times \cos\theta$$

Equation 17.1

where LAI = leaf area index, θ = sun zenith angle in degrees (this can be calculated from time, date and position), k = canopy light extinction coefficient, which is a function of the angle and spatial arrangement of the leaves. The derivation of this formula is given in English *et al.* (1997). Nel and Wessman (1993) and Clough *et al.* (1997) should be consulted for a full discussion of the assumptions on which this model is based. For each field site, $\log_e(I_c/I_o)$ was calculated for pairs of simultaneous readings and averaged (Plate 22). A value for k of 0.525 was chosen as being appropriate to mangrove stands (B. Clough, personal communication).

☞ **Most spreadsheet packages require angles to be entered in radians, in which case θ should be multiplied by (π/180).**

Table 17.1 Band settings used on the Compact Airborne Spectrographic Imager (CASI) for applications in this chapter. We found the bands shown in bold to be the most useful for establishing a relationship between mangrove LAI and image NDVI.

Band	Part of electromagnetic spectrum	Wavelength (nm)
1	Blue	402.5–421.8
2	Blue	453.4–469.2
3	Green	531.1–543.5
4	Green	571.9–584.3
5	Red	630.7–643.2
6	**Red**	**666.5–673.7**
7	**Near Infra-red**	**736.6–752.8**
8	Near Infra-red	776.3–785.4

BOX 17.3

Direct and diffuse radiation

The radiation at any point under the canopy is a mixture of direct and diffuse radiation. Direct radiation has arrived at that point directly, passing through only the canopy perpendicularly above the position, and the atmosphere. Diffuse light has arrived at that point by being reflected off clouds, 'sideways' through adjacent areas of canopy and back up from the soil or understory vegetation. Direct light predominates on clear days and at high sun angles (i.e. when the sun is more or less directly overhead). Diffuse light predominates on cloudy days and at low sun angles. Measurement of canopy transmittance using the field survey method described below, assumes that all the light recorded at a point under the canopy is direct radiation. It is possible to measure and correct for diffuse radiation. Two more sensors inside collimating tubes would have been required to measure diffuse radiation simultaneously with the readings of I_c and I_o (see Chason *et al.* 1991, Nel and Wessman 1993). No attempt was made to correct for diffuse radiation however because (i) uncorrected measurements have been shown to yield estimates of LAI similar to direct methods, (ii) uncorrected measurements would under-estimate, rather than over estimate LAI, (iii) the cost of extra detectors was high, and (iv) a reduction in mobility through the mangroves would result from a system of four inter-connected detectors and cables.

Field survey methods

Measurements were taken on clear sunny days between 1000 and 1400 hours, local time. The solar zenith angle was judged to be sufficiently close to normal (i.e. more or less vertically overhead) two hours either side of noon for directly transmitted light to dominate the radiation spectrum under the canopy (Box 17.3). At other times the sun is too low and diffuse light predominates. Photosynthetically active radiation (PAR) was measured using two MACAM™ SD101Q-Cos 2π PAR detectors connected to a MACAM™ Q102 radiometer. One detector was positioned vertically outside the mangrove canopy on the end of a 40 m waterproof cable and recorded I_o. The other detector recorded I_c and was connected to the radiometer by a 10 m waterproof cable. If the mangrove prop roots and trunks were not too dense to prevent a person moving around underneath the canopy this detector (which was attached to a spirit level) was hand-held. If the mangroves were too dense then the I_c detector was attached to a 5.4 m extendible pole and inserted into the mangrove stand. The spirit level was attached to the end of the pole to ensure the detector was always vertical (Plate 22). All recordings of I_c were taken at waist height, approximately 0.8 m above the substrate. Eighty pairs of simultaneous readings of I_c and I_o were taken at each site, the ratio I_c/I_o calculated and averaged for input into Equation 17.1.

The estimation of LAI from remotely sensed data

SPOT XS

Figure 17.2 is a scatter plot of *in situ* LAI values against the NDVI information derived from the SPOT XS data for 29 field sites surveyed in 1995. A linear regression was fitted to these data and a good coefficient of determination obtained ($r^2 = 0.74$, $P < 0.001$, $n = 29$). The F-test for the model and *t*-test for the slope estimate were both significant at the 0.001 level of confidence, indicating a strong relationship which can be used to convert NDVI values to LAI.

The NDVI model was then used to estimate values of mangrove LAI for the entire image. LAI ranged from 0.83 to 8.51, with a mean value of 3.96. This produced a thematic image of LAI for the mangrove areas of the Caicos Bank. Field data from 32 accuracy sites surveyed in 1996 were used to test the accuracy of this image. The 95% confidence intervals of values of LAI predicted from NDVI were calculated for the regression model (Zar 1996). The accuracy of the LAI image was defined as the proportion of 1996 sites at which the value of LAI estimated from *in situ* measurements of canopy transmittance lay within the 95% confidence interval for that value of NDVI (Figure 17.3). Figure 17.3 shows that the thematic maps of LAI are adequately accurate; 88% of the LAIs predicted from NDVI were within the 95% confidence interval (Table 17.2). The mean difference between predicted LAI and the value estimated from field measurements was only 13% for the accuracy test sites.

An estimate of the precision of the LAI prediction was obtained by expressing the standard error of a predicted LAI value as a percentage of LAI estimated from *in situ* measurements of canopy transmittance. Figure 17.4 shows that the precision is high (< 10%) for values of LAI of 1.5 and greater.

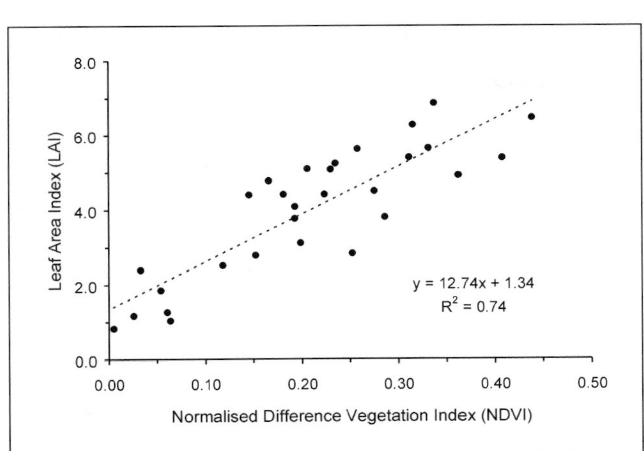

Figure 17.2 The relationship between leaf area index (LAI) estimated from *in situ* measurements of mangrove canopy transmittance and normalised difference vegetation index (NDVI) derived from SPOT XS data for 29 sites near South Caicos, Turks and Caicos Islands. A linear regression has been fitted to the data and is significant at the 0.001 level.

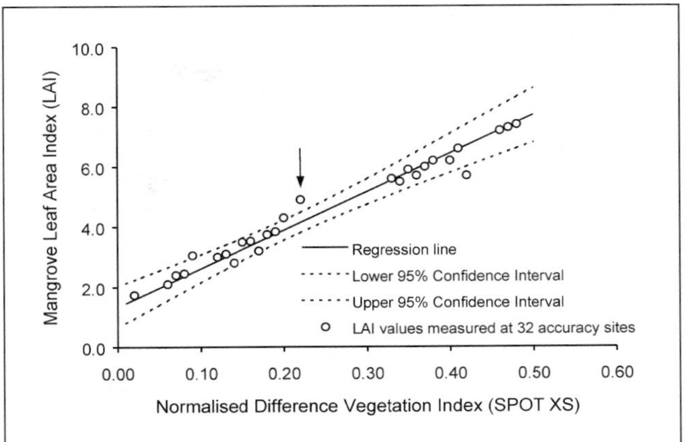

Figure 17.3 A plot of the 95% confidence intervals of the LAI:NDVI regression model for SPOT XS imagery. *In situ* measured values of LAI for 32 accuracy sites have been superimposed over this plot. The accuracy of the LAI image was defined as the proportion of 1996 accuracy sites where the LAI value lay within the 95% confidence interval for that value of NDVI. For example, accuracy site number 16 (indicated by an arrow) has a NDVI of 0.22. At this NDVI the 95% confidence interval of a predicted value of LAI is 3.81–4.48. However the LAI at that site was calculated from *in situ* measurements of canopy transmittance to be 4.9. Therefore, the estimation of LAI at accuracy site number 16 was deemed inaccurate. LAIs of 28 of the 32 accuracy sites do lie between the appropriate confidence intervals. The accuracy of the SPOT XS derived thematic map of LAI is thus deemed to be 88%.

Table 17.2 The accuracy of LAI and percentage canopy closure prediction using three image types.

	Landsat TM	SPOT XS	CASI
Leaf area index	71%	88%	94%
Canopy closure	65%	76%	80%

Landsat TM

The accuracy and precision of the LAI image was assessed in the same manner as for SPOT XS but was lower, 71% (Table 17.2).

CASI

The relationship between NDVI calculated from bands 8 (near IR, 776.3–785.4 nm) and 5 (red, 630.7–643.2 nm), and values of LAI estimated from *in situ* measured canopy transmittance, was not significant (Table 17.3). Neither was a model using NDVI calculated from bands 8 and 6. However, there was a significant relationship when LAI was regressed against NDVI calculated either from bands 7 and 6 or 7 and 5 (Table 17.3). The former model was deemed more appropriate for the prediction of LAI because (i) it accounts for a much higher proportion of the total variation in the dependent variable, and (ii) the accuracy with which the model predicts the dependence of LAI on NDVI is higher (the standard error of the estimate is lower).

The accuracy and precision of the LAI image was assessed in the same manner as for SPOT XS. Some 94% of the LAIs predicted from NDVI were within the 95% confidence interval (Table 17.2). In other words, anyone using this thematic image knows that there is a 94% probability that the 95% confidence interval of any value of LAI predicted from CASI data includes the value of LAI which would be obtained by field measurements of canopy transmittance. The mean difference between pre-

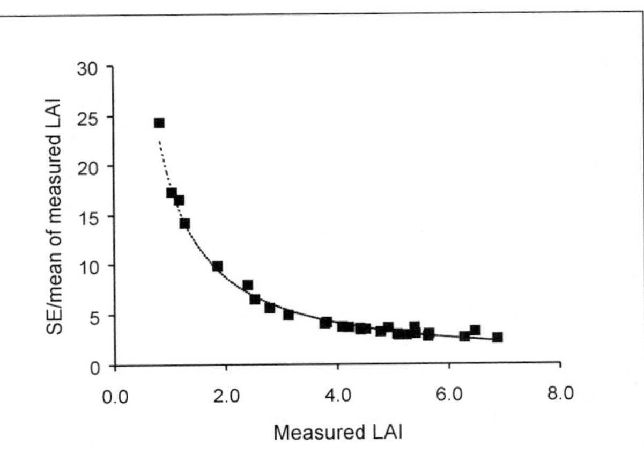

Figure 17.4 The precision of the LAI prediction. Precision is defined as the standard error/mean (see Equations 16.1–16.3). Good precision is indicated by low values of the standard error/mean, that is the error of measurement is small in relation to the mean. The standard errors of predicted values of LAI have been calculated as a percentage of mean measured LAI and are plotted against measured LAI. The precision of LAI prediction is high (< 10%) for LAI values greater than about 1.5–2.0.

Table 17.3 A summary of the four ways used to calculate NDVI from CASI bands using combinations of bands 5 to 8. NDVI calculated from CASI data was regressed against LAI estimated from *in situ* measurements of canopy transmittance. Bands 5 and 6 are in the visible red portion of the electromagnetic spectrum, and bands 7 and 8 in the near infra-red (for exact wavelengths refer to Table 17.1). r^2 = coefficient of determination, P = probability of the F-test for the model, NS = not significant at the 5% level, SE = standard error of the estimate. Degrees of freedom = 1, 29 in all cases. For models marked * the F-test and t-test for the slope estimate were both significant at the 0.1% level, indicating a strongly significant relationship which could be used to convert NDVI values to LAI.

NDVI equation	r^2	Intercept	Slope	P	SE
(Band 8 - Band 5)/(Band 8 + Band 5)	0.22	0.19	0.05	NS	1.93
(Band 8 - Band 6)/(Band 8 + Band 6)	0.12	3.68	2.29	NS	2.18
(Band 7 - Band 6)/(Band 7 + Band 6)*	0.77	0.32	9.76	<0.001	0.73
(Band 7 - Band 5)/(Band 7 + Band 5)*	0.43	1.81	5.71	<0.001	1.69

dicted LAI and the value estimated from *in situ* measurements of canopy transmittance was only 9% for the accuracy sites.

A thematic map of LAI predicted for mangroves on South Caicos using CASI imagery (using bands 6 and 7) is shown in Figure 17.5 (Plate 21). Comparison with Figure 10.9 (Plate 11) shows that the highest LAI occurred in areas of tall *Rhizophora mangle*.

The use of indices other than NDVI in semi-empirical models of LAI

The results of modelling the relationship between LAI and more atmospherically robust indices (Angular Vegetation Index – AVI and Global Environment Monitoring Index – GEMI, see Box 13.1) are given in Table 17.4 for SPOT XS and CASI. In the interests of simplicity only the results of the model using NDVI are discussed further although readers should note that these indices would be preferable to NDVI if LAI is to be monitored over time with several images (see Box 13.1 for more details on the use of different vegetation indices).

Table 17.4 Modelling the relationship between LAI and atmospherically robust vegetation indices; Angular Vegetation Index (AVI) and Global Environment Monitoring Index (GEMI), calculated from SPOT XS and CASI data. Refer to Box 13.1 for more information on AVI and GEMI.

Index	Sensor	r^2	Intercept	Slope	SE
AVI	SPOT XS	0.78	2.66	5.01	0.99
AVI	CASI	0.78	3.45	6.77	0.39
GEMI	SPOT XS	0.72	5.76	7.01	0.87
GEMI	CASI	0.80	3.81	5.89	1.03

Canopy closure

The canopy closure (expressed on a percentage scale) of a mangrove stand is highly correlated with NDVI and other indices (Jensen 1991). Canopy closure can therefore be measured from remotely sensed data in the same way as LAI (i.e. field measurements can be regressed against NDVI derived from the imagery and the resulting model used to calibrate the entire scene). Jensen (1991) proceeded to use canopy closure as a surrogate measure of tree density and determine the sensitivity of different mangrove areas to oil spill on the basis that oil would penetrate further into less dense stands.

Field techniques for measuring canopy closure

Wherever access underneath the mangrove canopy was possible percentage canopy structure was measured using a hand held semi-hemispherical mirror (a spherical densiometer) which had a grid graticule engraved on its surface. Percentage canopy closure was then estimated by computing the proportion of the mirror area covered by the reflection of leaves and stems (Plate 22). Eighty readings were taken at each site, converted into percentage canopy closure and averaged. A hemi-spherical densiometer was purchased from: Forest Densiometers, 5733 SE Cornell Drive, Bartlesville, OK 74006, USA (Tel: +1 918 333 2830).

Summary of results

NDVIs of approximately 0.60 or above were obtained from sites with 100% canopy closure. Below 0.60 the relationship between NDVI and percentage canopy closure was linear for all three image types. NDVI calculated from CASI bands 6 and 7 was again a superior predictor of percentage canopy closure (r^2 = 0.92, P <0.001, n = 19) to NDVI calculated from other band combinations. Regression models were fitted to scatter plots of percentage canopy closure against NDVI. The equations of these regression models were then used to predict mangrove canopy closure over the entire mangrove area.

Accuracy of this estimation of percentage canopy closure was assessed in the same manner as the LAI (see the estimation of LAI from remotely sensed data). The most accurate estimation was obtained from CASI, the least from Landsat TM (Table 17.2). The mean difference between predicted canopy closure and the *in situ* measured value was 4% for CASI, 11% for SPOT XS and 17% for Landsat TM.

Discussion

Comparison of LAI derived from remotely sensed data with other methods

Clough *et al.* (1997) have published LAI values for mangroves from the west coast of Peninsular Malaysia. They obtained indices ranging from 2.2 to 7.4 (mean 4.9) by direct measurement and a mean value of 5.1 when LAI was estimated indirectly from light transmission measurements over four transects. Our values of LAI derived from CASI and satellite data of Caribbean mangroves are similar to their findings and other published values for mangrove LAI (Table 17.5). The LAI of surrounding vegetation is usually quite different from mangrove. Tropical rainforest typically has a higher LAI than mangrove (~10). The difference has been attributed to the shade intolerance of mangroves, the dense aggregation of foliage in the upper portion of the mangrove canopy and the absence of an understory (Cintrón and Schaeffer-Novelli 1985). In contrast the acacia scrub in the Turks and Caicos Islands has a lower LAI than mangroves (mean 2.0), a feature of the more open canopy of that type of vegetation.

Table 17.5 Values for mangrove leaf area index estimated using airborne and satellite data in the Turks and Caicos Islands compared with *in situ* studies in other areas.

Reference	Leaf Area Index Minimum	Maximum	Mean
Araújo *et al.* (1997)	3.0	5.7	-
Clough *et al.* (1997)	2.2	7.4	4.9
Cintrón and Schaeffer-Novelli (1985)	0.2	5.1	-
Cintrón *et al.* (1980)	-	-	3.8
Landsat TM	0.7	7.2	3.8
CASI	1.0	8.5	5.3
SPOT XS	0.8	7.0	4.0

The lower LAI values reported from the Turks and Caicos Islands reflect the inclusion of some sites with sparse mangroves and more open canopies. Care should be taken when interpreting low (< ~1.5) values of LAI derived from remotely sensed data, especially if the understory vegetation is not uniform. The spectral signature of such sites will contain a larger proportion of light that has been reflected from the understory than if the canopy was denser. Ground cover beneath sparse mangrove sites was variable in the Turks and Caicos Islands. White sand, dark organic detritus and dense green mats of the succulent *Salicornia perennis* were all recorded, ground covers with presumably very different optical properties. The LAI of mangroves with relatively open canopies might be under-estimated if they were growing over white sand, or over-estimated if dark organic detritus covered the sediment. LAI values higher than previously published probably reflect the effectiveness of using remote sensing because they include especially dense mangrove areas which others might not have been able to access.

Above an LAI of 1.5–2.0 satellite data can be used to estimate mangrove LAI with considerable precision. The extent to which LAI obtained in this way can be used to model ecological processes in mangrove forests may only be limited by the availability of supporting data. If appropriate data exist then LAI may be used to produce thematic maps of, for example, rates of photosynthesis, transpiration, respiration and nutrient uptake. Such thematic maps could then be combined with other spatial data in a GIS for the management of mangrove areas.

☞ **Maps of mangrove LAI could be incorporated in a GIS and presented in 3-D in the same fashion as seagrass standing crop (see *Selection of methods for mapping and monitoring seagrass beds, p. 243,* and Figure 16.8, Plate 20).**

The selection of an appropriate sensor for monitoring mangrove LAI

LAI can be estimated from CASI data at a significantly greater level of accuracy than is possible from either SPOT XS or Landsat TM (Table 17.2) but this must be interpreted in the context of the area covered. The area of CASI coverage used in this Chapter was slightly more than 0.5 km^2. Although calibration of CASI imagery can be carried out at higher accuracies than satellite data, the latter cover an area that is approximately 10^4 or 10^5 times as large. There appears to be a trade-off between accuracy and coverage. CASI is also relatively expensive in both financial cost and processing time For example, in the Turks and Caicos study CASI costs (£ sterling km^{-2}) were approximately 400 times as much as SPOT XS, whilst acquisition and correction of imagery took about twice as long. Although CASI offers extremely high spatial resolution, great care should be taken to decide whether high resolution is really necessary. Reducing pixel size from (say) 3 m to 1 m will have a direct affect on the width of the area surveyed along each flight line (approx. 1.5 km to 0.5 km). At a resolution of 1 m the error present in each position fix, even with a DGPS, means that site-specific information has to be analysed

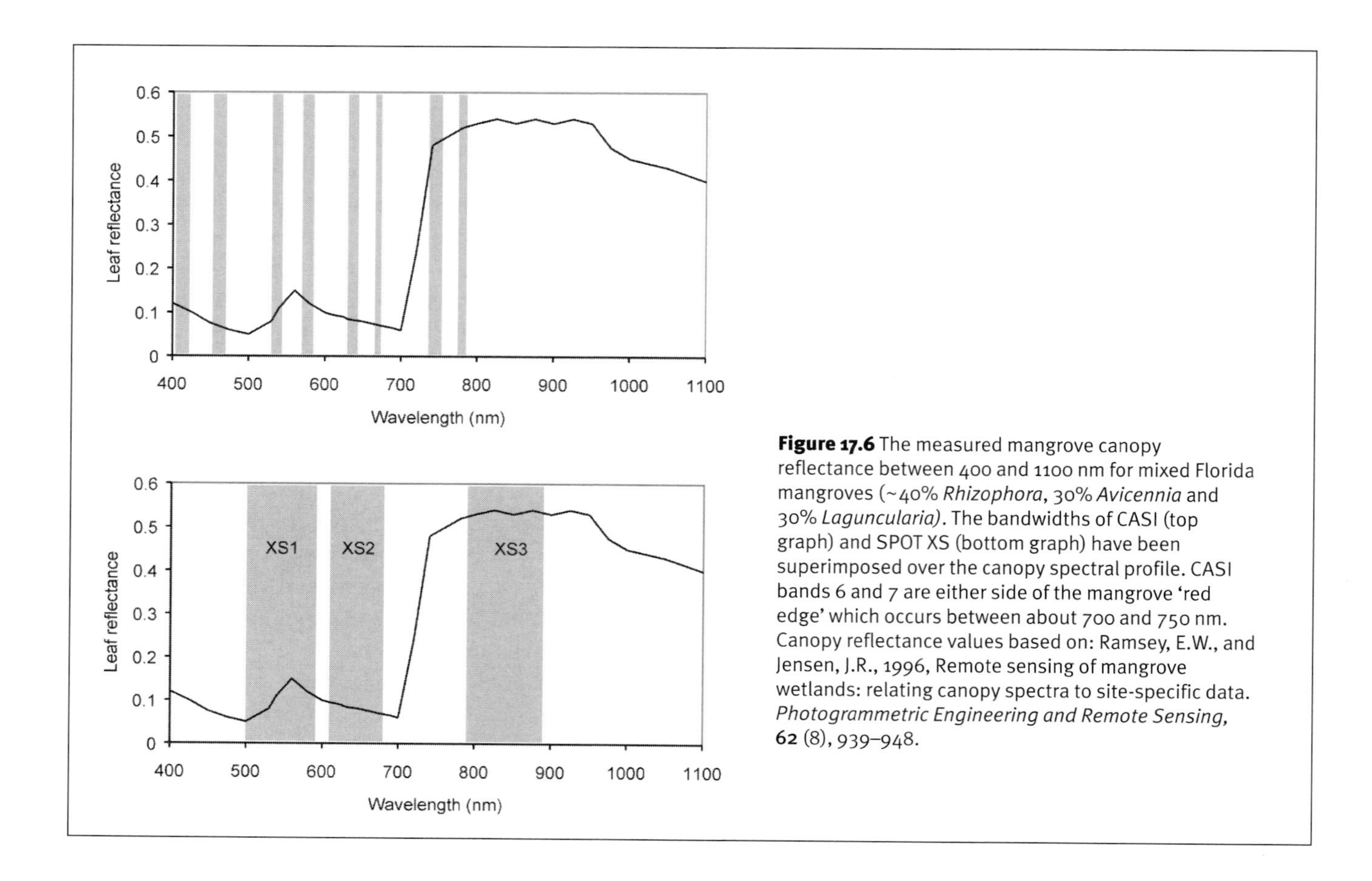

Figure 17.6 The measured mangrove canopy reflectance between 400 and 1100 nm for mixed Florida mangroves (~40% *Rhizophora*, 30% *Avicennia* and 30% *Laguncularia)*. The bandwidths of CASI (top graph) and SPOT XS (bottom graph) have been superimposed over the canopy spectral profile. CASI bands 6 and 7 are either side of the mangrove 'red edge' which occurs between about 700 and 750 nm. Canopy reflectance values based on: Ramsey, E.W., and Jensen, J.R., 1996, Remote sensing of mangrove wetlands: relating canopy spectra to site-specific data. *Photogrammetric Engineering and Remote Sensing,* **62** (8), 939–948.

from a 5 x 5 block of pixels. Other considerations include (i) the ratio of useful signal to noise (Box 5.1), which tends to be less for small pixel sizes, (ii) the frequent need to smooth images through filtering to facilitate visual interpretation of the final product, and (iii) reduced spectral resolution (8 rather than 18 bands for CASI). Weighed against these considerations is the ability to detect small, subtle features in CASI data.

High spectral resolution and the ability to select the location and width of the bands are considerable advantages to CASI, though ancillary data will frequently be necessary to exploit this feature fully. It is clear from the work of Ramsey and Jensen (1995, 1996), who have modelled mangrove canopy reflectance in Florida at species compositions very similar to the Turks and Caicos, that there is a sharp increase in reflectance at wavelengths of 710–720 nm (the mangrove 'red edge', Figure 17.6). With hindsight a better configuration for CASI might have been to place two red bands and two infra-red bands either side of this red edge (i.e. at approximately 680–690 nm, 700–710 nm, 720–730 nm and 740–750 nm). Bands 7 and 6 were either side of the red edge and this probably explains why NDVI calculated from them was a better predictor of LAI and canopy closure than the more spectrally distant Bands 8 and 5 (Figure 17.6).

An inevitable consequence of high spectral resolution is that image processing can be complicated and considerably extended as a result of the various combinations of bands which can be used to generate signatures, calculate indices etc. Undoubtedly, high spectral resolution can be immensely useful but readers need to be aware of these drawbacks.

Advantages of remote sensing for estimating LAI

The major advantage of remote sensing methods is that estimates of LAI for large areas of mangrove (a SPOT XS scene covers 3600 km^2) can be obtained without the need for extensive field effort in areas where logistical and practical problems can be severe. *Rhizophora mangle*, in particular, can grow extremely densely and access to the interior of mangrove thickets is physically impossible in many cases. Remote sensing is therefore the only way that LAI can be estimated non-destructively. In addition, the LAI data is obtained relatively quickly; the field measurements presented here required a team of two working for a total of 34 person-days; processing time involved another 22 person-days.

References

Araújo, R.J, Jaramillo, J.C., and Snedaker, S.C. 1997, Leaf area index and leaf size differences in two red mangrove forest types in South Florida. *Bulletin of Marine Science*, **60** (3), 643–647.

Botkin, D.B., 1986, Remote sensing of the biosphere. National Academy of Sciences, Report of the Committee on Planetary Biology, National Research Council, Washington D.C., USA.

Chason, J.W., Baldocchi, D.D., and Huston, M.A., 1991, A comparison of direct and indirect methods for estimating forest canopy leaf area. *Agricultural and Forestry Meteorology*, **57**, 107–128.

Cintrón, G., and Schaeffer-Novelli, Y., 1985, Características y desarrollo estructural de los manglares de Norte y Sur America. *Ciencia Interamericana*, **25**, 4–15.

Cintrón, G., Lugo, A.E., and Martínez, R., 1980, Structural and functional properties of mangrove forests. A symposium signalling the Completion of the 'Flora of Panama'. Universidad de Panamá Monographs in Systematic Botany, Missouri Botanical Garden.

Clough, B.F., Ong, J.E., and Gong, G.W., 1997, Estimating leaf area index and photosynthetic production in canopies of the mangrove *Rhizophora apiculata. Marine Ecology Progress Series*, **159**, 285–292.

English, S., Wilkinson, C., and Baker, V., 1997, *Survey Manual for Tropical Marine Resources*. 2nd Edition. (Townsville: Australian Institute of Marine Science).

Gholz, H.L., 1982, Environmental limits on above-ground net primary production, leaf area and biomass in vegetation zones of the Pacific Northwest. *Ecology*, **63**, 469–481.

Gholz, H.L., Vogel, S.A., Cropper, W.P., McKelvey, K., Ewel, K.C., Teskey, R.O., and Curran, P.J., 1991, Dynamics of canopy structure and light interception in *Pinus elliottii* stands, North Florida. *Ecological Monographs*, **6**, 33–51.

Green, E.P., Mumby, P.J., Edwards, A.J., and Clark, C.D., 1996, A review of remote sensing for the assessment and management of tropical coastal resources. *Coastal Management*, **24**, 1–40

Jensen, J.R., Ramset, E., Davis, B.A., and Thoemke, C.W., 1991, The measurement of mangrove characteristics in south-west Florida using SPOT multispectral data. *Geocarto International*, **2**, 13–21.

Kaufmann, M.R., Edminster, C.B., and Troendle, C., 1982, Leaf area determinations for subalpine tree species in the central Rocky Mountains. US Department of Agriculture and Rocky Mountains Forestry Rangers Experimental Station General Technical Report, RM–238.

Meyers, T.P., and Paw, U.K.T., 1987, Modelling the plant canopy micrometeorology with higher-order closure principles. *Agricultural and Forestry Meteorology*, **41**, 143–163.

Meyers, T.P., and Paw, U.K.T., 1986, Testing of a higher-order closure model for modeling airflow within and above plant canopies. *Boundary-Layer Meteorology*, **37**, 297–311.

McNaughton, K.G., and Jarvis, P.G., 1983, Predicting effects of vegetation changes on transpiration and evaporation. In *Water Deficits and Plant Growth*. Vol. 7., edited by T.T. Kozlowski, (London: Academic Press), pp. 1–47.

Monteith, J.L., 1972, Solar radiation and productivity in tropical ecosystems. *Journal of Applied Ecology*, **9**, 747–766.

Nel, E.M., and Wessman, C.A., 1993, Canopy transmittance models for estimating forest leaf area index. *Canadian Journal of Forestry Research*, **23**, 2579–2586.

Norman, J.M., 1980, Photosynthesis in Sitka Spruce (*Picea sitchensis*). Radiation penetration theory and a test case. *Journal of Applied Ecology*, **12**, 839–878.

Pierce, L.L., and Running, S.W., 1988, Rapid estimation of coniferous forest leaf area index using a portable integrating radiometer. *Ecology*, **69**, 1762–1767.

Ramsey, E.W., and Jensen, J.R., 1995, Modelling mangrove canopy reflectance by using a light interception model and an optimisation technique. In *Wetland and Environmental Applications of GIS*, (Chelsea, Michigan: Lewis Publishing), pp. 61–81.

Ramsey, E.W., and Jensen, J.R., 1996, Remote sensing of mangrove wetlands: relating canopy spectra to site-specific data. *Photogrammetric Engineering and Remote Sensing*, **62** (8), 939–948.

Waring, R.H., 1985, Estimates of forest growth and efficiency in relation to canopy leaf area. *Advances in Ecological Research*, **13**, 327–354.

Zar, J.H., 1996, *Biostatistical Analysis*. Third Edition. (New Jersey, USA: Prentice Hall).

18

Assessment of Coastal Marine Resources: a Review

Summary *Remote sensing can augment traditional methods of assessing the stocks of some marine animal and plant resources. A few resources, such as seaweeds (harvested for alginates, agar and carrageenan), can be remotely sensed directly. However, most marine animal resources can only be assessed indirectly once quantitative relationships between features which the remote sensor can detect (e.g. ocean colour and temperature, or habitat type) and the resources (e.g. Albacore tuna or conch) have been established. Establishing such relationships generally requires considerable costly boat-based survey.*

The assessment of phytoplankton biomass and production is briefly reviewed because the ability to monitor primary production or phytoplankton biomass in coastal waters would in theory seem useful for management. However, this is a specialist field and techniques work least well in coastal waters where silt and dissolved organic matter from river outflows are present. The background and limitations of the techniques are presented. The assessment of coastal fisheries resources by both direct and indirect remote sensing is reviewed. Airborne remote sensing can be useful in certain circumstances in aiding assessments of spawning stocks or censusing schools of a few schooling fish species. Satellite remote sensing of ocean colour and sea surface temperature (SST) can be used to identify areas where certain commercial species (e.g. Albacore tuna, Californian anchoveta, Gulf menhaden) are likely to be congregated and guide commercial fishing fleets to them. However, even for such very valuable stocks, the remotely sensed fisheries support products have not been found to be economically viable despite clearly demonstrated financial benefits to fishermen.

*The assessment of benthic resources such as conch (*Strombus gigas*),* Trochus *and lobster depends on finding clear associations between the densities of these species and marine habitat characteristics which can be accurately mapped from remotely sensed imagery. Considerable experimental research on conch and* Trochus *resources has indicated that the integration of remote sensing into field-based stock assessment surveys can extend the scope of the surveys, making them more cost-effective and probably more accurate. Seaweed resources can be assessed directly and monitoring of major commercially exploited species in the wild (e.g.* Macrocystis*) or large areas of seaweed culture appears both practicable and economically viable.*

Introduction

Although marine plant resources such as seaweeds can be assessed directly using remote sensing, most animal marine resources cannot be directly assessed, with rare exceptions, such as some near-surface schooling fish and marine mammals which can be remotely sensed from aircraft (e.g. Borstad *et al.* 1992). However, there is potential for assessing stocks indirectly using a combination of field survey and remote sensing technologies. This chapter briefly considers how remotely sensed characteristics of coastal waters, such as chlorophyll concentration and sea-surface temperature, and the ability to discriminate habitats may help in assessing potential levels of economic resources such as finfish and shellfish. It also outlines how seaweed and phytoplankton resources may be mapped (Figure 18.1).

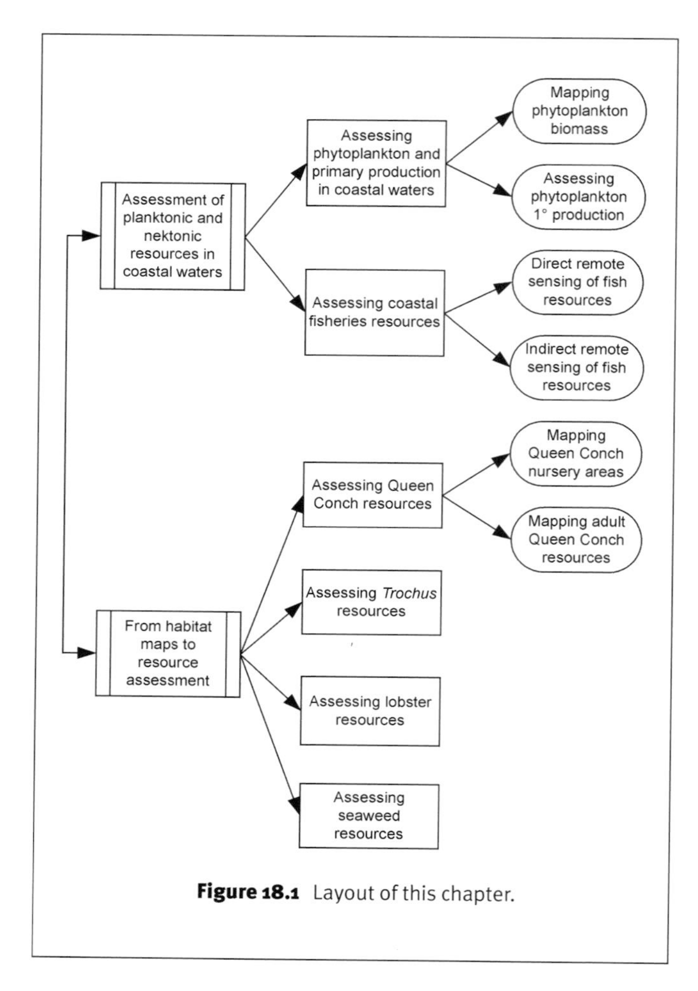

Figure 18.1 Layout of this chapter.

The progression from mapping of habitats or ocean colour to assessment of living aquatic resources requires the establishment of significant quantitative relationships between features which the remote sensor can reliably detect and levels of the living resource being studied. Establishing such relationships generally requires a large amount of field survey data and thus any attempts to assess resources using remote sensing are likely to be costly.

In terms of cost-effectiveness, the question which needs to be asked is whether the costs of the addition of a remote sensing component to the stock assessment study are outweighed by the benefits. Remote sensing does not replace traditional methods of stock assessment; it only augments existing methods. It may allow a) a more accurate determination of stocks, and b) some reduction in the need for field survey by allowing extrapolation to a wider study area. However, this is only possible once a clear relationship between a feature which can be distinguished on the imagery and levels of the stock being assessed has been established empirically or using analytical models.

Aerial surveys involving direct counting and/or counting of individuals from photographs have been widely used to monitor populations of manatees, dugong, dolphins and other coastal marine mammals (e.g. Heinsohn 1981, Irvine *et al.* 1982, Packard *et al.* 1986, Lefebvre and Kochman 1991). Methods are well documented in these references and will not be discussed further here.

Assessment of planktonic and nektonic resources in coastal waters

In this section the ways in which remote sensing can contribute to the assessment in coastal waters of planktonic (passively drifting) resources, notably phytoplankton, and nektonic (free-swimming) resources, notably fish, are examined.

Assessing phytoplankton and primary production in coastal waters

The sea-based primary production which supports coastal marine food webs has two main components which are susceptible to assessment by remote sensing: phytoplankton in the water column and attached macrophytes such as seaweeds and seagrasses on the intertidal shore and shallow seabed. This section briefly considers what can be achieved in assessing phytoplankton biomass and primary production using remote sensing. Assessment of seaweed resources is reviewed later in this chapter, mapping of seagrass is explored in detail in Chapter 12 and the measurement of seagrass standing crop using remote sensing is discussed in Chapter 16. For coastal resource managers the ability to assess coastal primary production over wide areas using a combination of field survey and remote sensing would in theory seem attractive.

Mapping phytoplankton biomass

Phytoplankton in surface water is readily detected by remote sensing due to the strong spectral signature of chlorophyll-*a* (and accessory pigments such as phycocyanin, phycoerythrin, fucoxanthin and peridinin). Remotely sensed measurement of chlorophyll-*a* and hence phytoplankton biomass is well established (Gordon and Morel 1983) for waters that do not have significant suspended sediment or yellow substance (very fine decayed organic matter) concentrations. Unfortunately, coastal waters often have significant concentrations of either one or both of these leading to considerable problems (Tassan 1988, Bagheri and Dios 1990, López García and Caselles 1990, Garcia and Robinson 1991, Parslow 1991, Han *et al.* 1994, Dierberg and Zaitzeff 1997). Notwithstanding the difficulties of obtaining accurate chlorophyll-*a* values it is fairly easy to map the location and pattern of chlorophyll-*a* anomalies. The use of such anomalies to detect cyanobacterial (blue-green algal) blooms and toxic algal blooms which can have adverse impacts on coastal waters (e.g. killing fish in cage aquaculture) of varying severity is discussed in Chapter 14. Here we will focus on the feasibility of firstly mapping phytoplankton biomass in coastal waters and secondly assessing its primary production.

As stressed above, silt from rivers or stirred up by waves and tidal streams, and yellow substance from terrestrial run-off make it more difficult to measure surface-layer chlorophyll and hence phytoplankton biomass concentrations in coastal waters. Even in the open ocean, where interference from these substances is minimal, phytoplankton pigment concentrations were only measurable to an accuracy of about ±35% using the Coastal Zone Color Scanner (Sathyendrath and Platt 1993). (This is perhaps not as bad as it seems, since concentrations range over 4 orders of magnitude). Accuracy should improve with SeaWiFS (Plate 23) and similar new sensors (e.g. Richardson and Ambrosia 1997) and such sensors and modelling developments are likely to lead to improved compensation for the effects of silt and yellow substance. Nevertheless, it seems probable that in all but very clear coastal waters accuracies are likely to remain fairly poor. Thus in reality the information available from an image is a loosely quantified synoptic view of relative phytoplankton biomass at a single point in time.

The spatial and temporal variability of coastal waters tends to be higher than that of the open ocean and the value to coastal natural resource managers of such an image needs to be considered critically. Perhaps the first point is that a single image is of very limited operational value but frequent (e.g. daily) images would allow patterns of phytoplankton biomass in the surface mixed layer to be built up. In the next section the use of such images together with AVHRR sea-surface temperature images for identifying good offshore fishing areas is discussed. Inshore, such images might allow areas of anomalously high phytoplankton resulting from, for example, fertiliser run-off or sewage outfalls, or anomalously low phytoplankton concentration resulting from, for example, herbicide run-off or toxic wastes, to be identified. Such anomalies would most likely be detectable in areas such as lagoons (e.g. López García and Caselles 1990, Ruiz-Azuara 1995) or large bays where dispersal processes are poor and could be further investigated by field survey.

At present the spatial resolution of sensors which can provide imagery cheaply at weekly or better frequency is in the order of 1 km. Such resolution will only be of use for impacts over large areas. Higher spatial resolution imagery such as Landsat TM and SPOT XS (20–30 m pixel size) is available at much lower frequency and higher cost and may allow seasonal or year-to-year changes in phytoplankton biomass to be inferred using multiple images from respectively one (e.g. López García and Caselles 1990) or several (e.g. Ruiz-Azuara 1995) years when coupled with extensive field survey. For environmental regulators, airborne instruments such as CASI (Vallis *et al.* 1996) or other airborne multispectral sensors (Zibordi *et al.* 1990) or LIDAR fluorosensors (e.g. Chekalyuk and Gorbunov 1994) could be used to map suspected impacts but at considerable expense. However, costs could perhaps be recouped via the legal system.

In summary, mapping phytoplankton biomass in coastal waters may in some circumstances be useful for coastal management, but users should be aware of the limitations and make sure their objectives are both clear and achievable.

Assessing phytoplankton primary production

At the next level of complexity is the possibility of remotely sensing planktonic primary production. As stated by Sathyendranath and Platt (1993), in their excellent succinct review of the philosophy and methodologies for estimating primary production, 'remote sensing of primary production is a science in its infancy'. However, it is growing up fast. In oceanic waters, satellite techniques and optical models can provide information on phytoplankton biomass and available light at depth. These data, together with *in situ* information on the vertical structure of the biomass and rate constants describing the relationship between light and photosynthesis, allow growth rates and thus primary production to be estimated (Sathyendranath *et al.* 1991). As photosynthesis-light models are improved and bio-geochemical provinces are better characterised (Sathyendranath *et al.* 1995), a combination of satellite observations of ocean colour and temperature (e.g. using SeaWiFS and AVHRR) and *in situ* ship-borne measurements seems likely to allow routine synoptic estimates of primary production in oceanic waters in the near future (Platt *et al.* 1995).

With regard to coastal waters where the situation is altogether more complex, routine estimates of primary production using remote sensing seem further off. Although good correlations between net primary production in a lagoon in Mexico measured *in situ* and outputs from processed Landsat TM imagery captured simultaneously were achieved by Ruiz-Azuara (1995), this is more an indication of theoretical possibility rather than operational practicality.

Assessing coastal fisheries resources

Recent reviews (e.g. Populus and Lantieri 1991, 1992) reveal that little progress has been made in the way of assessment of tropical coastal finfish resources using remote sensing. This is not surprising given the capabilities of remote sensors and the ecology of the fish important to these fisheries. However, some success has been achieved in direct remote sensing of near-surface schooling fish from aircraft (Borstad *et al.* 1992). Indirect

Table 18.1 Some of the fish species which have been deemed amenable to direct (D) or indirect (I) remote sensing. Direct sensing involves location and measurement of spawning aggregations or schools (examples cited used CASI or aerial photography). Indirect remote sensing involves determining areas where there is a high probability of finding the species using ocean colour and/or temperature data from satellite-borne sensors (examples cited used CZCS and/or AVHRR data except for the study of the Gulf menhaden which utilised Landsat MSS imagery).

Common name	Family	Species	Direct/Indirect
Jack mackerel	Carangidae	*Trachurus murphyi*	I
Pacific herring	Clupeidae	*Clupea pallasii*	D
Chilean herring	Clupeidae	*Strangomera bentincki*	I
Gulf menhaden	Clupeidae	*Brevoortia patronus*	I
Japanese pilchard	Clupeidae	*Sardinops melanostictus*	D, I
Californian anchoveta	Engraulididae	*Engraulis mordax*	I
Peruvian anchoveta	Engraulididae	*Engraulis ringens*	I
Capelin	Osmeridae	*Mallotus villosus*	D
Chinook salmon	Salmonidae	*Oncorhynchus tshawytscha*	I
Albacore	Scombridae	*Thunnus alalunga*	I
Skipjack tuna	Scombridae	*Katsuwonus pelamis*	I
Yellowfin tuna	Scombridae	*Thunnus albacares*	I
Swordfish	Xiphiidae	*Xiphias gladius*	I

remote sensing of parameters such as sea-surface temperature (SST) and ocean colour, has been used operationally to aid commercial fishermen to locate pelagic stocks of tuna, swordfish, mackerel, anchovy and sardine, but these techniques do not tend to be useful close to the coast.

Butler *et al.* (1988) provided a good review of the application of remote sensing technology to marine fisheries in general, including a useful series of case studies. Le Gall (1989) reviewed the use of satellite remote sensing to detect tuna, Simpson (1992) reviewed the use of remote sensing and GIS in marine fisheries, and Fernandes (1993) provided a general review of the role of remote sensing in aquatic resource assessment and management. The reader is referred to these comprehensive publications for details of the subject. Here, the use of both direct and indirect remote sensing of finfish resources will be outlined with special reference to their potential for tropical coastal seas.

Direct remote sensing of fish resources

Fish species which shoal in near-surface waters (e.g. sardines, anchovies, capelin, herring, mackerel, menhaden, horse mackerel and tuna – see Table 18.1) can be detected by airborne remote sensors. Visual spotting of valuable fisheries species such as tuna and menhaden from aircraft and helicopters has been used to direct commercial fishing fleets for some time. A more systematic approach using aerial photography to study the distribution and abundance of such stocks (e.g. Bullis 1967, Kemmerer 1979, Nakashima 1985, Hara 1985a) has led on to the use of airborne digital scanners such as CASI to census Pacific herring (Borstad *et al.* 1992) and capelin (Nakashima *et al.* 1989) off Canada.

Hara (1985a, b) studied the use of aerial surveys (visual observation and aerial photographs) to provide information of the size, shape and moving direction of schools of Japanese pilchard (*Sardinops melanostictus*). This species forms an important seasonal (July to October) commercial fishery off south-east Hokkaido when schools of the pilchard migrate along the coast from south to north. The results obtained from aerial surveys were compared to acoustic survey data from the area. Acoustic survey indicated that the schools occurred at a modal depth of around 10 m, although there was variation from day to day and site to site. More than 70% of schools were found in the upper 20 m of the water column and over 90% in the top 30 m. Initial aerial surveys identified almost 300 oval or crescent shaped schools. The crescent shaped schools were 100–200 m in length and around 10 m wide, whilst the oval schools were 20–30 m wide at their widest point. The technique appeared to offer potential both for censusing the pilchard and perhaps guiding fishing boats to high densities of the fish. However, further aerial studies by Hara (1989) with more or less contemporaneous acoustic survey, indicated that airborne surveys spotted less than 25% of the schools recorded in the area by acoustic survey. Analysis showed that schools below about 6 m depth did not seem to be detected by aerial survey. In conclusion, for Japanese pilchard it appears that aerial survey does not penetrate deep enough to provide reliable estimates of school density or act as a census tool.

In Newfoundland, huge shoals of capelin (*Mallotus villosus*) are observed in shallow coastal waters near beaches and bays where they spawn typically in 0–5 m depth water. Since 1982 aerial surveys have been used to produce an index of relative abundance (school surface area index) as an indicator of the status of capelin stocks, initially using aerial photography and since 1989 using CASI (Nakashima *et al.* 1989). CASI had several advantages over colour aerial photography. It allowed data: 1) to be collected in light conditions unfavourable for aerial photography, 2) to be stored digitally in real time and then rapidly analysed after each flight (school surface area index typically took several months to process from aerial photography), and 3) to be easily incorporated into a GIS. These surveys assist monitoring of capelin abundance from year to year and have been used to tune Virtual Population Analyses of the stock.

Borstad *et al.* (1992) used CASI flown at an altitude of 975 m (spatial resolution 4 m) to detect Pacific herring (*Clupea pallasii*) schools in shallow water as they prepared for spawning off the coast of British Columbia, Canada. The dark dorsal surfaces of the fish made the schools appear dark relative to the surrounding sea or, in shallower water, against the sea bottom between wavelengths of 475–575 nm. This reduced reflectance was detected by two of the wavebands to which they had set the CASI instrument (467–476 nm, 540–550 nm) and allowed the area of herring schools to be calculated.

The CASI-based technology is being further developed with the long-range objective of digital assessment of Atlantic capelin stocks. This involves obtaining a measure of density as well as area from the imagery and developing software to automatically detect schools on the imagery (rather than visually interpreting the processed imagery). Fernandes (1993) reported the cost of CASI for this type of survey in Canada to be in the order of £80–320 per 100 km^2 covered (depending on requirements) with the daily rate including aircraft charter and post-flight data processing being about £800.

These novel technologies have, as far as we are aware, not been applied to near-surface schooling stocks in tropical coastal waters but clearly could be applied to some commercially important fisheries for such stocks. Until such time as the techniques are shown to be fully operational and generating cost-effective information of real use to fisheries managers, their adoption should be approached with caution.

In terms of exploitation of coastal fisheries resources, remote sensing from aircraft using visual assessment and aerial photography has been used to assess the distribution and intensity of inshore fishing off the west coast of India and to map intensity of lobster fishing (by mapping buoys marking traps) off the coast of Canada (Butler *et al.* 1988: 77–79, Sharp *et al.* 1989). Such techniques, coupled with appropriate field survey, can have a useful role in mapping resource use and providing relative measures of effort for coastal and fisheries managers.

Indirect remote sensing of fish resources

For a brief review of this area see Laurs (1989) and Petit *et al.* (1994). Pelagic fishes tend to be congregated where there is an abundant supply of food and also show water temperature preferences. This food supply is on the whole ultimately dependent on there being a supply of nutrients to support phytoplankton. The phytoplankton is fed on by zooplankton and by some commerically important fish (e.g. menhaden which filter-feed on diatoms and dinoflagellates). The zooplankton is in turn consumed by many commercially important fishes such as sardines, herrings, anchovies and mackerel. Larger fish such as the tunas (*Thunnus* spp.) may take larger zooplankton, smaller fish (e.g. anchovies) and other nekton. The swordfish (*Xiphias gladius*) takes squid, small fish such as herrings and anchovies, as well as large fish such as tuna. Remote sensing can detect two key features which relate to phytoplankton abundance – sea-surface temperature and ocean colour. This allows remote sensing to demarcate patches of sea (or lake) where food is likely to be more abundant and which fall within known temperature preferences for species and hence where there is an enhanced probability of finding fish (e.g. Stretta 1991, Petit *et al.* 1994). Nutrients tend to be most abundant at upwelling areas and fronts, where cooler nutrient rich water is rising to the surface or different water masses are meeting. Well-known upwelling areas that support large fisheries include those off the coast of Peru, Namibia, and the Cape Verde region of West Africa. Fronts tend to occur where different currents meet and near the continental shelf edge where coastal waters meet oceanic ones.

Such mesoscale phenomena (over tens to hundreds of kilometres) can be readily mapped from the associated changes in sea-surface temperatures (SST) by NOAA's polar-orbiting satellites' AVHRR sensor (or more recently by the ERS Along Track Scanning Radiometer). Phytoplankton possess chlorophyll and accessory pigments, the concentrations of which have been mapped with considerable success using the Nimbus-7 Coastal Zone Color Scanner (CZCS: operational from 1979-1986) and which are amenable for more accurate mapping using the SeaWiFS sensor (launched in July 1997) and a range of other new sensors (e.g. the Ocean Color and Temperature Scanner (OCTS) on the Japanese Advanced Earth Observing Satellite (ADEOS) launched in August 1996).

AVHRR temperature data alone has been used operationally to direct tuna and salmon fishing fleets by supplying near real-time maps by direct Automatic Picture Transmission (APT) of two AVHRR wavebands at reduced spatial resolution (typically 4.4 km), or by High Resolution Picture Transmission (HRPT) facsimile charts at full resolution (1.1 km) from land-based stations. Such charts can reduce tuna fleet search times by 25–40%. An 18 month experiment was also made with mailing AVHRR based charts to fishermen off the north-eastern US. The charts had a mixed reception with problems with spatial resolution, time delay and cloud cover being noted (Cornillon *et al.* 1986). Although fishermen estimated savings in excess of US$5000 per year from using the charts, most were not prepared to pay more than US$50 per year for them. This meant that it would be uneconomic to provide the charts.

Lasker *et al.* (1981) showed a correlation between AVHRR derived SST and Californian anchoveta (*Engraulis mordax*) spawning in the Southern California Bight but noted that other factors were also determining the distribution of spawning. Laurs (1989) noted that a combination of AVHRR SST data and CZCS phytoplankton pigment data defined the spatial distribution of the spawning stock nearly completely.

Since phytoplankton patches are not always associated with temperature changes, a combination of ocean colour and SST is likely to be more fruitful, providing closer correlations with fish distributions. Laurs *et al.* (1984) used CZCS and AVHRR to study temperature and phytoplankton distributions simultaneously, and found a clear association of *Thunnus alalunga* with oceanic fronts off California with fish congregating in particular on the ocean side of oceanic water intrusions into coastal water masses. Montgomery *et al.* (1986) used CZCS data combined with SST and weather data to provide charts to guide tuna, swordfish and salmon fishermen off the west coast of the United States. Although successful in terms of improved fish location and decreased fuel costs, the high cost of chart production (approximately $1000 per chart) seemed to preclude development of a viable commercial market.

There are also similar reports of the use of NOAA AVHRR sea surface temperature (SST) imagery to assist fishermen in China, India, and Chile. In the East China Sea, Qinsheng (1990) reported potential for SST data to delimit commercial ('hairtail' and 'filefish') fishing grounds whilst Shixing *et al.* (1990) reported its use along with fishery vessel data to show concentration of the Japanese pilchard (*Sardinops melanostictus*) on the warm side of frontal zones and in warm-core eddies. The latter authors concluded that it would be possible to forecast the formation and stability of fishing grounds and the timing of the pilchard fishing season from satellite SST data. Xingwei and Baide (1990) reported supply of fisheries charts for the East China Sea and Yellow Sea based on SST data by fax to fishermen in 1987–89 with a delay of about 48 hours from overpass. One fishing company estimated time and fuel savings of about 20% from using these charts. In the Indian EEZ, Narain *et al.* (1990) used AVHRR data to map temperature fronts and generate 'fishery prospect' charts within 55–60 hours of an overpass. These charts appeared to generate higher than expected catches by fishermen using them and thus may have increased fishing efficiency. In central Chile, hard-copy fish harvesting probability charts, based on AVHRR SST data and indicating areas of high, medium and low probability of catching albacore (*Thunnus alalunga*), were used successfully for 1986–89 to support small-scale fishermen. The charts appeared to significantly reduce search times and fuel costs (Barbieri *et al.* 1989). Recently, Yáñez *et al.* (1997) have indicated the potential for using SST data also to support fisheries for the Chilean herring (*Strangomera bentincki*), the Peruvian anchoveta (*Engraulis ringens*) and the jack mackerel (*Trachurus murphyi*).

Overall these studies have shown that satellite remote sensing can increase the probability of locating certain major commercial species which are found offshore (as opposed to in nearshore truly coastal waters). Unfortunately, even for these very valuable commercial stocks off the United States the remotely sensed fisheries support products have not been considered economically viable despite demonstrable benefits to fishermen worth thousands of dollars per year per boat. If a self-financing system cannot be set up in these fisheries, it seems unlikely that it can elsewhere!

However, remote sensing can also provide environmental information which may help with traditional stock assessment programmes. Combinations of ocean colour and SST sensors are likely to be most effective and use of SeaWiFS together with AVHRR or ATSR (or possibly the Visible and Thermal Infra-red Radiometer (VTIR) on the Japanese Marine Observation Satellite (MOS); see Chong, 1990), or OCTS (which has both) offers the potential for near real-time operational support for fishing fleets exploiting these valuable resources. The biggest drawback is cloud cover, which prevents both ocean colour measurement in the visible and SST measurement in the infra-red. This may render large areas of ocean effectively unviewable for weeks at a time.

With respect to species fished inshore the phytoplankton signal is likely to be masked by suspended matter and yellow substance and the coastal waters are likely to be well-mixed so that temperature gradients are most likely to be associated with river outflows. However, Landsat MSS was found useful for identifying areas of the Northern Gulf of Mexico where the Gulf menhaden (*Brevoortia patronus*) were most likely to be located (Kemmerer 1980). Their distribution was correlated with water turbidity and colour, and charts indicating areas of high and low probability of occurrence were produced from Landsat MSS images. Both cloud cover and repeat time were a drawback, but for this major fishery (with catches of approximately 500,000 tonnes a year in the 1970s and early 1980s) the charts were considered worthwhile. Despite this example, in general satellite remote sensing seems at best to offer limited support to either fish location or inshore stock assessment in terms of synoptic views of water movements and environmental features which may assist in modelling.

From habitat maps to resource assessment

This section reviews how remote sensing researchers have tried to move beyond just mapping habitats to assessing the stocks of selected key resources (conch, *Trochus*, lobster and seaweeds) associated with particular habitats. Such studies are few and should be regarded as experimental.

Assessing Queen Conch resources

Mapping Queen Conch nursery areas
In the Caribbean and Bahamas the Queen Conch, *Strombus gigas*, is an important marine resource with some 4000 tonnes being harvested each year with a value of some US$ 40 million. Conch are shallow water species which are rarely found at depths greater than 30 m. Juvenile conch (1–2 years old) are found in aggregations, commonly at densities of 1–2 m^{-2}, associated with seagrass meadows in very shallow water (< 4 m deep in the Bahamas) where there is abundant macroalgal food. From a management viewpoint, protection of these potentially vulnerable nursery stocks, which tend to be found in well-defined, localised areas, would be a key precautionary measure. Stoner *et al.* (1996) studied how remote sensing could help to identify such nursery areas in an 11,300 ha region of the Great Bahama Bank. Jones (1996) extended this work using a Geographical Information System (GIS) to study further the spatial association of the juvenile conch with biological and physical habitat features.

Stoner *et al.* (1996) used a Landsat Thematic Mapper image to map both seagrass biomass (to four densities: < 5, 5–30, 30–80, and > 80 g dry weight m^{-2}) and bathymetry (to six depth intervals: 0–0.5, 0.5–1.0, 1.0–1.5, 1.5–2.5, 2.5–4.0, and > 4 m deep at mean low water – see Chapter 15 for discussion of whether such precise mapping of bathymetry is realistic from optical imagery). Field survey, involving collection of seagrass and algae at three 0.25 x 0.25 m quadrats at each of 29 stations, was used to calibrate the biomass image (Armstrong 1993). Extensive local knowledge and ground truthing subsequent to image analysis helped correct errors caused by dark sediment and patches of macroalgae in seagrass free areas. The bathymetry map was constructed using the TM ratio algorithm of Spitzer and Dirks (1987). Mapping using the TM image was facilitated by the shallowness (about 75 % of the Bank is < 2.5 m deep) and relatively simple habitats of the Great Bahama Bank, which is composed almost exclusively of sand and single species stands of the seagrass *Thalassia testudinum*, with few coral reefs.

The distribution of juvenile conch was mapped by field surveys carried out annually between 1989 and 1995. Comparison of image derived data and distribution of conch nursery areas indicated that juvenile conch were associated with seagrass densities of 30–80 $g.m^{-2}$ and depths of 1.5–4 m (Stoner *et al.* 1996). Jones (1996) found optimal nursery areas (those occupied by juvenile conch in 4 out of the 7 years) at seagrass densities of 18–75 $g.m^{-2}$ and depths of 2–4 m. Although over 90% of persistent aggregations of juvenile conch were found in this optimal habitat, only about 10% of the optimal habitat defined by depth and seagrass biomass was occupied by juvenile conch. Thus, image-derived information would grossly overestimate conch nursery areas. However, it could be useful (if the correlations established in the Bahamas apply elsewhere) in focusing field surveys and thus reducing the costs of these. The data of Stoner *et al.* (1996) suggest that it would have focused field survey on about 10% of their study area, limiting surveys to about 1000 ha of the 11,000 ha area covered by the sub-scene, but at the risk of missing some juvenile conch aggregations.

In an attempt to refine their prediction of juvenile conch habitat other factors were considered, in particular the relationship of the nursery areas to the degree of tidal flushing of the Bank by cooler clear oligotrophic oceanic water. Using sea surface temperature (measured by field survey, not remote sensing) as a proxy, Jones (1996) mapped the extent of tidal flushing and found that juvenile conch were restricted to areas within 5–6 km of cuts between islands through which oceanic water moved onto the Bank with each tide. This allowed the extent of the optimal habitat and overall suitable habitat (occupied by juvenile conch in any year) to be further refined. However, even with this additional criterion optimal habitat remained at about 760 ha out of the 11,000 ha, and on average only about 80 ha of this habitat (~10%) was occupied. Whilst for overall suitable habitat, some 2350 ha were identified (only about 3% of which was utilised). The authors conclude that either the Bank was undersaturated with conch or that the ecological requirements of the nursery areas were not fully understood.

There are perhaps two lessons we can learn from these studies. Firstly, the bulk of the resource assessment was achieved using painstaking traditional field survey methods. The remote sensing element primarily allowed extension of these survey data to broader areas of the

Bank, but relied strongly on the high quality field data to achieve this. Secondly, relationships between resources and biological and physical features may be complex and multifaceted. In such cases, if even one critical feature is not measurable by remote sensing then it may not be feasible to predict the occurrence of the resource from imagery with sufficient accuracy to be useful. In the present case study, remote sensing data and additional tidal excursion information would at best allow a manager to predict conch nursery habitat with a 10% accuracy. Thus, assessment of juvenile conch resources would clearly not be possible using remote sensing. However, it does have the potential to assist field survey by focusing ground studies in areas where juvenile conch are most likely to occur.

The question of whether the Great Bahama Bank relationship between conch nursery areas and features discernable on remotely sensed imagery is applicable elsewhere determines how cost-effective remote sensing might be in other areas. If the lessons learned at one site can be applied elsewhere then the scale of field survey can be significantly reduced. If on the other hand the relationship is site-specific, then all the field survey has to be repeated for each site. Unfortunately, the classic report of Randall (1964) on the biology of the Queen Conch at St John in the US Virgin Islands suggests that the relationship is site-specific. He found juvenile conch to be very abundant on a coral rubble bottom in 9–12 m of water.

Mapping adult Queen Conch resources

We have not seen any published accounts of attempts to do this using remote sensing; however, given that conch habitat is generally observable by remote sensors, it should in theory be possible to use remotely sensed imagery as an aid to *Strombus gigas* (Plate 23) stock assessment. The basis for any such stock assessment lies in establishing the average density and size of conch in the different shallow water habitats which can be discriminated by remote sensing. Thus, considerable field survey would need to be undertaken at each site to establish, firstly, what the main habitats distinguishable on imagery are, and secondly, the nature of the relationship between conch and these habitats. Remote sensing would primarily facilitate assessment of the area of each habitat type and thus allow the densities/biomasses assessed in the field survey to be used to estimate stocks in the whole region covered by the image.

The studies of Stoner and Waite (1990) in the Bahamas, which show a highly significant correlation of Queen Conch biomass and density with the biomass (or shoot density) of the seagrass *Thalassia testudinum* and with water depth, suggest that such an approach might be successful, as both these parameters can be inferred from remotely sensed imagery. In a detailed study in Belize, where annual landings of conch meat have averaged 180 tonnes in recent years, Appeldoorn and Rolke (1996) studied the relationship between conch (juvenile, legal – defined as > 15 cm shell length, and adult) density and seven different habitats. Legal (exploitable) conch showed a preference for three habitats (sparse seagrass and algae, moderately dense seagrass and algae, and gorgonian plain) with adults preferring the the former two habitats. Exploitable populations occurred mainly at depths of 4.8–7.6 m, well within the depth range for optical remote sensing, and our experience indicates that the different habitats can be distinguished from remotely sensed imagery. Although their extensive field survey data allowed densities of exploitable conch in each habitat to be estimated, they did not have remote sensing derived habitat maps which would allow the areas of each habitat to be calculated and so they used an average density value (approx. 15 ha^{-1}) to estimate the total exploitable population in the 1,500 km^2 of conch grounds present in water < 15 m deep. This gave an estimated exploitable population of 2.3 million conch (95% confidence limits, 1.6–3.8 million) and a maximum sustainable yield (MSY) of approximately 190 tonnes per year. Detailed habitat maps would have allowed better estimates of the conch population but perhaps more usefully (given the uncertainties inherent in calculating MSY) would reduce subsequent survey costs by guiding stratified sampling of the key conch habitats.

Such stock assessments will always be site (and time) specific as local fishing pressure will be the other key determinant. Nevertheless, the integration of remote sensing techniques into stock assessment surveys for conch seems likely to be a cost-effective approach.

Assessing Trochus *resources*

Large top shells (Trochidae) of the genera *Trochus* and *Tectus* are widely fished in the Indo-Pacific for the valuable mother-of-pearl of the inner nacreous layers of their shells. *Trochus niloticus* is important commercially to many Pacific islanders and Bour *et al.* (1986) reported that 2000 tons of mother-of-pearl were exported from New Caledonia alone in 1978. In many areas *Trochus* is being overexploited and management is urgently required.

This gastropod mollusc mainly lives on coral reefs at depths of less than 10 m and so its habitat is accessible to remote sensing as well as human predation (Plate 23). Juveniles and small (< 5 cm) individuals favour the outer reef-flat rubble zone at depths < 3 m, whereas larger commercial-sized (> 8 cm) individuals are generally distributed along the windward edge of reefs both intertidally and subtidally to depths of about 8 m (Long *et al.* 1993). *Trochus* prefers exposed, gently sloping, structurally complex substrates rich in crevices, with abundant coralline and filamentous algae. Being confined to shallow water, easily located and collected, long-lived (10–14 years), slow-maturing (2–3 years), with limited dispersal abilities and high value, it is very susceptible to overfishing with reports of overfishing throughout its range.

Against this background, use of remote sensing to assist in estimating standing stocks would clearly be welcome. The rationale is that, if the *Trochus* habitat can be mapped accurately using remotely sensed imagery, and both the density of *Trochus* per unit area of habitat and the mean weight of individuals determined by field surveys, then the stock of *Trochus* can be estimated. Such information can be used to determine the annual catches of *Trochus niloticus* which can be taken on a sustainable basis. It would also provide fisheries/coastal managers with useful spatial information on which to base conservation measures should the stock be shown to be overfished.

Two such studies have been undertaken, one in the Bourke Isles in the Torres Strait using a Landsat Thematic Mapper (TM) image (Long *et al.* 1993), and the other in New Caledonia using a SPOT High Resolution Visible (HRV) image (Bour *et al.* 1986, Bour 1988).

In New Caledonia *Trochus* habitat can be found on fringing reefs, inner lagoon reefs and barrier reefs and represents a large fraction of the 20,000 km^2 of shallow water habitat surrounding New Caledonia. Bour *et al.* (1986) noted the impracticality of surveying such a large area to assess *Trochus* resources by conventional field survey methods and explored the use of remote sensing using simulated SPOT HRV multispectral (20 m spatial resolution) imagery to map *Trochus* biotope. (The simulation used a Daedalus airborne multispectral scanner set to the SPOT wavebands and flown at an altitude to give a spatial resolution of 20 m).

Bour *et al.* (1986) carried out their experimental study on part of Tetembia Reef on the south-west of the island. They were able to use simulated green (XS1) and red (XS2) bands of the SPOT HRV sensor to distinguish 5 reef types within the general hard bottom category of reef flat at 0–2 m depth. One of these was *Trochus* habitat – 'reef flat, suitable for *Trochus* (slabs with dead coral rubble)' – which represented almost 50% of the 278 ha of shallow reef flat surveyed and about 17% of all bottom types which covered a total of 820 ha in the image.

Bour (1988) extended this work to the whole of the Tetembia Reef once real SPOT data became available in 1986. Again he used SPOT XS1 and XS2 bands to classify the shallow water habitats (< 25–30 m deep). A different classification scheme was adopted with 6 categories being distinguished (3 being hard coral bottom types and 3 soft sandy bottom types). Two of the hard bottom categories were identified as *Trochus* biotope. The total area of these *Trochus* habitats was estimated from the image at 1038 ha of a total shallow water area of 4401 ha. Previous studies had shown the mean density of *Trochus* on reefs around New Caledonia to be 45 ha^{-1}. To determine the mean weight of the gastropods on Tetembia Reef sampling was carried out in the field and indicated a mean mass of 418 g. These figures were multiplied together to provide an estimated biomass of 18.8 kg.ha^{-1} and a total biomass of *Trochus niloticus* on Tetembia Reef of 19.5 tonnes.

Unfortunately data which would allow the efficacy of this approach to be assessed were not presented. The kind of information that would be needed is: a) the accuracy of *Trochus* habitat identification using the SPOT image, b) the variability of *Trochus* densities within the two image identified habitats, and c) the variability of mean weight from one area to another. With such data the likely error of biomass estimates and the amount of field survey needed to achieve useful estimates could be evaluated. Without statistical analysis of the relationship between *Trochus* biomass predicted from the image and *Trochus* biomass estimated by field survey in, for example, a sample of 0.36–1.0 ha (9–25 pixel) units, the practical usefulness of the method must remain equivocal.

Long *et al.* (1993) reported the methodology and results of their study in much more detail and provide most of the information managers might wish in planning such a stock assessment. *Trochus* habitat was mapped from the Landsat TM image using relatively simple image processing. Band 2 (green) was divided by Band 3 (red) and the resultant ratio image was contrast stretched; pixels with brightness values above 115 (on a scale of 0–255), if located on the windward sides of reefs exposed to the south-east trade winds, were classified as *Trochus* habitat. The element of contextual editing was needed to exclude the lee sides of reefs and submerged shoals where *Trochus* are not normally found.

Long *et al.* give a detailed account of their field survey strategy (including time costs) which is very instructive and recommended reading for anyone embarking on such a study. They carried out field sampling at 89 sites (each approximating in area to one Landsat TM pixel) distributed over 8 reefs (out of a total of 28 in the Bourke Isles) using band transects, surveying a total area of just over 2 ha out of 275 ha of *Trochus* habitat on these reefs (total area 4105 ha). Habitat data and samples of *Trochus* shells were also collected at each site; the first to define objectively the characteristics of the shell's habitat, the second to allow the proportion and biomass of commercial-sized *Trochus* to be calculated.

The average width of the *Trochus* habitat ranged from only 24 m on some reefs to 136 m on the largest reef studied. This indicates that on some reefs at least the width of the habitat is less than the spatial resolution (*ca.* 30 m) of the Landsat TM sensor and will be represented as 'mixels' on the image and lead to error in estimates. Three principal habitats were characterised as being associated with *Trochus* in the field sampling (algal pavement, rubble/algal pavement, and rubble zone) and average densities of the shell in these were estimated at 445–590 ha^{-1}. Combining their field and satellite data they were able to calculate that the 8 study reefs held stocks of about 28,500 commercial-sized *Trochus* with a biomass of around 6.3 tonnes. Extending this to the area of habitat estimated from the Landsat image for the 28 reefs of the Bourke Isles, they estimated a standing stock of 13.7 tonnes (±24%).

This case study is useful in that it outlines how remote sensing can be integrated with field based stock assessment techniques and, with careful design, extend the scope of the assessment in a potentially cost-effective way. Assumptions have to be made but, where these are clearly stated and the likely errors of estimates stated, such assessments can be of considerable benefit to managers. A major assumption here is that the 20 reefs not subject to field survey have similar *Trochus* densities to the sample of 8 which were studied. In the present example the accuracy of the final commercial stock estimate is stated as ±24%. This gives a manager the option to base management decisions on both the best estimate of a standing stock of 13.7 tonnes and a 'worst-case scenario' of only 10.4 tonnes. Given the uncertainties indicated by the authors, a precautionary approach using the lower estimate would be recommended.

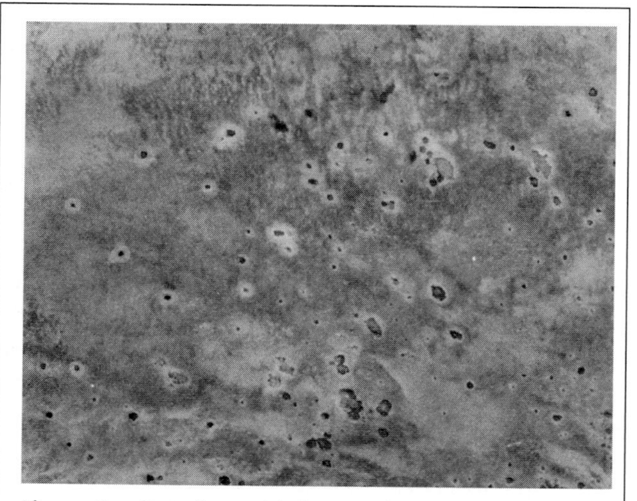

Figure 18.2 Part of an aerial photograph showing an approximately 1 km^2 area of the Caicos Bank south-east of Little Ambergris Cay which is rich in patch reefs harbouring lobsters (*Panulirus argus*). Around the patch reefs are halos (grazed areas) which are clearly visible. The widths of the halos and sizes of the patch reefs can be measured from digitised aerial photographs.

Assessing lobster resources

Spiny lobsters (e.g. species of *Panulirus*) are associated with coral reef habitats. In the Turks and Caicos Islands extensive field survey of *Panulirus argus* resources indicated a correlation between patch reef size and halo width (Figure 18.2) and the numbers of lobsters on the reef (Ninnes and Medley 1995). Aerial imagery allows the locations and sizes of patch reefs to be mapped and thus could be used to augment field survey in providing estimates of unexploited biomass, a useful parameter for a number of models used to manage stocks (e.g. Medley and Ninnes 1997). Patch reef maps may also be of use in modelling the migrations of spiny lobster and thus help management to predict the effect of closed areas and casitas which could lead to fishery enhancement (Medley pers. comm. 1997). Satellite images such as SPOT Pan and Landsat TM (for large patches) allow areas of complex habitat with lots of patches to be demarcated for possible inclusion in closed areas but can contribute little else to lobster resource management.

As for the other resources looked at, all that remote sensing can really do is to aid stock assessment, allowing detailed field survey to be used more cost-effectively on the basis that relationships between habitat and resource density established in 'training areas' apply throughout the image. With integration of resource data into a GIS, lobster resource maps could allow study of populations in space as well as time and ultimately improve management. However, it is likely to be many years yet before we see such data being used operationally in management. At present, key management options in lobster fisheries are quotas, effort controls, size limits, closed areas and closed seasons. Remote sensing can provide some useful inputs to quota setting (if it improves stock assessment) and into siting of closed areas, but otherwise is of little practical value.

Assessing seaweed resources

Seaweeds harvested from the wild and cultured on racks (Plate 23) are an increasingly valuable coastal resource. The Food and Agriculture Organisation (FAO) of the United Nations estimated that some 7.8 million tonnes (Mt) of seaweeds were harvested worldwide in 1995. These included 5.3 Mt of brown seaweeds (Phaeophyta), 1.8 Mt of red seaweeds (Rhodophyta) and 0.3 Mt of green seaweeds (Chlorophyta). The seaweeds are primarily collected for the alginates, agar or carrageenan they contain.

For remote sensing purposes, seaweed resources can be notionally divided into four categories: 1) giant brown kelps with fronds which float on the sea surface (e.g. species of *Macrocystis*), 2) intertidal seaweeds (e.g. the brown Fucales such as *Fucus* spp. and *Sargassum*), which are collected from the shore at low tide, 3) subtidal kelps such as the Laminariales, which have to be sensed through the water column, and 4) cultured seaweeds such as *Euchema* growing on racks in shallow water. For each category the appropriate wavebands and spatial resolutions required to map and monitor the resources will differ. The studies reported below are primarily from temperate waters but the principles should apply in the tropics.

In California alone the giant brown kelp *Macrocystis pyrifera* supported an industry, primarily based around the extraction of algin from the seaweed, worth US$ 20 million a year in 1975 (Jensen *et al.* 1980). This kelp, which grows on average 45 cm per day, is also used as meal in animal feed and in small quantities for human consumption and has been proposed as a renewable energy source (as a substrate for producing methane by anaerobic digestion). The fronds of this kelp are buoyed up by gas-filled bladders and so float on the sea surface making the kelp readily detectable by colour infra-red aerial photography, a range of satellite (Landsat MSS and TM, SPOT XS) and airborne multispectral scanners, and radar (Jensen *et al.* 1980, Augenstein *et al.* 1991, Deysher 1993). Except when the water was turbid the kelp covered sea surface was readily distinguished from surrounding sea in the near infra-red and could be mapped. In turbid conditions other wavebands allowed discrimination of the kelp. Due to the relatively large areas being mapped (for example, Jensen *et al.* (1980) were mapping areas in the order of 275 ha) satellite sensors were generally found to be adequate in terms of spatial resolution. Distinguishing kelp areas using radar relies on a) kelp beds giving greater backscatter than the surrounding sea in very calm sea conditions due to the kelp fronds sticking through the glassy surface, or b) the kelp beds damping wave choppiness in rougher sea conditions resulting in less backscatter from the beds than the surrounding sea.

To quantify the resource (in terms of surface canopy density) Augenstein *et al.* (1991) were able to demonstrate a relationship between near infra-red (SPOT XS band 3) radiance and canopy density. This allowed beds to be separated into low, medium and high density. Deysher (1993), using airborne multispectral imagery, was also able to map subsurface *Macrocystis* not detectable using colour infra-red aerial photography. His conclusion that SPOT imagery cannot resolve *Macrocystis* beds smaller than 10 ha (equivalent to some 250 SPOT XS pixels!) seems improbable and is contradicted by the findings of Augenstein *et al.* (1991).

The primary objective of monitoring the *Macrocystis* resource is to quantify how much resource is harvestable (floating on the surface in dense mats) and how much has been harvested. Harvesting involves trimming the fronds off to about one metre's depth at which point the kelp's infra-red reflectance will not be detectable. Thus change through time in the area of infra-red reflectance gives a measure of both kelp regeneration and the areas harvested. Such information for management is relatively easily obtained by remote sensing using either satellite or aerial sensors with infra-red capability.

Seaweeds growing in the intertidal zone are also harvested for alginate, agar and carrageenan extraction or use as animal feed on both temperate and tropical coasts. For example, in southern India the Gulf of Mannar is estimated to have 125 km^2 of intertidal wild seaweed beds where the browns *Sargassum* and *Turbinaria* are collected for alginates and the reds *Gracilaria*, *Gelidium* and *Gelidiella* are collected for agar with a total of some 3000 tonnes dry-weight being landed per year (Coppen and Nambiar 1991).

A pioneering study of the use of black-and-white and colour aerial photography to map seaweed resources was that of Cameron (1950) for the main seaweed harvesting area in Nova Scotia, Canada. Intertidal *Fucus serratus* and subtidal kelp (*Laminaria*) beds could be distinguished and were mapped successfully. Although biomass estimates still required field survey, the large savings in boat time as a result of knowing the boundaries of the seaweed beds was deemed to make the aerial survey cost-effective.

Belsher and Viollier (1984) used simulated SPOT multispectral imagery (flown from an aircraft but with 20 m pixel size) to map intertidal seaweeds around Roscoff, France and produced a thematic map showing five categories of seaweed (*Fucus* spp., *Ascophyllum*, *Sargassum*, *Ulva* and *Enteromorpha*, and *Pelvetia*). Unfortunately, the user accuracy of the map was not stated and so it is unclear how well the SPOT sensor could distinguish between the four brown and one green seaweed category. Later, Belsher (1986) reported the use of large scale colour and colour infra-red aerial photography flown at 275 m altitude to map intertidal brown seaweeds (Fucales, mainly *Fucus serratus*) and green seaweeds (Ulvales) on the French coast. Using a combination of field survey data and densitometric analysis of the photographs he then estimated the number of tonnes of harvestable seaweed of each type. Meulstee *et al.* (1986) in the Netherlands used a similar approach to map the coverage of the green seaweeds *Enteromorpha* and *Ulva* and the brown *Fucus vesiculosus* on intertidal estuarine mudflats and estimate the biomass of each seaweed type. The main problem with such intertidal surveys is that they must coincide with low tide. Appropriate satellite images are thus likely to be too rare to be useful and, unless the shores are very wide and seaweed patch sizes measured in many hundreds of square metres, may also be of insufficient spatial resolution. Commissioning aerial surveys to coincide with spring low tides would solve these problems but it is unclear whether the economics of intertidal seaweed collection (unlike the multi-million dollar *Macrocystis* industry) would support the expense of such studies.

There are also a range of subtidal exploitable kelp resources which do not float on the surface (although some fronds may break the surface at low tide) such as species of *Laminaria* which can form extensive subtidal kelp forests in temperate waters. These tend to occur as relatively narrow strips in shallow water (< 5–20 m deep depending on average water clarity) along rocky coasts and are amenable to high-resolution airborne remote sensing when water clarity is good. A study of such beds in Newfoundland using a multispectral digital airborne scanner (CASI) flown at 2 m resolution allowed areas of bare substrate, open kelp canopy and dense kelp canopy to be mapped (Simms 1997). Field survey provided biomass per unit area estimates that allowed the standing crop of kelp in the *ca.* 4.5 km^2 study area to be estimated. Results indicated around 13.6 ha of dense kelp beds with a standing crop in the order of 1000 tonnes. Research is currently underway in the UK to look at airborne multispectral and boat-based acoustic methods of both identifying exploitable subtidal kelp resources and monitoring their harvesting.

Remote sensing of sites of seaweed culture in order to monitor development of the industry or to assess compensation claims subsequent to oil spills or cyclones is another area where airborne digital imagery would clearly be useful.

In summary, remote sensing of seaweed resources is practicable and for major commercially exploited species appears economically justifiable. Existing multispectral airborne digital technologies coupled with field survey could routinely provide quantitative monitoring information of greater detail than seems to be currently collected.

References

Armstrong, R.A., 1993, Remote sensing of submerged vegetation canopies for biomass estimations. *International Journal of Remote Sensing*, **14**, 621–627.

Appeldoorn, R.S., and Rolke, W., 1996, Stock abundance and potential yield of the Queen Conch resource in Belize. Report to CARICOM Fisheries Resource Assessment and Management Program (CFRAMP) and Belize Fisheries Department, Belize City, 15 pp.

Augenstein, E.W., Stow, D.A., and Hope, A.S., 1991, Evaluation of SPOT HRV-XS data for kelp resource inventories. *Photogrammetric Engineering and Remote Sensing*, **57** (5), 501–509.

Bagheri, S., and Dios, R.A., 1990, Chlorophyll-*a* estimation in New Jersey's coastal waters using Thematic Mapper data. *International Journal of Remote Sensing*, **11**, 289–299.

Barbieri, M.A., Yáñez, E., Farias, M., and Aguilera, R., 1989, *IGARSS '89 Remote Sensing: an Economic Tool for the Nineties, Vancouver, Canada July 10–14, 1989* (Vancouver, Canada: IGARSS '89 12th Canadian Symposium on Remote Sensing), **4**, 2447–2450.

Belsher, T., 1986, Measuring the standing crop of intertidal seaweeds by remote sensing, Ch. 28, in *Land and its Uses – Actual and Potential: an Environmental Appraisal*, edited by F.T. Last, M.C.B. Hotz and A.G. Bell (New York: Plenum Press), pp. 453–456.

Belsher, T., and Viollier, M., 1984, Thematic study of the 1982 SPOT simulation of Roscoff and the west coast of the Contentin peninsula (France). *Proceedings of the 18th International Symposium on Remote Sensing of Environment*, Paris, France (Ann Arbor: Environment Research Institute of Michigan), pp. 1161–1166.

Borstad, G.A., Hill, D.A., Kerr, R.C., and Nakashima, B.S., 1992, Direct digital remote sensing of herring schools. *International Journal of Remote Sensing*, **13** (12), 2191–2198.

Bour, W., 1988, SPOT images for coral reef mapping in New Caledonia. A fruitful approach for classic and new topics. *Proceedings of the 6th International Coral Reef Symposium*, **2**, 445–448.

Bour, W., Loubersac, L., and Rual, P., 1986, Thematic mapping of reefs by processing of simulated SPOT satellite data: application to the *Trochus niloticus* biotope on Tetembia Reef (New Caledonia). *Marine Ecology Progress Series*, **34**, 243–249.

Bullis, H.R., 1967, A program to develop aerial photo technology of surface fish schools. *Proceedings of the 20th Annual Session. Gulf and Caribbean Fisheries Institute* (Miami, Florida: Gulf and Caribbean Fisheries Institute), pp. 40–43.

Butler, M.J.A., Mouchot, M.-C., Barale, V., and LeBlanc, C., 1988, The application of remote sensing technology to marine fisheries: an introductory manual. *FAO Fisheries Technical Paper*, No. 295, 165 pp.

Cameron, H.L., 1950, The use of aerial photography in seaweed surveys. *Photogrammetric Engineering*, **16** (4), 493–501.

Chekalyuk, A.M., and Gorbunov, M.Y., 1994, Pump-and-probe LIDAR fluorosensor and its applications for estimates of phytoplankton photosynthetic activity. *Proceedings of the Second Thematic Conference on Remote Sensing for Marine and Coastal Environments*, **1**, 381–400.

Chong, Y.J., 1990, Satellite measurement of sea surface temperature in a humid atmosphere. *Remote Sensing for Marine Studies. Report of the Seminar on Remote Sensing Applications for Oceanography and Fishing Environment Analysis, Beijing, China, 7 to 11 May 1990* (Bangkok: ESCAP/UNDP Regional Remote Sensing Programme (RAS/86/141) ST/ESCAP/897), pp. 263–273.

Coppen, J.J.W., and Nambiar, P., 1991, Agar and alginate production from seaweed in India. *Bay of Bengal Programme Working Paper*, No. 69, 24 pp.

Cornillon, P., Hickox, S., and Turton, H., 1986, Sea surface temperature charts for the southern New England fishing community. *Marine Technology Society Journal*, **20** (2), 57–65.

Deysher, L.E., 1993, Evaluation of remote sensing techniques for monitoring giant kelp populations. *Hydrobiologia*, **260/261**, 307–312.

Dierberg, F.E., and Zaitzeff, J., 1997, Assessing the application of an airborne intensified multispectral video camera to measure chlorophyll a in three Florida estuaries. *Proceedings of the*

Fourth International Conference on Remote Sensing for Marine and Coastal Environments, **2**, 80–87.

Fernandes, P.G., 1993, *The Role of Remote Sensing in Aquatic Resource Assessment and Management*, (London: Overseas Development Administration), 47 pp.

Garcia, C.A.E, and Robinson, I.S., 1991, Chlorophyll–*a* mapping using Airborne Thematic Mapper in the Bristol Channel (South Gower Coastline). *International Journal of Remote Sensing*, **12**, 2073–2086.

Gordon, H.R., and Morel, A.Y., 1983, *Remote Assessment of Ocean Color for Interpretation of Satellite Visible Imagery: a Review* (New York: Springer-Verlag).

Han, L., Rundquist, D.C., Liu, L.L., and Fraser, R.N., 1994, The spectral responses of algal chlorophyll in water with varying levels of suspended sediment. *International Journal of Remote Sensing*, **15** (18), 3707–3718.

Hara, I., 1985a, Shape and size of Japanese sardine school in the waters of southeastern Hokkaido on the basis of acoustic and aerial surveys. *Bulletin of the Japanese Society of Scientific Fisheries*, **51** (1), 41–46.

Hara, I., 1985b, Moving direction of Japanese sardine school on the basis of aerial surveys. *Bulletin of the Japanese Society of Scientific Fisheries*, **51** (12), 1939–1945.

Hara, I., 1989, Comparison of ship and aerial surveys of sardine school. *IGARSS '89 Remote Sensing: an Economic Tool for the Nineties, Vancouver, Canada July 10–14, 1989* (Vancouver, Canada: IGARSS '89 12th Canadian Symposium on Remote Sensing), **4,** 2041–2043.

Heinsohn, G.E., 1981, Aerial survey techniques for dugongs. In *The Dugong: Proceedings of a Seminar/Workshop held at James Cook University, 8–13 May 1979*, edited by H. Marsh, pp. 125–129.

Irvine, A.B., Caffin, J.E., and Kochman, H.I., 1982, Aerial surveys for manatees and dolphins in western peninsular Florida. *Fishery Bulletin*, **80** (3), 621–630.

Jensen, J.R., Estes, J.E., and Tinney, L., 1980, Remote sensing techniques for kelp surveys. *Photogrammetric Engineering and Remote Sensing*, **46** (6), 743–755.

Jones, R.L., 1996, Spatial analysis of biological and physical features associated with the distribution of Queen Conch, *Strombus gigas*, nursery habitats. M.S. thesis, Florida Institute of Technology, ix + 98 pp.

Kemmerer, A.J., 1979, Remote sensing of living marine resources. *Proceedings of the 13th International Symposium on Remote Sensing of the Environment, 23–27 April 1979* (Ann Arbor, Michigan: Environmental Research Institute of Michigan), pp. 729–738.

Kemmerer, A.J., 1980, Environmental preferences and behavior patterns of Gulf menhaden (*Brevoortia patronus*) inferred from fishing and remotely sensed data. *ICLARM Conference Proceedings*, No. 5, 345–370.

Lasker, R., Peláez, J., and Laurs, R.M., 1981, The use of satellite infrared imagery for describing ocean processes in relation to spawning of the northern anchovy (*Engraulis mordax*). *Remote Sensing of Environment*, **11**, 439–453.

Laurs, R.M., 1989, Review of satellite applications to fisheries. *IGARSS '89 Remote Sensing: an Economic Tool for the Nineties, Vancouver, Canada July 10–14, 1989* (Vancouver, Canada: IGARSS '89 12th Canadian Symposium on Remote Sensing), **4**, 2037–2039.

Laurs, R.M., Fiedler, P.C., and Montgomery, D.R., 1984, Albacore tuna catch distributions relative to environmental features observed from satellites. *Deep Sea Research*, **31** (9), 1085–1099.

Lefebvre, L.W., and Kochman, H.I., 1991, An evaluation of aerial survey replicate count methodology to determine trends in manatee abundance. *Wildlife Society Bulletin*, **19**, 298–309.

Le Gall, J.-Y., 1989, Télédétection satellitaire et pêcheries thonières océaniques. *FAO Document Techniques sur les Pêches*, No. 302, 148 pp.

López García, M.J. and Caselles, V., 1990, A multi-temporal study of chlorophyll-*a* concentration in the Albufera lagoon of Valencia, Spain, using Thematic Mapper data. *International Journal of Remote Sensing*, **11** (2), 301–311.

Long, B.G., Poiner, I.R., and Harris, A.N.M., 1993, Method of estimating the standing stock of *Trochus niloticus* incorporating Landsat satellite data, with application to the trochus resources of the Bourke Isles, Torres Strait, Australia. *Marine Biology*, **115**, 587–593.

Medley, P.A.H., and Ninnes, C.H., 1997, A recruitment index and population model for spiny lobster (*Panulirus argus*) using catch and effort data. *Canadian Journal of Fisheries and Aquatic Sciences*, **54**, 1414–1421.

Meulstee, C., Nienhuis, P.H., and Van Stokkom, H.T.C., 1986, Biomass assessment of estuarine macrophytobenthos using aerial photography. *Marine Biology*, **91**, 331–335.

Montgomery, D.R., Wittenberg-Fay, R.E., and Austin, R.W., 1986, The applications of satellite-derived ocean color products to commercial fishing operations. *Marine Technology Society Journal*, **20** (2), 72–86.

Nakashima, B.S., 1985, The design and application of aerial surveys to estimate inshore distribution and relative abundance of capelin. Northwest Atlantic Fisheries Organization Scientific Council Research Document 85/54, pp. 1–11.

Nakashima, B.S., Borstad, G.A., Hill, D.A., and Kerr, R.C., 1989, Remote sensing of fish schools: early results from a digital imaging spectrometer. *IGARSS '89 Remote Sensing: an Economic Tool for the Nineties, Vancouver, Canada July 10–14, 1989* (Vancouver, Canada: IGARSS '89 12th Canadian Symposium on Remote Sensing), **4**, 2044–2047.

Narain, A., Dwivedi, R.M., Solanki, H.U., Kumari, B., Chaturvedi, N., James, P.S.B.R., Subbaraju, G., Sudarsan, D., Sivaparkasam, T.E., and Somvanshi, V.S., 1990, The use of NOAA AVHRR data for fisheries exploration in the Indian EEZ. *Remote Sensing for Marine Studies. Report of the Seminar on Remote Sensing Applications for Oceanography and Fishing Environment Analysis, Beijing, China, 7 to 11 May 1990* (Bangkok: ESCAP/UNDP Regional Remote Sensing Programme (RAS/86/141) ST/ESCAP/897), pp. 226–232.

Ninnes, C.H., and Medley, P.A.H., 1995, *Sector Guidelines for the Management and Development of the Commercial Fisheries of the Turks and Caicos Islands.* Report to the UK Department for International Development.

Packard, J.M., Siniff, D.B., and Cornell, J.A., 1986, Use of replicate counts to improve indices of trends in manatee abundance. *Wildlife Society Bulletin*, **14**, 265–275.

Parslow, J.S., 1991, An efficient algorithm for estimating chlorophyll from Coastal Zone Color Scanner data. *International Journal of Remote Sensing*, **12**, 2065–2072.

Petit, M., Dagorn, L., Lena, P., Slepoukha, M., Ramos, A.G., and Streta, J.M., 1994, Oceanic landscape concept and operational fisheries oceanography. *Mémoires de l'Institut Océanographique*, Monaco, **18**, 85–97.

Platt, T., Sathyendranath, S., and Longhurst, A., 1995, Remote sensing of primary production of the ocean: promise and fulfilment. *Philosophical Transactions of the Royal Society of London B*, **348**, 191–202.

Populus, J., and Lantieri, D., 1991, Use of high resolution satellite data for coastal fisheries. *FAO Remote Sensing Centre Series*, No. 58, (*High Resolution Satellite Data Series*, No. 5), 43 pp.

Populus, J., and Lantieri, D., 1992, High resolution satellite data for assessment of tropical coastal fisheries. Case study in the Philippines. In *Pix'Iles 90. Remote Sensing and Insular Environments in the Pacific: Integrated Approaches*, edited by W. Bour and L. Loubersac (Nouméa: ORSTOM), pp. 523–536.

Qinsheng, M., 1990, Satellite observation and analysis of winter thermal fronts in the East China Sea. *Remote Sensing for Marine Studies. Report of the Seminar on Remote Sensing Applications for Oceanography and Fishing Environment Analysis, Beijing, China, 7–11 May 1990* (Bangkok: ESCAP/UNDP Regional Remote Sensing Programme (RAS/86/141) ST/ESCAP/897), pp. 185–198.

Randall, J.E., 1964, Contributions to the biology of the Queen Conch, *Strombus gigas*. *Bulletin of Marine Science of the Gulf and Caribbean*, 14, 246–295.

Richardson, L.L., and Ambrosia, V.G., 1997, Remote sensing of algal pigments to determine coastal phytoplankton dynamics in Florida Bay. *Proceedings of the Fourth International Conference on Remote Sensing for Marine and Coastal Environments*, **1**, 75–81.

Ruiz-Azuara, P., 1995, Multitemporal analysis of 'simultaneous' Landsat imagery (MSS and TM) for monitoring primary production in a small tropical coastal lagoon. *Photogrammetric Engineering and Remote Sensing*, **61** (2), 187–198.

Sathyendranath, S., Longhurst, A., Caverhill, C.M., and Platt, T., 1995, Regionally and seasonally differentiated primary production in the North Atlantic. *Deep-Sea Research*, **42** (10), 1773–1802.

Sathyendranath, S., and Platt, T., 1993, Remote sensing of water-column primary production. *ICES Marine Science Symposia*, **197**, 236–243.

Sathyendranath, S., Platt, T., Horne, E.P.W., Harrison, W.G., Ulloa, O., Outerbridge, R., and Hoepffner, N., 1991, Estimation of new production in the ocean by compound remote sensing. *Nature*, **353**, 129–133.

Sharp, G., Pringle, J., and Duggan, R., 1989, Assessing fishing effort by remote sensing in the Scotia-Fundy region of Fisheries and Oceans Canada. *IGARSS '89 Remote Sensing: an Economic Tool for the Nineties, Vancouver, Canada July 10–14, 1989* (Vancouver, Canada: IGARSS '89 12th Canadian Symposium on Remote Sensing), **4**, 2056–2060.

Shixing, H., Jianhua, S., and Zhongming, H., 1990, The use of satellite information for the exploitation of sardine stocks. *Remote Sensing for Marine Studies. Report of the Seminar on Remote Sensing Applications for Oceanography and Fishing Environment Analysis, Beijing, China, 7 to 11 May 1990* (Bangkok: ESCAP/UNDP Regional Remote Sensing Programme (RAS/86/141) ST/ESCAP/897), pp. 220–225.

Simms, É.L., 1997, Submerged kelp biomass assessment using remote sensing. *CoastGIS'97*, not paginated (10 pp.).

Simpson, J.J., 1992, Remote sensing and geographical information systems: Their past, present and future use in global marine fisheries. *Fisheries Oceanography*, **1** (3), 238–280.

Spitzer, D., and Dirks, R.W.J., 1987, Bottom influence on the reflectance of the sea. *International Journal of Remote Sensing*, **8**, 279–290.

Stoner, A.W., Pitts, P.A., and Armstrong, R.A., 1996, Interaction of physical and biological factors in the large-scale distribution of juvenile Queen Conch in seagrass meadows. *Bulletin of Marine Science*, **58** (1), 217–233.

Stoner, A.W., and Waite, J.M., 1990, Distribution and behavior of the Queen Conch *Strombus gigas* relative to seagrass standing crop. *Fishery Bulletin, U.S.*, **88**, 573–585.

Stretta, J.-M., 1991, Forecasting models for tuna fishery with aerospatial remote sensing. *International Journal of Remote Sensing*, **12** (4), 771–779.

Tassan, S., 1988, The effect of dissolved 'yellow substance' on the quantitative retrieval of chlorophyll and total suspended sediment concentrations from remote measurements of water colour. *International Journal of Remote Sensing*, **9**, 787–797.

Vallis, R., Currie, D., and Dechka, J., 1996, Assessment of casi data for monitoring chlorophyll–*a* and suspended solids in the Northumberland Strait, P.E.I. *Proceedings of the Sixth (1996) International Offshore and Polar Engineering Conference*, **2**, 380–383.

Xingwei, S., and Baide, X., 1990, Quick reporting of sea state information and fishing ground forecasts. *Remote Sensing for Marine Studies. Report of the Seminar on Remote Sensing Applications for Oceanography and Fishing Environment Analysis, Beijing, China, 7 to 11 May 1990* (Bangkok: ESCAP/UNDP Regional Remote Sensing Programme (RAS/86/141) ST/ESCAP/897), pp. 233–238.

Yáñez, E., Barbieri, M.A., and Catasti, V., 1997, Sea surface thermal structure associated to the small pelagic fish resources distribution in central Chile. *Proceedings of the Fourth International Conference on Remote Sensing for Marine and Coastal Environments*, **1**, 583–592.

Zibordi, G., Maracci, G., and Schlittenhardt, P., 1990, Ocean colour analysis in coastal waters by airborne sensors. *International Journal of Remote Sensing*, **11**, 705–725.

Part 6

Cost-effectiveness of Remote Sensing

19

Cost-effectiveness of Remote Sensing for Coastal Management

Summary *Habitat mapping is an expensive undertaking and using remote sensing to augment field survey is the most cost-effective means of achieving outputs for scientific and management purposes. Four types of cost are encountered when undertaking remote sensing: (1) set-up costs, (2) field survey costs, (3) image acquisition costs, and (4) the time spent on analysis of field data and processing imagery. The largest of these are set-up costs, such as the acquisition of hardware and software, which may comprise 40–72% of the total cost of the project depending on specific objectives.*

For coarse-level habitat mapping with satellite imagery, the second most important cost is field survey which can account for ca 25% of total costs and over 80% of costs if a remote sensing facility already exists (i.e. in the absence of set-up costs). Field survey is a vital component of any habitat mapping programme and may constitute ca. *70% of the time spent on a project.*

Detailed habitat mapping should be undertaken using digital airborne scanners or interpretation of colour aerial photography. The cost of commissioning the acquisition of such imagery can be high (£15,000– £26,000 even for small areas of 150 km²) and may constitute 33–45% of total costs (64–75% if set-up costs are excluded). For a moderate-sized study larger than ca. *150 km², about a month will be spent deriving habitat classes from field data and processing digital image data – irrespective of the digital data used (though, if staff need to be trained, this will increase time requirements considerably).*

Since the set-up costs, field survey costs and analyst time do not vary appreciably between satellite sensors, the selection of cost-effective methods boils down to map accuracy and the cost of imagery, the latter of which depends on the size of the study area and choice of sensor. Using a case study of mapping coarse-level coastal habitats of the Caicos Bank, it appears that SPOT XS is the most cost-effective satellite sensor for mapping sites whose size does not exceed 60 km in any direction (i.e. falls within a single SPOT scene). For larger areas, Landsat TM is the most cost-effective and accurate sensor.

The relative cost-effectiveness of digital airborne scanners and aerial photography are more difficult to ascertain because they are case-specific. Where possible, we recommend that professional survey companies be approached for quotes. In our experience, the acquisition of digital airborne imagery such as CASI is more expensive than the acquisition of colour aerial photography. However, this must be offset against the huge investment in time required to create maps from aerial photograph interpretation (API). If habitat maps are needed urgently, say in response to a specific impact on coastal resources, API might take too long and therefore be inappropriate. For small areas of say 150 km², a map could be created within 120 days using CASI but might take almost twice this time to create using API. We estimate that, in such a scenario, API is only cheaper if the staff costs for API are less than £75 per day. If consultants are used, this is unlikely to be the case. Further, as the area that needs to be covered by the survey increases, the cost of API is likely to rise much faster than the cost of a digital airborne scanner survey, making API progressively less cost-effective as area increases. In cases where the costs of API and digital airborne scanners are similar, the latter should be favoured because they are likely to yield more accurate results than API.

Ball-park figures of the costs involved to map an area of ca 150 km^2 are given below:

Habitat detail	Method	Cost £ (incl. set-up)	Cost £ (excl. set-up)	Time taken (person days)
Coarse	SPOT XS	33,020	10,170	97
Fine	Digital airborne scanner (e.g. CASI)	57,620	34,770	117

Introduction

The cost-effectiveness of a remote sensing survey is perhaps best assessed in relation to alternative means of achieving the same management objectives. The three primary objectives of remote sensing surveys identified by end-users were: providing a background to management planning, detecting change in habitats over time, and planning monitoring strategies (see Chapter 2). In all cases the expected outputs are habitat/resource maps of varying detail. Once the political decision has been taken that coastal resource management needs to be strengthened and that, as part of this process, an inventory of coastal ecosystems is required, the question becomes 'Does a remote sensing approach offer the most cost-effective option to achieve this objective?'

The only alternative to remote sensing for mapping marine and shoreline habitats is use of boat-based or land-based surveys where the habitat type is recorded at each point in a grid and boundaries are fitted using interpolation methods (Figure 19.1; for more details see Box 19.1). Unlike remote sensing methods that sample the entire seascape, errors arise from sampling a grid of points because of the possibility of overlooking some habitats between adjacent survey sites. The probability of missing habitats decreases if a finer grid is surveyed, but a finer grid requires much greater survey effort. If say a 10% risk of missing boundaries is tolerated, the appropriate sampling density is approximately half the mean distance between habitat boundaries (Burgess and Webster 1984a, b).

By calculating the spatial frequency (average patch size) of habitats on the Caicos Bank we were able to estimate the costs of using boat survey to achieve maps of similar detail to those obtained using remote sensing. The analysis is conservative with respect to boat-survey because an additional reconnaissance survey of boundary spacings would be required to plan the sampling strategy for a purely boat-based survey. Using remotely sensed data as a surrogate reconnaissance survey makes the seemingly fair assumption that remotely sensed data reveal the boundaries between habitats although this does not include the unrealistic notion that the habitats are identified correctly. At the scale of the Caicos Bank (15,000 km^2), coarse descriptive resolution (corals, seagrass, algae, sand) is more feasible than detailed habitat mapping (e.g. coral assemblages, seagrass standing crop) and data from the sensor Landsat TM were used to estimate mean boundary spacings from 2650 polygons on the image.

For a 10% risk of missing boundaries, the interval for a synoptic boat survey of the Caicos Bank mapping at a coarse descriptive resolution would be 152 m, which translates to *ca* 190,000 sites at an estimated cost of *ca* £380,000. This would take a survey team of three more than 8½ years to complete!

At the scale of marine protected areas (MPA median size 16 km^2; Kelleher *et al.* 1996), where detailed habitat mapping is more feasible, CASI would be a more appropriate remote sensing method than Landsat TM (Mumby *et al.* 1998). Although the mean boundary spacing of habitats will vary according to the location of the MPA, a mean boundary spacing of 20 m was obtained from CASI data of a representative fringing reef and lagoon in the TCI (Cockburn Harbour). However, a corresponding grid spacing of 10 m would barely be possible given the 5 m positional errors of differential global positioning systems (August *et al.* 1994). A more realistic grid spacing of 25 m (half the mean boundary spacing of dense seagrass, which had the largest patches) would still require 25,600 points taking a team of three surveyors 170 days and would cost *ca* £21,250 in boat charges. CASI hire for such a small area would be approximately half this cost (£12,000) and field survey would be reduced to 1 day. Although image analysis may take about a month, boat-based survey would still be less accurate, more expensive, and involve an extra 16 months of person-days of effort!

In summary, while almost 70% of questionnaire respondents considered the cost of remote sensing to be a

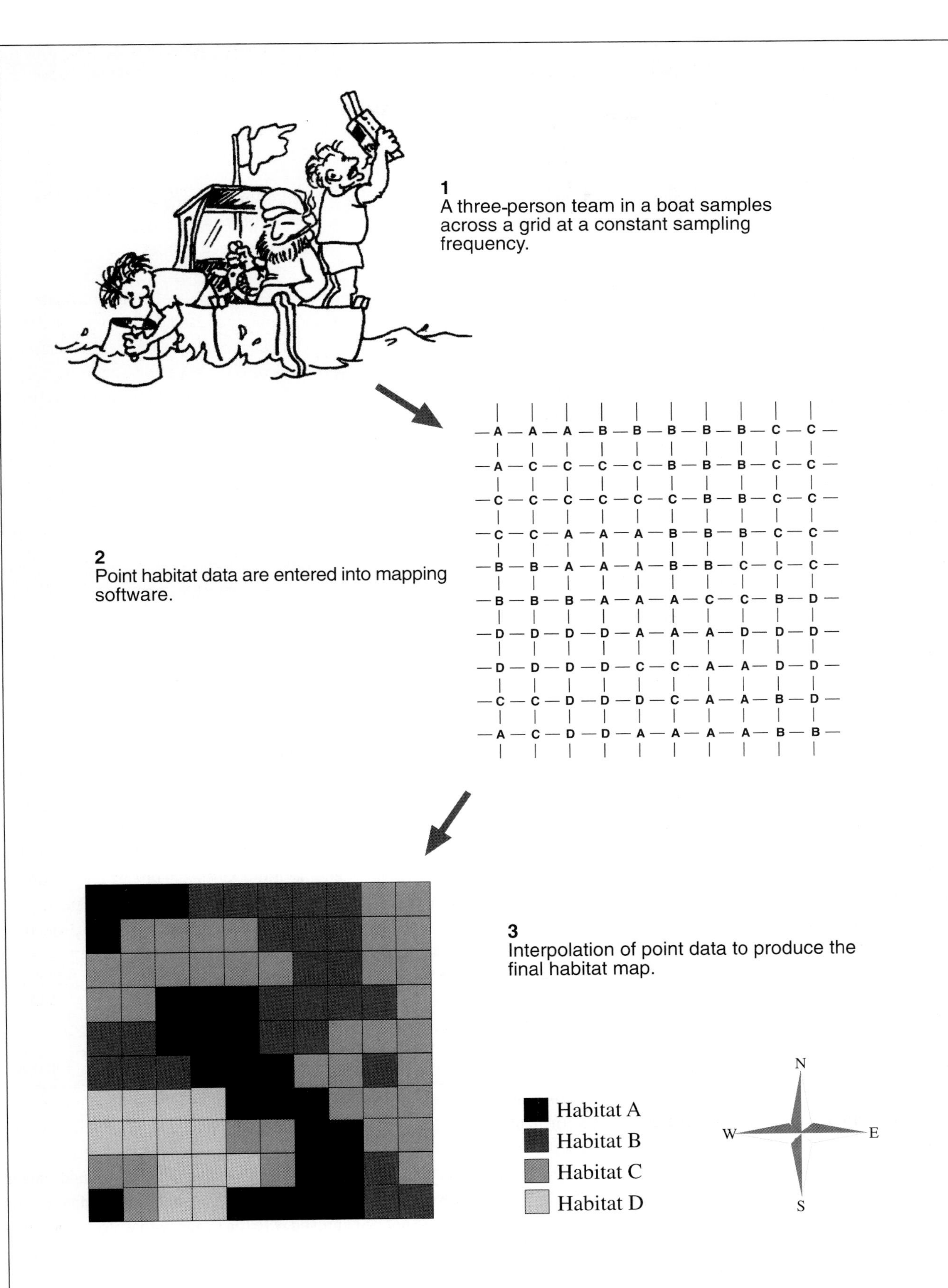

Figure 19.1 Overview of boat-based habitat mapping methods. A three-person team collects field data (1) across a grid and notes the habitat type at each point on the grid (2). The point samples are interpolated to create a thematic habitat map (3).

BOX 19.1

Estimating survey effort for a boat-based survey of marine habitats

It is possible to create a marine habitat map purely from field survey. The habitat type is recorded at each point within a large grid and habitat boundaries are fitted using interpolation methods. Unlike remote sensing methods, which sample the entire seascape, sampling a grid of points entails error and there is a possibility of overlooking some habitats between adjacent survey sites. The probability of missing habitats decreases if a finer grid is surveyed, but a finer grid requires much greater survey effort. In practice, a 10% risk of missing boundaries tends to be adopted (Burgess and Webster 1984a) and the appropriate sampling density for this risk of error is approximately half the mean distance between habitat boundaries (Burgess and Webster 1984b).

Ideally the mean boundary spacing would be determined in a prior reconnaissance survey, but we may obtain a surrogate measure directly from remote sensing. This entails the seemingly fair assumption that remotely sensed data can be used to identify the boundaries between habitats although this does not require that the habitats be identified correctly. Therefore, Landsat TM data were used to estimate the mean boundary spacing of habitats of the Caicos Bank based on a coarse descriptive resolution (coral, sand, seagrass, and algal habitats). To estimate the spatial frequency of habitats described in greater detail (i.e. 9 habitats), measurements were taken from CASI data of Cockburn Harbour, which encompassed a variety of lagoon and reef environments. Mean boundary spacings are listed in Table 19.1.

In the light of guidelines produced by Burgess and Webster (1984b), Table 19.1 suggests that, for a constant boundary omission risk of 10%, the interval for a synoptic boat survey mapping at a fine descriptive resolution should be 10 m and 152 m for a survey at coarse descriptive resolution. It is also clear from Table 19.1 that there are considerable differences in the average sizes of different habitats. Large areas of the Caicos Bank are covered with bare sand (mean boundary spacing by class = 951 m), whereas dense seagrass, for example, occurs in comparatively small beds (mean boundary spacing by class in Cockburn Harbour = 51 m). Therefore, a survey conducted at a sampling frequency of 105 m would miss more of the dense seagrass boundaries than the sand boundaries, thus under-sampling smaller habitats and producing a map biased in favour of larger habitats.

main hindrance to using it for coastal habitat mapping (Chapter 2), the issue is not that remote sensing is expensive but that habitat mapping is expensive. Remote sensing is just a tool that allows habitat mapping to be carried out at reasonable cost. Therefore, the main issue facing practitioners is: 'Which is the least expensive remote sensing method to achieve a given habitat mapping task with acceptable accuracy?'

Any generic discussion of the costs and cost-effectiveness of remote sensing is somewhat limited by the large number of variables to take into account such as hardware availability, the size of the study site, the technical expertise of staff, the quotations from aerial survey companies for obtaining airborne data for a given region, and so on. Therefore, the chapter begins by setting out the costs incurred during remote sensing in as generic a format as possible, thus allowing readers to estimate costs to suit their own requirements.

Costs are divided into four main categories:

1. Set-up costs (e.g. hardware and software requirements).
2. Field survey costs.
3. The time required for image processing and derivation of habitat classes.
4. Cost of imagery (discussed in detail in Chapter 5 and implicit in latter sections of this chapter)

Table 19.1 Boundary spacing for fine and coarse habitats of Cockburn Harbour and the Caicos Bank respectively with appropriate grid spacing for habitat mapping.

	Descriptive resolution	
	Fine	Coarse
Mean boundary spacing (m)	20	304
Number of habitat polygons used in calculation	490	2650
Range of mean boundary spacing by class (m)	4–51	194–951
Appropriate grid spacing (m)	10	152
Number of points to cover 16 km^2 at fine and the whole Caicos Bank at coarse descriptive resolution	25,600	190,000

☞ **The term 'processing' relates to the combined time of the operator and computer when manipulating imagery and should not be confused with the speed with which a computer will undertake computations.**

The chapter then draws on the Turks and Caicos Islands case study of coastal habitat mapping in the Caribbean to make a comparison of the cost-effectiveness of various sensors. We point out upfront, however, that

Table 19.2 The cost of mapping submerged marine habitats based on a case study of 150 km² in the Turks and Caicos Islands. The relative proportions of set-up (S), field survey (F), and image acquisition (I) costs are given for two scenarios – the costs required to start from scratch (i.e. including set-up costs) and the cost given existing remote sensing facilities (excluding set-up costs). API = aerial photographic interpretation. To convert costs and time estimates to those pertinent to mapping a similar area of mangrove habitats, subtract 57 days from each time estimate and subtract £2200 from each total cost (i.e. remove boat hire costs and net difference in field requirements – Table 19.4). If required, add the cost of either purchasing or hiring a small craft (e.g. kayak) for transport with the mangal. The main assumptions of the table are described below.

	Landsat		SPOT		Airborne	API
All costs in £	MSS	TM	XS	Pan	Digital Scanners	
Total costs incl. set-up	31,870	33,570	33,020	33,420	57,620	47,120
Cost component	S F I	S F I	S F I	S F I	S F I	S F I
% total cost	72 28 1	68 26 6	69 27 4	68 26 5	40 15 45	48 19 33
Total costs excl. set-up	8,930	10,720	10,170	10,570	34,770	24,270
Cost component	F I	F I	F I	F I	F I	F I
% total cost	98 2	82 18	86 14	83 17	25 75	36 64
Time taken (person-days)	97	98	97	95	117	229

Set-up costs assume that a full commercial image processing software licence is purchased and that the software is run on a UNIX workstation. Field survey costs assume that a differential GPS is included (see following section for further explanation), that 170 ground-truthing sites and 450 accuracy sites are visited, and that boat costs are £125 day^{-1}. Estimates of time are based on a 3-person survey team and include the time required to derive habitat classes from field data and process imagery / photo-interpret and digitise photographs into a geographic information system. Field survey time is estimated at 69 person-days in each case with additional time being that required for image acquisition and processing (Table 19.6) or digitisation in the case of API. Total costs (including set-up) for API: some remote sensing software (e.g. Erdas Imagine 8.3 Professional) is particularly user-friendly for hand digitising polygons from aerial photographs and thus the cost of the platform and software has been included in these costs.

this comparison is case-specific and that our conclusions must not be accepted blindly without considering the similarity between our study and that of the reader. The Turks and Caicos Islands are an ideal site for remote sensing of coral reefs and seagrass beds because the banks are relatively shallow (average depth *ca* 10 m) and the water clarity is high (horizontal Secchi distance 30–50 m). Therefore, the accuracies quoted for habitat maps are likely to represent the maximum accuracies possible for such habitats. However, the mangrove areas in the TCI are fairly small and higher accuracies may be expected in areas with a better developed mangal where patterns of species zonation would be larger and more appropriate for the larger pixels of satellite sensors.

In an attempt to consolidate the cost information, the chapter ends with a cost checklist that should help practitioners budget for a remote sensing facility or campaign.

Costs incurred during remote sensing

This part of the chapter discusses the costs associated with set-up, field survey and the analysis of field data, image acquisition and processing. Although each type of cost will be dealt with separately, it is worth beginning with an overview to compare the relative importance of set-up, field survey and analysis costs. All monetary costs are given in pounds sterling (£) and time is quoted in person-days (pd). Costs are based on 1999 prices and are subject to change (particularly the costs of computing equipment where a given level of performance costs less each year). Time requirements assume that staff are proficient in field survey, field data analysis and image processing. Estimates of the additional time required for training in remote sensing can be obtained from the appropriate chapters of the book.

Overview of costs incurred during remote sensing

The overall costs of remote sensing have been simulated for an area of *ca* 150 km² based on the costs and time spent conducting remote sensing in the Turks and Caicos Islands (Table 19.2).

Although set-up costs could be avoided (to some extent) by contracting consultants to undertake the work, this may prove to be a false economy given that much of the set-up equipment would be needed if an institution is to make effective use of the output habitat maps in a geographic information system or image processing environment. Looking at the costs including set-up, it can be seen that imagery would represent a small proportion of the total cost (4–6%) if coarse-level habitat mapping using appropriate satellite sensors (Landsat TM and SPOT XS) was the objective. This rises to about 33–45% of project costs if accurate fine-level habitat mapping is required, in which case only digital airborne scanners (e.g. CASI) or API of colour aerial photography are adequate. If set-up costs are not included, the total costs fall dramatically and the acquisition of SPOT or Landsat TM satellite images account for between 10–20% of total costs.

Field survey constitutes a significant proportion of remote sensing costs even if set-up costs are included. However, the time required to undertake processing of digital remotely sensed images is largely independent of the sensor used (whether using satellite-borne or airborne scanners). Even for a small area of 150 km², aerial photograph interpretation (API) is at least twice as time consuming as digital remote sensing – and this disparity in time would grow as the study site becomes larger.

Set-up costs

Set-up costs are defined here as the cost of those items that have to be purchased only once to enable a mapping campaign to be instigated. Under this definition, set-up costs are independent of the type of imagery used, the duration of fieldwork and the number and distribution of sites. The most significant costs are hardware and software but training and reference materials should also be considered (see Table 19.3).

A UNIX workstation is recommended in preference to a high-specification PC because data processing is faster. This is a considerable bonus when conducting processing that requires many stages, or if using whole satellite images or digital airborne data that have large data volumes.

☞ **The rapid development of affordable high-specification PCs is narrowing the advantages of UNIX based workstations: however, at the same time, UNIX workstations are steadily becoming more and more affordable.**

Field survey costs

Field survey costs can be divided into fixed (e.g. equipment) and variable categories, the latter of which depend on the number of field sites required.

Fixed costs of field survey

A major equipment cost for field survey is a global positioning system (GPS). Hand-held stand-alone instruments can be purchased cheaply for several hundred pounds but their spatial accuracy is poor (Chapter 4). August *et al.* (1994) found that 95% of position fixes from GPS were within 73 m of the true position – equivalent to several pixels of most satellite sensors! A differential facility (DGPS) reduced this error margin considerably to just 6 m. While DGPS is *ca* £1800 more expensive than GPS (Table 19.4), the extra expenditure is well worth the gains in accuracy. After all, field survey is expensive and, although the use of GPS may reduce capital expenditure, any saving is likely to be outweighed by the costs of discarding some field data due to poor positional accuracy (i.e. wasting information that was expensive to collect).

Table 19.3 Example breakdown of remote sensing set-up costs for mapping submerged tropical habitats. Prices are approximate.

Set-up Requirement	Cost (£)
Hardware	
SUN Ultra 10 with 21" monitor, 256MB RAM and 18GB hard disk	6,300
HP DesignJet Printer (A4–A1 capacity) with stand and extras	2,175
Ink and paper consumables	200
8 mm tape drive (7GB EXABYTE with SCSI card)	1,625
8 mm tapes to backup and store data	100
sub-total	**10,400**
Software	
ERDAS Imagine Professional 8.3 licence†	12,000
Training and reference materials	
Image processing textbooks and ERDAS Imagine training course	220
Charts (1:60,000 and 1:200,000)	50
Ordnance survey maps (1:25,000) (@ *ca.* £10 per map)	130
Ancillary aerial photographs to aid image interpretation (@ *ca.* £25 per photo)	50
sub-total	**450**

† This quote is for a commercial UNIX-based licence. A commercial PC-based licence would cost about £7,000. Academic and non-profit users can usually negotiate a substantial discount from ERDAS.

Variable costs of field survey

The number of ground-truthing sites required per habitat is difficult to quantify and depends on the size of the area and distribution (i.e. complexity) of habitats: smaller study areas and areas of relatively uniform habitat (e.g. sand banks) require less ground-truthing. For example, 10 sites per habitat may be acceptable for a single bay whereas 30 may be required to map an entire coastline. However, whereas ground-truthing requirements may vary, it is imperative that an adequate number of sites are visited for accuracy assessment – ideally at least 50 independent sites per habitat.

Table 19.4 Example of fixed costs for field survey equipment

Equipment	Cost (£)
General survey equipment	
GPS	150
DGPS (preferred option)	2,000
Notebook computer for data entry and analysis (10 GB hard-disk and 128 MB RAM)	2,480
Laminated colour prints of imagery (@ *ca* £3 per print at 1:80,000 for Landsat TM)	100
Recording equipment (water-resistant paper)	190
Alkaline batteries for DGPS and sonar	140
Identification guides	100
sub-total (including DGPS)	**5,010**
Additional equipment for marine surveys	
Diving equipment (per set)	800
Hand-held depth sonar	110
Quadrats	80
sub-total	**990**
Additional equipment for mangrove surveys	
Hemi-spherical densiometer (to measure percent canopy closure)	140
Pair of PAR incident light sensors (to measure leaf area index)	1,300
Telescopic measuring pole (to measure mangrove height)	120
sub-total	**1,560**

The following section concerns the costs of visiting field sites, usually by boat. First, the number of sites must be estimated and expressed in terms of the number of boat days. The cost is then calculated for boat hire/fuel costs and the corresponding staff time.

Apart from stating that adequate accuracy assessment is required, it is difficult to generalise how much field survey should be undertaken. To provide some insight, however, we use the Turks and Caicos Islands case-study to examine the affect of different amounts of fieldwork on the accuracy of marine habitat maps using Landsat TM. A total of 157 sites were used to ground-truth this image. To assess the effect of varying amounts of field data on accuracy, the supervised classification procedure was repeated using a sample of 25%, 50%, and 75% of these data. Each simulation was repeated three times with a random selection of field data and the accuracy of the resulting habitat maps was assessed using independent field data.

Figure 19.2 shows that the amount of field survey used to direct supervised classification profoundly influenced the accuracy of outputs although the increase in accuracy between 75% and 100% of field survey inputs was not sig-

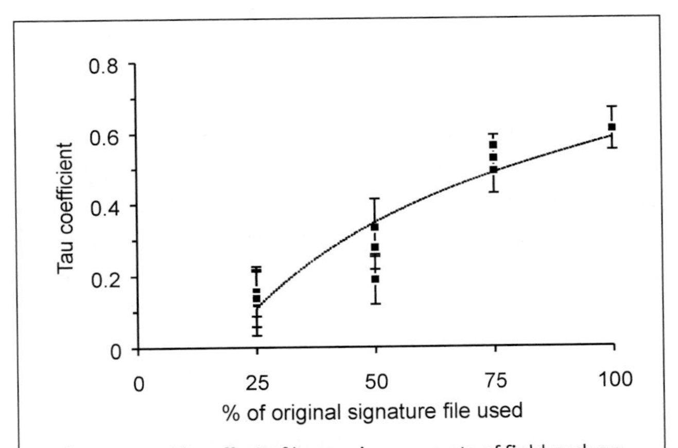

Figure 19.2 The effect of increasing amounts of fieldwork on classification accuracy (expressed as the tau coefficient) for Landsat TM supervised classification (coarse-level habitat discrimination). The vertical bars are 95% confidence limits. Three simulations were carried out at each level of partial field survey input (25%, 50% and 75% of signature file).

Table 19.5 Estimated survey rates for conducting surveys in the Turks and Caicos Islands where 1 boat day consisted of *ca* 8 hours and the overall study site measured *ca* 1100 km². The survey rate for smaller areas (i.e. the size of a single harbour of 1 km²) are given in parentheses. Boat hire, operator and fuel were charged at £125 day⁻¹. The distribution of non-mangrove sites is presented in Figure 19.3.

Type of survey	Survey rate (No. sites day^{-1})	Average distance travelled (km day^{-1})	Cost of boat hire, 1 operator & fuel (£ site^{-1})
Detailed ground-truthing of coral reefs and seagrass beds (quadrat surveys)	11 (25)	25	11
Detailed ground-truthing of mangroves* (using a Kayak not a boat)	10 (10)	1	n/a
Accuracy assessment of coral reefs (using glass-bottomed bucket)	60 (150)	40	2

* Based on a 4 hour survey period around noon to measure leaf area index (see Chapter 17)

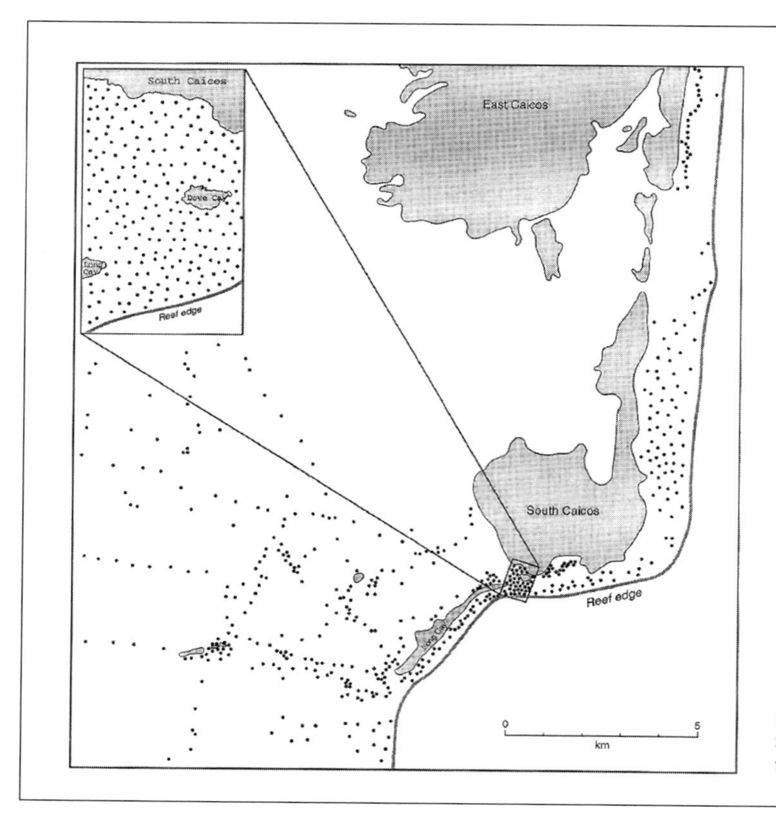

Figure 19.3 Study area in the Turks and Caicos Islands showing the locations of all sampling sites. CASI was flown across Cockburn Harbour (inset).

nificant. 75% of the signature file in Figure 19.2 would give about 30 sites per class at coarse-level (4 habitat classes). An additional 50 sites per habitat class should be surveyed for accuracy assessment (i.e. a total of 80 sites per habitat).

Once the number of sites has been estimated they can be grouped into geographical areas. The number of boat days needed to survey these areas depends on:

- the distance of each area from harbour,
- the number of sites within each area,
- the distance between sites in each area, and
- the speed of boat travel – itself a function of boat specifications and navigational hazards.

To obtain an accurate estimate of the number of boat days required and their cost (i.e. based on the rate of fuel consumption and the time required to visit particular areas) it is best to seek advice from the boat operator(s). As an extremely general guideline, ball-park estimates of survey rates are listed in Table 19.5 based on the Turks and Caicos Islands case study. In calculating staff time one should bear in mind that, in addition to the boat operator, 2–3 members of staff are required for marine surveys (Figure 19.1) and a minimum of 2 staff are required to conduct mangrove surveys. For surveys requiring SCUBA diving, 3–4 staff may be required for safety reasons.

The time required for image processing and derivation of habitat classes

The time required to undertake data analysis and image processing depends on the level of expertise of the operator (i.e. depending on whether training is required), the volume of data, and the speed of the hardware used to support image processing. Given the range of these variables, it is difficult to be specific on such time requirements. We have therefore continued our use of the Turks and Caicos Islands case-study so readers should remember that the figures in Table 19.6 are simply meant to provide approximate estimates of time requirements. The estimates assume: (i) that image processing is conducted on a UNIX workstation, (ii) that operators are proficient in image processing methods (but see individual chapters for details on the time required to learn each method), (iii) that only a single satellite image is processed per sensor, (iv) that five CASI strips were processed covering an area of 50 km^2, (v) that field data encompassed 170 sites for derivation of habitat classes, and (vi) that the accuracy assessment used *ca* 450 sites per satellite image and 200 sites for CASI.

The combined time for analysis of field data and image processing was approximately one month irrespective of the type of satellite imagery used in the case study. CASI imagery took longer to process because of the extra time

Table 19.6 Estimated time requirements in person-days to derive habitat classes from field data and conduct image processing. Habitat classes were derived using the software PRIMER (see Chapter 9). See text for assumptions of time estimates.

Activity	Landsat MSS	Landsat TM	SPOT XS	SPOT Pan	CASI
IMAGE ACQUISITION					
Searching archives / choosing CASI operator	2	2	2	2	4
Choosing appropriate images from archive	1	1	1	1	n/a
Negotiation with *SPOT Image* (programming request) and CASI operator	n/a	n/a	2	2	2
Selection of CASI bandwidths and location	n/a	n/a	n/a	n/a	2
Delivery time	14	14	60	60	180
Acquisition sub-total (excluding delivery time)	**3**	**3**	**5**	**5**	**8**
CORRECTIONS TO IMAGERY					
Importing data	0.5	1	0.3	0.3	6
Geometric correction	2	3	1	1	16
Radiometric correction	4	4	3.5	3.5	n/a
Water column correction	2	2	2	n/a	2
Correction sub-total	**8.5**	**10**	**6.8**	**4.8**	**24**
DERIVATION OF HABITAT CLASSES FROM FIELD DATA					
Entry of data into database	4	4	4	4	4
Interpretation of database	2	2	2	2	2
Hierarchical cluster analysis and SIMPER analysis	6	6	6	6	6
Derivation of habitat classes sub-total	**12**	**12**	**12**	**12**	**12**
IMAGE CLASSIFICATION					
Differential correction of GPS (to locate field data on imagery)	4	4	4	4	1
Unsupervised classification	0.5	0.5	0.5	0.5	0.5
Supervised classification step 1– collecting signatures	1	1	1	1	0.3
Supervised classification step 2 – editing signatures	1	1	1	1	0.3
Contextual editing	1	1	1	1	1
Accuracy assessment	1	1	1	1	1
Image classification sub-total	**4.5**	**4.5**	**4.5**	**4.5**	**4.1**
Total time required (excluding delivery time)	**28**	**29**	**28.3**	**26.3**	**48.1**

required to import large data volumes and the complexity of geometric correction which required added field survey (i.e. the location of *ca* 30 ground control points in the field). Extra ground control points were needed to geometrically correct individual flight lines but this was a fairly site-specific requirement and might not be necessary where a substantial part of each flight line encompasses terrestrial habitats which have been mapped in detail elsewhere (e.g. in large scale Ordnance Survey maps). Readers should note that only a few CASI flight lines were mosaiced in this case study and time requirements could rise considerably for larger surveys where mosaicing is more complex.

A wide variety of image-processing methods exist and the effects on accuracy of conducting some of these are presented in Figure 19.4. Two key points emerge from this figure. Firstly, to achieve acceptable accuracy a considerable investment in staff time is required to carry out image acquisition, correction and subsequent processing; approximately one month for Landsat TM (and other satellite imagery such as SPOT XS), and about 1.5 months for CASI or similar airborne imagery. Secondly, increased image processing effort generally leads to increasing accuracy of outputs. A notable exception to this rule is the merging of Landsat TM data with SPOT Pan to increase spatial resolution. Although the merge

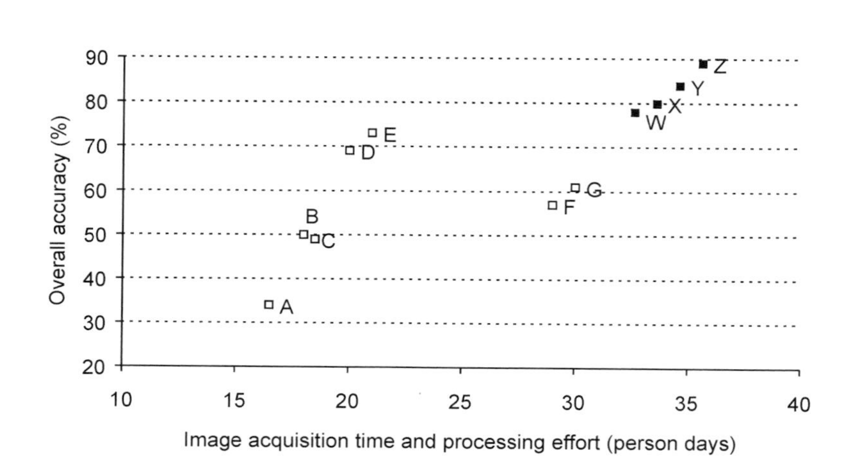

Figure 19.4 The relationship between the overall accuracy of marine habitat maps of the TCI and the time required in preparation from Landsat TM (and TM/SPOT Pan merge) and CASI data at **coarse** (□) and **fine** (■) levels of habitat discrimination respectively. Time includes image acquisition, correction, image classification, and the merging of SPOT Pan with Landsat TM data.

Five levels of image processing effort were applied to Landsat TM alone:
A = unsupervised classification of raw (DN) data
B = supervised classification of raw (DN) data
C = unsupervised classification of depth-invariant data
D = supervised classification of depth-invariant data, without contextual editing
E = supervised classification of depth-invariant data, with contextual editing

Depth-invariant Landsat TM data were combined with SPOT Pan to increase the spatial resolution, and the resultant merged image classified in two ways:
F = supervised classification of merged image, without contextual editing
G = supervised classification of merged data, with contextual editing

Four levels of image processing effort were applied to CASI:
W = supervised classification of raw (reflectance) data, without contextual editing
X =supervised classification of raw (reflectance) data, with contextual editing
Y = supervised classification of depth-invariant data, without contextual editing
Z = supervised classification of depth-invariant data, with contextual editing

produces visually pleasing outputs (Figure 19.5, Plate 24) it produced no benefits in terms of improved accuracy (in fact reducing it). Given the considerable costs of inputs such as field survey data and set-up costs extra effort devoted to image processing is clearly very worthwhile.

☞ Time requirements will rise if multiple images are used and although some time may be saved if corrections are performed concurrently on several images, additional time will be necessary to mosaic the classified images. Readers are therefore advised to multiply the time estimates in Table 19.5 by the number of images to be used.

The relative cost-effectiveness of different remote sensing methods

Whether a practitioner's objective is to map marine habitats or terrestrial habitats, the choice of satellite versus airborne sensors depends on the level of habitat detail required. If coarse descriptive resolution (e.g. corals versus seagrasses; mangrove versus non-mangrove) is all that is required, satellite sensors will, almost certainly, be the most cost-effective option and reasonable accuracy (~ 60–80%) might be expected. However, Chapters 11–13 concluded that for detailed habitat mapping, airborne (particularly digital airborne) remote sensing methods are far more likely to provide results of high accuracy. Therefore, the comparison between satellite and airborne methods is somewhat irrelevant because they are used for achieving different descriptive resolutions. (Although airborne remote sensing could be used to achieve coarse descriptive resolution, it would be highly cost-inefficient given that satellite imagery is an order of magnitude cheaper and may attain similar overall accuracy). Thus, the cost-effectiveness of satellite sensors and airborne sensors are compared separately. Once again, the Turks and Caicos Islands case study is used to examine issues of cost-effectiveness. Data are presented for mapping marine habitats but a similar conclusion would be reached had we used data for mangrove mapping (see Chapter 13 for estimates of map accuracy with various sensors).

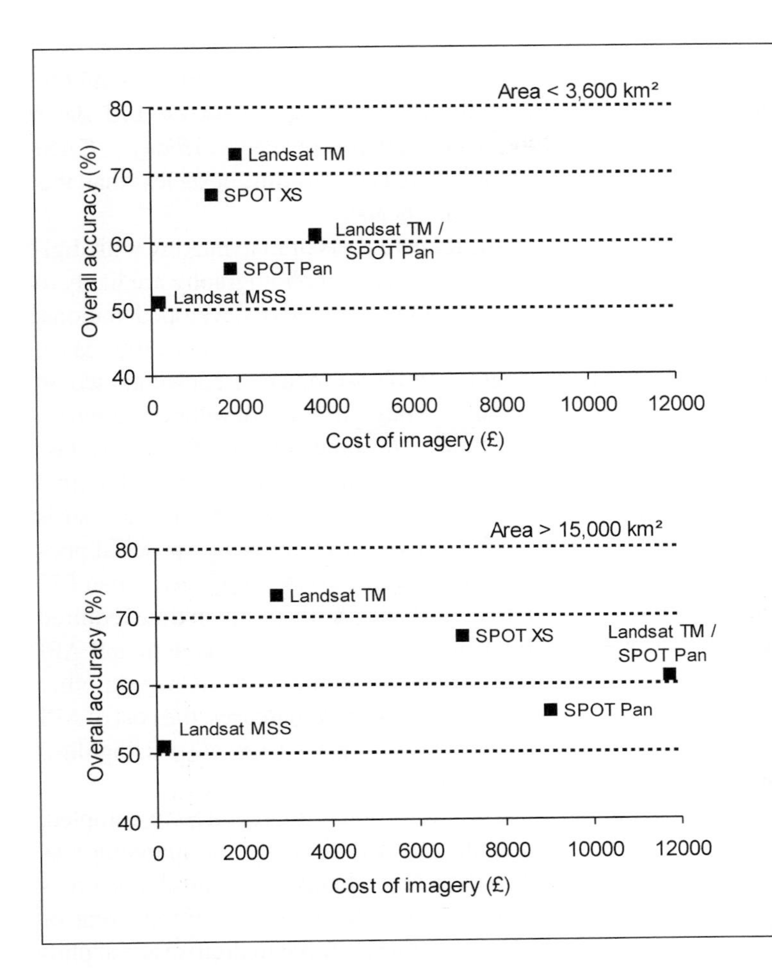

Figure 19.6 Cost-effectiveness of various satellite sensors for mapping coastal habitats of the Caicos Bank with coarse detail. Upper figure shows imagery costs for a relatively small area, such as would be contained within a single SPOT scene (3,600 km²). Lower figure shows costs for a larger area (about the size of the Caicos Bank), which would fall within a single Landsat scene but would require five SPOT scenes.

Cost-effectiveness of various optical satellite-borne sensors

The time required to process an individual satellite image does not vary substantially from sensor to sensor (see Table 19.6). The most cost-effective solution therefore depends on the cost of each image and the accuracy achievable. Figure 19.6 compares the cost and expected map accuracy resulting from various satellite sensors for study sites of various sizes. For areas less than 3,600 km^2, SPOT XS is possibly slightly more cost-effective than Landsat TM: the small (6%) rise in accuracy from SPOT XS to Landsat TM would cost *ca* £550 (i.e. just under £100 per 1% rise in accuracy). However, once the study site is too large to fall within a single SPOT scene (60 km x 60 km), the cost-effectiveness of SPOT is drastically reduced. For example, an area of say 15,000 km^2 (such as the Caicos Bank) would fit easily within a single Landsat TM scene while requiring five SPOT XS scenes. The superior cost-effectiveness of Landsat TM then becomes clear even before the added time required to process multiple SPOT scenes is taken into account (Figure 19.6).

☞ **The cost of Landsat TM imagery for areas < 10,000 km² is reduced by purchasing a sub-scene (see Chapter 5). This assumption has been made in preparing Figure 19.6 – the upper graph uses a sub-scene and the lower graph requires a full scene.**

In this case-study, Landsat MSS would only be considered cost-effective if an overall accuracy of *ca* 50% were considered acceptable. However, given that half the pixels might be incorrectly assigned, this seems unlikely to be the case.

Cost-effectiveness of digital airborne scanners versus aerial photography

It is exceedingly difficult to assess the relative cost-effectiveness of airborne remote sensing methods because the cost of data acquisition is so variable and case-specific; it is best to obtain quotes from professional survey companies (refer to Chapter 5 for advice on obtaining such quotes). In addition, the accuracies of tropical coastal habitat maps resulting from digital scanners and aerial photographs have not been compared directly. The comparisons made in Chapter 11 between the results of the Turks and Caicos Islands case study using CASI and the results of Sheppard *et al.* (1995) in Anguilla using 1:10,000 colour aerial photography are not conclusive because of disparity in the size of areas mapped and some differences in habitats between studies. Nevertheless, given their high spectral versatility and resolution, digital airborne scanners like CASI are likely to be at least as accurate as aerial photography and probably more so. Therefore, if the costs of acquiring and processing digital airborne data and aerial photography are similar for a given study, we recommend that the scanner is chosen because of the likely improvement in effectiveness.

To provide a tentative insight into the relative cost of commissioning new aerial photography and digital airborne scanners, we obtained four independent quotes for mapping a coastal area of approximately 150 km^2. The prices are based on a remote coastal area so the survey aircraft had to be specially mobilised (in this case, leased from the United States). Specifically, quotes were sought for 1:10,000 colour aerial photography and CASI imagery with 3 m spatial resolution (Table 19.7). Readers should bear in mind that this is a fairly small area and that the time required to photo-interpret and digitise aerial photographs would be disproportionately greater for larger study sites (i.e. if the reader's area of interest is much greater than 150 km^2, do not use our estimates of time – these will be highly conservative). Although scanning of aerial photographs should increase the speed of map production by allowing digital classification techniques to be used, the effect of scanning on accuracy and processing time has not yet been evaluated.

This comparison between CASI and aerial photography assumes that set-up costs and fieldwork costs do not differ between studies. This seems to be a realistic assumption, particularly if the map derived from aerial photography is to be used within a geographic information system such as ARC/Info which has similar hardware requirements and software costs to image processing software. Assuming that the costs and mapping rates given in Table 19.7 are fair, CASI is more expensive to acquire but map production from API requires a greater investment in staff time. The overall cost of the two methods would be equal if the photo-interpreter was paid *ca* £75 day^{-1} (£10,500/140 pd). If staff costs exceed £75 day^{-1}, CASI would be cheaper whereas if staff time costs less than this figure, API would be cheaper.

In practice, the relative costs of map making with digital airborne scanners and aerial photography are likely to differ between developed and less developed nations. A consultant might charge in excess of £300 day^{-1}, whereas the average staff cost for a non-consultant would probably be about £75 day^{-1} in some developed countries and less than £75 day^{-1} in some developed and most less developed countries. Given that the accuracy of the final map is likely to be greater if a digital airborne scanner is used, it is probably only cost-effective to use aerial photography when staff costs are appreciably lower than £75 day^{-1}. As pointed out above, the disparity in time required to create habitat maps from airborne scanners and API will increase as the study site increases in size. In other words, as the scope of the survey increases, the cost of API is likely to rise much faster than survey costs using digital airborne scanners.

A final consideration is the time required to complete and deliver outputs, particularly if habitat mapping was carried out for detecting change in coastal resources (e.g. if investigating the effects of a pollution event or cyclone) – if the extra time required to digitise aerial photographs is prohibitive, airborne scanners may be the only feasible solution.

Cost check-list

The costs of undertaking remote sensing have been outlined in some detail and segregated into major types of expense and time requirement. Unfortunately, it is difficult to extrapolate from specific case-studies to provide

Table 19.7 Cost of mapping a coastal area of 150 km^2 using CASI and 1:10,000 colour aerial photography interpretation (API). CASI is more expensive to acquire but, being digital, requires much less processing time post-acquisition. Processing time for CASI assumes that mosaicing and geometric correction are carried out by the contractor but includes time for negotiations with CASI operator, selection of CASI bands, etc. Processing time for API assumes that polygons are digitised by hand using conventional cartographic methods and a geographic information system. pd = person-days

Method	Cost of acquisition (£) quotes	mean	CASI - API	Staff time required for processing (pd)	API - CASI
CASI – 3 m pixels	27,000 25,000	26,000		20	
			10,500		**140**
1:10,000 colour API	16,000 15,000	15,500		160	

generic figures or to anticipate all requirements, but we hope that the chapter serves as a guide to help readers create their own budgets for remote sensing. To this end, Table 19.8 is included to act as a checklist of the main costs which may (or may not) be necessary for a particular study.

Conclusion

Habitat mapping is an expensive undertaking and using remote sensing to augment field survey is the most cost-effective means of achieving outputs for scientific and management purposes. To help practitioners match their survey objectives to appropriate remote sensing methods, a quick-reference summary of the main conclusions is provided in Figure 19.7.

Satellite imagery is the most cost-effective method for producing habitat maps with coarse descriptive resolution (e.g. corals, sand, seagrass and algae). Using a case study of mapping coarse-level coastal habitats of the Caicos Bank, it appears that SPOT XS is the most cost-effective satellite sensor for mapping sites whose size does not exceed 60 km in any direction (i.e. falls within a single SPOT scene). For larger areas, Landsat TM is the most cost-effective and accurate sensor among the satellite sensors tested.

Table 19.8 Checklist of main cost and time considerations when planning a remote sensing facility or campaign. Direct capital expenditure divided into four orders of magnitude (£): XXXX (over ten thousand) XXX (thousands), XX (hundreds), X (tens). Staff time expressed as XXX (> 50 days), XX (10–50 days), X (< 10 days).

Activity		Direct capital expenditure	Staff time required
SET-UP COSTS			
	UNIX workstation	XXX	
	Colour printer, ink, paper	XXX	
	8 mm tape drive and tapes	XXX	
	Software	XXXX	
	Reference books	XX	
	Charts and maps of area	XX	
	Archived aerial photographs	X	
FIELD SURVEY			
	Boat hire, operator time, fuel	XXX	
	Staff time (2–3 persons / day)		XXX
	DGPS (GPS)	XXX	
	Notebook computer	XXX	
	Laminated prints of imagery	XX	
	Water-proof recording equipment	XX	
	Depth sonar and batteries	XX	
	Identification guides	X	
	Diving equipment	XX	
	Quadrats	X	
	Hemi-spherical densiometer	XX	
	2 PAR light sensors and data logger	XXX	
	Telescopic measuring pole	XX	
IMAGE ACQUISITION			
	Purchase of satellite data	XXX	
	Seeking quotes for airborne data		X
	Commissioning of airborne data	XXXX	
IMAGE PROCESSING AND DERIVATION OF HABITAT CLASSES			
	Corrections to imagery		X or XX
	Derivation of habitat classes		XX
	Image classification		X
	Mosaicing of aerial photographs		XX
	Aerial photograph interpretation		XXX
	Digitising polygons from photographs		XX
	Contextual editing		X
	Accuracy assessment		X

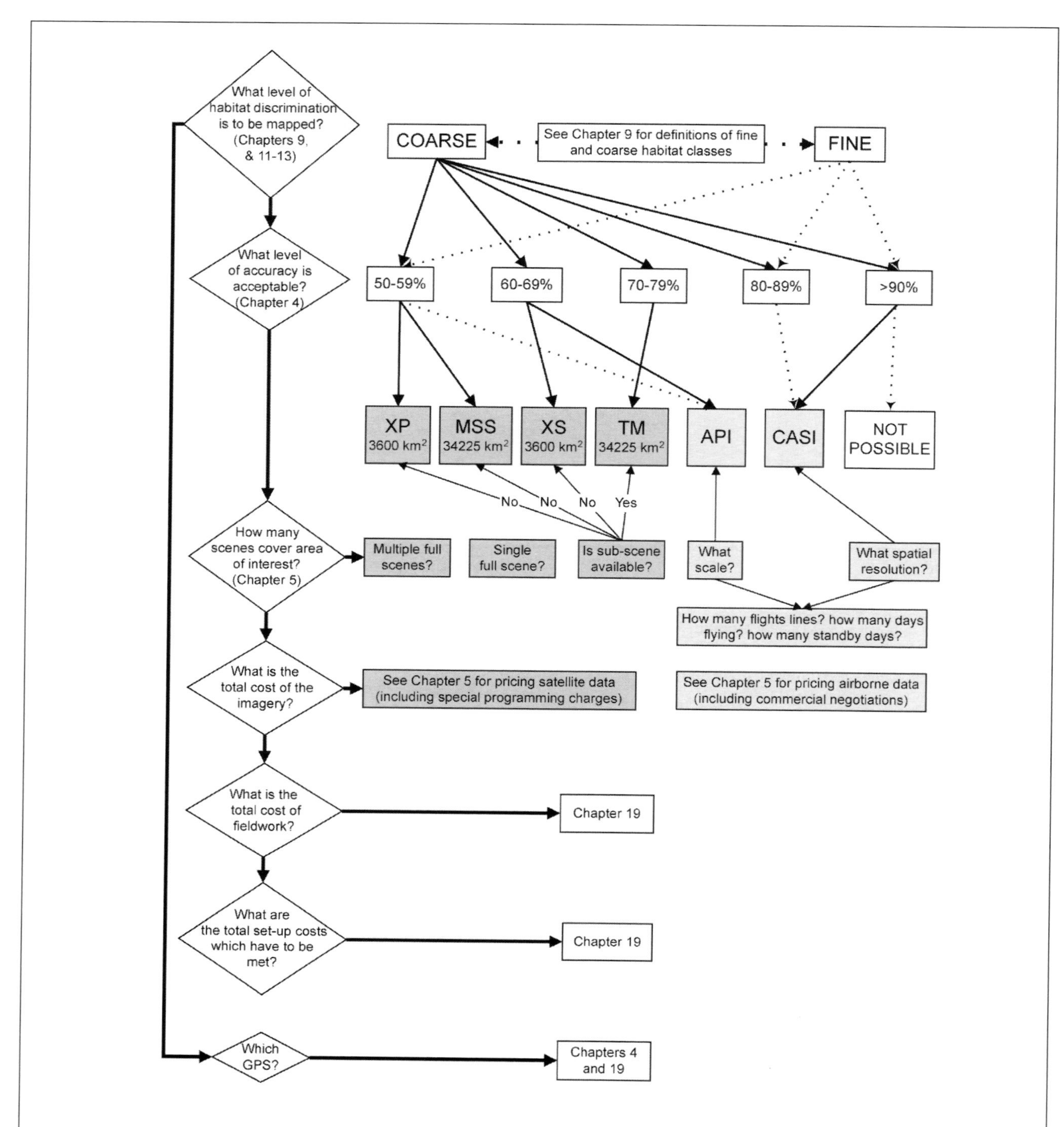

Figure 19.7 A schematic diagram illustrating the stages in selecting the most cost-effective imagery for habitat mapping. Key decisions are listed on the left. For example, if coarse habitats are to be mapped and an accuracy of 60–69% is acceptable then either SPOT XS or Landsat TM would be suitable (it would be a waste of resources to use aerial imagery in this case for reasons discussed in the text). However the most cost-effective option would depend on the area to be mapped: if less than 3600 km² then SPOT XS is likely to be the most cost-effective, if greater than 3600 km² then either a subscene or full image of Landsat TM data. References to other chapters in the *Handbook* are provided to assist the complete costing of a mapping project.

Detailed habitat mapping should be undertaken using digital airborne scanners or interpretation of colour aerial photography. However the relative cost-effectiveness of these methods is more difficult to ascertain because quotes are case-specific. In our experience, while the acquisition of digital airborne imagery such as CASI is more expensive than the acquisition of colour aerial photography, its high cost must be offset against the huge investment in time required to create maps from aerial photograph interpretation (API). If habitat maps are needed urgently API might take too long and therefore be inappropriate. For small areas of say 150 km^2, a map could be created within 120 days using CASI but might take almost twice this time to create using API. We estimate that, in such a scenario, API is only cheaper if the staff costs for API are less than £75 day^{-1} (Mumby *et al.* 1999). If consultants are used, this is unlikely to be the case. Further, as the area which needs to be covered by the survey increases, the cost of API is likely to rise much faster than the cost of a digital airborne scanner survey, making API progressively less cost-effective as area increases. In cases where the costs of API and digital airborne scanners are similar, the latter should be favoured because they are likely to yield more accurate results than API.

References

August, P., Michaud, J., Labash, C., and Smith, C., 1994, GPS for environmental applications: accuracy and precision of locational data. *Photogrammetric Engineering and Remote Sensing*, **60**, 41–45.

Burgess, T.M., and Webster, R., 1984a, Optimal sampling strategies for mapping soil types I. Risk Distribution of boundary spacings. *Journal of Soil Science*, **35**, 641–654.

Burgess, T.M., and Webster, R., 1984b, Optimal sampling strategies for mapping soil types II. Risk functions and sampling intervals. *Journal of Soil Science*, **35**, 655–665.

Kelleher, G., Bleakely, C., and Wells S., 1996, A global representative system of marine protected areas. (Washington DC: World Bank Publications).

Mumby, P.J., Green, E.P., Clark, C.D., and Edwards, A.J., 1998, Digital analysis of multispectral airborne imagery of coral reefs. *Coral Reefs*, **17**, 59–69.

Mumby, P.J., Green, E.P., Edwards, A.J., and Clark, C.D., 1999, The cost-effectiveness of remote sensing for tropical coastal resources assessment and management. *Journal of Environmental Management,* **55**, 157–166.

Sheppard, C.R.C., Matheson, K., Bythell, J.C., Murphy, P., Blair-Myers, C., and Blake, B., 1995, Habitat mapping in the Caribbean for management and conservation: use and assessment of aerial photography. *Aquatic Conservation: Marine and Freshwater Ecosystems*, **5**, 277–298

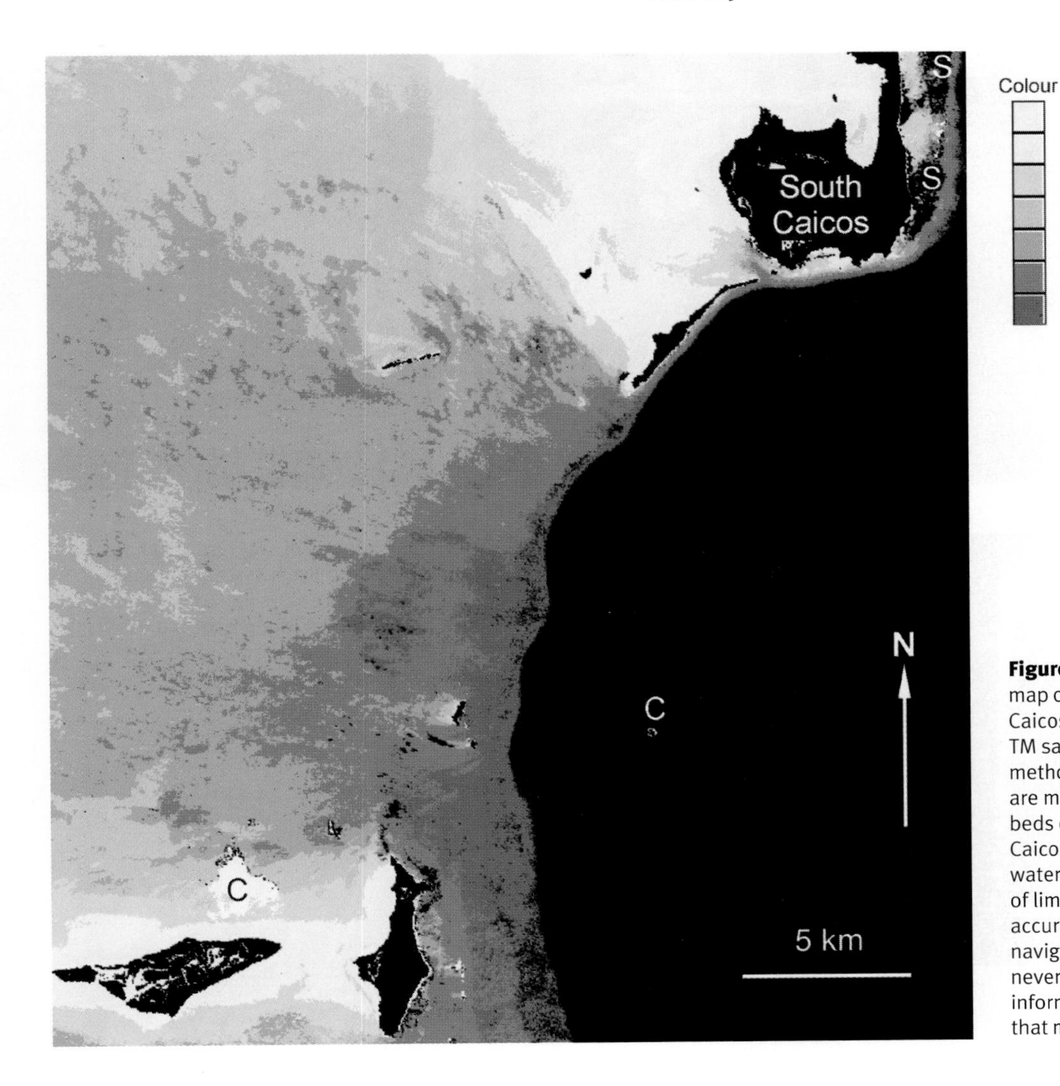

Figure 15.10 A crude bathymetric map of the south-east side of the Caicos Bank prepared from Landsat TM satellite imagery using the method of Jupp (1988). Two clouds are marked 'C', and dense seagrass beds on the east side of South Caicos which classify with deep water are marked 'S'. Such maps are of limited use because of their accuracy (i.e. should not be used for navigation! – Figure 15.6), but nevertheless convey broadscale information on depth distributions that may be useful to managers.

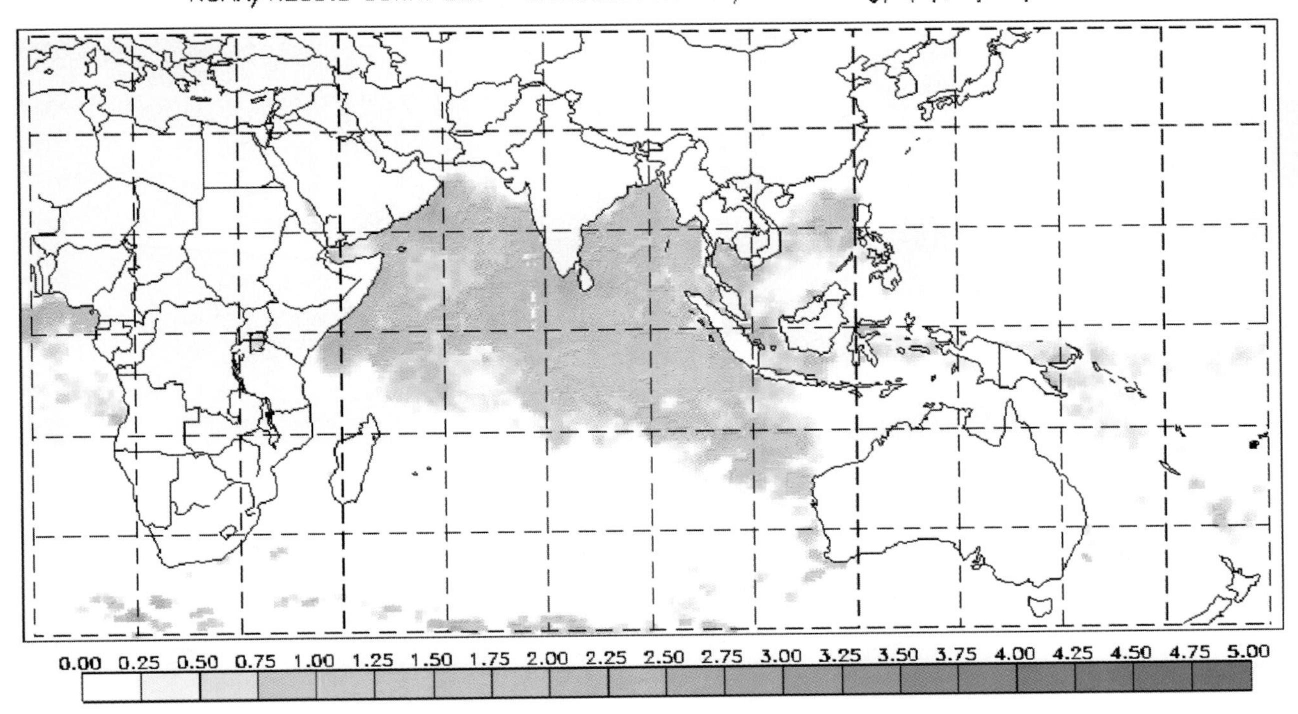

Example of the sea surface temperature (SST) anomaly maps (coral reef 'hotspots') available over the internet from the US National Oceanic and Atmospheric Administration (NOAA) based on Advanced Very High Resolution Radiometer (AVHRR) satellite imagery. The example presented is from 12 May 1998 at the height of the 1998 coral bleaching in the central Indian Ocean. Severe coral bleaching was occurring in Maldives, Sri Lanka, Seychelles and Chagos at around this time.

Varying levels of seagrass standing crop on the visual assessment scale used on the Caicos Bank, showing respectively above-ground biomasses (standing crops) of 1, 2, 3, 4 and 6. [*Photographs:* E. Green]

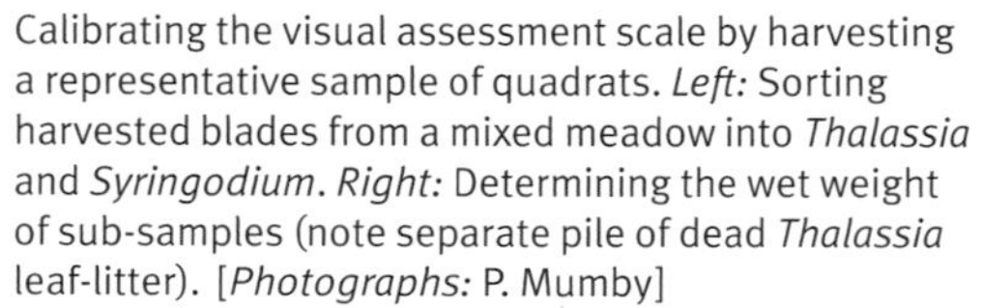

Calibrating the visual assessment scale by harvesting a representative sample of quadrats. *Left:* Sorting harvested blades from a mixed meadow into *Thalassia* and *Syringodium*. *Right:* Determining the wet weight of sub-samples (note separate pile of dead *Thalassia* leaf-litter). [*Photographs:* P. Mumby]

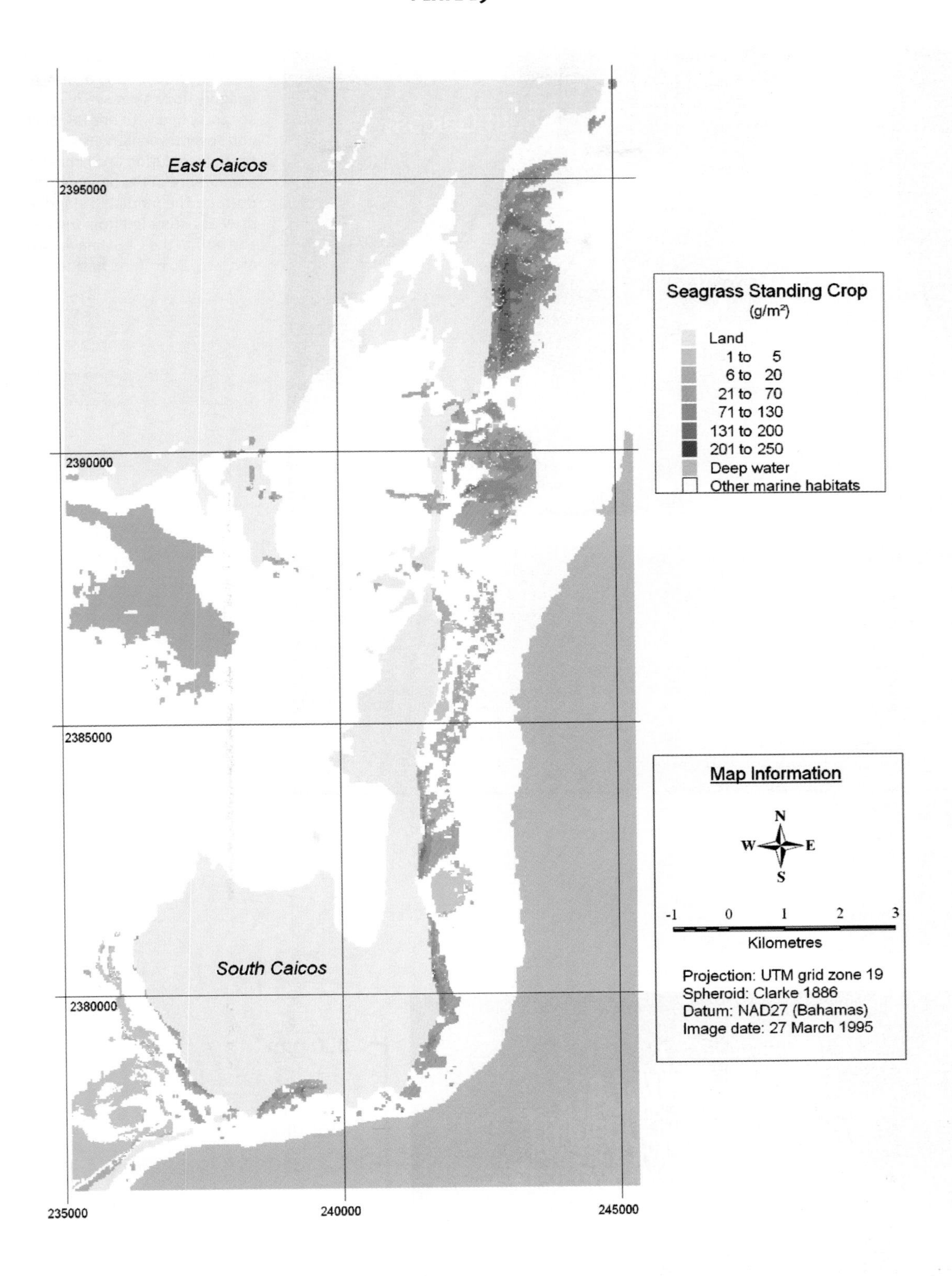

Figure 16.4 Seagrass resource map for the area around South Caicos derived from SPOT XS imagery. Different shades of green show seagrass standing crop (g.m^{-2}) on a pixel by pixel basis with an accuracy comparable to that achieved using visual assessment.

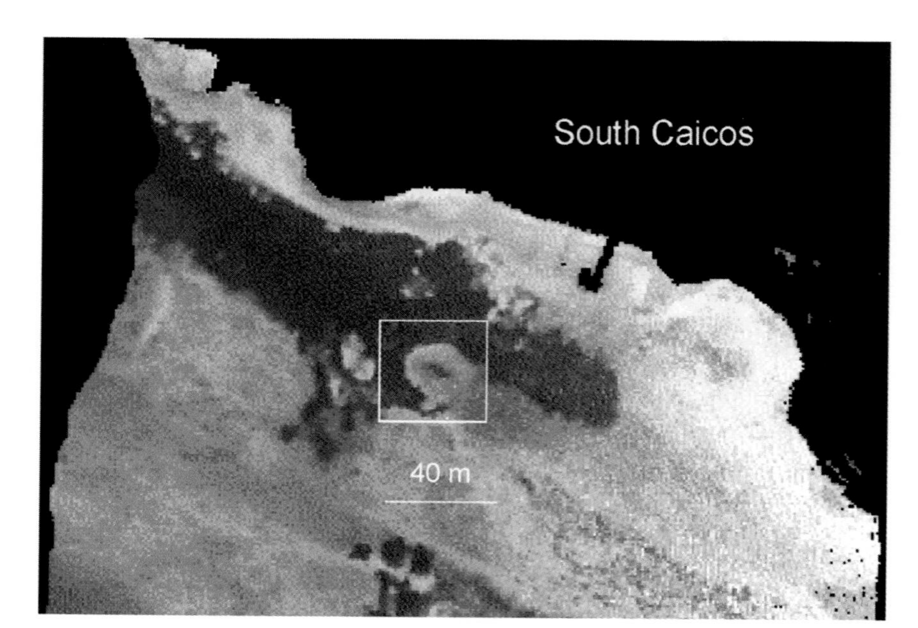

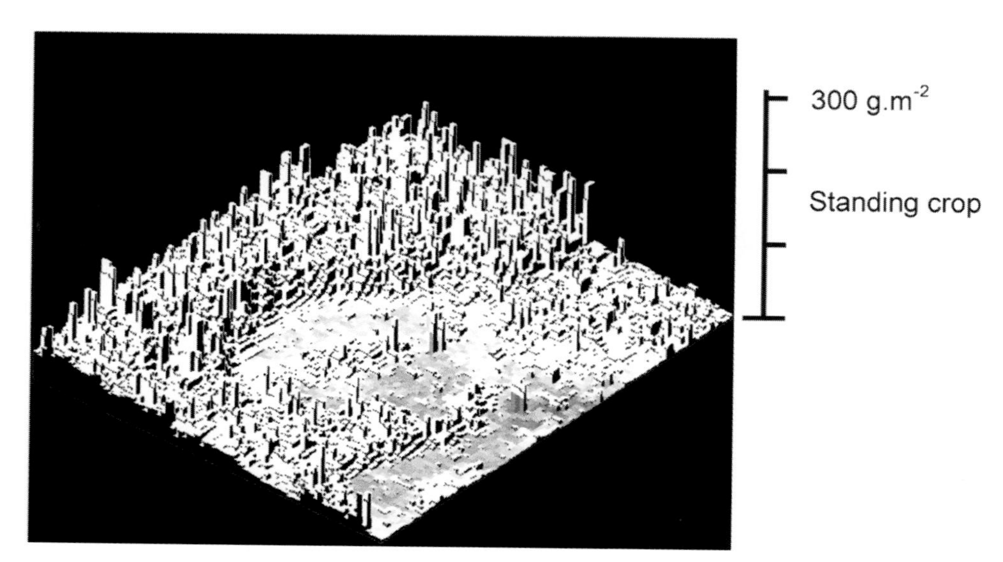

Figure 16.8 Mapping of seagrass standing crop around a blow-out using CASI imagery. *Top:* A dense *Thalassia*-dominated seagrass bed with one large blow-out and several smaller ones, as shown on 1 m spatial resolution CASI imagery. *Middle:* The appearance of the blow-out underwater. *Bottom:* The predicted standing crop in each pixel as estimated from the CASI data. The box on the CASI image covers an area of *c.* 40 x 40 m; land has been masked out.

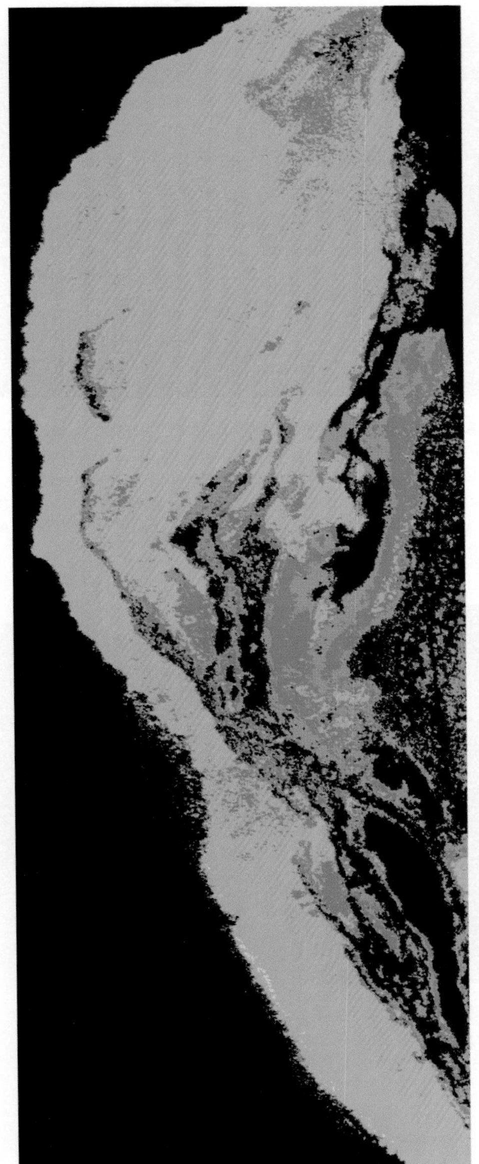

Colour	Leaf Area Index
	< 2
	2-4
	4-6
	6-8
	> 8

Figure 17.5 Thematic map of mangrove leaf area index (LAI) for mangroves on the north-west of South Caicos Island. Mangrove LAI has been categorised into 5 categories with progressively brighter shades representing progressively higher LAIs. Comparison of this figure with Figure 10.9 (Plate 11) indicates that LAI values of >8 were only found in a few patches of tall *Rhizophora mangle*, and that most of the tall *Rhizophora* and tall mixed mangrove community had LAIs of 6–8. Short, high density *Rhizophora* and much of the *Avicennia germinans* stands had LAIs of 4–6.

Oblique aerial view of most of the mangrove area on north-west South Caicos shown on the LAI map in Figure 17.5 above and on Plate 11. [*Photograph:* C. Clark]

View of the seaward fringe of Red mangrove, *Rhizophora mangle*, on South Caicos at a point where there is progradation of the mangrove (foreground) – see point A on Plate 11. [*Photograph*: P. Mumby]

A fairly tall stand of the Black mangrove, *Avicennia germinans*. [*Photograph*: P. Mumby]

Recording measurements of photosynthetically active radiation (PAR) from two MACAM™ SD101Q-Cos 2π PAR detectors connected to a MACAM™ Q102 radiometer. Eighty pairs of simultaneous readings of I_c (light flux density beneath the canopy) and I_o (light flux density outside the canopy) were taken at each site to determine leaf area index (LAI). [*Photograph:* P. Mumby]

Where the mangroves were too dense or low to enter then the I_c detector (measuring light flux density beneath the canopy) was attached to a 5.4 m extendible pole and inserted into the mangrove stand. A spirit level was attached to the end of the pole to ensure the detector was always vertical. All recordings of I_c were taken at waist height, approximately 0.8 m above the substrate. [*Photograph:* P. Mumby]

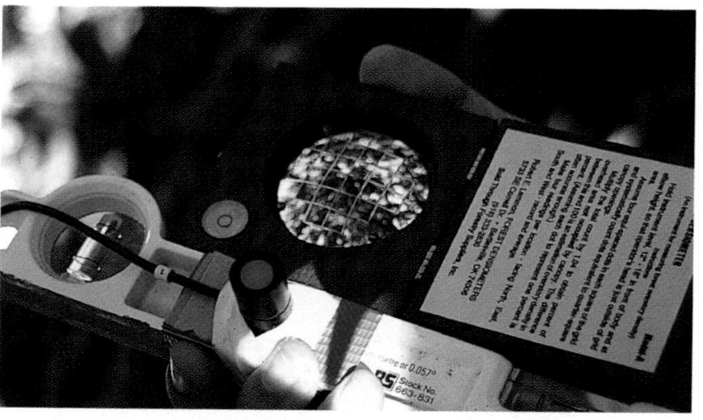

Percentage canopy cover was measured by estimating the proportion of the mirror area covered by the reflection of leaves and stems using a hand held semi-hemispherical mirror (a spherical densiometer) with a grid graticule engraved on its surface. Eighty readings were taken at each site and averaged. The spirit level was used to ensure the mirror was horizontal. [*Photograph:* P. Mumby]

Using a differential global positioning system (DGPS) with its aerial mounted in a backpack during accuracy assessment surveys of mangrove habitats. [*Photograph:* P. Mumby]

Two-man kayak set up for mangrove surveys where very shallow water prevented access by powered boats. On board are a differential global positioning system (DGPS), two PAR detectors connected to a radiometer for measuring LAI, laminated hard copy of a geometrically corrected SPOT XS image for the area and printed sheets for recording LAI. [*Photograph:* C. Clark]

Climbing into the mangrove canopy to get a DGPS fix and place the I_o detector (on the end of a 40 m waterproof cable) outside the canopy. [*Photograph:* P. Mumby]

Conch (*Strombus gigas*) midden. This species is extensively exploited for food in the Caribbean and has been overexploited in most areas. However, stocks are currently well managed in the Turks and Caicos Islands. Attempts have been made to use remote sensing to aid stock assessment of this species by mapping its habitat (Chapter 18). [*Photograph:* E. Green]

A pile of trochus shells (*Trochus niloticus*) being left to rot in Micronesia prior to being shipped to be made into buttons. Mapping of *Trochus* habitat using remote sensing has been used to help in stock assessment (Chapter 18). [*Photograph:* A. Edwards]

Seaweed mariculture in East Asia. Such areas and natural intertidal algal beds may be amenable to remote sensing (Chapter 18). They are also at risk from oil pollution events (Chapter 14). [*Photographs:* ITOPF]

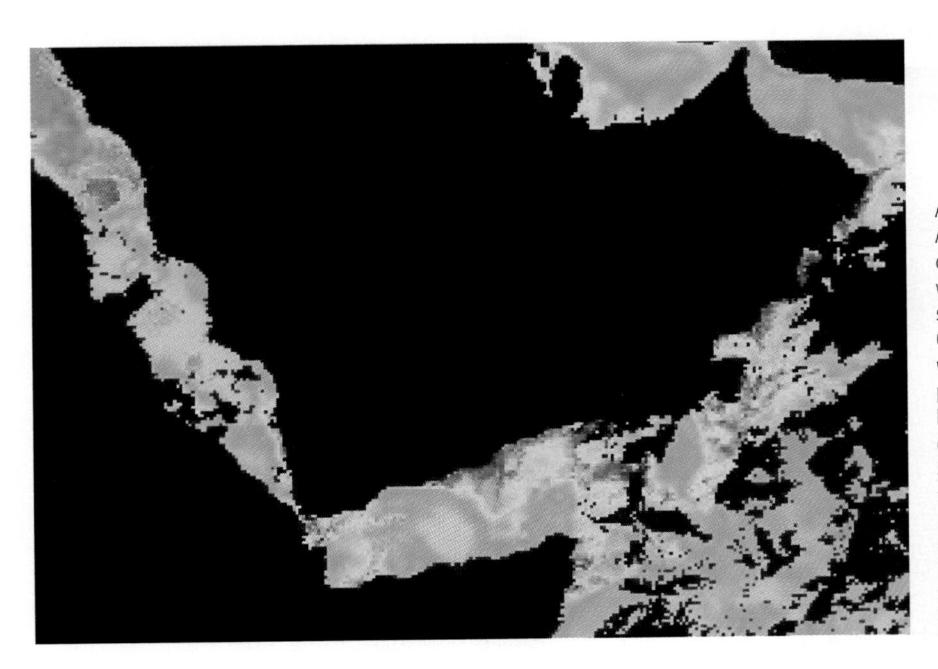

A National Aeronautics and Space Administration (NASA) SeaWiFS chlorophyll-*a* map of the north-west Indian Ocean in August 1998 showing upwelling off Yemen, Oman and Somalia associated with high phytoplankton production (Chapter 18). Red areas have the highest chlorophyll-*a* concentrations (10–50 $mg.m^{-3}$) and blue coloured areas (e.g. in the northern Red Sea) the lowest (0.03–0.3 $mg.m^{-3}$). Orange, yellow and green tones denote intermediate levels. See http://seawifs.gsfc.nasa.gov/SEAWIFS.html for further details.

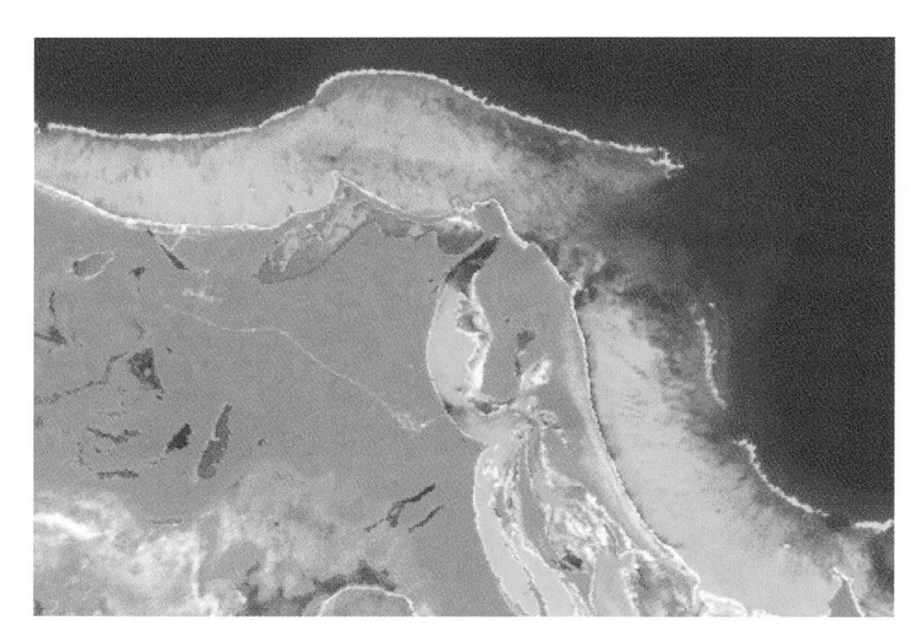

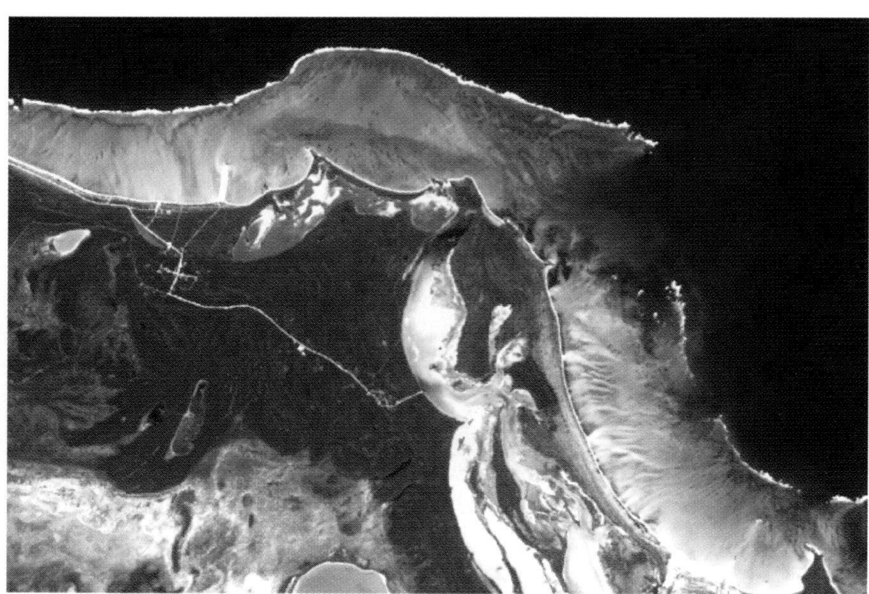

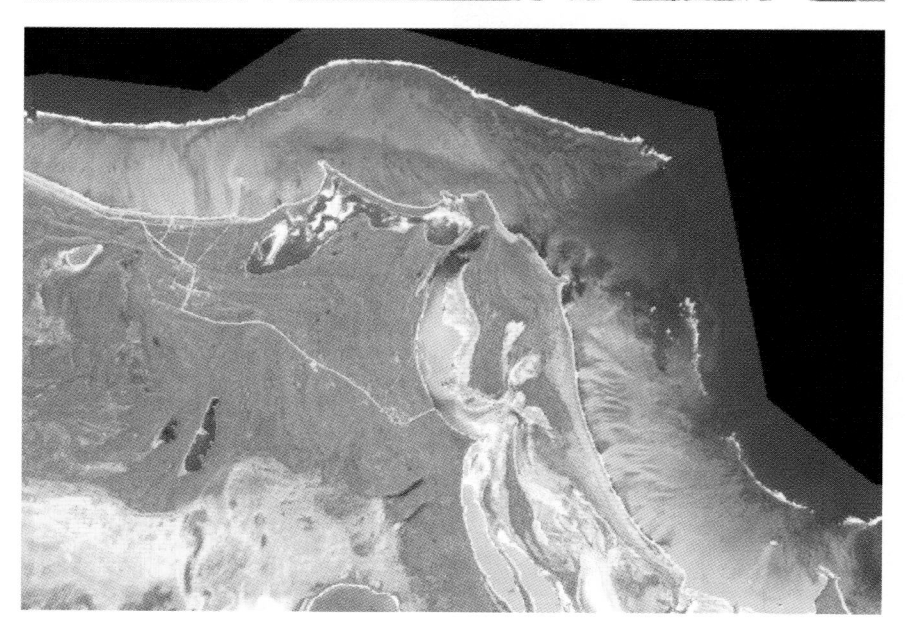

Figure 19.5 Merging Landsat TM and SPOT Pan satellite imagery. Clarity is limited by printer resolution but it can be seen that merging 30 m resolution Landsat TM (top) with 10 m resolution SPOT Pan (middle) increases the detail which is revealed (bottom). Unfortunately although visual interpretation is improved, supervised classification becomes less accurate.

Future Prospects

Introduction

Remote sensing, or 'Earth Observation' as it is more recently being called, is making rapid advances both in terms of the range and number of sensors that are put into orbit and in the way in which we extract information from the data they produce. For these reasons it is useful to comment briefly on the future prospects of remote sensing.

Proliferation of new sensors

The main current sensors of use for coastal resources assessment and management were described in Chapter 3. Most of these are likely to continue providing data in the near future, and so our evaluations of these should serve well. In addition however, there will soon be a large number of new sensors that may be of use. These are likely to modify some of our recommendations, improve accuracy and open up new applications.

There has been a major shift in the way in which new satellites are financed, planned, built and launched. Up and until now, most sensors have been on large and costly platforms designed and built by space agencies of the USA, Europe, Japan, Canada and India. Greatly reduced costs within the space industry, including that of satellite-launching, has now led to the emergence of commercial remote sensing. We are at the beginning of an era where private companies are building and launching their own remote sensing satellites and doing so for the financial return on the images they will sell. This moves us away from large multi-sensor platforms which seek to address a wide range of applications, towards smaller platforms carrying sensors which may be highly specialised and with the imagery marketed at particular target applications. The reduced scale of these projects permits a much faster implementation phase such that satellites may be designed, built and launched within a few years rather than closer to a decade. It may be that the space industry will be faster in responding to the needs of the user. The space agencies are also likely to enter this new phase of development and produce a range of sensors more quickly and more tightly focused on specific applications. The main implications are therefore:

- faster response to user needs,
- greater specialism in sensor design, and therefore suitability to a required application,
- greater number of sensors in orbit and therefore better temporal resolution,
- and perhaps, if successful, a reduction in image costs.

A down-side to these developments is that users will be faced with a bewildering array of sensors to choose from and an increased element of marketing, which may include mis-information and oversell of the potential of the imagery. The combination of these factors will make it much harder for the coastal manager or remote sensing scientist to make the best decisions about a cost-effective strategy to meet their objectives. It is likely that future exercises similar to the concept that forms the basis of this *Handbook* (i.e. which sensors and methods are most cost-effective for coastal management) will be in even higher demand in the future.

By the year 2000 it is estimated that there will be about 30 remote sensing satellites in orbit. Some may not be launched for financial reasons and some may fail on launch or in orbit, but a large number of these are likely to succeed and have an impact on coastal applications. In order to try to make sense of the range of new sensors they are grouped into five basic types, although some cut across these categories (see Table 1):

1. Sensors designed specifically for marine applications

These are primarily aimed at oceanographic rather than coastal applications and the spatial resolution is therefore in the range of 700–1000 m. They will mostly be used for measuring ocean colour and sea surface temperatures, with each swath covering as much as 1000 km.

2. Medium spatial resolution, multispectral

These are sensors with resolutions of the order of 50 m and with a number of multispectral wavebands. They are best thought of as of filling the familiar Landsat-type niche. They typically produce images with a swath width of about 150 km every 16 days.

3. High spatial resolution
This is the main growth area for new satellites. Sensors with high spatial resolutions will produce images capable of mapping features 1–10 m in size. This is achieved at the expense of spectral resolution, with the highest resolution sensors acquiring just a single panchromatic waveband. Images cover only small areas (4–40 km across) and will mostly require user requests for their acquisition.

4. Hyperspectral
Sensors of this type will acquire information in numerous (e.g. 50 to 200) narrow wavebands permitting detailed characterisation of the spectra of targets of interest to be made from the data. This contrasts with the crude characterisation from the usual 3 or 4 broad wavebands of Landsat-type sensors.

5. Radar
Utilising microwave, rather than optical/infra-red wavelengths, and doing so by transmitting their own energy and measuring the return, these sensors provide a very different view of the surface. As microwaves do not penetrate the sea surface, applications are limited to the terrestrial portion of coasts, and may become of use for mangrove mapping and biomass assessment, and for change detection. Clouds are transparent to radar sensors, making them the only type of data that is acquirable for particularly cloudy areas.

Discontinuation of Landsat MSS

From 1993 onwards the collection of Landsat multispectral scanner (MSS) ceased. These data had a spatial resolution of around 80 m and 4 wavebands and proved useful for coarse-level habitat assessment. Technically, it was superseded by the much better Thematic Mapper (TM) (30 m and 7 wavebands), but the intention was to continue MSS acquisition as it was regarded by many as the 'workhorse' of remote sensing. The lack of continuation represents a great loss to the user community. This is particularly so for long term monitoring studies as MSS has the longest archive which goes back to 1972. If commercial remote sensing only produces satellite/sensors of short duration and of frequently changing specification then this will seriously hinder our ability to monitor change. It is our hope that space agencies will continue to co-ordinate their activities such that a full range of capabilities is covered and standard or compatible specifications are maintained. Demise of Landsat TM (archived since 1984) for example, would be most unfortunate, especially as we have noted its great utility for assessment and monitoring

Another factor in the loss of MSS is that it was the only medium resolution data that could be obtained relatively cheaply. Images typically cost about £200, a whole order of magnitude less than the cost of a TM or SPOT image (see Chapter 5).

Developments in information extraction

Essentially we attempt two types of information extraction from remotely sensed data; mapping and measurement. Mapping is a descriptive procedure that allows us to produce thematic maps of habitats, and measurement permits us to estimate quantitative values such as seagrass standing crop, for example. As research continues to grow in both of these areas it is likely that accuracy and effectiveness of remote sensing will improve above the levels outlined in this *Handbook*.

There is always a large range of new techniques under development, some of which may prove useful enough, or have enough promise, to be included in remote sensing software packages. We anticipate that two techniques in particular, artificial neural networks and spectral mixture modelling, will soon be incorporated in widely available packages and may prove of use for coastal applications.

Artificial neural networks (ANNs) utilise a computer architecture analogous to the biological nervous system, in that they employ multiple interconnectivity and a high level of parallelism. The network is built as it 'learns' the data structures which can then be used for the purpose in hand. The main use so far has been as an alternative to statistically-based multispectral classification (e.g. Benediktsson *et al.* 1990, Bishof *et al.* 1992). The advantage of ANNs in this context is that they do not assume the data to be normally distributed (see Chapter 10) and instead develop a modelling approach that is dependent upon the nature of the actual data. It remains to be seen whether neural networks will outperform statistical multispectral classification as an operational tool for habitat mapping.

Spectral mixture modelling can be used to estimate the proportions of a cover type within individual pixels (e.g. Settle and Drake 1993). This is of most use if there are two brightly contrasting cover types (e.g. coral heads and sand) that are known to exist in patches smaller in size than the pixel dimensions. The technique assumes that there is a linear mix of cover-type proportions in relation to their mixing of spectral response. By collecting spectra of, for example, pure reef and pure sand, and assuming a linear mix between them the technique computes the constituent proportions within all pixels in the image.

The major change in information extraction is a move away from simple empirical research which seeks statis-

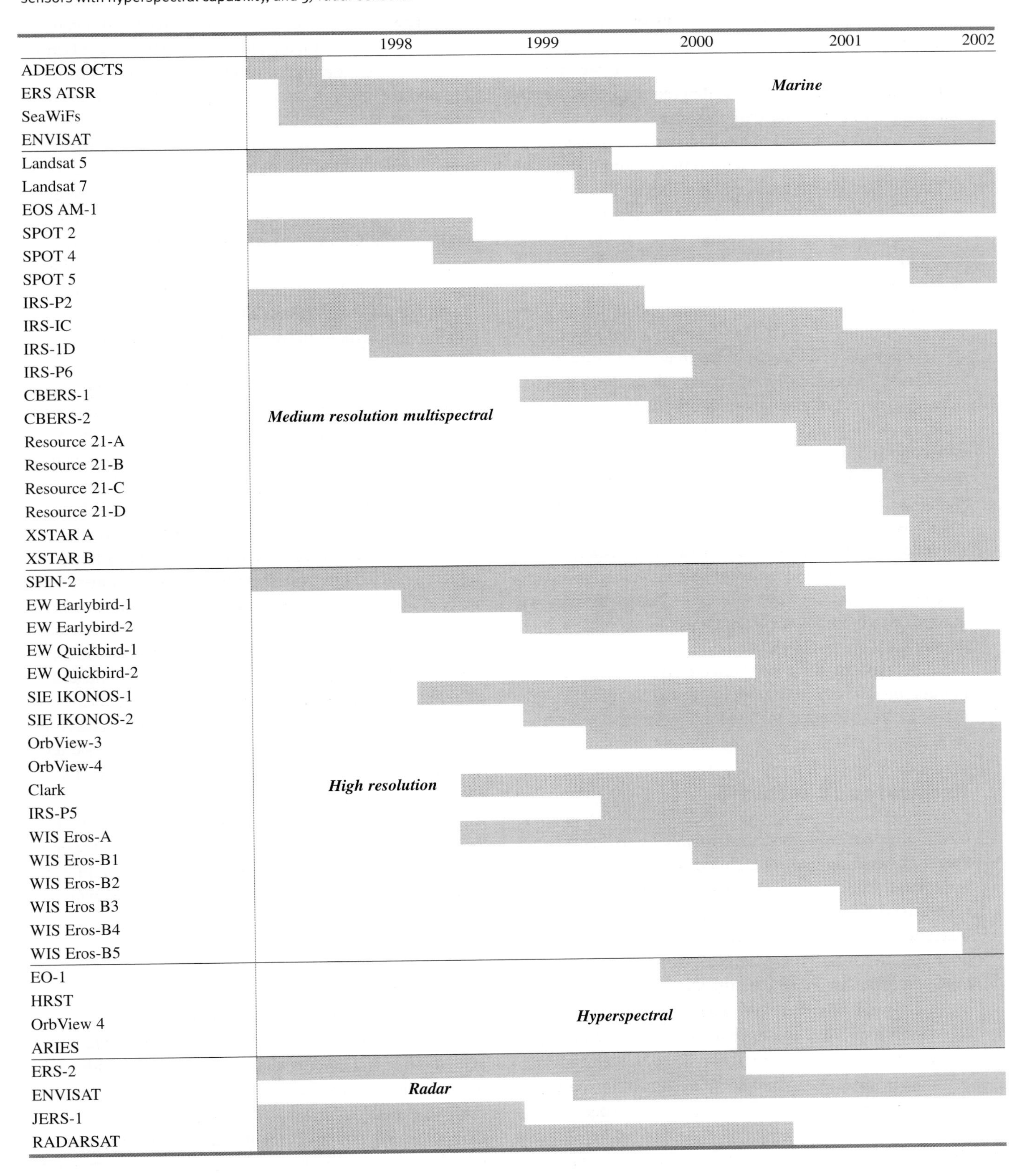

Table 1 The main remote sensing satellites of the immediate future. These are split into 5 categories: 1) those designed for marine and oceanographic applications, 2) satellites offering medium resolution multispectral data, 3) those promising high resolution data, 4) satellite sensors with hyperspectral capability, and 5) radar sensors.

☞ If further information is required, most of these sensors have their own web pages (Appendix 1).

tical relationships between reflectance and the target, towards a theoretical approach which seeks to actually understand and model the interactions of radiation down through the atmosphere (and water) to the target and back to the satellite. The ideology here, is that if we can produce a radiative transfer model that includes all scattering and attenuation of radiation, and how it interacts with its target then we can use the model in inverse mode to convert our remotely sensed measurements into an estimate of target biomass for example. In Chapter 16 we demonstrated the empirical approach for estimating seagrass standing crop. The calibration relationship between standing crop and our remotely sensed measurements do not imply that we understand the physics of interactions, but simply that a good relationship exists that we can exploit. The specific relationship can only be used for those seagrass beds at the time that the fieldwork and imagery was acquired. The aim of analytical modelling is to produce a technique that is robust to changes in area and time, so that it could be used elsewhere and without field calibration. Modelling also has the advantage of being able to make simulations so that it is possible to answer questions without conducting expensive field experiments such as those reported in this handbook. If a good model existed we would be able to answer questions such as 'can remote sensing measure seagrass standing crop in water of depth x with water quality y'; or 'can the new sensor z distinguish between seagrass and algae at a water depth of x?'.

At present there are very few invertible (i.e. of use for extracting information) models of proven reliability, but this is likely to change as this is a rapidly advancing field.

Hardware and software

Computing hardware continues to evolve at a fast pace such that data handling and image processing tasks that today appear as tiresome or prohibitively computer-intensive will soon become routine and easy. This will ease the workloads that we have described, but it is likely that there will be a commensurate increase in data volumes (e.g. from high resolution or hyperspectral sensors) and image processing complexity that will slow us down again! While all our research was conducted using UNIX workstations, because existing PCs would have made our work too time-consuming, this necessity is set to change as the performance of these two platforms is steadily narrowing and PCs continue to get faster. In a few years time it is probable that all the computational tasks conducted for this *Handbook* could be adequately performed on a PC.

Researchers close to the application end of remote sensing rarely have the skills or time for writing computer programs of their own and so are reliant on commercially available software. Over the last ten years the main improvements in software have been: a greater range of image processing techniques; a move from command line to 'point and click' menus; better image handling (zooming, panning, mosaicing, overlaying, etc.); faster processing; and the inclusion of non-programming methods for implementing algorithms. While they have made our work much easier and faster, it is this latter improvement that has been of greatest benefit. It is now possible to use a visual or flow diagram approach to implement your own algorithms without recourse to programming. All of our work on radiometric, atmospheric and water column correction for example was performed within the software package by using a flow diagram of operations to build up the equations we required. This is important, as it allows users to implement newer or highly specific techniques that the software has not yet incorporated.

The future direction of software is likely to continue as above, but with an emphasis on making specific tasks easier by requiring the user to know less about them. An example may be that rather than having to understand the theory of water column correction (Chapter 8) and being able to implement it by building your own algorithm, that in the future there may be a simple menu or range of buttons that will execute the task for you. It is likely that packages will proliferate with this kind of application-specific task. While it is in the interests of the vendor to convey how easy it may be to conduct a variety of complex tasks, it is in the interest of the user to be sure that the methods are appropriate and valid for the objective in hand and for the limitations of the data. There is a conflict here that in our view can only be resolved by the vendors providing good manuals describing the physical basis of the techniques and an extensive bibliography which can be followed up when required.

The main remote sensing software packages that we know of are Erdas Imagine, PCI, Envi and ER-mapper, all of which can be found via their web pages on the Internet. All of the image processing for this *Handbook* was conducted using Erdas Imagine.

Conclusions

As remote sensing is a fast moving field in terms of the sensors involved and the techniques we use to understand the data they produce, it is tempting to say that we are entering a new era in which all current difficulties will be overcome. We should be cautious of such statements as they have been said many times before. While new sensors and techniques may solve current problems, they also make new ones. If you were to have a hyperspectral image of your field area, does the insight, knowledge and software already exist with which to analyse it? Probably not.

It is usual that there be a time lag of at least 5 years after the launch of a new type of sensor before application users can really get on and use the imagery for routine rather than experimental tasks. As an example, ERS-1 was launched in 1991 collecting radar images, but it was not until the mid-1990s that appropriate software tools became widely available and there were enough research papers to guide new users.

How the proliferation of new sensors is going to impact on coastal habitat mapping and management is hard to predict. It will certainly be much harder to decide on a suitable strategy, and actual acquisition of imagery will be more complicated as a wide range of data suppliers will need to be contacted. Over-marketing and misinformation may lead to some practitioners making expensive mistakes in the imagery they buy. Basic research leading to handbooks such as this, or demonstration case-studies using the new image types will alleviate these problems. If radiative transfer modelling advances sufficiently fast, the best approach may be to move away from case-study testing of individual sensors to using model simulations based on the sensor specifications.

Imagery with spatial resolutions of less than 10 m will compete with aerial photography and may prove easier and less expensive to acquire. Aerial photography however, is also undergoing a steady transformation towards digital rather than photographic capture. How this type of data will compare in effectiveness and cost to the new satellite data is as yet unknown. A key problem to consider is the signal to noise ratio of the new satellite and airborne high resolution sensors. Reducing the pixel size reduces the amount of signal that is measured (there is less upwelling radiation from a smaller area), and so for a given sensor sensitivity, smaller pixel sizes will result in greater problems with background and sensor noise. As the high resolution sensors are designed for terrestrial use where signals are usually much higher, this may be particularly acute when operating over water as the signal levels may be too low for meaningful measurement.

A bonus of more sensors in orbit is the greater chance of obtaining cloud free imagery. This may be useful if techniques are developed which allow us to be less dependent upon always using the same sensor type and which will allow valid comparisons to be made between data from different sensors.

Increased use of radiative transfer modelling should result in better estimations of quantitative parameters and importantly should permit simulations to tell us when not to use remote sensing. Modelling should be able to tell us, for example, under what water quality (e.g. suspended sediment) conditions seagrass standing crop prediction ceases to become viable. At the moment, however, the empirical approach is still the best method for operational application, and this will remain the case until modelling investigations are fully tested against field data and proven to work.

The future prospects of remote sensing look good, in that higher spatial and spectral resolution sensors are being put into orbit. Careful use of the data they produce should be of great benefit producing better accuracies and opening up new applications. Unfortunately, and due to physical limitations, there will not be any spaceborne high spatial and high spectral resolution sensors, which are what we really want. Users will have to use a combination of satellite image types or go the extra expense of acquiring airborne multispectral data.

The future remote sensing practitioner with a list of objectives to achieve will require very good skills in distinguishing actual from potential applications, and operational from experimental techniques. It is hoped that this *Handbook* aids in this purpose for current sensors and techniques, and helps frame the critical questions that should be asked, but for the future we wish you the best of luck.

Postscript

Recent news (*Nature*, 19 August 1999, p. 702) indicates that Landsat Thematic Mapper imagery (from the new ETM+ sensor) is to be made available at a greatly reduced price. Images from the recently launched Landsat 7 will cost just US$475, much less than the previous price of around US$4000. As Landsat TM imagery is of great utility, this is a welcome change and may affect the decisions made in image acquisition (Chapters 5 and 19).

References

Benediktsson, J.A., Swain, P.H., and Ersoy, O.K., 1990, Neural network approaches versus statistical methods in classification of multisource remote sensing data. *IEEE Transactions on Geoscience and Remote Sensing*, **28** (4), 540–552.

Bishof, H., Schneider, W., and Pinz, A.J., 1992, Multispectral classification of Landsat images using neural networks. *IEEE Transactions on Geoscience and Remote Sensing*, **30** (3), 482–490.

Settle, J.J., and Drake, N.A., 1993, Linear mixing and the estimation of ground cover proportions. *International Journal of Remote Sensing*, **14** (6), 1159–1177.

Glossary of Terms and Acronyms

The Canada Centre for Remote Sensing (CCRS) offers a very useful on-line Remote Sensing Glossary at the following URL: http://www.ccrs.nrcan.gc.ca/ccrs/eduref/ref/glosndxe.html. Readers are referred to this for items not listed here.

Absorption A process by which radiation is converted to other types of energy (especially heat) by a material. Reduction in strength of an electromagnetic wave propagating through a medium.

Absorptivity Ratio of the absorbed to incident electromagnetic radiation on a surface.

Active remote sensing Remote sensing methods that provide there own source of electromagnetic radiation, e.g. radar.

Additive primary colours The colours blue, green and red. Filters of these colours transmit the primary colour of the filter and absorb the other two colours.

ADEOS Advanced Earth Observing Satellite. A Japanese Earth observation satellite. Carries the OCTS sensor. See Chapter 3.

Adjacency effect The change in the digital number of a pixel caused by atmospheric scattering of radiance that originates outside of the sensor element's field of view.

Aerial photography Photography from airborne platforms.

Albedo The ratio of the amount of electromagnetic energy reflected by a surface to the amount of energy incident upon it.

Algorithm A statement of predefined steps to be followed in the solution of a problem, such as a set of image processing steps (each a mathematical manipulation of the image data) to bring about a desired outcome.

Altimeter Instrument which determines the altitude of an object with respect to a fixed level such as sea level.

All weather Refers to the capability of a sensing system to operate under any weather condition.

AMS (Daedalus) Airborne Multispectral Scanner. See Chapter 3.

Analogue A form of data display in which values are shown in graphic form such as curves. Also a form of computing in which values are represented by directly measurable quantities such as voltages of resistances. Analog computing methods contrast with digital methods in which values are treated numerically.

Analogue recorder Data recorder in which data are stored in continuous form as contrasted with digital data having discrete values.

Angle of incidence (1) The angle between the direction of incoming EMR and the normal to the intercepting surface; (2) In SLAR systems this is the angle between the vertical and a line connecting the antenna and the target.

Ångstrom (Å) A measurement of length (10^{-10} m). Equivalent to 10 nm.

Angular field of view The angle subtended by lines from a remote sensing system to the outer margins of the strip of terrain that is viewed by the system.

Antenna The device that transmits and/or receives microwave and radio energy.

API	Aerial photographic interpretation. The process of visually interpreting aerial photographs.
APT	Automatic Picture Transmission (TIROS, ATS, ESSA, NOAA-n).
ATM (Daedalus)	Airborne Thematic Mapper sensor. See Chapter 3.
Atmospheric absorption	The process whereby some or all of the energy of sound waves or electromagnetic waves is transferred to the constituents of the atmosphere. See Chapter 7.
Atmospheric reflectance	Ratio of reflected radiation from the atmosphere to incident radiation.
Atmospheric scattering	Process whereby part or all of the energy of electromagnetic radiation is dispersed when traversing the atmosphere. See Chapter 7.
Atmospheric windows	Wavelength intervals at which atmosphere transmit most electromagnetic radiation.
ATSR	Along Track Scanning Radiometer. Sensor carried on the European Remote Sensing satellites. See Chapter 3.
AVHRR	Advanced Very High Resolution Radiometer. Carried on NOAA polar orbiting satellites. Spatial resolution 1.1 x 1.1 km.
AVI	Angular Vegetation Index. An index of vegetation cover. See Chapter 13.
AVIRIS	Airborne Visible/Infra-Red Imaging Spectrometer. Produces multispectral data with 224 narrow (10 nm) bands between 400 and 2400 nm. See Chapter 3.
Azimuth	The geographic orientation of a line given as an angle measured clockwise from north.
Backscatter	(1) The scattering of radiant energy into the hemisphere of space bounded by a plane normal to the direction of the incident radiation and lying on the same side as the incident ray; (2) In SLAR usage this refers to the portion of the microwave energy scattered by the terrain surface that is directed back towards the antenna.
Band	A wavelength interval in the electromagnetic spectrum. For example, in Landsat sensors the bands designate specific wavelength intervals at which images are required.
BIL	Band Interleaved by Line. A format for data storage. See Chapter 5.
BIP	Band Interleaved by Pixel. A format for data storage. See Chapter 5.
Bit	In digital computer terminology, this is a binary digit that is an exponent of the base 2.
Brightness	Magnitude of the response produced in the eye by light.
Byte	A group of eight bits of digital data.
Calibration	The process of comparing measurements, made by an instrument, with a standard.
CASI	Compact Airborne Spectrographic Imager. A digital airborne multispectral sensor.
CCD	Charge Coupled Device. A light sensitive solid-state detector that generates a voltage which is proportional to the intensity of illumination. Arrays of CCDs make up pushbroom scanners.
CCRS	Canada Centre for Remote Sensing.
Classification	The process of assigning individual pixels of a digital image to categories, generally on the basis of spectral reflectance or radiometric characteristics.
Colour composite image	A colour image prepared by combining individual band images. Each band (up to a maximum of 3) is assigned one of the three additive primary colours: blue, green and red.
Computer Compatible Tape (CCT)	The magnetic tape upon which the digital data for remotely sensed images are sometimes distributed.
Conformal	A map projection that has the property of true shape (conformality). See Chapter 6.
Contextual editing	The use of non-spectral information, e.g. location or depth, to improve classification of spectrally overlapping classes.
Contrast	The difference between highlights and shadows in a photographic image. The larger the difference in density the greater the contrast.

Contrast stretching	Improving the contrast of images by digital processing. The original range of digital values is expanded to utilize the full contrast range of the recording film or display device.
CZCS	Coastal Zone Color Scanner. See Chapter 3.
DEM	Digital Elevation Model. See Chapter 6.
Density slicing	The process of converting the continuous grey tone of an image into a series of density intervals or slices, each corresponding to a specific digital range.
Descriptive resolution	The level of ecological/geomorphological detail to which a sensor can map a given area. Equivalent to habitat discrimination.
Detector	The component of a remote sensing system that converts the electromagnetic radiation into a signal that is recorded.
DGPS	Differential Global Positioning System.
Digital data	Data displayed, recorded or stored in binary notation.
Digital image processing	Computer manipulation of the digital values for picture elements of an image.
Digitizer	A device for scanning an image and converting it into numerical picture elements.
Electromagnetic Radiation (EMR)	Energy propagated through space or through material media in the form of an advancing interaction between electric and magnetic fields.
Emission	With respect to EMR, the process by which a body emits EMR usually as a consequence of its temperature only.
Emissivity	The ratio of radiant flux from a body to that from a blackbody at the same kinetic temperature.
EMR	See Electromagnetic Radiation.
Enhancement	The process of altering the appearance of an image so that the interpreter can extract more information. Enhancement may be done by digital or photographic methods.
Equidistance	The property of a map projection to represent true distances between points. See Chapter 6.
Equivalence	The property of a map projection to represent all areas in true proportion to one another. See Chapter 6.
EOSAT	Earth Observation Satellite Company. A private company contracted to the US government to market Landsat data. Now superseded by Space Imaging.
ERDAS	Earth Resources Data Analysis System. An image processing and GIS software package now called ERDAS Imagine and produced by ERDAS Inc.
ERS	European Remote Sensing Satellite
ESA	European Space Agency
ETM	Enhanced Thematic Mapper. Sensor carried on Landsat 7. See Chapter 3.
EXABYTE	A tape format on which satellite imagery may be supplied.
Fluorescence	The emission of light from a substance caused by exposure to radiation from an external source.
Gain	An increase in signal power in transmission from one point to another and usually expressed in decibels.
GCP	Ground Control Point. A point on the ground whose position is accurately known and which can be used with other GCPs to geometrically correct an image. See Chapter 6.
Gelbstoffe	Dissolved yellow substances which absorb light in natural waters.

GEMI	Global Environment Monitoring Index. An index of vegetation cover. See Chapter 13.
Geoid	The figure of the earth considered as a sea-level surface extended continuously over the entire earth's surface.
Geometric correction	The correction of errors of skew, rotation, and perspective in raw, remotely sensed data. See Chapter 6.
GIS	Geographic Information System.
GPS	Global Positioning System. A network of 24 radio transmitting satellites (NAVSTAR) developed by the US Department of Defense to provide accurate geographical position fixing.
Grey scale	A calibrated sequence of grey tones ranging from black to white.
Groundbusters	Elite SWAT team of ground-truthing professionals with total accuracy dedication; reaching the habitats that other survey teams can't. Motto: Groundbusters are go!
Ground receiving station	A facility that records image data transmitted by airborne or spaceborne sensors.
Ground resolution cell	The area on the terrain that is covered by the instantaneous field of view of a detector. Size of the ground resolution cell is usually determined by the altitude of the remote sensing system and the instantaneous field of view of the detector.
Hertz (Hz)	Cycles per second. Measure of frequency.
HRV	Haute Resolution Visible. The visible and very near infra-red pushbroom sensor carried by SPOT 1–3.
Hue	The attribute of a colour that differentiates it from grey of the same brilliance and that allows it to be classed as blue, green, red or intermediate shades of these colours.
IFOV	See Instantaneous Field of View.
Image	The representation of a scene as recorded by a remote sensing system. Although image is a general term, it is commonly restricted to representations acquired by non-photographic methods.
IMU	Inertial Measurement Unit. See Chapter 6.
Infra-red (IR)	Portion of the electromagnetic spectrum lying between the red end of the visible spectrum and microwave radiation (700 nm to 1000 μm).
INS	Inertial Navigation System. See Chapter 6.
Instantaneous Field Of View (IFOV)	A measure of the area viewed by a single detector on a scanning system at a given moment in time.
Intensity	A measure of energy reflected from a surface.
Interactive processing	The method of data processing by which the operator views preliminary results and can alter the instructions to the computer to achieve optimum results.
Interpretation	The extraction of information from an image.
IR	See Infra-red.
IR colour film	A colour film consisting of three layers in which the red-imaging layer responds to photographic infra-red radiation ranging in wavelength from 700 to 1300 nm. The green-imaging layer responds to red light and the blue-imaging layer responds to green light.
Irradiance	The radiant power density incident on a surface (W cm^{-2}).
IRS	Indian Remote Sensing satellite. IRS-1A to 1D and IRS-P2 to P4 have been launched since 1988. See Chapter 3.
ISODATA	Iterative Self-Organizing Data Analysis Technique. A method of clustering used in image classification.
ITRES	A Canadian company who developed the Compact Airborne Spectrographic Imager (CASI).

LAI Leaf Area Index. Single-side leaf area per unit ground area; a unit-less measure. See Chapter 17.

Landsat A series (6 successfully launched since 1972) of unmanned earth-orbiting NASA satellites (formerly called Earth Resources Technology Satellite – ERTS).

Laser Light Amplification by Stimulated Emission of Radiation.

L-Band Radar wavelengths region from 15 to 30 cm.

Lidar Light Detection and Ranging.

Light EMR within 400–700 nm in wavelength that is detectable by the human eye.

LISS Linear Imaging Self-Scanning Sensor. A type of sensor carried on Indian Remote Sensing satellites. IRS-1C and 1D carry the LISS-III sensor. See Chapter 3.

Microwave The region of the electromagnetic spectrum in the wavelength range from 1 mm to beyond 1 m.

Mie scattering Multiple reflection of light waves by atmospheric particles that have the approximate dimensions of the wavelength of light. See Chapter 7.

MOS Marine Observation Satellite. MOS-1 was launched by Japan in 1987.

MSS See Multispectral Scanner.

Multispectral Scanner (MSS) A scanner system that simultaneously acquires images of the same scene in various wavelength bands. Landsat MSS was one such system which was operational on Landsat series satellites from 1972–1993.

Nadir The point on the ground vertically beneath the centre of a remote sensing system.

NASA National Aeronautics and Space Administration (USA).

NDVI Normalised Difference Vegetation Index. An index of vegetation biomass. See Chapter 13.

NIR Near Infra-Red. Wavelengths around 700–3000 nm.

NOAA National Oceanic and Atmospheric Administration (USA).

Noise Random or repetitive events that obscure or interfere with the desired information.

NRSC National Remote Sensing Centre (UK).

Oblique photograph A photograph acquired with the camera axis intentionally directed between the horizontal and vertical orientations.

OCTS Ocean Color and Temperature Scanner. Sensor carried on the Advanced Earth Observing Satellite (ADEOS). See Chapter 3.

Okta Unit of measurement for cloud cover. One okta = 1/8 or 12.5% cover.

Orbit The path of a satellite around a body under the influence of gravity.

Overall accuracy The overall probability that pixels on the image have been classified correctly.

Overlap The extent to which adjacent images or photographs cover the same terrain, expressed in percentages.

Panchromatic Sensitive to entire visible part of EMR spectrum.

Passive remote sensing Remote sensing of energy naturally reflected or radiated from the terrain.

PCA Principal Components Analysis. A statistical technique for reducing the number of bands in an image by generating a series of components, each of which is an uncorrelated combination of raw bands. The first few components will contain most of the information found within all the raw bands.

PCI EASI/PACE An image processing and GIS software package produced by the PCI Remote Sensing Corporation.

Photogrammetry	The making of maps using aerial photographs.
Photograph	A representation of targets formed by the action of light on silver halide grains of an emulsion.
Photon	The elementary quantity of radiant energy.
Picture element	In a digitized image this is the area on the ground represented by each digital value. Because the analog signal from the detector of a scanner may be sampled at any desired interval, the picture element may be smaller than the ground resolution cell of the detector. Commonly abbreviated as pixel.
Pitch	Rotation of an aircraft about the horizontal axis normal to its longitudinal axis, that causes a nose-up or nose-down attitude.
Pixel	A contraction of picture element.
Producer accuracy	The probability that a pixel in a given habitat category will have been classified correctly on the image.
Pushbroom scanner	An imaging system consisting of a linear array of many detectors (CCDs) that record the brightness of a line of pixels directly below the sensor in the across-track direction.
Radar	Radio Detection and Ranging.
RADARSAT	A Canadian radar satellite launched in 1995.
Radar shadow	A dark area of no return on a radar image that extends in the far-range direction from an object on the terrain that intercepts the radar beam.
Ratio image	An image prepared by processing digital multispectral data. For each pixel the value for one band is divided by that of another. The resulting digital values are displayed as an image.
Rayleigh scattering	Selective scattering of light in the atmosphere by particles that are small compared to the wavelength of light. See Chapter 7.
RBV	See Return Beam Vidicon.
Real time	Time in which reporting of events or recording of events is simultaneous with the event.
Reflectance	The ratio of the radiant energy reflected by a body to that incident upon it.
Reflectance, spectral	Reflectance measured at a specific wavelength interval.
Refraction	The bending of electromagnetic rays as they pass from one medium into another.
Remote sensing	The collection of information about an object or event without being in physical contact with the object or event. Remote sensing is restricted to methods that record the electromagnetic radiation reflected or radiated from an object, which excludes magnetic and gravity surveys that record force fields.
Resolution	The ability to distinguish closely spaced objects on an image or photograph. Commonly expressed as the spacing, in line-pairs per unit distance, of the most closely spaced lines that can be distinguished.
Return Beam Vidicon (RBV)	A little-used imaging system on Landsat that consists of three cameras operating in the green, red and photographic infra-red spectral regions. Instead of using film, the images are formed on the photosensitive surface of a vacuum tube. The image is scanned with an electron beam and transmitted to an earth receiving station.
RGB	Red, green blue. The primary additive colours which are used on most display hardware (computer monitors) to display images.
RMS error	Root Mean Square error. The distance between the input (source) location of a GCP and the retransformed location for the same GCP. Measure used to assess the accuracy of geometric correction. See Chapter 6.
Roll	Rotation of an aircraft about the longitudinal axis to cause a wing-up or wing-down attitude.

SAR Synthetic Aperture Radar. A radar imaging system in which high resolution in the azimuth direction is achieved by using Doppler shift of backscattered waves to identify waves from ahead and behind the platform, thereby simulating a very long antenna. Available on both satellite (ERS, RADARSAT) and airborne platforms.

Scale The ratio of the distance on an image to the equivalent distance on the ground.

Scan line The narrow strip on the ground that is swept by the instantaneous field of view of a detector in a scanner system.

Scan skew Distortion of scanner images caused by forward motion of the aircraft or satellite during the time required to complete a scan.

Scanner (1) Any device that scans, and thus produces an image. (2) A radar set incorporating a rotatable antenna for directing a searching radar beam through space and imparting target information to an indicator.

Scanner distortion The geometric distortion that is characteristic of scanner images. The scale of the image is constant in the direction parallel with the aircraft or spacecraft flight direction. At right angles to this direction, however, the image scale becomes progressively smaller from the nadir line outward toward either margin of the image. Linear features, such as roads, that trend diagonally across a scanner image are distorted into S-shaped curves. Distortion is imperceptible for scanners with a narrow angular field of view but becomes more pronounced with a larger angular field of view. Also called panoramic distortion.

Scattering Multiple reflection of electromagnetic waves by gases or particles in the atmosphere.

Scene The area on the ground that is covered by an image or photograph.

SeaWiFS Sea-viewing Wide Field-of-view Sensor carried on the SeaStar satellite. See Chapter 3.

Sensitivity The degree to which a detector responds to electromagnetic energy incident upon it.

Sensor A device that receives electromagnetic radiation and converts it into a signal that can be recorded and displayed as numerical data or as an image.

Signature A characteristic, or combination of characteristics, by which a material or an object may be identified on an image or photograph.

Skylight The component of light that is scattered by the atmosphere and consists predominantly of shorter wavelengths of light.

SLAR Side-Looking Airborne Radar. Airborne systems that use real-aperture radar to generate images. Cheaper than synthetic aperture radar (SAR).

Spectral envelope Boundaries of the spectral properties of a feature, expressed as the range of brightness values in a number of spectral bands.

Spectral sensitivity The response or sensitivity of a film or detector to radiation in different spectral regions.

Spectrometer Device for measuring intensity of radiation absorbed or reflected by a material as a function of wavelength.

Spectrum (1) In physics, any series of energies arranged according to wavelength (or frequency); (2) The series of images produced when a beam of radiant energy is subject to dispersion. A rainbow-coloured band of light is formed when white light is passed through a prism or a diffraction grating. This band of colours results from the fact that the different wavelengths of light are bent in varying degrees by the dispersing medium and is evidence of the fact that white light is composed of coloured light of various wavelengths.

SPOT Satellite Pour l'Observation de la Terre. A French satellite carrying two pushbroom imaging systems. (Originally Système Probatoire de l'Observation de la Terre).

SST Sea Surface Temperature.

Subtractive primary colours Yellow, cyan and magenta. When used as filters for white light these colours remove blue, red, and green, respectively.

Sun synchronous An Earth satellite orbit in which the orbit plane is near polar and the altitude such that the satellite passes over all places on earth having the same latitude twice daily at the same local sun time. Also known as heliosynchronous.

Supervised classification A classification method whereby thematic classes are defined by the spectral characteristics of pixels within an image that correspond to training areas in the field chosen to represent known features (e.g. habitats). Each pixel within the image is then assigned to a thematic class using one of several decision rules (see Chapter 10).

Surface phenomenon Interaction between electromagnetic radiation and the surface of a material.

Swath width The linear ground distance in the across-track direction which is covered by a sensor on a single overpass.

Synthetic Aperture Radar A radar imaging system in which high resolution in the azimuth direction is achieved by utilizing the Doppler principle to give the effect of a very long antenna.

Systematic distortion Predictable geometric irregularities on images that are caused by the characteristics of the imaging system.

Tau coefficient (accuracy) A coefficient which describes the accuracy of a thematic map (see Chapter 4).

Thermal capacity The ability of a material to store heat, expressed in $J\ kg^{-1}\ K^{-1}$.

Thermal conductivity The measure of the rate at which heat passes through a material, expressed in $W\ m^{-1}\ K^{-1}$.

Thermal IR The portion of the infra-red region from approximately 3–15μm wavelength that corresponds to heat radiation. This spectral region spans the radiant power peak of the Earth.

TIROS Television Infrared Observational Satellite.

TM Thematic Mapper.

Ultraviolet (UV) radiation The region of the electromagnetic spectrum consisting of wavelengths from 1 to 400 nm.

UNDP/GEF United Nations Development Programme/Global Environment Facility

Unsupervised classification Using a computer to automatically generate a thematic map from digital remotely sensed imagery by statistically clustering pixels on the basis of spectral similarity. The clusters may then be assigned labels (e.g. habitat names) using the operator's field knowledge.

User accuracy The probability that a pixel classified as a particular habitat on the image is actually that habitat.

UTM Universal Transverse Mercator. A widely used geographical coordinate system. See Chapter 6.

Visible radiation Energy at wavelengths from 400 to 700 nm that is detectable by the eye.

Wavelength The distance between successive wave crests or other equivalent points in a harmonic wave.

X-Band Radar wavelength region from 2.4 to 3.8 cm.

XS Multispectral SPOT imagery from the HRV sensor. See Chapter 3.

Yaw Rotation of an aircraft about its vertical axis that causes the longitudinal axis to deviate from the flight line.

Appendices

APPENDIX 1

Information Networks and Internet Resources Directory

Sources of data on the World Wide Web

The following list of data sources provides URLs (Uniform Resource Locators) or 'web addresses' of a range of sites we have found particularly useful. These relate primarily to remote sensing but with the emphasis on sensors and satellites of use in coastal management or products of use to coastal managers.

Low spatial resolution sensors (> 100 m) and products

AVHRR sensor characteristics
http://edcwww.cr.usgs.gov/landdaac/1KM/avhrr.sensor.html

NOAA Oceanic Research and Applications Division home page
http://manati.wwb.noaa.gov/orad/

Experimental 'daily' SST anomaly charts (NOAA AVHRR)
http://psbsgi1.nesdis.noaa.gov:8080/PSB/EPS/SST/climo.html

Sea Surface Temperatures (SST) from NOAA AVHRR
http://manati.wwb.noaa.gov/orad/sub/sstlink.html

Coral bleaching hotspots (NOAA AVHRR)
http://psbsgi1.nesdis.noaa.gov:8080/PSB/EPS/SST/climohot.html

Coral reef 'hotspots' (NOAA/NESDIS)
http://manati.wwb.noaa.gov/orad/sub/noaarsrc.html

NERC Dundee Satellite Receiving Station (UK). Up-to-date archive of images from NOAA and SeaStar polar orbiting satellites.
http://www.sat.dundee.ac.uk/

Coastal Zone Color Scanner (CZCS) description
http://daac.gsfc.nasa.gov/WORKINPROGRESS/UDAY/czcs.html

Getting started with Coastal Zone Color Scanner (CZCS) data
http://daac.gsfc.nasa.gov/CAMPAIGN_DOCS/OCDST/CZCS_Starter_kit.html#8

Advanced Earth Observing Satellite (ADEOS)
http://mentor.eorc.nasda.go.jp/ADEOS/index.html

Ocean Color and Temperature Scanner (OCTS) on ADEOS
http://www.eorc.nasda.go.jp/ADEOS/UserGuide/

SeaWiFS Project home page
http://seawifs.gsfc.nasa.gov/SEAWIFS.html

Medium spatial resolution sensors (10–100 m)

NASA Earth Science Enterprise missions homepage (USA)
http://www.earth.nasa.gov/missions/index.html

SPOT Image home page
http://www.spotimage.fr/spot-us.htm

Landsat program
http://geo.arc.nasa.gov/esd/esdstaff/landsat/landsat.html

USGS EROS Data Center: Earthshots (Satellite images of environmental change: Landsat)
http://edcwww.cr.usgs.gov/earthshots/slow/tableofcontents

National Remote Sensing Centre (UK)
http://www.nrsc.co.uk/

Canada Centre for Remote Sensing home page
http://www.ccrs.nrcan.gc.ca/ccrs/homepg.pl?e

European Space Agency home page
http://www.esrin.esa.it/

ESA (European Space Agency) earth observation programmes including ERS and ENVISAT
http://www.esrin.esa.it/esa/progs/eo.html

European Space Agency ERS project
http://earth1.esrin.esa.it/ERS/ or http://www.asf.alaska.edu/source_documents/ers1_source.html

European Space Agency ENVISAT project
http://envisat.estec.esa.nl/

Indian Space Research Organisation
http://ww.ipdpg.gov.in/

Airborne digital remote sensing and aerial photography

ITRES Research Limited (CASI)
http://www.itres.com/

Borstad Associates Ltd. (CASI)
http://www.borstad.com

Hyperspectral Data International Inc. (CASI)
http://www.brunnet.net/hdi/casi%20SERVICES.htm

Technical specification of the York University Compact Airborne Spectrographic Imager (CASI)
http://www.eol.ists.ca/projects/boreas/plan/A-casi-specs.html

Aerial photography and remote sensing (University of Texas at Austin)
http://www.utexas.edu/depts/grg/gcraft/notes/remote/remote.html

GIS, GPS and mapping

GPS receiver database (GPS World magazine)
http://www.gpsworld.com

Global Positioning System overview (University of Texas at Austin)
http://www.utexas.edu/depts/grg/gcraft/notes/gps/gps.html

Review of GIS and image processing software by UNEP/GRID programme
http://grid2.cr.usgs.gov/publications/gissurvey97/flyer2.html

Remote sensing and geographic information systems (University of Minnesota)
http://www.gis.umn.edu/rsgisinfo/rsgis.html

Map Scales. A useful site explaining map scales. (US Geological Survey)
http://info.er.usgs.gov/fact-sheets/map-scales/index.html

Cartographic reference resources (University of Texas at Austin)
http://www.lib.utexas.edu/Libs/PCL/Map_collection/Cartographic_reference.html

Useful links pages and miscellaneous information sources

Remote Sensing Society, Modelling and Advanced Techniques Special Interest Group - Hot Links page
http://www.soton.ac.uk/~pma/matsiglinks.html

CSIRO Marine Research: Remote Sensing Project (Australia)
http://www.marine.csiro.au/~lband

Satellite related World Wide Web sites (Dundee, UK)
http://www.sat.dundee.ac.uk/web.html

Remote Sensing Laboratory, Department of Marine Sciences, University of South Florida (USA)
http://paria.marine.usf.edu

The Remote Sensing Society (UK)
http://www.the-rss.org/index.htm

The Satellite Imagery FAQ
http://www.geog.nottingham.ac.uk/remote/satfaq.html

Remote sensing and geoscience home pages (Analytical Imaging and Geophysics LLC, Boulder, Colorado)
http://shell.rmi.net/~kruse/rs_pages.htm

The Virtual Geography Department home page
http://www.utexas.edu/depts/grg/virtdept/contents.html

OCMA - Ocean and Coastal Management Archives home page (University of Genoa, Italy)
http://www.polis.unige.it/ocma98

UNESCO Bilko project: training modules in remote sensing
http://www.unesco.bilko.org/

RAPIDS: Real-time Acquisition and Processing - Integrated Data System. PC-based transportable groundstation for real-time, local reception and local processing of high resolution satellite data.
http://neonet.nlr.nl/rapids/

CCRS Remote Sensing Glossary
http://www.ccrs.nrcan.gc.ca/ccrs/eduref/ref/glosndxe.html

APPENDIX 2

Suppliers of Remotely Sensed Data

Satellite data

A vast number of data receiving and data distribution centres exist throughout the world. To acquire remotely sensed data, seek the advice of the nearest centre. While some centres receive satellite data, they might not be able to distribute it, in which case, contact the appropriate owners of the data – EOSAT (Landsat TM), EROS (Landsat MSS), SPOT Image (SPOT).

The following list of data centres is not exhaustive but should provide a reasonable distribution of information sources.

Satellite data owners

Landsat TM	EOSAT, 9351 Grant Street, Suite 500, Thornton, Colorado 80229, USA http://www.spaceimage.com/index.html
Landsat MSS	US Department of the Interior, Geological Survey, EROS Data Center, Sioux Falls SD 57198, USA http://edcwww.cr.usgs.gov/content_about.html
SPOT XS and Pan	SPOT Image, 5 rue des Satellites, BP 4359, F 31030 Toulouse cedex, France http://www.spotimage.fr/
Indian Remote Sensing Satellite	Indian Space Research Organisation, National Remote Sensing Agency, Data Centre, Hyderabad, India. http://www.ipdpg.gov.in/

Satellite data centres

AFRICA

Algeria	INC - National Institute of Cartography, 123 Rue De Tripoli, BP 430, Hussein Dey, Alger, Algeria
Benin	Centre National de Télédétection et de Surveillance du Couvert Forestier (CE.NA.TEL), B.P. 06-711, Cotonou Benin
Egypt	GeoMAP consultants, 13 El Obour Buildings, Salah Salem Street, PO Box 85, Saray El Kobba, Cairo 11371, Egypt
Ghana	GIANT METRO COMPANY Ltd., North Industrial Area, PO Box 12923, Accra-North, Ghana
Israel	Interdisciplinary Center for Technological Analysis & Forecasting (ICTAF), Tel-Aviv University, Ramat Aviv, 69978 Tel-Aviv, Israel Israel Space Agency (ISA), 26a Chaim Levanon Street, PO Box 17185, Tel-Aviv 61171, Israel PAIMEX Co. Ltd., 3 Wohlman Street, Tel-Aviv 69400, Israel
Jordan	Research & Consulting Services Co. Ltd., PO Box 5013, Amman, Hashemite Kingdom of Jordan

Kenya	Andrews Aeronautical & Allied Equipment Ltd., Kasuku Road, Kilimani, PO Box 1152, Nairobi, Kenya
Kuwait	COMET International Co., PO Box 29606, Safat, Code 13157, Kuwait
Libya	Libyan Center for Remote Sensing and Space Science, PO Box 397, Tripoli, Libya
Morocco	Centre Royal de Télédétection Spatiale, 16 bis, avenue de France, Agdal Rabat, Morocco
Mozambique	CENACARTA, 537 Josina Machel Av., PO Box 83, Maputo, Mozambique
Nigeria	Information Research Ltd., 3, Gerrard Road, Ikoyi, PO Box 53386, Ikoyi, Lagos, Nigeria
Oman	INTERTEC SYSTEMS L.L.C., PO Box 345, Postal Code 118, Sultanate of Oman
South Africa	Satellite Applications Centre, Division of Microelectronics & Communications Technology, PO Box 395, Pretoria 0001, South Africa

CANADA and USA

Canada	Canada Centre for Remote Sensing, 588 Booth Street, Ottawa, Ontario K1A 0Y7, Canada http://www.ccrs.nrcan.gc.ca/ccrs/home.html
	Radarsat International Inc., 275 Slater Street, Suite 1203, Ottawa, Ontario, K1P 5H9, Canada
USA	Earth Satellite Corporation, 6011 Executive Blvd., Suite 602, Rockville, MD 20852, USA
	Environmental Research Institute of Michigan, PO Box 134001, Ann Arbor, MI 48113, USA
	SPOT IMAGE Corporation, 1897 Preston White Drive, Reston, VA 20191-4368, USA

SOUTH AMERICA

Argentina	AEROESPACIO S.R.L., Piedras 77 Piso 10, Capital Federal C.P. 1070, Buenos Aires, Argentina
Brazil	INPE, Av. dos Astronautas, 1758, Caixa Postal 515, São José dos Campos, SP Brazil
	INTERSAT, Alameda José Alves Siqueira Filho, 139 Vila Bethânia Sao José dos Campos, CEO 12245 260 SP Brazil
Chile	COESPA, Av. Libertador O'Higgins 580, Of. 603, 13716 Santiago, Chile
Colombia	HERINDSER LTDA, Carrera 11 Nº 82.38 Oficina 402, A.A. 101 348, Santafe de Bogota DC, Colombia
Ecuador	CLIRSEN, Apartado 8216, Quito, Ecuador
	DEVECOM, Avenida Amazonas 477, Of. 905-907, Quito, Ecuador
Mexico	COSMOCOLOR, Castellanos Quinto 393, Col. Educacion, Coyoacan 04400, Mexico DF, Mexico
Paraguay	PARECO Representaciones, Iturbe 1023-1, Casilla de Correo 2031, Asunción, Paraguay
Peru	EKODES Consultores SCRL, Av. Nicolas de Rivera 544, San Isidro, Lima 27, Peru

SOUTH-EAST ASIA AND MIDDLE EAST

China	RSGS China Remote Sensing Satellite Station, 45 Beisanhvanxi Road, Beijing 100086, China
Hong Kong	STARVISION Ltd., Shop 106, Eastern Plaza, 111 Chai Wan Road, Chai Wan, Hong Kong
India	National Remote Sensing Agency, Balanagar, Hyderabad 500 037, Andhra Pradesh, India
Indonesia	LAPAN, National Aeronautics & Space Inst., Jl. Pemuda Persil No 1, Rawamangun, PO Box 1020/JAT, Jakarta 13220, Indonesia
Iran	JAZAYERI & ASSOCIATES, Tenth Street, Gandhi Avenue, Tehran, 15178 Iran
Japan	CRC Research Institute Inc., 2-7-5 Minamisuna Koto-Ku, Tokyo 136, Japan
	ORION PRESS, 1-55 Kanda-Jimbocho, Chiyoda-ku, Tokyo 101, Japan
	RESTEC, Roppongi First Building, 12F, 1-9-9, Roppongi, Minato-ku, Tokyo 106, Japan
Korea	KAIST-SaTRec, 373-1 Kusung-dong, Yusong-gu, Taejon 305-701, Republic of Korea
	CHANG WOO Inc., Keum Young Bldg, 15-11, Yeo Eui Do-Dong, Young Deung Po-gu, Seoul, Korea
Laos	Lao Survey & Exploration Services, 206 Thaduea Road, Sissatanak, Vientiane, Lao People's Democratic Republic
Macau	I.F.S., Beco do Gonçalo, 18 R/C, Macau
Nepal	LYRA International, PO Box 5057, Annapurna Arcade - Durbar Marg, Kathmandu, Nepal
Pakistan	SUPARCO, Off University Road, PO Box 8402, Karachi 75270, Pakistan
Philippines	CERTEZA Surveying & Aerophoto Inc., 795 E. Delos Santos Avenue, Quezon City, Philippines
Singapore	CRISP, Lower Kent Ridge Road, Singapore 0511, Republic of Singapore
	SPOT ASIA, 73 Amoy Street, Singapore 069892, Republic of Singapore
Taiwan	Centre for Space & Remote Sensing Research (CSRSR), National Central University, Chung-Li, Taiwan, Republic of China
Thailand	National Reasearch Council of Thailand (NRCT), 196 Phahonyothin Road, Chatuchak, Bangkok 10900 Thailand

EUROPE

Austria	Oesterreichisches Fernerkundungs – Datenzentrum (OFD), Jakob Haringer Strasse 1, A-5020 Salzburg, Austria
Belgium	Services Fédéraux des Affaires Scientifiques, Techniques et Culturelles (SSTC), Rue de la Science 8, B-1040 Bruxelles, Belgium
Bulgaria	Bulgarian Academy of Science (BAS), Space Research Institute, 6 Moskovska Street, PO Box 799, 1000 Sofia, Bulgaria
Croatia	GISDATA, Svetice 15, 41000 Zagreb, Croatia
Czech Rep. & Slovakia	GISAT, Charkovska 7, 10000 Praha 10, Czech Republic
France	ESRI France S.A., 21 rue des Capucins, 92190 Meudon, France
	SCIENCES IMAGES, 10 rue de l'Industrie, B.P. 220, 74105 Annemasse, France
Germany	Gesellschaft Fuer Angewandte, Fernerkundung mbH (GAF), Leonrodstr. 68, D-80636 Muenchen 19, Germany
Greece	GEOMET Ltd., Praduna 7 & Odissou, GR-11525 Athens, Greece

Hungary	FÖMI Remote Sensing Center, Guszev U19, PO Box 546, H-1373 Budapest, Hungary
Ireland	ERA - MAPTEC Ltd., 36 Dame Street, Dublin 2, Ireland
Italy	NUOVA TELESPAZIO SPA, Via Tiburtina, 965, 00156 Roma, Italy
The Netherlands	National Lucht-En Ruimtevaart Laboratorium NLR, PO Box 153, 8300 AD Ammeloord, The Netherlands
Portugal	INFORGEO, Avenida Marquês de Tomar, 100 R/C Dto, 1050 Lisboa, Portugal
Romania	Centre Roumain pour l'Utilisation de la Télédétection en Agriculture (CRUTA), 35-37 Oltenitei, 79 656 Bucarest IV, Romania
Russia	RPA Planeta, Bolshoy Predtechensky per., Moscow 123242, Russia
	DERSI, Schmitorski Proezd, Maison 13, Moscow 123100, Russia
Slovenia	GISDATA, Saranoviceva 12, 61000 Ljubljana, Slovenia
Spain	AURENSA TELEDETECCION S.A., C/San Francisco de Sales 38, 28003 Madrid, Spain
Sweden	SSC Satellitbild, PO Box 816, S-98128 Kiruna, Sweden
Switzerland	Institut für Kommunikations Technik (ETH), Gloriastrasse 35, CH-8092 Zurich, Switzerland
Turkey	DATASEL, Mesrutiyet, cad 31/8-9-11, Yenisehir 06420, Ankara, Turkey
United Kingdom	National Remote Sensing Centre Ltd. (NRSC), Arthur Street, Barwell, Leicestershire LE9 8G2, United Kingdom

AUSTRALIA AND SOUTH PACIFIC

Australia	SPOT IMAGING SERVICES, 156 Pacific Highway, St Leonards 2065, Australia
	Mapping & Monitoring Technology Pty Ltd, PO Box 5704 MC, 37 Tully Street, South Townsville, QLD 4810, Australia
	Remote Sensing Services, Department of Land Administration, PO Box 471, 65 Brockway Road, Florear, Wembley WA 6014, Australia
New Zealand	Manaaki Whenva Landcare Research, PO Box 38491, Wellington Mail Centre, Wellington, New Zealand
Pacific Area	SCOOP, Immeuble Fong, B.P. 5670, Pirae, Tahiti
Papau New Guinea	Arman Larmer Surveys, Varahe Street, Gordon, PO Box 372, Port Moresby, PNG

Airborne multispectral imagery

CASI	Hyperspectral Data International Inc., One Research Drive, Dartmouth, Nova Scotia B2Y 4M9 Canada. Tel: +1 (902) 465-8877. Fax: +1 (902) 466-6889. E-mail: hdiinfo@hdi.ns.ca
Dr Gary Borstad	114–9865 West Saanich Road, Sidney, British Columbia, Canada V8L 5Y8 Tel: +1 (250) 656-5633. Fax: +1 (250) 656-3646. gary@Sparc2.Borstad.com
Daedalus ATM	Daedalus Enterprises Inc., PO Box 1869, Ann Arbor, Michigan 48106-1869 USA Tel: +1 (313) 769 5649. Fax: +1 (313) 769 0429
Multisensor airborne monitoring of oil pollution	Roger Stockham, Air Atlantique, Coventry Airport, Coventry CV8 3AZ United Kingdom. Fax: +44 (1203) 882633

Index